AF761723

VOLUME ONE HUNDRED AND THIRTY FOUR

PROGRESS IN MOLECULAR BIOLOGY AND TRANSLATIONAL SCIENCE

Molecular Biology of Eye Disease

VOLUME ONE HUNDRED AND THIRTY FOUR

PROGRESS IN MOLECULAR BIOLOGY AND TRANSLATIONAL SCIENCE

Molecular Biology of Eye Disease

Edited by

J. FIELDING HEJTMANCIK

Ophthalmic Molecular Genetics Section, Ophthalmic Genetics and Visual Function Branch, National Eye Institute, Rockville, Maryland, USA

JOHN M. NICKERSON

Department of Ophthalmology, Emory University, Atlanta, Georgia, USA

ELSEVIER

AMSTERDAM • BOSTON • HEIDELBERG • LONDON
NEW YORK • OXFORD • PARIS • SAN DIEGO
SAN FRANCISCO • SINGAPORE • SYDNEY • TOKYO

Academic Press is an imprint of Elsevier

Academic Press is an imprint of Elsevier
225 Wyman Street, Waltham, MA 02451, USA
525 B Street, Suite 1800, San Diego, CA 92101-4495, USA
125 London Wall, London, EC2Y 5AS, UK
The Boulevard, Langford Lane, Kidlington, Oxford OX5 1GB, UK

First edition 2015

© 2015 Elsevier Inc. All rights reserved.

No part of this publication may be reproduced or transmitted in any form or by any means, electronic or mechanical, including photocopying, recording, or any information storage and retrieval system, without permission in writing from the publisher. Details on how to seek permission, further information about the Publisher's permissions policies and our arrangements with organizations such as the Copyright Clearance Center and the Copyright Licensing Agency, can be found at our website: www.elsevier.com/permissions.

This book and the individual contributions contained in it are protected under copyright by the Publisher (other than as may be noted herein).

Notices
Knowledge and best practice in this field are constantly changing. As new research and experience broaden our understanding, changes in research methods, professional practices, or medical treatment may become necessary.

Practitioners and researchers must always rely on their own experience and knowledge in evaluating and using any information, methods, compounds, or experiments described herein. In using such information or methods they should be mindful of their own safety and the safety of others, including parties for whom they have a professional responsibility.

To the fullest extent of the law, neither the Publisher nor the authors, contributors, or editors, assume any liability for any injury and/or damage to persons or property as a matter of products liability, negligence or otherwise, or from any use or operation of any methods, products, instructions, or ideas contained in the material herein.

ISBN: 978-0-12-801059-4
ISSN: 1877-1173

For information on all Academic Press publications
visit our website at store.elsevier.com

ELSEVIER Book Aid International
Working together
to grow libraries in
developing countries

www.elsevier.com • www.bookaid.org

CONTENTS

CONTRIBUTORS

Zsolt Ablonczy
Department of Ophthalmology, Albert Florens Storm Eye Institute, Medical University of South Carolina, Charleston, South Carolina, USA

Leopold Adler IV
Department of Ophthalmology, Albert Florens Storm Eye Institute, Medical University of South Carolina, Charleston, South Carolina, USA

S. Amer Riazuddin
Wilmer Eye Institute, Johns Hopkins University School of Medicine, Baltimore, Maryland, USA

Jeffrey H. Boatright
Department of Ophthalmology, Emory University School of Medicine, Atlanta, and Center for Visual and Neurocognitive Rehabilitation, Atlanta VA Medical Center, Decatur, Georgia, USA

Hannah E. Bowrey
Department of Ophthalmology, Albert Florens Storm Eye Institute, Medical University of South Carolina, Charleston, South Carolina, USA

Nicholas P. Boyer
Department of Ophthalmology, Albert Florens Storm Eye Institute, Medical University of South Carolina, Charleston, South Carolina, USA

Barbara M. Braunger
Institute of Human Anatomy and Embryology, University of Regensburg, Regensburg, Germany

Ranjay Chakraborty
Department of Ophthalmology, Emory University School of Medicine, Atlanta, Georgia, USA

Chunhe Chen
Department of Ophthalmology, Albert Florens Storm Eye Institute, Medical University of South Carolina, Charleston, South Carolina, USA

Seung-il Choi
Corneal Dystrophy Research Institute, Yonsei University College of Medicine, Seoul, Republic of Korea

Micah A. Chrenek
Department of Ophthalmology, Emory University School of Medicine, Atlanta, Georgia, USA

Rosalie K. Crouch
Department of Ophthalmology, Albert Florens Storm Eye Institute, Medical University of South Carolina, Charleston, South Carolina, USA

Ales Cvekl
Departments of Genetics and Ophthalmology and Visual Sciences, Albert Einstein College of Medicine, Bronx, New York, USA

Lucian V. Del Priore
Department of Ophthalmology, Albert Florens Storm Eye Institute, Medical University of South Carolina, Charleston, South Carolina, USA

Allen O. Eghrari
Wilmer Eye Institute, Johns Hopkins University School of Medicine, Baltimore, Maryland, USA

J. Fielding Hejtmancik
Ophthalmic Genetics and Visual Function Branch, National Eye Institute, National Institutes of Health, Bethesda, Maryland, USA

Mark A. Fields
Department of Ophthalmology, Albert Florens Storm Eye Institute, Medical University of South Carolina, Charleston, South Carolina, USA

Rudolf Fuchshofer
Institute of Human Anatomy and Embryology, University of Regensburg, Regensburg, Germany

James L. Funderburgh
Department of Ophthalmology, University of Pittsburgh, Pittsburgh, Pennsylvania, USA

Eldon E. Geisert
Department of Ophthalmology, Emory University School of Medicine, Atlanta, Georgia, USA

Jie Gong
Department of Ophthalmology, Albert Florens Storm Eye Institute, Medical University of South Carolina, Charleston, South Carolina, USA

John D. Gottsch
Wilmer Eye Institute, Johns Hopkins University School of Medicine, Baltimore, Maryland, USA

Hans E. Grossniklaus
Department of Ophthalmology, Emory University School of Medicine, Atlanta, Georgia, USA

Andrew J. Hertsenberg
Department of Ophthalmology, University of Pittsburgh, Pittsburgh, Pennsylvania, USA

Shengping Hou
The First Affiliated Hospital of Chongqing Medical University, Chongqing Key Lab of Ophthalmology, Chongqing Eye Institute, Chongqing, PR China

Winston Whei-Yang Kao
Edith Crawley Ophthalmic Research Laboratory, Department of Ophthalmology, College of Medicine, University of Cincinnati, Cincinnati, Ohio, USA

Aize Kijlstra
University Eye Clinic Maastricht, Maastricht, The Netherlands

Eung Kweon Kim
Department of Ophthalmology, Vision Research Institute, Severance Hospital; Corneal Dystrophy Research Institute, and BK21 Plus Project for Medical Science and Severance Biomedical Science Institute, Yonsei University College of Medicine, Seoul, Republic of Korea

Masahiro Kono
Department of Ophthalmology, Albert Florens Storm Eye Institute, Medical University of South Carolina, Charleston, South Carolina, USA

Yiannis Koutalos
Department of Ophthalmology, Albert Florens Storm Eye Institute, Medical University of South Carolina, Charleston, South Carolina, USA

Hun Lee
Department of Ophthalmology, Vision Research Institute, Severance Hospital, and Corneal Dystrophy Research Institute, Yonsei University College of Medicine, Seoul, Republic of Korea

Chia-Yang Liu
Edith Crawley Ophthalmic Research Laboratory, Department of Ophthalmology, College of Medicine, University of Cincinnati, Cincinnati, Ohio, USA

Wei Liu
Departments of Genetics and Ophthalmology and Visual Sciences, Albert Einstein College of Medicine, Bronx, New York, USA

Peter Y. Lwigale
Department of Biosciences, Rice University, Houston, Texas, USA

Caitlin E. Mac Nair
Ophthalmology and Visual Sciences, and Cellular and Molecular Pathology Graduate Program, University of Wisconsin—Madison, Madison, Wisconsin, USA

Rebecca McGreal
Departments of Genetics and Ophthalmology and Visual Sciences, Albert Einstein College of Medicine, Bronx, New York, USA

Pia R. Mendoza
Department of Ophthalmology, Emory University School of Medicine, Atlanta, Georgia, USA

Ravi Metlapally
UC Berkeley School of Optometry, Berkeley, California, USA

T. Michael Redmond
Laboratory of Retinal Cell and Molecular Biology, National Eye Institute, NIH, Bethesda, Maryland, USA

Robert S. Molday
Department of Biochemistry and Molecular Biology, Centre for Macular Research, University of British Columbia, Vancouver, British Columbia, Canada

Robert W. Nickells
Ophthalmology and Visual Sciences, University of Wisconsin—Madison, Madison, Wisconsin, USA

John M. Nickerson
Department of Ophthalmology, Emory University School of Medicine, Atlanta, Georgia, USA

Machelle T. Pardue
Department of Ophthalmology, Emory University School of Medicine, Atlanta, Georgia, USA

Kevin Schey
Department of Biochemistry, Vanderbilt University, Nashville, Tennessee, USA

Robin H. Schmidt
Department of Ophthalmology, Emory University School of Medicine, Atlanta, Georgia, USA

Daniel Schorderet
IRO - Institute for Research in Ophthalmology, Sion; Faculty of Life Sciences, Swiss Federal Institute of Technology, and Department of Ophthalmology, University of Lausanne, Lausanne, Switzerland

Alan Shiels
Department of Ophthalmology and Visual Sciences, Washington University School of Medicine, St. Louis, Missouri, USA

Deborah L. Stenkamp
Department of Biological Sciences, University of Idaho, Moscow, Idaho, USA

Felix L. Struebing
Department of Ophthalmology, Emory University School of Medicine, Atlanta, Georgia, USA

Ernst R. Tamm
Institute of Human Anatomy and Embryology, University of Regensburg, Regensburg, Germany

Janey L. Wiggs
Harvard Medical School, and Massachusetts Eye and Ear Infirmary, Boston, Massachusetts, USA

Christine F. Wildsoet
School of Optometry, University of California, Berkeley, California, USA

Charles B. Wright
Department of Ophthalmology and Visual Sciences, University of Kentucky, Lexington, Kentucky, USA

Peizeng Yang
The First Affiliated Hospital of Chongqing Medical University, Chongqing Key Lab of Ophthalmology, Chongqing Eye Institute, Chongqing, PR China

Qingjiong Zhang
State Key Lab of Ophthalmology, Zhongshan Ophthalmic Center, Sun Yat-sen University, Guangzhou, PR China

Yan Zhang
School of Optometry, University of California, Berkeley, California, USA

PREFACE

The only thing worse than being blind is having sight but no vision.

Helen Keller

The visual process is complex, depending on a combination of precise functions of the anterior segment, which serve to focus light precisely on the retina, and the posterior segment, which receives light signals, transforms them into electrical signals, and performs preliminary processing before transmitting them through the optic nerve and pathways to the visual cortex. Each component of this system must function precisely and dependably for correct vision. This requires that each part of the visual system undergoes appropriate developmental regulation, that each molecule in the various metabolic and functional pathways functions correctly, and that they all interact seamlessly in carrying out visual perception. Finally, the biological systems that support and maintain homeostasis of the cells making up the visual system are required to protect and preserve vision over the lifetime of the individual to prevent age-related causes of blindness such as age-related cataract and macular degeneration.

One way in which to gain insight into the intricate processes supporting vision is through the examination of inherited diseases affecting vision, and from these the proteins and pathways which they affect. In addition to obvious candidates, such as rhodopsin for retinal degenerations and lens crystallins for cataracts, the study of inherited visual diseases has identified previously unsuspected pathways and processes critical for visual function, including the role of complement and other immune regulators in the retina and processes such as message sequestration and autophagy in the lens. However, the process works both ways: in order to study or even understand the molecular genetics of vision, one must have a firm foundation in the basic biochemistry, cell and developmental biology, and molecular biology of its component parts. It is this interrelationship between the basic sciences and the study of inherited diseases, and eventually the clinical application of knowledge derived from both, that this book aspires to delineate.

In order to accomplish this intertwining of basic biology, genetics, and clinical application, coverage of each component of the eye begins with its developmental biology and progresses through its biochemistry and molecular biology before finishing with the molecular genetics of its inherited diseases. The material aims at being approachable by a graduate student or even an advanced undergraduate, while at the same time being of sufficient depth to provide even an advanced researcher a concise overview of each area. In that vein, while avoiding being cumbersome each chapter is sufficiently referenced to provide access to the original literature in the area which it covers. Finally, the chapters are written to maintain the interest of the reader and hopefully will inspire young scientists to pursue a career in vision research.

J. Fielding Hejtmancik
John M. Nickerson

CHAPTER ONE

Overview of the Visual System

J. Fielding Hejtmancik*, John M. Nickerson†,[1]
*Ophthalmic Genetics and Visual Function Branch, National Eye Institute, National Institutes of Health, Bethesda, Maryland, USA
†Department of Ophthalmology, Emory University School of Medicine, Atlanta, Georgia, USA
[1]Corresponding author: e-mail address: litjn@emory.edu

Abstract

This introduction provides an overview of the retina, in which we survey the fundus, layers of the retina, retinal cell types, visual transduction cascade, vitamin A cycle, neuronal wiring of the retina, and blood supply of the retina.

The visual system includes the optical components of the anterior segment of the eye: in order, the cornea, the aqueous humor, and the lens; and the posterior segment, including the vitreous body, the retina, and the optic nerve. Finally, the visual system includes the optic tracts and optic radiations, transmitting neural signals to the visual cortex, and several additional nuclei of the brain (Fig. 1). Each of these components is critical in receiving, transmitting, and interpreting visual information.

The optical components in the anterior segment of the eye focus light onto the retina, which then transduces the light signal into neural signals. In addition, the retina also carries out initial processing of the neural signals before passing them through the optic nerves and tracts to central nervous system components that carry out their elaborate processing and integration with other senses. In addition, the oculomotor system, basically the efferent arm of the visual system, controls stability of position of the eyes as well as directing and coordinating movements of the eyes to objects of interest.

Light initially traverses the anterior chamber where it first passes through the transparent cornea, aqueous humor, lens, and vitreous body (Fig. 1). The speed with which light travels through each of these components is inversely proportional to its density, with the ratio of the velocity of light in a vacuum to the velocity in medium being the refractive index. Thus, light waves striking the surface of the cornea at an angle are slowed differentially, so that the light entering the cornea first is slowed more than that which travels longer through air. This bends the direction of the light called refraction. If the components of the anterior segment, especially the cornea and lens, develop

Progress in Molecular Biology and Translational Science, Volume 134
ISSN 1877-1173
http://dx.doi.org/10.1016/bs.pmbts.2015.05.006
© 2015 Elsevier Inc.
All rights reserved.

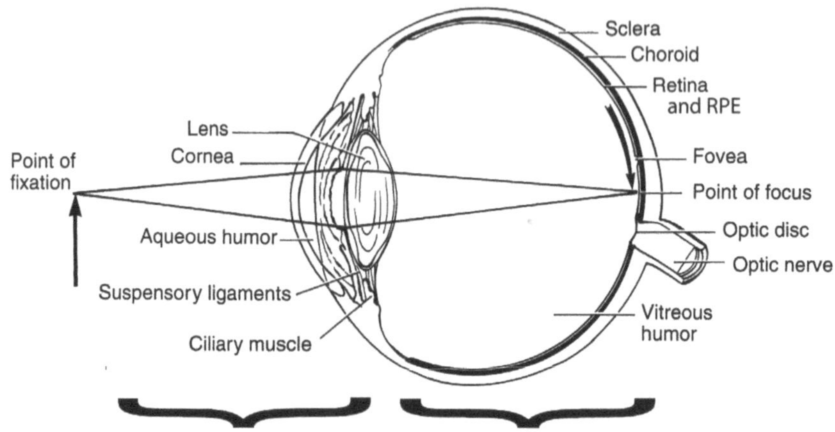

Figure 1 Overview of the eye including the anterior and posterior segments. The refraction of light rays from the fixation point at the tip of the arrow to the focal point on the retina represents the summed effects of the anterior and posterior surfaces of the cornea and lens. The image of the arrow is projected in an inverted orientation on the retina.

into the proper shape and optical density, and are appropriately transparent, light rays originating from point source similarly are focused onto a single point on the retina. This results in the image of an object being projected in an inverted fashion onto the retina, so that the inferior visual field is projected onto the superior region of the retina, and the nasal visual fields are projected onto the temporal retina. The entire biology of the anterior segment is oriented toward accomplishing the clear transmission and sharp focusing of light onto the retina, and many of the genetic lesions of this part of the eye interfere with this task.

In the retina, the light signals transmitted by the anterior segment are converted to neural signals that undergo some initial processing before being transmitted through the optic nerve and radiations to the brain. The retina comprises two functional and structural parts: the retinal pigment epithelium or RPE, which is the nonneural component, and the adjacent but distinct neural or sensory retina. RPE cells contain melanin granules, which absorb light passed through the retina, preventing its reflection by the sclera, which would and degrade the quality of vision. In addition, the cells of the RPE aid the photoreceptors by recycling visual pigments and phagocytizing shed photoreceptor outer-segment tips. This requires that the outer segments including the visual pigments are physically close to the RPE layer.

Conversely, the neural processing networks of the retina are the anterior-most structures, through which light passes before stimulating the photoreceptor cells in the posterior layer of the retina adjacent to the RPE (Fig. 2).

The neural retina is composed of six neuronal-cell types as well as non-neuronal glial Müller cells. These exist in three nuclear layers: from anterior to posterior the ganglion cell, inner nuclear, and outer nuclear layers, separated by the inner plexiform and outer plexiform layers, in which synapses occur. The photoreceptor cells, whose cell bodies lie in the outer nuclear layer and whose outer segments lie adjacent to the RPE, carry out phototransduction, the biochemical process of transforming light to the electrical energy of neural signals. There are two types of these highly specialized cells: rod and cone cells. Rods contain rhodopsin and occur at greater density in the peripheral retina. They mediate black and white vision and are able to detect light under dim illumination, important for night vision. Cones are densely packed in the central retina, especially the macula, and carry out precision and color vision under strong illumination, for example in daylight. The photoreceptor cells synapse with horizontal and bipolar cells in the outer plexiform layer. Bipolar cells correspondingly synapse with amacrine and ganglion cells in the inner plexiform layer. The cell bodies of the amacrine, bipolar, and horizontal, as well as interplexiform cells lie in the inner nuclear layer, while the cell bodies of the ganglion cells lie in the ganglion cell layer. Finally, axons of the ganglion cells traverse the nerve fiber

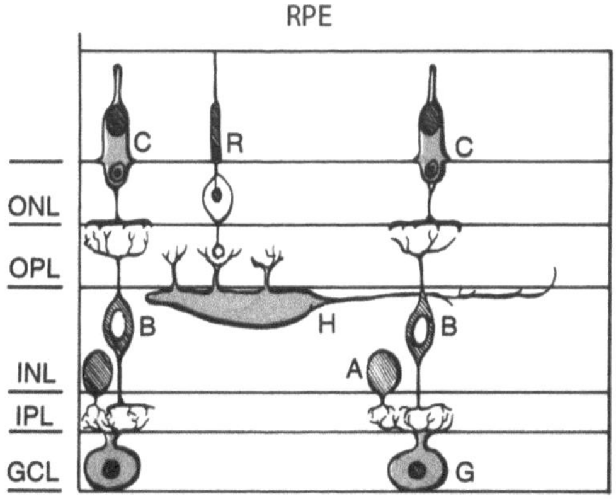

Figure 2 The vertebrate retina. Schematic of the cells in the retina. ONL, outer nuclear layer; OPL, outer plexiform layer; INL, inner nuclear layer; IPL, inner plexiform layer; GCL, ganglion cell layer; R, rod cell; C, cone cell; B, bipolar cell; H, horizontal cell; A, amacrine cell; G, ganglion cell; RPE, retinal pigmented epithelium.

layer of the retina, collecting to form optic nerve, which leads through the optic tracts to the optic radiations and visual cortex of the brain.

Each of these components including, its biology, function, and genetics, is considered in detail in the following chapters. Because of the philosophical approach of this volume as well as the limitations of space, general principles are emphasized rather than the experiments and the results through which our knowledge has been accrued. Students who wish additional information regarding the experimental approach to visual science or a more detailed description of the biology or genetics of the visual system should be able to find it in the references, which are plentiful for each chapter.

SECTION I

Cornea

CHAPTER TWO

Overview of the Cornea: Structure, Function, and Development

Allen O. Eghrari, S. Amer Riazuddin, John D. Gottsch[1]
Wilmer Eye Institute, Johns Hopkins University School of Medicine, Baltimore, Maryland, USA
[1]Corresponding author: e-mail address: jgottsch@jhmi.edu

Contents

Abstract

The cornea is a transparent tissue with significant refractive and barrier functions. The epithelium serves as the principal barrier to fluid and pathogens, a function performed through production of tight junctions, and constant repopulation through differentiation and maturation of dividing cells in its basal cell layer. It is supported posteriorly by basement membrane and Bowman's layer and assists in maintenance of stromal dehydration. The stroma composes the majority of corneal volume, provides support and clarity, and assists in ocular immunity. The posterior cornea, composed of Descemet membrane and endothelium, is essential for stromal dehydration, maintained through

Progress in Molecular Biology and Translational Science, Volume 134
ISSN 1877-1173
http://dx.doi.org/10.1016/bs.pmbts.2015.04.001
© 2015 Elsevier Inc.
All rights reserved.

tight junctions and endothelial pumps. Corneal development begins with primitive formation of epithelium and lens, followed by waves of migration from cells of neural crest origin between these two structures to produce the stroma and endothelium. Descemet membrane is secreted by the latter and gradually thickens.

1. STRUCTURE

The human cornea is an avascular tissue that measures approximately 11.5 mm horizontally and 10.5 mm vertically. Its relative transparency, with average refractive index of 1.3375, and anterior radius of curvature centrally of 7.8 mm makes this tissue responsible for three-fourths of the total refractive power of the human eye.

1.1 Epithelium

The corneal epithelium is composed of four to six layers of nonkeratinized, stratified squamous epithelial cells, and in humans, it measures approximately 50 μm in thickness.

The most superficial two to three layers are flat and polygonal in shape[1] with apical microvilli and microplicae, and covered by a charged glycocalyx,[2] which maximizes surface area with the mucinous layer of the tear film. At the cell periphery, tight junctions provide a watertight seal and assist in the prevention of pathogenic organisms from entering the cornea.

Directly posterior, the wing or suprabasal cells contribute a two- to three-cell thick layer and also demonstrate tight junction complexes between cells.

Basal epithelial cells represent the posterior-most layer of the corneal epithelium. Perilimbal basal epithelial cells differentiate and migrate anteriorly to repopulate the cornea; microvilli appear on the surface gradually during this process of maturation. Basal epithelial cells utilize hemidesmosomes to adhere to the underlying basement membrane and underlying stroma. The hemidesmosome, anchoring fibril, and anchoring filament complex produce an anchoring complex, which represents a common link between the intracellular cytoskeleton of the basal epithelial cell and the stroma posteriorly.[3]

1.2 Epithelial Basement Membrane

The epithelial basement membrane lies posterior to the epithelium and anterior to the corneal stroma and is laid down by basal epithelial cells.

Transmission electron microscopy reveals an anterior lamina lucida and posterior lamina densa, visible through transmission electron microscopy.[4]

While the lamina lucida is structured with laminins, the lamina densa is largely composed of collagens, laminin, heparan sulfate proteoglycans, and nidogens.[5] Immunohistochemical studies reveal a heterogeneously distributed regional variation of collagen IV subtypes dependent on its location in the central or peripheral cornea; these subtypes are composed of heterotrimers from six alpha chains.[5,6]

Laminin is the most frequent protein besides collagen and is also composed of heterotrimers, with one alpha, one beta, and one gamma chain. Structurally, it self-assembles into sheets and contributes to the embryological development of the epithelial basement membrane. Expression of laminin subunits varies over time during development,[7] and knockouts demonstrate severe dysfunction.[8]

The major heparan sulfate proteoglycan is perlacan, a protein distributed in basement membranes throughout the body which mediates migration, proliferation, and differentiation of other cells.[4] Keratinocyte survival and differentiation are regulated by perlacan, which has been shown to be critical for the formation of epidermis,[9] and upregulated after corneal stromal injury.[10]

Nidogens are sulfated glycoproteins with three globular domains connected with rodlike or thin segments[11] and are distributed throughout the basement membrane with strong affinity to laminin and collagen IV.[4] Nidogen-1 and Nidogen-2 each demonstrate distinct binding sites to collagen IV and laminin, respectively, reflected in inhibition assays and studies of recombinant fragment binding.[11]

1.3 Bowman's Layer

Bowman's layer is an acellular, nonregenerating layer posterior to the epithelial basement membrane, approximately 8–12 μm in depth and decreases in thickness over time.

Its collagen fibrils are distributed such that their posterior surfaces merge with the anterior stroma, leaving a smooth anterior surface. These fibers are only half to two-thirds the thickness of collagen fibrils in the stroma.[12]

1.4 Stroma

The stroma contributes the majority of the cornea's structural framework, measuring approximately 500 μm in humans and representing approximately 90% of the corneal anterior–posterior axis, as seen in Fig. 1. This layer

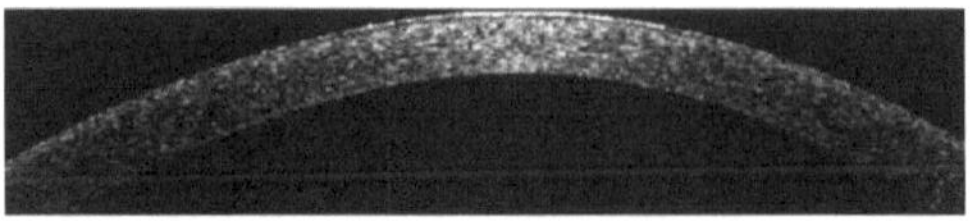

Figure 1 Normal human corneal structure as illustrated through anterior segment optical coherence tomography. The stroma composes over 90% of corneal volume. The epithelium, appreciated as a white band at the superior margin of the cornea, contributes less than 10% of overall thickness. Descemet membrane is seen as an even thinner band at the inferior edge.

is organized with a network of collagen fibers and ground substance, with an extracellular matrix composed of water, inorganic salts, proteoglycans, and glycoproteins.[13]

Stromal collagen is composed of a heterodimeric complex of Type I and Type V collagen with a narrow diameter.[13] Lumican and keratocan are the major keratan sulfate proteoglycans in the stroma, which also include mimecan. These demonstrate a bifunctional role to contribute structural support; protein moieties bind to collagen fibrils to modulate their diameters, while glycosaminoglycan chains are highly charged and promote interfibrillar spacing.[14] Loss of lumican in a mouse model results in corneal opacity,[15] and in addition to its structural role, lumican regulates neutrophil migration during bacterial infections of the stroma.[16] Decorin is a major proteoglycan associated with dermatan sulfate and similarly surrounds collagen fibrils. The concentration of proteoglycans varies with depth in the anterior–posterior axis, such that there is greater hydration in the posterior stroma.

Patterns of collagen lamellae also vary with depth. Anteriorly, the stroma is marked by short, narrow sheets with extensive interweaving, while collagen posteriorly demonstrates long, wide, thick lamellae extending from limbus to limbus, and without significant interlamellar connections.

Keratocytes, the major cells of the stroma, maintain the integrity of this layer, producing collagen, glycosaminoglycans, and matrix metalloproteinases.[1]

The corneal stroma represents one of the most highly innervated tissues in the human body. Sensory nerves from the nasociliary branch of the ophthalmic division of the trigeminal nerve course radially toward the central cornea through the anterior stroma. Between Bowman's layer and the anterior stroma, these nerves form the subepithelial nerve plexus, and then perforate through Bowman's layer to become the subbasal epithelial nerve plexus, innervating the basal epithelial layer.[17]

1.5 Descemet Membrane

Descemet membrane is the basement membrane of the corneal endothelium and measures approximately 3 μm in thickness in children, gradually thickening to 10 μm in adults. This membrane is composed of two layers: an anterior banded layer which is developed by collagen lamellae and proteoglycans and detected in fetal corneas as early as 12 weeks of gestation,[18] and a posterior nonbanded layer which is laid down by endothelial cells and thickens over decades.

Structurally, Descemet membrane contains collagen type IV and VIII fibrils. In contrast to other basement membranes throughout the body in which the type IV subtype is common, collagen type VIII is relatively specific to Descemet membrane and forms ladder-like structures visible under electron microscopy. In the anterior banded layer, collagen fibrils demonstrate a lattice-like configuration with periodic banding at 110 nm intervals; an additional 0.3–0.4 μm extension of this layer borders the stroma and is electron dense and homogenous.[18,19]

Posteriorly, the nonbanded layer is relatively homogenous with a fine granular appearance and increases in thickness over time as it is laid down by the corneal endothelium.

Similar to the stroma, laminin and fibronectin, as well as keratan sulfate, heparan sulfate, and dermatan sulfate are also present, and appropriate hydration is required to maintain clarity.

1.6 Endothelium

The corneal endothelium is composed of a single layer of flat, polygonal cells which line the posterior surface of the cornea, demonstrated in Fig. 2. These cells, approximately 5 μm in depth and with diameter of 20 μm, maintain the relatively dehydrated status of the stroma through ionic pumps in basolateral plasma membranes.

Actin filaments mediate cell migration and maintain cell shape and are located as dense peripheral bands apically.[20] Cadherin forms a thin pericellular band at apical cell junctions and extends along basolateral cell borders; alpha-catenin, beta-catenin, and plakoglobulin follow a similar pattern.[21] Tight junctions are also located near the apical aspect of cell membranes, and corneal endothelial cells express the cytoplasmic complex ZO-1,[22] which is associated with tight junctions.

Anterior to tight junctions, gap junctions are also present in corneal endothelium and contribute to the electrical coupling of endothelial

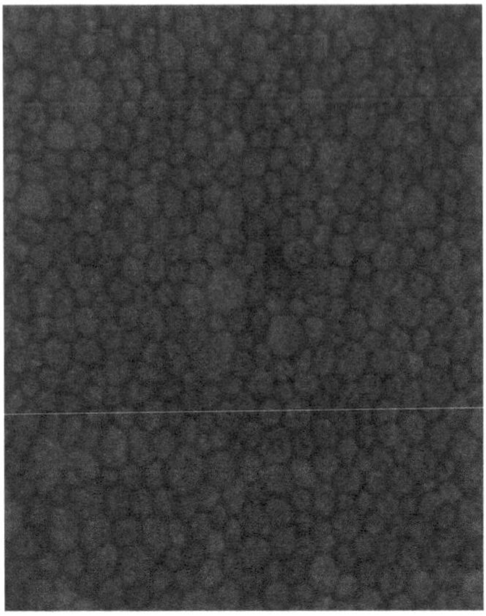

Figure 2 Confocal microscopy of the human corneal endothelium reveals a monolayer of flat cells distributed with a hexagonal pattern. Several larger cells contribute to slight variability in size (polymegethism) and shape (polymorphism).

cells.[23] The gap junction protein connexin-43 is expressed in the corneal endothelium.[24]

2. FUNCTION

2.1 Epithelium

At the surface, the epithelium plays a central role in vision as the majority of the refractive power of the eye occurs at the interface of air and the tear film. Among the anterior layers of flat squamous cells, tight junctions provide a surface barrier and assist to maintain the dehydrated state of the cornea, as well as inhibiting entry of pathogens into the stroma.

Corneal epithelial cells regularly undergo apoptosis and desquamation and are replenished by division from stem cells at the limbus, contributing to a net movement of epithelial cells in an apical direction. On average, the lifespan of epithelial cells is 7–10 days.[25] Genetic expression varies in basal epithelial cells as they shift from a perilimbal location to the central cornea, with mouse studies demonstrating 100 differentially expressed genes between basal epithelial cells in the two locations.[26] Distinct patterns of expression are evidenced by limbal-specific staining for ABCG2, K19, vimentin, KGF-R, metallothionein, and integrin alpha9, whereas basal cells of the corneal epithelium specifically stain for K3 and K12, Connexin 43, involucrin, P-cadherin, nestin, and integrins alpha2, alpha6, and beta4.[27]

The expression of Keratin-12 is specific to the corneal epithelium.[28] Dysfunction of this protein is appreciated with heterozygous mutations in KRT12, which are causative for Meesman's corneal dystrophy.

The barrier function of the epithelium is additionally important for immune regulation and prevention of pathogen entry into the cornea. Epithelial cells maintain toll-like receptors and secrete proinflammatory cytokines IL-1β, IL-6, IL-8, and TNF-α.[29] Langerhans cells exist at the basal layer of the corneal epithelium and subbasal nerve plexus[30] and demonstrate chemotaxis to the central cornea in response to epithelial-secreted IL-1.[31]

2.2 Epithelial Basement Membrane

The epithelium releases factors that modulate cell differentiation and apoptosis, including TGF-β1 and PDGF. As such, the epithelial basement membrane serves as a physical barrier to modulate the effect of these factors on keratocyte function. Removal of the epithelial basement membrane has been shown to accelerate processes of stromal wound healing[32] and carries implications for scar and haze formation in the stroma.

2.3 Bowman's Layer

Bowman's layer may play a role in protection of the subepithelial nerve plexus which courses through from the anterior stroma. Its absence, however, which occurs most commonly after phototherapeutic keratectomy in humans, does not appear to result in loss of vision or significant structural changes of the cornea overall. Moreover, among mammals, this layer is specific to primates.

2.4 Stroma

The stroma provides both structural support to the cornea and transparency by facilitating the passage of light through the framework of collagen fibrils in a manner that prevents scattering. The bifunctional role of proteoglycans promotes a small distance between collagen fibrils, decreasing the opportunity for light to scatter. The refractive index of the cornea decreases as light passes through the anterior–posterior axis due to a more hydrated posterior cornea. Disruptions in this pattern of lamellae, whether through hydration or abnormal deposits, disturb transparency and result in loss of vision.

For proper stromal function in the passage of light, the cornea must overcome scattering from keratocytes, from which the scattering of light is clearly appreciated during *in vivo* confocal microscopy. Similar to the lens, in which water-soluble crystallins contribute to clarity, corneal crystallins are water-soluble proteins which decrease light scattering. These include aldehyde dehydrogenase class 3,[33] transketolase,[34] and alpha-enolase.[35] The former two comprise 50% and 10% of water-soluble proteins in the cornea, respectively, and may play a role in reducing variability in the index of refraction as light passes through the stroma.[36]

Avascularity of the corneal stroma is also needed to facilitate transparency, and the production of antiangiogenic factors assists to maintain a balance with proangiogenic factors. These include the angiogenic properties of the molecule vascular endothelial growth factor (VEGF), which are effectively nullified by corneal secretion of soluble VEGF receptor-1 (sVEGFR-1 or sflt-1), trapping VEGF and blocking its activity. Supportive evidence of this neutralizing effect is seen through suppression of the soluble receptor by neutralizing antibodies or RNA interference resulting in angiogenesis, and the fact that patients with aniridia are deficient in this molecule.[37] Additional angiogenic factors include basic fibroblast growth factor (bFGF) and membrane type-1-MMP (MT1-MMP).[38] Thrombospondins may additionally play a role in regulating neovascularization.[39]

Stromal clarity is also dependent on appropriate keratocyte cell differentiation. Keratocyte-derived myofibroblast formation, which may occur with exposure to TGF-β1 and PDGF, is associated with development of corneal haze and disordered extracellular matrix production.[40]

In addition to surface immune regulation by the epithelium, the stroma plays a pivotal role in corneal immunity. The central cornea contains immature and precursor-type dendritic cells, while the peripheral cornea demonstrates resident bone marrow-derived dendritic cells.[41] Keratocytes may be incited by TNF-α and epithelial release of IL-1α to produce IL-6 and defensins.[31] Macrophages are present in the posterior stroma.[42]

2.5 Descemet Membrane

As the basement membrane for the corneal endothelium, Descemet membrane assists in the maintenance of corneal dehydration. This is evident in corneal hydrops, focal acute episodes of corneal edema experienced in keratoconic corneas in which breaks occur in Descemet membrane.

Iatrogenic Descemet membrane tear or detachment after intraocular surgery also results in corneal edema.

The posterior nonbanded layers continue to thicken throughout life through endothelial secretion, in contrast to the anterior descemet membrane (DM) which remains relatively constant.

2.6 Endothelium

The corneal endothelium plays an essential role in preserving stromal dehydration, thereby maximizing the fidelity of light passing through the cornea. Maintenance of this gradient of hydration depends on tight junctions among endothelial cells and pump function associated with Na^+/K^+-ATPase and bicarbonate-dependent Mg^{2+}-ATPase. Endothelial failure is, in turn, marked by corneal edema and represents a net influx of aqueous fluid into the cornea at a higher rate than the amount pumped out during a given period of time.

Aquaporin-1 is expressed in corneal endothelial cells, and mice lacking AQP1 demonstrate a reduced ability to counter corneal edema secondary to hypotonic saline exposure.[43]

A minimum number of endothelial cells are required to provide adequate pump function, and endothelial cell density decreases from approximately 3000–4000 cells/mm^2 at birth to 2500 cells/mm^2 in late adulthood. Confocal microscopy reveals endothelial cell loss over a lifetime at a rate of 10.92 cells/mm^2 per year.[44] A loss of corneal endothelial density to several hundred cells per mm^2, which may occur from intraocular surgery or other trauma, generally results in corneal edema.

Unlike the epithelium, endothelial cells do not demonstrate mitosis *in vivo*. Replacement of endothelial cell function from damage and cell death occurs not through cell division but migration. A focal loss of cells centrally, therefore, results in centripetal migration of adjacent endothelial cells, subsequent formation of tight junctions and restoration of pump function, and then a graduation remodeling of endothelial structure from enlarged, irregularly shaped endothelial cells to a more hexagonal pattern.[1]

A study of central, peripheral, and paracentral endothelial cell density in normal eyes and eye bank corneas demonstrated that endothelial cell density in paracentral and peripheral zones was higher by 5% and 10%, respectively, compared to the central cornea.[45] The exact causes of this distinction are unclear and may be due to patterns of migration or aqueous flow dynamics, or a process of peripheral endothelial cell proliferation yet to be understood.

3. DEVELOPMENT

The eye originates from the somatic ectoderm and neural tube and induction by the chordamesoderm. Following development of the diencephalon from induction of the anterior neural tube by the chordamesoderm, the optic vesicle begins to protrude. As the optic vesicle extends and meets the head ectoderm, it induces the thickening of surface ectoderm and formation of the lens placode, which invaginates to develop the lens. This process correlates with expression of the transcription factors Six3, Pax6, and Rx1 in the anterior neural plate. Rx1 is essential for eye formation, and Rx1-null mice do not develop optic cups and therefore do not produce eyes.[46] Six3 activation precedes Pax6 in the mouse, and in early inactivation of Six3, the surface ectoderm does not thicken to produce the lens placode.[66]

Pax6 serves as a master regulator of eye development. In the mouse, Pax6 expression is detected in head surface ectoderm at day 8 of embryogenesis (E8), and spatial differences in Pax6 expression are noted in surface ectoderm by E9.5[47]: a band of surface ectoderm in which Pax6 expression is switched off gives rise to epidermis, and an adjacent narrow band in which Pax6 is expressed commits to ocular development, subsequently resulting in formation of the lens and corneal epithelium.[48]

3.1 Epithelium

As the surface ectoderm above the neuronal optic cup invaginates, developing the crystalline lens, this primitive lens becomes round and induces formation of the overlying ectoderm into corneal epithelium. This process occurs in humans at approximately day 33 of gestation, and a bilayered epithelium is present by the fifth week of gestation.

In the mouse, commitment of the ectoderm to corneal epithelium begins at days E11–E12, when epithelial progenitor cytokeratins K5 and K14 replace ectodermal cytokeratins K8 and K18. At approximately days E15.5–E17.5, cytokeratins K12 and K3 are detected and are specific to the corneal epithelium.[49]

The mature corneal epithelium, which is composed of six to seven layers of cells, develops upon opening of the eyelids, which in humans is at approximately 24 weeks of gestation. The posterior-most basal epithelial cells morph from a flattened, ovoid shape to cuboidal and later a columnar formation. These cells flatten as they migrate anteriorly into the suprabasal or wing cells and then to the anterior corneal epithelium.

Corneal epithelial development is regulated at the molecular level by dynamic changes in genetic expression throughout development. In mice, microarray analysis of the expression of 8666 genes between postnatal day 10 and days 49–56 revealed 442 genes with distinct levels of expression between the two time points.[50]

MicroRNA (miRNA) function is essential for epithelial development, and targeted deletion of a ribonuclease necessary for miRNA function, Dicer, results in poorly stratified corneal epithelium and microphthalmia.[51] In human-induced pluripotent stem cells, miR-450b-5p serves as a switch for Pax6, inhibiting Pax6 expression and directing corneal epithelial fate; miR-184 is expressed in the corneal epithelium, and knockdown results in decreased Pax6 expression.[52] Mutation of a single base pair in miR-184 is responsible for EDICT syndrome in humans.[53]

3.2 Stroma

At approximately the seventh week of gestation in humans, a subpopulation of neural crest cells, which have originated at the junction of the closing neural tube and ectoderm and populated the periocular space,[54] develop mesenchymal cells and migrate into the space between lens and primitive corneal epithelium and develop into the stroma, endothelium, iris, and trabecular meshwork.

During this seventh week, migration of mesenchymal cells occurs in three waves. In humans, as well as reptiles, birds, and other primates, the first wave contributes to development of corneal and trabecular meshwork endothelium, and the second wave differentiates into keratocytes. In contrast, a single wave is present in rodents, cats, rabbits, and cattle.[55] A third wave contributes to iris development. Both TGF-β2 and FOXC contribute to differentiation of neural crest cells into corneal stroma.

Forkhead box (FOX) proteins share homology with the Drosophila forkhead transcription factor, which regulates expression of genes associated with cell growth, proliferation, and differentiation. In the mouse, Foxc1-null mice demonstrate disrupted stromal development.[56] In humans, autosomal dominant mutations in FOXC1 are associated with anterior segment dysgenesis resembling Axenfeld–Rieger syndrome and Peters anomaly.[57]

TGF-β2 modulates stromal development, and in knockout mice targeting the three TGF-β isoforms, only mice lacking the TGF-β2 isoform demonstrated corneal abnormalities; these included thin corneal stroma, absent corneal endothelium, corneolenticular fusion, and excessive hyaline cells in the vitreous.[58] Although this did not appear to involve apoptosis, stromal thinning was associated with fewer keratocytes and lumican, keratocan, and collagen I were significantly decreased.

During stromal development, glycosaminoglycans bind to proteins and fill spaces between collagen fibers, and keratan sulfate proteoglycans accumulate in this space. Lumican is essential for corneal clarity, and mice homozygous for a mutation in lumican develop bilateral corneal opacification and thick collagen fibrils visualized with transmission electron microscopy.[15] Keratocan expression during development correlates directly with lumican expression,[59] and mutations in KERA are associated with cornea plana in humans.

Migration of neural crest cells appears to be modulated by a combination of factors expressed by the lens and neural crest-derived cells that have initiated migration. Removal of the lens results in premature invasion and abnormal corneal differentiation. Semaphorin3A (Sema3A) is expressed in the lens placode and epithelium throughout eye development and inhibits periocular neural crest migration *in vitro*. Periocular neural crest expresses neuropilin-1 (Npn-1), and levels of expression decrease in cells that migrate to become corneal endothelium and stroma. A peptide which inhibits Sema3A/Npn-1 signaling results in early entry of neural crest cells over the lens.[54]

3.3 Descemet Membrane

The collagen of Descemet membrane is secreted by corneal endothelial cells and detected *in utero* as early 12 weeks of gestation, at which point an electron-lucent zone bordering the corneal endothelium lies adjacent to an electron-dense zone bordering the stroma.[18] These measure less than 40 nm in thickness, which is comparable at that age of gestation to the epithelial basement membrane. Descemet membrane becomes thicker with the addition of multiple layers, resulting in a multilayered structure. This process continues throughout gestation, from 1 layer at 12 weeks to 10 layers at approximately 27 weeks, and 30–40 layers by birth.[18]

By 16 weeks, early signs of differentiation are present, with the development of linear filaments measuring 170 nm in length and 40 nm in width, aligned perpendicularly between the electron-dense layers. These become uniformly distributed in a 110 nm banding pattern characteristic to the anterior banded layer.[18] At the time of birth, the membrane is largely composed of the anterior banded layer, measuring 3 μm in thickness.

In contrast, the posterior nonbanded layer continues to thicken over one's lifetime, at a slower rate and lacking the striated appearance of the anterior layer.

Mutations in COL8A2 are associated with a markedly thickened anterior banded layer of Descemet membrane, more than three times thicker than normal.[60]

3.4 Endothelium

Formation of the corneal endothelium, derived from the neural crest, occurs alongside development of the stroma, and factors which disturb stromal development frequently affect both layers. Foxc1-null mice demonstrate a lack of corneal endothelium in addition to disruption of stromal development,[56] and TGF-β2-null mice demonstrate an absent corneal endothelium associated with stromal thinning. In humans, miR-184-associated EDICT syndrome results in both stromal thinning and a beaten-metal appearance to the posterior cornea.[53]

The corneal endothelium develops as a monolayer and maintains this status through development and adult life, arrested in the G1 phase of development. Prevention of progression through the cell cycle is modulated by p27kip1, a G1-phase inhibitor, and cell–cell contact inhibits cell proliferation.[23] Contact inhibition is likely signaled from adherens junctions to the nucleus,[61] and disruption of adherens junctions may affect proliferation through β-catenin or p120 catenin.

Reversal or modulation of processes of corneal endothelial development may carry implications for wound healing and potentially for future therapies in repopulating corneal endothelium. In response to wound formation, fibroblast growth factor-2 plays a key role in an endothelial-to-mesenchymal transition, degrading p27, facilitating synthesis of type I collagen, inducing a change in endothelial cell shape into a fibroblastic morphology, and resulting in loss of the endothelial monolayer,[62] modulating this process through IL-1β and NF-κB.[63] In cultured human endothelial cells, exposure to EDTA alone disrupts cell junctions, and addition of bFGF results in both proliferation and endothelial-to-mesenchymal transition, with a fibroblastic-type appearance, and relocation of N-cadherin to the cytoplasm from its original location in the cell junction; addition of TGF-β1 induces a fibroblastic state but without proliferation.[64] Use of a selective inhibitor of TGF-β, SB431542, counteracts the fibroblastic phenotype in cultured human corneal endothelial cells.[65]

ACKNOWLEDGMENTS

This work was supported by NIH Grants K12 EY015025 (A.O.E.) and R01 EY016835 (J.D.G.).

REFERENCES

1. DelMonte DW, Kim T. Anatomy and physiology of the cornea. *J Cataract Refract Surg.* 2011;37(3):588–598.
2. Mochizuki H, Fukui M, Hatou S, Yamada M, Tsubota K. Evaluation of ocular surface glycocalyx using lectin-conjugated fluorescein. *Clin Ophthalmol.* 2010;4:925–930.
3. Gipson IK, Spurr-Michaud SJ, Tisdale AS. Anchoring fibrils form a complex network in human and rabbit cornea. *Invest Ophthalmol Vis Sci.* 1987;28(2):212–220.
4. Torricelli AA, Singh V, Santhiago MR, Wilson SE. The corneal epithelial basement membrane: structure, function, and disease. *Invest Ophthalmol Vis Sci.* 2013;54(9):6390–6400.
5. Tuori A, Uusitalo H, Burgeson RE, Terttunen J, Virtanen I. The immunohistochemical composition of the human corneal basement membrane. *Cornea.* 1996;15(3):286–294.
6. Ljubimov AV, Burgeson RE, Butkowski RJ, Michael AF, Sun TT, Kenney MC. Human corneal basement membrane heterogeneity: topographical differences in the expression of type IV collagen and laminin isoforms. *Lab Invest.* 1995;72(4):461–473.
7. Miner JH, Yurchenco PD. Laminin functions in tissue morphogenesis. *Annu Rev Cell Dev Biol.* 2004;20:255–284.
8. Smyth N, Vatansever HS, Murray P, et al. Absence of basement membranes after targeting the LAMC1 gene results in embryonic lethality due to failure of endoderm differentiation. *J Cell Biol.* 1999;144(1):151–160.
9. Sher I, Zisman-Rozen S, Eliahu L, et al. Targeting perlecan in human keratinocytes reveals novel roles for perlecan in epidermal formation. *J Biol Chem.* 2006;281(8):5178–5187.
10. Vittitow J, Borras T. Genes expressed in the human trabecular meshwork during pressure-induced homeostatic response. *J Cell Physiol.* 2004;201(1):126–137.
11. Fox JW, Mayer U, Nischt R, et al. Recombinant nidogen consists of three globular domains and mediates binding of laminin to collagen type IV. *EMBO J.* 1991;10(11):3137–3146.
12. Jacobsen IE, Jensen OA, Prause JU. Structure and composition of Bowman's membrane. Study by frozen resin cracking. *Acta Ophthalmol (Copenh).* 1984;62(1):39–53.
13. Torricelli AA, Wilson SE. Cellular and extracellular matrix modulation of corneal stromal opacity. *Exp Eye Res.* 2014;129:151–160.
14. Kao WW, Liu CY. Roles of lumican and keratocan on corneal transparency. *Glycoconj J.* 2002;19(4–5):275–285.
15. Chakravarti S, Magnuson T, Lass JH, Jepsen KJ, LaMantia C, Carroll H. Lumican regulates collagen fibril assembly: skin fragility and corneal opacity in the absence of lumican. *J Cell Biol.* 1998;141(5):1277–1286.
16. Wu F, Vij N, Roberts L, Lopez-Briones S, Joyce S, Chakravarti S. A novel role of the lumican core protein in bacterial lipopolysaccharide-induced innate immune response. *J Biol Chem.* 2007;282(36):26409–26417.

17. Oliveira-Soto L, Efron N. Morphology of corneal nerves using confocal microscopy. *Cornea.* 2001;20(4):374–384.
18. Murphy C, Alvarado J, Juster R. Prenatal and postnatal growth of the human Descemet's membrane. *Invest Ophthalmol Vis Sci.* 1984;25(12):1402–1415.
19. Johnson DH, Bourne WM, Campbell RJ. The ultrastructure of Descemet's membrane. I. changes with age in normal corneas. *Arch Ophthalmol.* 1982;100(12):1942–1947.
20. Olsen EG, Davanger M, Moen T. The role of microfilaments in the healing of the corneal endothelium. *Acta Ophthalmol (Copenh).* 1985;63(1):104–108.
21. Petroll WM, Hsu JK, Bean J, Cavanagh HD, Jester JV. The spatial organization of apical junctional complex-associated proteins in feline and human corneal endothelium. *Curr Eye Res.* 1999;18(1):10–19.
22. Barry PA, Petroll WM, Andrews PM, Cavanagh HD, Jester JV. The spatial organization of corneal endothelial cytoskeletal proteins and their relationship to the apical junctional complex. *Invest Ophthalmol Vis Sci.* 1995;36(6):1115–1124.
23. Joyce NC. Proliferative capacity of the corneal endothelium. *Prog Retin Eye Res.* 2003;22(3):359–389.
24. Mohay J, McLaughlin BJ. Corneal endothelial wound repair in normal and mitotically inhibited cultures. *Graefes Arch Clin Exp Ophthalmol.* 1995;233(11):727–736.
25. Hanna C, Bicknell DS, O'brien JE. Cell turnover in the adult human eye. *Arch Ophthalmol.* 1961;65:695–698.
26. Zhou M, Li XM, Lavker RM. Transcriptional profiling of enriched populations of stem cells versus transient amplifying cells. A comparison of limbal and corneal epithelial basal cells. *J Biol Chem.* 2006;281(28):19600–19609.
27. Schlotzer-Schrehardt U, Kruse FE. Identification and characterization of limbal stem cells. *Exp Eye Res.* 2005;81(3):247–264.
28. Tanifuji-Terai N, Terai K, Hayashi Y, Chikama T, Kao WW. Expression of keratin 12 and maturation of corneal epithelium during development and postnatal growth. *Invest Ophthalmol Vis Sci.* 2006;47(2):545–551.
29. Kumar A, Zhang J, Yu FS. Toll-like receptor 2-mediated expression of beta-defensin-2 in human corneal epithelial cells. *Microbes Infect.* 2006;8(2):380–389.
30. Zhivov A, Stave J, Vollmar B, Guthoff R. In vivo confocal microscopic evaluation of Langerhans cell density and distribution in the normal human corneal epithelium. *Graefes Arch Clin Exp Ophthalmol.* 2005;243(10):1056–1061.
31. Niederkorn JY, Peeler JS, Mellon J. Phagocytosis of particulate antigens by corneal epithelial cells stimulates interleukin-1 secretion and migration of Langerhans cells into the central cornea. *Reg Immunol.* 1989;2(2):83–90.
32. Pal-Ghosh S, Pajoohesh-Ganji A, Tadvalkar G, Stepp MA. Removal of the basement membrane enhances corneal wound healing. *Exp Eye Res.* 2011;93(6):927–936.
33. Alexander RJ, Silverman B, Henley WL. Isolation and characterization of BCP 54, the major soluble protein of bovine cornea. *Exp Eye Res.* 1981;32(2):205–216.
34. Jester JV, Moller-Pedersen T, Huang J, et al. The cellular basis of corneal transparency: evidence for 'corneal crystallins'. *J Cell Sci.* 1999;112(pt 5):613–622.
35. Cuthbertson RA, Tomarev SI, Piatigorsky J. Taxon-specific recruitment of enzymes as major soluble proteins in the corneal epithelium of three mammals, chicken, and squid. *Proc Natl Acad Sci USA.* 1992;89(9):4004–4008.
36. Piatigorsky J. Gene sharing in lens and cornea: facts and implications. *Prog Retin Eye Res.* 1998;17(2):145–174.
37. Ambati BK, Nozaki M, Singh N, et al. Corneal avascularity is due to soluble VEGF receptor-1. *Nature.* 2006;443(7114):993–997.
38. Han KY, Fahd DC, Tshionyi M, et al. MT1-MMP modulates bFGF-induced VEGF-A expression in corneal fibroblasts. *Protein Pept Lett.* 2012;19(12):1334–1339.

39. Cursiefen C, Masli S, Ng TF, et al. Roles of thrombospondin-1 and -2 in regulating corneal and iris angiogenesis. *Invest Ophthalmol Vis Sci.* 2004;45(4):1117–1124.
40. Wilson SE, Liu JJ, Mohan RR. Stromal-epithelial interactions in the cornea. *Prog Retin Eye Res.* 1999;18(3):293–309.
41. Hamrah P, Dana MR. Corneal antigen-presenting cells. *Chem Immunol Allergy.* 2007;92:58–70.
42. Hamrah P, Liu Y, Zhang Q, Dana MR. The corneal stroma is endowed with a significant number of resident dendritic cells. *Invest Ophthalmol Vis Sci.* 2003;44(2):581–589.
43. Thiagarajah JR, Verkman AS. Aquaporin deletion in mice reduces corneal water permeability and delays restoration of transparency after swelling. *J Biol Chem.* 2002;277(21):19139–19144.
44. Gambato C, Longhin E, Catania AG, Lazzarini D, Parrozzani R, Midena E. Aging and corneal layers: an in vivo corneal confocal microscopy study. *Graefes Arch Clin Exp Ophthalmol.* 2014;253(2):267–275.
45. Amann J, Holley GP, Lee SB, Edelhauser HF. Increased endothelial cell density in the paracentral and peripheral regions of the human cornea. *Am J Ophthalmol.* 2003;135(5):584–590.
46. Mathers PH, Grinberg A, Mahon KA, Jamrich M. The Rx homeobox gene is essential for vertebrate eye development. *Nature.* 1997;387(6633):603–607.
47. Swamynathan SK. Ocular surface development and gene expression. *J Ophthalmol.* 2013;2013:103947.
48. Ashery-Padan R, Gruss P. Pax6 lights-up the way for eye development. *Curr Opin Cell Biol.* 2001;13(6):706–714.
49. Wolosin JM, Budak MT, Akinci MA. Ocular surface epithelial and stem cell development. *Int J Dev Biol.* 2004;48(8–9):981–991.
50. Wu F, Lee S, Schumacher M, Jun A, Chakravarti S. Differential gene expression patterns of the developing and adult mouse cornea compared to the lens and tendon. *Exp Eye Res.* 2008;87(3):214–225.
51. Li Y, Piatigorsky J. Targeted deletion of dicer disrupts lens morphogenesis, corneal epithelium stratification, and whole eye development. *Dev Dyn.* 2009;238(9):2388–2400.
52. Shalom-Feuerstein R, Serror L, De La Forest Divonne S, et al. Pluripotent stem cell model reveals essential roles for miR-450b-5p and miR-184 in embryonic corneal lineage specification. *Stem Cells.* 2012;30(5):898–909.
53. Iliff BW, Riazuddin SA, Gottsch JD. A single-base substitution in the seed region of miR-184 causes EDICT syndrome. *Invest Ophthalmol Vis Sci.* 2012;53(1):348–353.
54. Lwigale PY, Bronner-Fraser M. Semaphorin3A/neuropilin-1 signaling acts as a molecular switch regulating neural crest migration during cornea development. *Dev Biol.* 2009;336(2):257–265.
55. Zieske JD. Corneal development associated with eyelid opening. *Int J Dev Biol.* 2004;48(8–9):903–911.
56. Kidson SH, Kume T, Deng K, Winfrey V, Hogan BL. The forkhead/winged-helix gene, Mf1, is necessary for the normal development of the cornea and formation of the anterior chamber in the mouse eye. *Dev Biol.* 1999;211(2):306–322.
57. Honkanen RA, Nishimura DY, Swiderski RE, et al. A family with Axenfeld-Rieger syndrome and peters anomaly caused by a point mutation (Phe112Ser) in the FOXC1 gene. *Am J Ophthalmol.* 2003;135(3):368–375.
58. Saika S, Saika S, Liu CY, et al. TGFbeta2 in corneal morphogenesis during mouse embryonic development. *Dev Biol.* 2001;240(2):419–432.
59. Carlson EC, Liu CY, Chikama T, et al. Keratocan, a cornea-specific keratan sulfate proteoglycan, is regulated by lumican. *J Biol Chem.* 2005;280(27):25541–25547.

60. Gottsch JD, Zhang C, Sundin OH, Bell WR, Stark WJ, Green WR. Fuchs corneal dystrophy: aberrant collagen distribution in an L450W mutant of the COL8A2 gene. *Invest Ophthalmol Vis Sci.* 2005;46(12):4504–4511.
61. Perez-Moreno M, Jamora C, Fuchs E. Sticky business: orchestrating cellular signals at adherens junctions. *Cell.* 2003;112(4):535–548.
62. Lee JG, Ko MK, Kay EP. Endothelial mesenchymal transformation mediated by IL-1beta-induced FGF-2 in corneal endothelial cells. *Exp Eye Res.* 2012;95(1):35–39.
63. Lee JG, Kay EP. NF-kappaB is the transcription factor for FGF-2 that causes endothelial mesenchymal transformation in cornea. *Invest Ophthalmol Vis Sci.* 2012;53(3):1530–1538.
64. Zhu YT, Chen HC, Chen SY, Tseng SC. Nuclear p120 catenin unlocks mitotic block of contact-inhibited human corneal endothelial monolayers without disrupting adherent junctions. *J Cell Sci.* 2012;125(pt 15):3636–3648.
65. Okumura N, Kay EP, Nakahara M, Hamuro J, Kinoshita S, Koizumi N. Inhibition of TGF-beta signaling enables human corneal endothelial cell expansion in vitro for use in regenerative medicine. *PLoS One.* 2013;8(2):e58000.
66. Liu W, Lagutin OV, Mende M, Streit A, Oliver G. Six3 activation of Pax6 expression is essential for mammalian lens induction and specification. *EMBO J.* 2006;25(22):5383–5395.

CHAPTER THREE

Stem Cells in the Cornea

Andrew J. Hertsenberg, James L. Funderburgh[1]
Department of Ophthalmology, University of Pittsburgh, Pittsburgh, Pennsylvania, USA
[1]Corresponding author: e-mail address: jlfunder@pitt.edu

Contents

Abstract

The cornea is the tough, transparent tissue through which light first enters the eye and functions as a barrier to debris and infection as well as two-thirds of the refractive power of the eye. Corneal damage that is not promptly treated will often lead to scarring and vision impairment. Due to the limited options currently available to treat corneal scars, the identification and isolation of stem cells in the cornea has received much attention, as they may have potential for autologous, cell-based approaches to the treatment of damaged corneal tissue.

The cornea is the tough, transparent tissue through which light first enters the eye. Serving as a barrier to debris and infection as well as two-thirds of the refractive power of the eye, the cornea is imperative to proper vision. Corneal damage that is not promptly treated will often lead to scarring and vision impairment. In fact, millions of people around the world suffer from corneal scars resulting in the loss of vision.[1]

Progress in Molecular Biology and Translational Science, Volume 134
ISSN 1877-1173
http://dx.doi.org/10.1016/bs.pmbts.2015.04.002
© 2015 Elsevier Inc.
All rights reserved.

The cornea is composed of the three cellular layers: the epithelium, stroma, and endothelium. The corneal epithelium is the most anterior layer and the first cellular barrier between the eye and environment. Like the epidermis of the skin, superficial corneal epithelium is continually sloughed off and replaced as it shields the eye from external insults. The stroma comprises roughly 90% of the cornea and is made primarily of highly organized collagen, making it both tough and transparent.[2] Damage to these layers by trauma or infection may result in corneal scarring, leading to visual impairment and often blindness. The corneal endothelium, the third and most posterior layer of the cornea, is a single-celled layer of epithelial cells responsible for maintaining deturgescence. The three cellular layers must function together to maintain transparency and, therefore, vision. Currently, the most common form of treatment for damage to any of these layers involves transplanting tissue, a procedure limited by the availability of donor tissue and complicated by the risk of immune-mediated rejection. In an attempt improve treatment options for corneal disorders and damage, research is being directed at bioprosthetics and stem cell biology.

Adult stem cells are characterized as slow-dividing cells with the ability to self-renew and give rise to differentiated progeny via mitosis. These adult stem cells are often found in specialized locations, or niches, in tissues throughout the body. When tissue is damaged (e.g., a flesh wound or blood loss), stem cell populations are often instrumental in replacing the lost cells to restore tissue function and integrity. Due to the devastating effects of corneal wounds and infections, and the limited options currently available to treat them, the identification and isolation of stem cells in the cornea has received much attention. The identification of stem cells in the cornea has the potential for autologous, cell-based approach to the treatment of damaged corneal tissue.

1. CORNEAL EPITHELIAL STEM CELLS

1.1 Anatomy

The corneal epithelium is a nonkeratinized, stratified squamous epithelium approximately 5–6 cells thick that covers the front of the cornea. The basal, columnar cell layer, is anchored to the basal lamina via hemidesmosomes and is covered by 2–3 layers of "wing" cells (Fig. 1). The outermost layer of cells is continuously sloughed off and replaced by the proliferation of wing and basal cells.[3] There is high corneal epithelial cell turnover due to blinking and both physical and chemical environmental insults. As such, there must be a

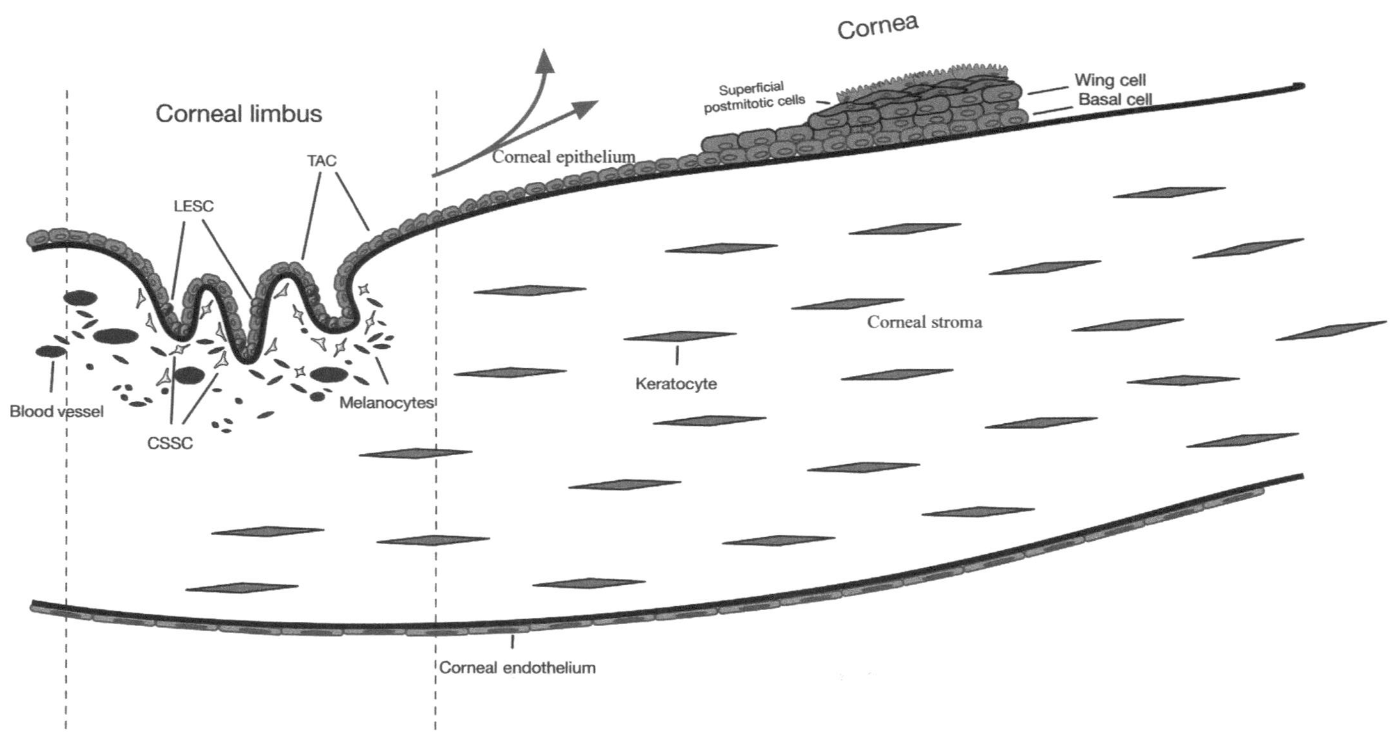

Figure 1 The cornea is composed of three cellular layers: the epithelium, stroma, and endothelium. The vascular limbal region is located at the peripheral cornea and is bordered by the conjunctiva—this region is the proposed niche for stem cell populations in each layer. LESC, limbal epithelial stem cell; TAC, transit-amplifying cell; CSSC, corneal stromal stem cell. (See the color plate.)

self-renewing source of corneal epithelial cells from which replacement cells can be drawn. It was suggested in 1971 that renewal of the corneal epithelium was maintained by the migration of epithelial cells in the basal layer of the epithelium.[4] We now know that this source is in the Palisades of Vogt at the limbal region that marks the transition zone between cornea and conjunctiva. A steady movement of epithelial cells in both human and mouse corneas from the limbal region toward the central cornea has been documented in a number of studies.[4–7] Located primarily at the superior and inferior corneal limbus, the Palisades are a vascularized series of crypts that provide a nutrient-rich, discrete, protected environment for limbal epithelial stem cells (LESCs) (Fig. 1). Cells here are protected from UV rays both by the upper and lower eyelids and by the presence of melanocytes. To support the hypothesis that this niche harbors LESC, the niche cells have been analyzed in a multitude of *in vitro* and *in vivo* studies for stem cell characteristics.

1.2 Characterization

DNA labeling of basal cells in the limbal region revealed them to be "slow cycling," a characteristic of stem cells. Basal limbal cells also show differences in protein expression when compared to basal cells of the central epithelium.[8] Keratin expression is notably distinct in the limbal basal cells with a lack of cytokeratins CK3 and CK12, and expression of CK14/CK5[9–11] (Fig. 2). Additionally, basal limbal cells grow clonally as holoclones, whereas clones from epithelial cells isolated from the central cornea are rare and do not proliferate extensively.[12] The adult stem cell marker ABCG2 has proven remarkably useful in the identification and isolation of stem cells due to its ability to efflux Hoechst 33342, a fluorescent dye that binds DNA. Expression of ABCG2 alters cellular fluorescence in the presence of Hoechst, allowing sorting of a cell population called the "side population" (SP) via fluorescence-activated cell sorting (FACS), separating the ABCG2$^+$ cells (stem cells) from a heterogeneous population. SP sorting of the limbal epithelium has indeed isolated a population of ABCG2$^+$ stem cells that have greater colony-forming efficiency and reduced K3/K12 expression compared to central corneal epithelium.[13,14] ABCB5, a member of the ATP-binding cassette family of proteins, has also been identified as a definitive LESC marker. Experiments in both mouse and human revealed that the ABCB5$^+$ limbal epithelium is necessary for proper wound healing, corneal development, and epithelial homeostasis.[15] Additional stem cells genes are

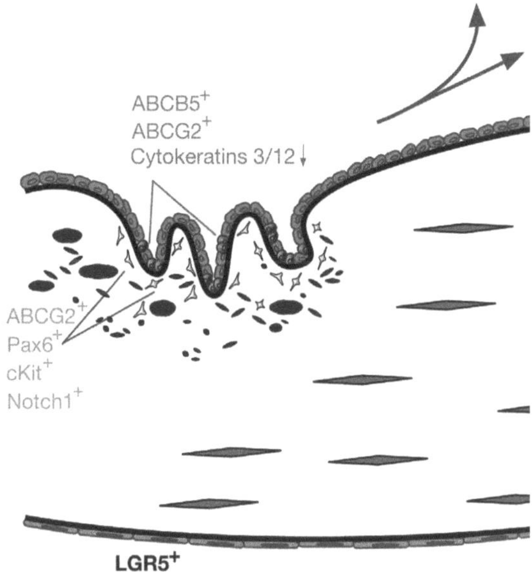

Figure 2 Stem cell markers/characteristics. Orange, limbal epithelial stem cell (LESC) markers, expression generally compared to central epithelial cells; green, human corneal stromal stem cell markers; and purple, proposed markers and characteristics corneal endothelial progenitor cells. (See the color plate.)

expressed by LESC, notable among them are C/EBP-delta and Notch1, proteins known to regulate stem cell self-renewal and differentiation.[16,17]

1.3 Wound Healing

In self-renewing epithelia, stem cells are essential for normal tissue maintenance; a second important characteristic is the ability of stem cells to regenerate tissue that has been damaged or lost.[18–20] Additional studies *in vivo* support the importance of LESC after epithelial debridement, concluding that limbal epithelial cells are involved in the rapid, functional healing of the epithelium.[21,22] Kruse *et al.* showed that when the basal limbal epithelium is completely removed before corneal wounding, vascularization, and conjunctivalization result. When the limbal cells were in place, however, wounds healed without vascularization and cells from the limbal region migrated into the central cornea to reform the epithelium.[23] In addition to replenishing the mitotic basal cells of the central epithelium, cells in the limbal epithelium appear to provide a biological barrier preventing conjunctivalization and vascularization of the corneal epithelium after wounding.[24,25]

Expansion of limbal cells *in vitro* and transplantation to central cornea can restore epithelial function for >10 years in human eyes.[26] In these corneas, there is no direct evidence that stem cells have repopulated a limbal niche, leaving open the question as to whether this tissue, like skin, can be supported by a distributed population of stem cells. The idea is supported by a study of destrin knockout mice, which show limited or no migration of epithelial cells.[27] Indeed, Dua observed human eyes with limbal damage to contain patches of normal epithelium in the central cornea, suggesting maintenance of the epithelial layer by local stem cells. In a 2008 study, limbal tissue expressing beta-galactosidase (βGal) was transplanted into normal mouse eyes but showed no centripetal migration of the labeled cells unless the central epithelium was wounded.[28] Both this and a more recent study found that cells from the central epithelium are capable of growing as holoclones.[28,29] These studies suggest that rare stem cells may occur in the central epithelium and that, where they do occur, they can maintain a stable column of epithelium in a manner similar to that of epidermis. A more recent study reconfirmed limbal stem cell-driven migration using lineage-tracing expression of fluorescent proteins driven by a K14 promoter. This gene is limited to stem and progenitor cells in the limbus, thus all fluorescent cells after induction will represent descendants of those progenitors. After induction, fluorescent cells were observed in radial streaks of cells moving toward the central cornea at about 11 μm/day in normal, unwounded eyes.[30] These experiments confirm the centripetal migration of cells observed in many earlier studies and support the idea that these cells in the central cornea are derived from limbal stem cells. The current state of our understanding leaves open the question as to whether stem cells in the central cornea can and do contribute to epithelial homeostasis on a long-term basis and whether the centripetal movement of progenitor-derived cells is a requirement for maintenance of healthy and stable corneal epithelium.

In spite of these unresolved questions, there remains no doubt in the clinical importance of LESC. This is manifest in the condition known as limbal stem cell deficiency (LSCD). In this relatively rare condition, loss of limbal cells, typically as a result of genetic disease or chemical burns, results in conjunctivalization of the cornea, inflammation, neovascularization, pain, and corneal opacity.[31] Standard penetrating keratoplasty involves only central cornea and allogenic tissue grafts that do not include limbal tissue experience an increased rate of failure.[32–34] In unilateral LSCD, limbal tissue can be harvested from the healthy eye without significant damage and cells transplanted on the affected cornea either directly or after expansion in culture.[35–37] Use of autologous LESC provides long-term stable epithelium to 50–70% of recipient eyes and regrafting can improve that result.[38,39] LESC allografts from cadaver eyes often fail[40]; however, use of biopsy tissue from living relatives and systemic antirejection drug therapy has led to survival rates for allogenic grafts similar to that of autologous tissue.[41] LESC treatment has been carried out on >1000 individuals, becoming the most common and successful of nonhematopoietic stem cell transplantation procedures, restoring vision to individuals with no alternative.

2. CORNEAL STROMAL STEM CELLS

2.1 Anatomy

The corneal stroma is a tough, collagenous tissue derived from the embryonic neural crest. Corneal collagen is arranged in sheets of fibrils called lamellae that are arranged orthogonally throughout the stroma. The precise spacing and arrangement of the fibrils and lamellae are essential to stromal transparency.[42] The stroma is sparsely populated by quiescent, mesenchymal cells called keratocytes which are responsible for collagen production and turnover.[43,44] Stromal scarring can occur as a result of ocular trauma, surgery, or infection. Loss of visual acuity due to stromal opacity affects more than 23 million individuals worldwide and 4.6 million are estimated to suffer bilateral corneal blindness. Although comprehensive data are limited, stromal opacity is the source of most corneal blindness, greatly exceeding the numbers of individuals affected by LSCD.[45–47] Although fully prosthetic corneal replacements are in limited use and acellular prosthetic tissue replacements are in clinical trials, replacement of scarred stromal tissue with cadaveric human stromal tissue is the current clinical method of choice.[48,49] Corneal scarring results from trauma or infectious keratitis that initiates an inflammatory response. Upon wounding, the normally quiescent keratocytes adjacent to the affected area become mitotically active fibroblasts, migrate to the wounded area, and subsequently differentiate to myofibroblasts. These cells are responsible for the production of fibrotic extracellular matrix components that contribute to light scatter by stromal scarring.[43,44,50,51]

2.2 Characterization

Over the past 15–20 years, a large number of reports have identified and characterized stem cells from various mesenchymal tissues. These stem cell populations are identified by self-renewal ability, differentiation into multiple cell types, and their ability to grow clonally. In an initial study, a small population of cells exhibiting clonal growth from bovine corneal stroma was found to express genes associated with mesenchymal stem cells.[52] Successful stem cell isolation from the bovine cornea led to investigation of human corneal tissue for a similar population. To achieve this, the well-documented adult stem cell marker ABCG2 was used.[53] ABCG2 is an ABC cassette membrane transporter, which has the ability to efflux the DNA-binding dye Hoechst 33342, allowing the $ABCG2^{(+)}$ cells to be sorted via FACS. $ABCG2^{+}$ cells from the limbal region of human cornea were found to represent less than 1% of the total cell population. These SP cells were shown to grow clonally and exhibited a multipotent differentiation potential, unlike $ABCG2^{(-)}$ cells isolated from the same region.[54] SP cells also expressed the stem cell markers Bmi1, Notch1, cKit, Six2, and Pax6 (Fig. 2). When cultured in serum-free medium supplemented with ascorbic acid and insulin, the human corneal SP cells upregulated keratocyte-specific markers, including the corneal stroma-specific proteoglycan keratocan.[54] Since the original report, cells isolated from corneal stroma with characteristics of mesenchymal stem cells have been described in a number of publications.[55–62] Similar properties and the location of these reported stem cells support the idea that each study is describing the same population of mesenchymal stem cells largely localized to the anterior limbal stroma.[63]

2.3 Niche Function of Stromal Stem Cells

Several studies have shown limbal mesenchymal cells with stem cell properties to be closely associated *in vivo* with limbal epithelial cells.[59,61,63–68] The two cell types are also coisolated in collagenase digests of the limbal tissue.[61,64] *In vivo* and *ex vivo*, the two associated cell types, exhibit different protein phenotypes, and both express stem cell genes.[61,64] A recent three-dimensional electron microscopic analysis of the limbal region found the epithelial basement membrane to be fenestrated, providing direct cell–cell contact between basal epithelial cells and elongated stromal cells in the limbus.[68] Melanocytes were also associated with the niche complex. Both LESC and the limbal mesenchymal stem cell express N-cadherin, suggesting this cell–cell junction protein provides interaction between these cell populations.[59,65,69,70] *In vitro*, coculture of LESC with coisolated stromal stem cells improved LESC expansion and clonogenicity.[59] Similarly, culture of limbal epithelial cells shows improved expansion, if it is carried out in the presence of limbal mesenchymal cells but less so in the presence of mesenchymal cells from the central stroma.[64] These findings lend credence to the idea that some or all of the mesenchymal stem cells in the stroma exist *in vivo* as a part of a multicellular limbal niche complex, and their presence supports the stem cell character of the LESC population.

2.4 Bioengineering Corneal Tissue with Stromal Stem Cells

Limited access to donated cadaveric tissue reduces the number of individuals with corneal blindness who benefit from corneal grafts. This shortage has led to widespread interest in development of alternative to penetrating keratoplasty. Alternatives include fully synthetic corneal prostheses as well as acellular polymeric replacements for scarred regions of the stroma.[48,49] Other investigators have examined the potential for bioengineering stromal tissue *in vitro* using a variety of cell types and scaffolding to generate stroma-like tissue.[32–34,38,40,41,71–103] Corneal stromal stem cells (CSSC) have also been shown to produce a collagenous matrix similar to that seen in the corneal stroma. When cultured on parallel, aligned nanofibers, CSSC secrete layers of organized collagen nearly identical in fibril diameter and spacing to that of the corneal stroma. This tissue construct also contains the unique corneal proteoglycans which are known to be required for corneal transparency.[74,75,88,94,95,100] Although the stem cell-based bioengineered constructs have yet to be tested *in vivo*, the stromal stem cells appear to be highly suited for generating tissue *in vitro* that may be useful for replacing scarred stroma.

2.5 Anti-Inflammatory Properties

In addition to their ability to generate corneal tissues *in vitro*, CSSC like many mesenchymal stem cells exhibit a potential to mediate immune response. In the lumican knockout mouse model of corneal haze, CSSC were shown to restore transparency after being injected directly into the stroma.[104] Even more striking, CSSC were shown to completely prevent stromal scarring in a mouse wound model.[105] Prevention of scarring appears to be the result of paracrine signaling, as tissue regeneration occurs both in the anterior stroma where the cells were present and the posterior where no cells were seen.[105] Importantly, in neither *in vivo* study was T-cell-mediated tissue rejection observed. This points to both the immune-privilege and immunomodulatory characteristics of CSSC.

3. CORNEAL ENDOTHELIUM STEM/PROGENITOR CELLS

3.1 Anatomy

The corneal endothelium is a simple, squamous epithelial sheet on the posterior side of the cornea. Serving as a "leaky" barrier, the corneal endothelium is responsible for pumping fluid out of the corneal stroma to prevent the development of edematous haze. The barrier "leaks" to let water and nutrients into the stroma from the anterior chamber, thus establishing two opposing forces that must function properly in order to maintain corneal transparency (Fig. 3).

The corneal endothelium is notable for its lack of mitotic activity after birth and inability to regenerate after damage.[106] Upon cell loss during aging, surrounding cells spread out and change shape. If enough cells are lost, the pump function is compromised and the cornea becomes edematous and cloudy. Perhaps, the most common disorder of this nature is Fuch's Dystrophy, leading to the thickening of Descemet's membrane and corneal edema causing corneal haze. Currently, the only treatment option to restore vision requires transplantation of donor tissue. While this is largely successful, it

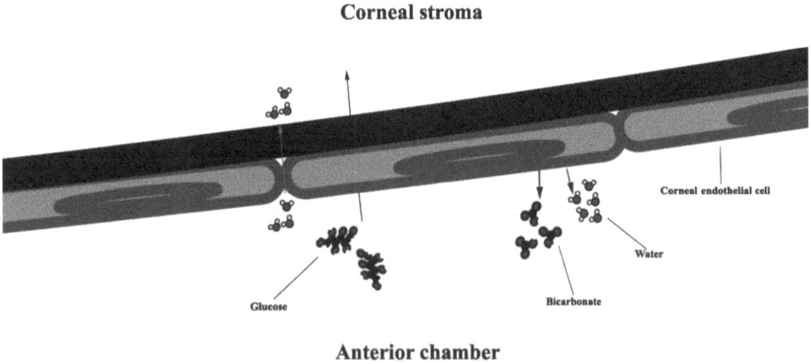

Figure 3 The corneal endothelium serves as a leaky barrier between the corneal stroma (anterior) and anterior chamber (posterior). Water passively moves from the ac into the stroma, while protein and other nutrients (such as glucose) are actively transported by corneal endothelial cells. To pump water from the stroma into the anterior chamber, bicarbonate is actively pumped out of the endothelial cells (along with other ions) and water follows. (See the color plate.)

does not avoid the problems often encountered with transplant operations, namely immune rejection and the declining availability of donor tissue.

3.2 Characterization

Many laboratories are searching for corneal endothelial stem cells for their potential in regenerative medicine. It has been shown that cells with clonogenic potential exist in the normal adult corneal endothelium by sphere formation.[107] While these cells did not express markers that would be indicative of a stem cell population, clonogenic sphere formation and their ability to form a hexagonal monolayer of cells with pump function may indicate that they are corneal endothelial progenitor cells.[56] As is seen in other ocular tissues, notably, the corneal epithelium, stroma, and retina, proposed stem cell niches are often found at the periphery of the tissue.[12,54,108,109] Further characterization of the sphere forming corneal endothelial cells revealed that the cells at the periphery of the endothelium have a greater propensity to form spheres than do those in the center, as has previously been described in rabbits.[110,111]

Recently, a progenitor cell population in human corneal endothelium was identified using the established stem cell marker LGR5.[112] $LGR5^+$ endothelial cells were shown to have greater proliferation capacity than $LGR5^-$ cells and were located at the periphery of the cornea, both expected characteristics of a stem/progenitor cell population. Another laboratory identified a progenitor cell population based on expression of the neural crest cell markers p75, SOX9, and FOXC2.[113] It was further shown that the progenitor cells had high proliferative capacity and demonstrated pump function. While there was no statistical difference between the number of $p75^+$ cells between the periphery and central corneal endothelium, the cells in the center were dispersed and those at the periphery much more concentrated, suggesting the presence of a niche.

3.3 Conclusion

Corneal wounding and disease often have the devastating consequence of visual impairment or blindness. While there are treatment options available when this occurs, tissue rejection and graft availability limit the ability to treat the numerous patients in need. Stem cells isolated from the cornea may offer an alternative to the current treatment options in that they can be used in an autologous fashion. This is already being used for trauma and corneal epithelial disorders such as LSCD and is widely successful.[39,114]

CSSC have successfully been isolated from human tissue and animal studies show encouraging results that these cells may soon be used to treat stromal scarring. The corneal endothelium appears to have a progenitor population that may provide a source of cells for regeneration, though more studies need to be completed to demonstrate this. Taken together, the progress made thus far in the isolation and use of corneal stem cells for use in the clinic is promising. Though much work still needs to be done, an alternative, cell-based treatment for corneal pathologies is achievable.

REFERENCES

1. Johnson GJ. Vision 2020: the right to sight: report on the sixth general assembly of the international agency for the prevention of blindness (IAPB). *Commun Eye Health.* 1999;12(32):59–60.
2. Hassell JR, Birk DE. The molecular basis of corneal transparency. *Exp Eye Res.* 2010;91(3):326–335.
3. Thoft RA, Friend J. The X, Y, Z hypothesis of corneal epithelial maintenance. *Invest Ophthalmol Vis Sci.* 1983;24(10):1442–1443.
4. Davanger M, Evensen A. Role of the pericorneal papillary structure in renewal of corneal epithelium. *Nature.* 1971;229(5286):560–561.
5. Nagasaki T, Zhao J. Centripetal movement of corneal epithelial cells in the normal adult mouse. *Invest Ophthalmol Vis Sci.* 2003;44(2):558–566.
6. Collinson JM, Morris L, Reid AI, et al. Clonal analysis of patterns of growth, stem cell activity, and cell movement during the development and maintenance of the murine corneal epithelium. *Dev Dyn.* 2002;224(4):432–440.
7. Buck RC. Measurement of centripetal migration of normal corneal epithelial cells in the mouse. *Invest Ophthalmol Vis Sci.* 1985;26(9):1296–1299.
8. Schlötzer-Schrehardt U, Kruse FE. Identification and characterization of limbal stem cells. *Exp Eye Res.* 2005;81(3):247–264.
9. Cotsarelis G, Cheng SZ, Dong G, Sun TT, Lavker RM. Existence of slow-cycling limbal epithelial basal cells that can be preferentially stimulated to proliferate: implications on epithelial stem cells. *Cell.* 1989;57(2):201–209.
10. Schermer A, Galvin S, Sun TT. Differentiation-related expression of a major 64K corneal keratin in vivo and in culture suggests limbal location of corneal epithelial stem cells. *J Cell Biol.* 1986;103(1):49–62.
11. Rodrigues M, Ben-Zvi A, Krachmer J, Schermer A, Sun TT. Suprabasal expression of a 64-kilodalton keratin (no. 3) in developing human corneal epithelium. *Differentiation.* 1987;34(1):60–67.
12. Lavker RM, Tseng SC, Sun TT. Corneal epithelial stem cells at the limbus: looking at some old problems from a new angle. *Exp Eye Res.* 2004;78(3):433–446.
13. Chen Z, de Paiva CS, Luo L, Kretzer FL, Pflugfelder SC, Li DQ. Characterization of putative stem cell phenotype in human limbal epithelia. *Stem Cells.* 2004;22(3):355–366.
14. de Paiva CS, Chen Z, Corrales RM, Pflugfelder SC, Li DQ. ABCG2 transporter identifies a population of clonogenic human limbal epithelial cells. *Stem Cells.* 2005;23(1):63–73.
15. Ksander BR, Kolovou PE, Wilson BJ, et al. ABCB5 is a limbal stem cell gene required for corneal development and repair. *Nature.* 2014;511(7509):353–357.

16. Barbaro V, Testa A, Di Iorio E, Mavilio F, Pellegrini G, De Luca M. C/EBPdelta regulates cell cycle and self-renewal of human limbal stem cells. *J Cell Biol.* 2007;177(6):1037–1049.
17. Thomas PB, Liu YH, Zhuang FF, et al. Identification of Notch-1 expression in the limbal basal epithelium. *Mol Vis.* 2007;13:337–344.
18. Barker N. Adult intestinal stem cells: critical drivers of epithelial homeostasis and regeneration. *Nat Rev Mol Cell Biol.* 2014;15(1):19–33.
19. Spradling A, Drummond-Barbosa D, Kai T. Stem cells find their niche. *Nature.* 2001;414(6859):98–104.
20. Tadeu AM, Horsley V. Epithelial stem cells in adult skin. *Curr Top Dev Biol.* 2014;107:109–131.
21. Chen JJ, Tseng SC. Corneal epithelial wound healing in partial limbal deficiency. *Invest Ophthalmol Vis Sci.* 1990;31(7):1301–1314.
22. Huang AJ, Tseng SC. Corneal epithelial wound healing in the absence of limbal epithelium. *Invest Ophthalmol Vis Sci.* 1991;32(1):96–105.
23. Kruse FE, Chen JJ, Tsai RJ, Tseng SC. Conjunctival transdifferentiation is due to the incomplete removal of limbal basal epithelium. *Invest Ophthalmol Vis Sci.* 1990;31(9):1903–1913.
24. Ahmad S. Concise review: limbal stem cell deficiency, dysfunction, and distress. *Stem Cells Transl Med.* 2012;1(2):110–115.
25. Sangwan VS. Limbal stem cells in health and disease. *Biosci Rep.* 2001;21(4):385–405.
26. Goldstein RS, Pomp O, Brokhman I, Ziegler L. Generation of neural crest cells and peripheral sensory neurons from human embryonic stem cells. *Methods Mol Biol.* 2010;584:283–300.
27. Zhang W, Zhao J, Chen L, Urbanowicz MM, Nagasaki T. Abnormal epithelial homeostasis in the cornea of mice with a destrin deletion. *Mol Vis.* 2008;14:1929–1939.
28. Majo F, Rochat A, Nicolas M, Jaoude GA, Barrandon Y. Oligopotent stem cells are distributed throughout the mammalian ocular surface. *Nature.* 2008;456(7219): 250–254.
29. Chang C-YA, McGhee JJ, Green CR, Sherwin T. Comparison of stem cell properties in cell populations isolated from human central and limbal corneal epithelium. *Cornea.* 2011;30(10):1155–1162.
30. Di Girolamo N, Bobba S, Raviraj V, et al. Tracing the fate of limbal epithelial progenitor cells in the murine cornea. *Stem Cells.* 2015;33(1):157–169.
31. Dua HS, Gomes JA, Singh A. Corneal epithelial wound healing. *Br J Ophthalmol.* 1994;78(5):401–408.
32. Reinhard T, Spelsberg H, Henke L, et al. Long-term results of allogeneic penetrating limbo-keratoplasty in total limbal stem cell deficiency. *Ophthalmology.* 2004;111(4): 775–782.
33. Miri A, Al-Deiri B, Dua HS. Long-term outcomes of autolimbal and allolimbal transplants. *Ophthalmology.* 2010;117(6):1207–1213.
34. Eberwein P, Böhringer D, Schwartzkopff J, Birnbaum F, Reinhard T. Allogenic limbo-keratoplasty with conjunctivoplasty, mitomycin C, and amniotic membrane for bilateral limbal stem cell deficiency. *Ophthalmology.* 2012;119(5):930–937.
35. Pellegrini G, Traverso CE, Franzi AT, Zingirian M, Cancedda R, De Luca M. Long-term restoration of damaged corneal surfaces with autologous cultivated corneal epithelium. *Lancet.* 1997;349(9057):990–993.
36. Tsai RJ, Tseng SC. Human allograft limbal transplantation for corneal surface reconstruction. *Cornea.* 1994;13(5):389–400.
37. Koizumi N, Inatomi T, Suzuki T, Sotozono C, Kinoshita S. Cultivated corneal epithelial stem cell transplantation in ocular surface disorders. *Ophthalmology.* 2001;108(9):1569–1574.

38. Sangwan VS, Jain R, Basu S, et al. Transforming ocular surface stem cell research into successful clinical practice. *Indian J Ophthalmol.* 2014;62(1):29–40.
39. Rama P, Matuska S, Paganoni G, Spinelli A, De Luca M, Pellegrini G. Limbal stem-cell therapy and long-term corneal regeneration. *N Engl J Med.* 2010;363(2):147–155.
40. Shortt AJ, Bunce C, Levis HJ, et al. Three-year outcomes of cultured limbal epithelial allografts in aniridia and Stevens-Johnson syndrome evaluated using the Clinical Outcome Assessment in Surgical Trials assessment tool. *Stem Cells Transl Med.* 2014;3(2): 265–275.
41. Basu S, Fernandez MM, Das S, Gaddipati S, Vemuganti GK, Sangwan VS. Clinical outcomes of xeno-free allogeneic cultivated limbal epithelial transplantation for bilateral limbal stem cell deficiency. *Br J Ophthalmol.* 2012;96(12):1504–1509.
42. Knupp C, Pinali C, Lewis PN, et al. The architecture of the cornea and structural basis of its transparency. *Adv Protein Chem Struct Biol.* 2009;78:25–49.
43. Funderburgh JL, Funderburgh ML, Mann MM, Corpuz L, Roth MR. Proteoglycan expression during transforming growth factor beta-induced keratocyte-myofibroblast transdifferentiation. *J Biol Chem.* 2001;276(47):44173–44178.
44. Funderburgh JL, Mann MM, Funderburgh ML. Keratocyte phenotype mediates proteoglycan structure: a role for fibroblasts in corneal fibrosis. *J Biol Chem.* 2003;278(46):45629–45637.
45. Oliva MS, Schottman T, Gulati M. Turning the tide of corneal blindness. *Indian J Ophthalmol.* 2012;60(5):423–427.
46. Shortt AJ, Tuft SJ, Daniels JT. Ex vivo cultured limbal epithelial transplantation. A clinical perspective. *Ocul Surf.* 2010;8(2):80–90.
47. Ontario HQ. Limbal stem cell transplantation: an evidence-based analysis. *Ont Health Technol Assess Ser.* 2008;8(7):1–58.
48. Tan DTH, Dart JKG, Holland EJ, Kinoshita S. Corneal transplantation. *Lancet.* 2012;379(9827):1749–1761.
49. Lagali N, Fagerholm P, Griffith M. Biosynthetic corneas: prospects for supplementing the human donor cornea supply. *Expert Rev Med Devices.* 2011;8(2):127–130.
50. Fini ME. Keratocyte and fibroblast phenotypes in the repairing cornea. *Prog Retin Eye Res.* 1999;18(4):529–551.
51. Cintron C, Schneider H, Kublin C. Corneal scar formation. *Exp Eye Res.* 1973;17(3):251–259.
52. Funderburgh ML, Du Y, Mann MM, Sundar Raj N, Funderburgh JL. PAX6 expression identifies progenitor cells for corneal keratocytes. *FASEB J.* 2005;19(10):1371–1373.
53. Golebiewska A, Brons NH, Bjerkvig R, Niclou SP. Critical appraisal of the side population assay in stem cell and cancer stem cell research. *Cell Stem Cell.* 2011;8(2):136–147.
54. Du Y, Funderburgh ML, Mann MM, SundarRaj N, Funderburgh JL. Multipotent stem cells in human corneal stroma. *Stem Cells.* 2005;23(9):1266–1275.
55. Yoshida S, Shimmura S, Nagoshi N, et al. Isolation of multipotent neural crest-derived stem cells from the adult mouse cornea. *Stem Cells.* 2006;24(12):2714–2722.
56. Amano S, Yamagami S, Mimura T, Uchida S, Yokoo S. Corneal stromal and endothelial cell precursors. *Cornea.* 2006;25(10 suppl 1):S73–S77.
57. Polisetty N, Fatima A, Madhira SL, Sangwan VS, Vemuganti GK. Mesenchymal cells from limbal stroma of human eye. *Mol Vis.* 2008;14:431–442.
58. Lu J-M, Zhou Z-Y, Zhang X-R, Li X-L, Wang H-F, Song X-J. A preliminary study of mesenchymal stem cell-like cells derived from murine corneal stroma. *Graefes Arch Clin Exp Ophthalmol.* 2010;248(9):1279–1285.
59. Xie H-T, Chen S-Y, Li G-G, Tseng SCG. Isolation and expansion of human limbal stromal niche cells. *Invest Ophthalmol Vis Sci.* 2012;53(1):279–286.

60. Branch MJ, Hashmani K, Dhillon P, Jones DRE, Dua HS, Hopkinson A. Mesenchymal stem cells in the human corneal limbal stroma. *Invest Ophthalmol Vis Sci*. 2012;53(9):5109–5116.
61. Li G-G, Zhu Y-T, Xie H-T, Chen S-Y, Tseng SCG. Mesenchymal stem cells derived from human limbal niche cells. *Invest Ophthalmol Vis Sci*. 2012;53(9):5686–5697.
62. Garfias Y, Nieves-Hernandez J, Garcia-Mejia M, Estrada-Reyes C, Jimenez-Martinez-MC. Stem cells isolated from the human stromal limbus possess immunosuppressant properties. *Mol Vis*. 2012;18:2087–2095.
63. Pinnamaneni N, Funderburgh JL. Concise review: stem cells in the corneal stroma. *Stem Cells*. 2012;30(6):1059–1063.
64. Chen S-Y, Hayashida Y, Chen M-Y, Xie HT, Tseng SCG. A new isolation method of human limbal progenitor cells by maintaining close association with their niche cells. *Tissue Eng Part C Methods*. 2011;17(5):537–548.
65. Hayashi R, Yamato M, Sugiyama H, et al. N-Cadherin is expressed by putative stem/progenitor cells and melanocytes in the human limbal epithelial stem cell niche. *Stem Cells*. 2007;25(2):289–296.
66. Higa K, Kato N, Yoshida S, et al. Aquaporin 1-positive stromal niche-like cells directly interact with N-cadherin-positive clusters in the basal limbal epithelium. *Stem Cell Res*. 2013;10(2):147–155.
67. Massie I, Dziasko M, Kureshi A, et al. Advanced imaging and tissue engineering of the human limbal epithelial stem cell niche. *Methods Mol Biol*. 2015;1235:179–202.
68. Dziasko MA, Armer HE, Levis HJ, Shortt AJ, Tuft S, Daniels JT. Localisation of epithelial cells capable of holoclone formation in vitro and direct interaction with stromal cells in the native human limbal crypt. *PLoS One*. 2014;9(4):e94283.
69. Higa K, Shimmura S, Miyashita H, et al. N-cadherin in the maintenance of human corneal limbal epithelial progenitor cells in vitro. *Invest Ophthalmol Vis Sci*. 2009;50(10):4640–4645.
70. Omoto M, Miyashita H, Shimmura S, et al. The use of human mesenchymal stem cell-derived feeder cells for the cultivation of transplantable epithelial sheets. *Invest Ophthalmol Vis Sci*. 2009;50(5):2109–2115.
71. Guan L, Ge H, Tang X, et al. Use of a silk fibroin-chitosan scaffold to construct a tissue-engineered corneal stroma. *Cells Tissues Organs*. 2013;198(3):190–197.
72. Karamichos D, Hutcheon AEK, Zieske JD. Transforming growth factor-β3 regulates assembly of a non-fibrotic matrix in a 3D corneal model. *J Tissue Eng Regen Med*. 2011;5(8):e228–e238.
73. Karamichos D, Rich CB, Hutcheon AEK, et al. Self-assembled matrix by umbilical cord stem cells. *J Funct Biomater*. 2011;2(3):213–229.
74. Du Y, Sundarraj N, Funderburgh ML, Harvey SA, Birk DE, Funderburgh JL. Secretion and organization of a cornea-like tissue in vitro by stem cells from human corneal stroma. *Invest Ophthalmol Vis Sci*. 2007;48(11):5038–5045.
75. Karamichos D, Funderburgh ML, Hutcheon AE, et al. A role for topographic cues in the organization of collagenous matrix by corneal fibroblasts and stem cells. *PLoS One*. 2014;9(1):e86260.
76. Karamichos D, Hutcheon AEK, Zieske JD. Reversal of fibrosis by TGF-β3 in a 3D in vitro model. *Exp Eye Res*. 2014;124:31–36.
77. Yoeruek E, Bayyoud T, Maurus C, et al. Reconstruction of corneal stroma with decellularized porcine xenografts in a rabbit model. *Acta Ophthalmol*. 2012;90(3):e206–e210.
78. Karamichos D, Zareian R, Guo X, Hutcheon AEK, Ruberti JW, Zieske JD. Novel in vitro model for keratoconus disease. *J Funct Biomater*. 2012;3(4):760–775.

79. Lai J-Y, Li Y-T, Cho C-H, Yu T-C. Nanoscale modification of porous gelatin scaffolds with chondroitin sulfate for corneal stromal tissue engineering. *Int J Nanomedicine.* 2012;7:1101–1114.
80. Guo X, Hutcheon AEK, Melotti SA, Zieske JD, Trinkaus-Randall V, Ruberti JW. Morphologic characterization of organized extracellular matrix deposition by ascorbic acid-stimulated human corneal fibroblasts. *Invest Ophthalmol Vis Sci.* 2007;48(9): 4050–4060.
81. Tanaka Y, Shi D, Kubota A, et al. Irreversible optical clearing of rabbit dermis for autogenic corneal stroma transplantation. *Biomaterials.* 2011;32(28):6764–6772.
82. Ionescu A-M, Alaminos M, de la Cruz Cardona J, et al. Investigating a novel nanostructured fibrin-agarose biomaterial for human cornea tissue engineering: rheological properties. *J Mech Behav Biomed Mater.* 2011;4(8):1963–1973.
83. Karamichos D, Hutcheon AEK, Rich CB, Trinkaus-Randall V, Asara JM, Zieske JD. In vitro model suggests oxidative stress involved in keratoconus disease. *Sci Rep.* 2014;4:4608.
84. Ren R, Hutcheon AEK, Guo XQ, et al. Human primary corneal fibroblasts synthesize and deposit proteoglycans in long-term 3-D cultures. *Dev Dyn.* 2008;237(10): 2705–2715.
85. Nakatsu MN, González S, Mei H, Deng SX. Human limbal mesenchymal cells support the growth of human corneal epithelial stem/progenitor cells. *Invest Ophthalmol Vis Sci.* 2014;55(10):6953–6959.
86. Karamichos D, Guo XQ, Hutcheon AEK, Zieske JD. Human corneal fibrosis: an in vitro model. *Invest Ophthalmol Vis Sci.* 2010;51(3):1382–1388.
87. Garzón I, Martín-Piedra MA, Alfonso-Rodríguez C, et al. Generation of a biomimetic human artificial cornea model using Wharton's jelly mesenchymal stem cells. *Invest Ophthalmol Vis Sci.* 2014;55(7):4073–4083.
88. Wu J, Du Y, Watkins SC, Funderburgh JL, Wagner WR. The engineering of organized human corneal tissue through the spatial guidance of corneal stromal stem cells. *Biomaterials.* 2012;33(5):1343–1352.
89. Then KY, Yang Y, Ahearne M, El Haj AJ. Effect of microtopographical cues on human keratocyte orientation and gene expression. *Curr Eye Res.* 2011;36(2):88–93.
90. Bray LJ, George KA, Hutmacher DW, Chirila TV, Harkin DG. A dual-layer silk fibroin scaffold for reconstructing the human corneal limbus. *Biomaterials.* 2012;33(13):3529–3538.
91. Saeidi N, Guo X, Hutcheon AEK, et al. Disorganized collagen scaffold interferes with fibroblast mediated deposition of organized extracellular matrix in vitro. *Biotechnol Bioeng.* 2012;109(10):2683–2698.
92. Katikireddy KR, Dana R, Jurkunas UV. Differentiation potential of limbal fibroblasts and bone marrow mesenchymal stem cells to corneal epithelial cells. *Stem Cells.* 2014;32(3):717–729.
93. Du L, Wu X. Development and characterization of a full-thickness acellular porcine cornea matrix for tissue engineering. *Artif Organs.* 2011;35(7):691–705.
94. Wu J, Du Y, Mann MM, Funderburgh JL, Wagner WR. Corneal stromal stem cells versus corneal fibroblasts in generating structurally appropriate corneal stromal tissue. *Exp Eye Res.* 2014;120:71–81.
95. Wu J, Rnjak-Kovacina J, Du Y, Funderburgh ML, Kaplan DL, Funderburgh JL. Corneal stromal bioequivalents secreted on patterned silk substrates. *Biomaterials.* 2014;35(12):3744–3755.
96. Arjamaa O. Corneal reconstruction by stem cells and bioengineering. *Clin Ophthalmol.* 2012;6:1407–1409.
97. Xiao J, Duan H, Liu Z, et al. Construction of the recellularized corneal stroma using porous acellular corneal scaffold. *Biomaterials.* 2011;32(29):6962–6971.

98. Ramachandran C, Basu S, Sangwan VS, Balasubramanian D. Concise review: the coming of age of stem cell treatment for corneal surface damage. *Stem Cells Transl Med.* 2014;3(10):1160–1168.
99. Grobe GM, Reichl S. Characterization of vitamin C-induced cell sheets formed from primary and immortalized human corneal stromal cells for tissue engineering applications. *Cells Tissues Organs.* 2013;197(4):283–297.
100. Wu J, Du Y, Mann MM, Yang E, Funderburgh JL, Wagner WR. Bioengineering organized, multilamellar human corneal stromal tissue by growth factor supplementation on highly aligned synthetic substrates. *Tissue Eng A.* 2013;19(17–18):2063–2075.
101. Boulze Pankert M, Goyer B, Zaguia F, et al. Biocompatibility and functionality of a tissue-engineered living corneal stroma transplanted in the feline eye. *Invest Ophthalmol Vis Sci.* 2014;55(10):6908–6920.
102. Espandar L, Bunnell B, Wang GY, Gregory P, McBride C, Moshirfar M. Adipose-derived stem cells on hyaluronic acid-derived scaffold: a new horizon in bioengineered cornea. *Arch Ophthalmol.* 2012;130(2):202–208.
103. Giasson CJ, Deschambeault A, Carrier P, Germain L. Adherens junction proteins are expressed in collagen corneal equivalents produced in vitro with human cells. *Mol Vis.* 2014;20:386–394.
104. Du Y, Carlson EC, Funderburgh ML, et al. Stem cell therapy restores transparency to defective murine corneas. *Stem Cells.* 2009;27(7):1635–1642.
105. Basu S, Hertsenberg AJ, Funderburgh ML, et al. Human limbal biopsy-derived stromal stem cells prevent corneal scarring. *Sci Transl Med.* 2014;6(266):266ra172.
106. Mishima S. Clinical investigations on the corneal endothelium. *Ophthalmology.* 1982;89(6):525–530.
107. Yokoo S, Yamagami S, Yanagi Y, et al. Human corneal endothelial cell precursors isolated by sphere-forming assay. *Invest Ophthalmol Vis Sci.* 2005;46(5):1626–1631.
108. Coles BL, Angenieux B, Inoue T, et al. Facile isolation and the characterization of human retinal stem cells. *Proc Natl Acad Sci USA.* 2004;101(44):15772–15777.
109. Tropepe V, Coles BL, Chiasson BJ, et al. Retinal stem cells in the adult mammalian eye. *Science.* 2000;287(5460):2032–2036.
110. Mimura T, Yamagami S, Yokoo S, Araie M, Amano S. Comparison of rabbit corneal endothelial cell precursors in the central and peripheral cornea. *Invest Ophthalmol Vis Sci.* 2005;46(10):3645–3648.
111. Yamagami S, Yokoo S, Mimura T, Takato T, Araie M, Amano S. Distribution of precursors in human corneal stromal cells and endothelial cells. *Ophthalmology.* 2007;114(3):433–439.
112. Hirata-Tominaga K, Nakamura T, Okumura N, et al. Corneal endothelial cell fate is maintained by LGR5 through the regulation of hedgehog and Wnt pathway. *Stem Cells.* 2013;31(7):1396–1407.
113. Hara S, Hayashi R, Soma T, et al. Identification and potential application of human corneal endothelial progenitor cells. *Stem Cells Dev.* 2014;23(18):2190–2201.
114. Satake Y, Yamaguchi T, Hirayama M, et al. Ocular surface reconstruction by cultivated epithelial sheet transplantation. *Cornea.* 2014;33(suppl 11):S42–S46.

CHAPTER FOUR

Corneal Development: Different Cells from a Common Progenitor

Peter Y. Lwigale[1]

Department of Biosciences, Rice University, Houston, Texas, USA
[1]Corresponding author e-mail address: lwigale@rice.edu

Contents

Abstract

Development of the vertebrate cornea is a multistep process that involves cellular interactions between various ectodermal-derived tissues. Bilateral interactions between the neural ectoderm-derived optic vesicles and the cranial ectoderm give rise to the presumptive corneal epithelium and other epithelia of the ocular surface. Interactions between the neural tube and the adjacent ectoderm give rise to the neural crest cells, a highly migratory and multipotent cell population. Neural crest cells migrate between the lens and presumptive corneal epithelium to form the corneal endothelium and the stromal keratocytes. The sensory nerves that abundantly innervate the corneal stroma and epithelium originate from the neural crest- and ectodermal placode-derived trigeminal ganglion. Concomitant with corneal innervation is the formation of the limbal vascular plexus and the establishment of corneal avascularity. This review summarizes historical and current research to provide an overview of the genesis of the cellular layers of the cornea, corneal innervation, and avascularity.

Progress in Molecular Biology and Translational Science, Volume 134
ISSN 1877-1173
http://dx.doi.org/10.1016/bs.pmbts.2015.04.003
© 2015 Elsevier Inc.
All rights reserved.

1. EMBRYONIC ORIGIN OF THE CORNEAL TISSUES

The vertebrate cornea is comprised of three cellular layers that originate from the cranial ectoderm (Fig. 1). A stratified epithelium (a direct descendant of the ectoderm) covers the outermost surface of the cornea. The middle layer (corneal stroma) and innermost surface (corneal endothelium) are both derived from neural crest cells. All these cellular layers initially combine in the anterior eye region to form the presumptive cornea in their undifferentiated states. Subsequent well-orchestrated cellular interactions and molecular signaling events induce their differentiation and concomitant formation of the transparent cornea.

During early development, the bilateral regions of the anterior cranial ectoderm become specified to form placodes (thickenings in the ectoderm) that later give rise to the olfactory and lens epithelium. The presumptive corneal epithelium is first identified as the ectoderm occupying the region adjacent to the lens placode. It becomes a continuous epithelial layer after the

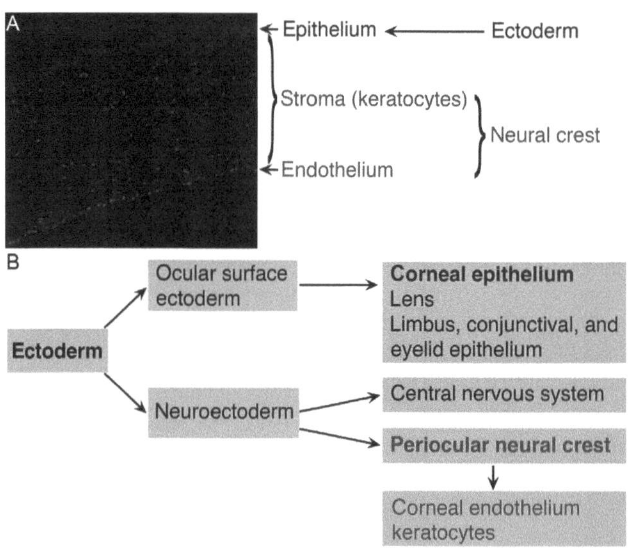

Figure 1 Ectodermal origin of the three cellular layers of the cornea. (A) Cross section through E8 chick cornea showing the ectoderm-derived corneal epithelium and neural crest cell-derived stromal keratocytes and corneal endothelium. (B) Schematic diagram showing the derivatives of the cranial ectoderm that contribute to the formation of the corneal epithelium and the neural crest cell progenitors of the corneal endothelium and keratocytes. (See the color plate.)

lens placode invaginates and detaches to form the lens vesicle. Although the specification of the placodes and the inductive signals between the optic vesicle and the ectoderm that lead to the formation of the lens have been well studied, little is known about the earliest origin of the corneal epithelium. Until recent, it was generally accepted that the corneal epithelium does not become specified until it receives signals from the lens vesicle.[1] However, recent studies indicate that the presumptive corneal and lens epithelium is specified in the ectoderm that expresses the paired box protein Pax6.[2] However, BMP signaling from the optic vesicles induces the formation of the lens placode, while the adjacent Pax6-expressing ectoderm forms the presumptive corneal epithelium. Microsurgical ablation of the lens placode at the time when the optic vesicles are signaling results in the induction of the lens vesicle from the regenerating adjacent presumptive epithelium. This indicates that the Pax6-expressing cranial ectoderm is competent to give rise to both lens and corneal epithelium and therefore, the common progenitor for both lens and corneal epithelium.

Neural crest cells are a multipotent, stem cell-like cell population that migrates into various regions of the embryo where they give rise to numerous derivatives.[3] The neural crest cell progenitors of the corneal endothelium and stromal keratocytes originate from the neural tube (presumptive central nervous system) located in midline region of the embryo. The surface ectoderm in the midline region takes on a neural identity following inhibition of Bmp and induction of Wnt signaling.[4–6] Signals from the neural tube and adjacent ectoderm are required for neural crest cell induction.[7,8]

The stream of neural crest cells that gives rise to the corneal endothelium and stromal keratocytes originates from the neural tube region between diencephalon and mesencephalon.[9–11] The optic stream of migratory neural crest cells also gives rise to other cranial structures including the cranial sensory and ciliary ganglia, smooth muscle, as well as the connective tissue, bone, and cartilage of the face.[3,12–14] The ocular component of the neural crest cells reside in the periocular region (also known as the periocular mesenchyme) prior to their migration into the presumptive corneal region (between the ectoderm and lens vesicle).[9] As the neural crest cells migrate into the ocular region, morphogenetic events transform the ectoderm and optic vesicles to generate the rudimentary eye, which at this time point is comprised of the presumptive corneal epithelium, the lens vesicle, and the optic cup (presumptive neuroretina and retinal pigment epithelium) surrounded by the periocular mesenchyme (Fig. 2A). This developmental stage occurs at about embryonic day (E)3 in chick,[15] E11.5 in mouse,[16] and after

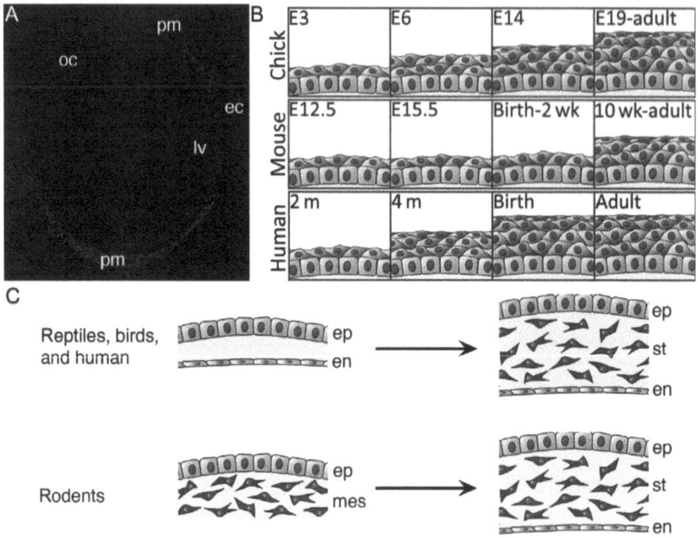

Figure 2 Cornea development. (A) Cross section through an E3 chick eye at the rudimentary stage showing the optic cup (oc) and lens vesicle (lv) being surrounded by the neural crest-derived periocular mesenchyme (pm), and the presumptive corneal epithelium (ec). (B) Schematic showing stages of stratification of the corneal epithelium and keratin expression in chick, mouse, and human. (C) Schematics showing the different modes of neural crest cell migration between the corneal epithelium (ep) and lens during the formation of the corneal endothelium (en) and stroma (st). mes, neural crest mesenchyme. (See the color plate.)

5 weeks of gestation in human.[17–19] This review will summarize the cellular and molecular events that lead to the formation of each corneal layer beginning at the rudimentary eye stage.

2. DEVELOPMENT OF THE CORNEAL EPITHELIUM

Shortly after the formation of the lens vesicle, the overlying ectoderm fuses together and becomes the presumptive corneal epithelium. Earlier studies using chick embryo models[15,20,21] indicated that interactions between the ectoderm and the lens vesicle resulted in synthesis of the primary stroma, an acellular extracellular matrix sandwiched by the two tissues. The primary stroma, which marks the beginning of corneal fibrillogenesis, is synthesized by the presumptive corneal epithelium, and it consists of orthogonally ordered layers of type I, II, IX collagen, and chondroitin sulfate proteoglycans.[22–24]

In chick, the corneal epithelium is two-cell layers thick between E3–E5. The basal cells have a cuboidal appearance that is distinct from the superficial cells. The corneal epithelium is specified by E5, given that it does not lose Pax6 expression when challenged by mesenchyme from the dorsal dermis.[2,25] By E6, the epithelium is three-layers thick and it increases to 4–5 cell layers at E19.[26] The corneal epithelium begins to express the keratin pair (K3/K12) at about E14,[27] and it is fully developed and consists of 6–8 stratified cell layers at the time of hatching (Fig. 2B).[28] The inner basal cells are columnar in shape and attach to the underlying basement membrane. They proliferate and produce the suprabasal layers of cells, which replenish the 1–3 layers of superficial squamous cells. Although the cellular arrangement in the adult mammalian corneal epithelium resembles that of chick, they are do not follow similar developmental patterns. In mouse the presumptive corneal epithelium expresses K12 at E15.5,[29] but it remains 1–2 cell layers thick throughout embryonic development and birth. It begins rapid proliferation when the eyelids open at postnatal day 12, giving rise to a stratified 6–8 cell layer thick mature corneal epithelium by 10 weeks after birth (Fig. 2B).[30] Stratification of the developing human corneal epithelium is similar to what has been observed in chick. The corneal epithelium is two-cell layers thick at about 2 months of gestation and it becomes 3–4 layers at 4 months, and it appears fully stratified at birth (Fig. 2B).[31]

3. DEVELOPMENT OF THE CORNEAL ENDOTHELIUM

Genesis of the corneal endothelium begins when periocular neural crest cells migrate between the presumptive corneal epithelium and lens vesicle. The neural crest cells undergo a mesenchyme-to-epithelial transition and form a monolayer that occupies the posterior surface of the cornea. In reptiles, birds, and human eyes, the cornea endothelium forms as a result of the first wave of neural crest cell migration into the region of the forming cornea (Fig. 2C).[21,32] This process differs in rodents, cats, rabbits, and cattle where the periocular neural crest cells migrate in a single wave of mesenchyme, after which only the cells closest to the lens vesicle undergo a mesenchyme-to-epithelial transition and coalesce to form a single-endothelial layer.[33]

Morphogenesis of the corneal endothelium is well studied in chick.[9,10,21] The first wave of periocular neural crest cell migration is initiated at E4.5, which results in the formation of a monolayer of cells between the lens and epithelium at E5. The earliest marker of the corneal endothelium, the cell adhesion molecule N-cadherin, is expressed in the periocular mesenchyme at E3,[10] and it is involved in the mesenchymal-to-epithelial transition of the neural crest cells and the formation of the monolayer. Studies have also shown that downregulation of N-cadherin is required for neural crest cell migration from the neural tube,[34–36] but its upregulation is required for the formation of tight junctions during morphogenesis.[37,38] In human, the presumptive corneal endothelium begin as a loosely arranged bilayer of cuboidal cells that thins out to form a flattened monolayer at about 8 weeks of gestation.[32] The zonular occluding junctions are well developed by week 17,[39] although the pump function is not yet fully established.[40] In mouse, the corneal endothelium is formed by the transition of the corneal mesenchyme adjacent to the lens into epithelium, which is completed by E15.5.[41] Tight junctions form and the developing corneal endothelium express N-cadherin.[42] The adult corneal endothelium expresses the tight junction marker ZO-1 and the functional pump marker Na^+/K^+-ATPase; however, the expression of these functional markers during mouse corneal development is not clear. In chick, the maturity of the endothelium is inferred from its pumping function evidenced by the transition from corneal opacity to transparency, which commences at E10.[43] Shortly after the corneal endothelium forms, it secrets the Descemet's membrane, a cell-free matrix that mostly consists of collagens.[44] In the adult cornea, the Descemet's membrane plays a crucial role in maintaining the integrity and function of the endothelium.

4. DEVELOPMENT OF THE CORNEAL STROMA

The stroma is the last cell layer to form during corneal development. In chick, formation of the corneal stroma is initiated by a second wave of neural crest migration between the presumptive corneal endothelium and epithelium.[10,11,21] This migration is underway in the periphery of the developing cornea by E6 and the cells are highly proliferative and have a mesenchymal morphology.[15] The presumptive stromal cells continue to proliferate and synthesize ECM components into which they become embedded and transform into the characteristic flattened morphology of the keratocytes.[15,45] The earliest expression of collagen and corneal proteoglycans (keratan and chondroitin/dermatan sulfate) is observed between E6 and E7,[46–49] which indicates the transformation from neural crest to keratocyte phenotype. The synthesis and distribution of collagen, a major component of the cornea, during development has been reviewed in detail by Quantock and Young.[50] Type I and type V collagens are synthesized by the stromal cells, whereas type II collagen is derived from the corneal epithelium. Other collagens including type VI, IX, XII, and XIV are present in the developing cornea where they are associated with its structural integrity.[50–54]

Interactions between the corneal proteoglycans and collagens are essential for fibrillogenesis and assembly of the corneal ECM, and its subsequent transparency.[55–57] Knockout of keratocan in the cornea results in increased diameter of collagen fibrils and thin corneas, although corneal transparency is not affected.[58] Mimecan and lumican deficient mice also have thicker collagen fibrils in the cornea but only the lumican knockout mice show corneal opacity.[56,59] Interactions between the proteoglycans and collagens entrap the keratocytes in the ECM as the cornea becomes compacted and transparent. Keratocytes cease to proliferate at about E14 in chick[21] or shortly after the opening of eyelids in mice.[30] Keratocytes become quiescent and remain in this state until when the normal corneal homeostasis is perturbed by trauma or infection.

5. CORNEAL INNERVATION

The cornea is highly innervated by sensory afferents from the trigeminal ganglion.[60–62] During ocular development in chick, trigeminal nerves arrive at the corneal periphery at about E4 and they form a pericorneal nerve ring before radially innervating the stroma at E9 (Fig. 3B and C).[63–65] The

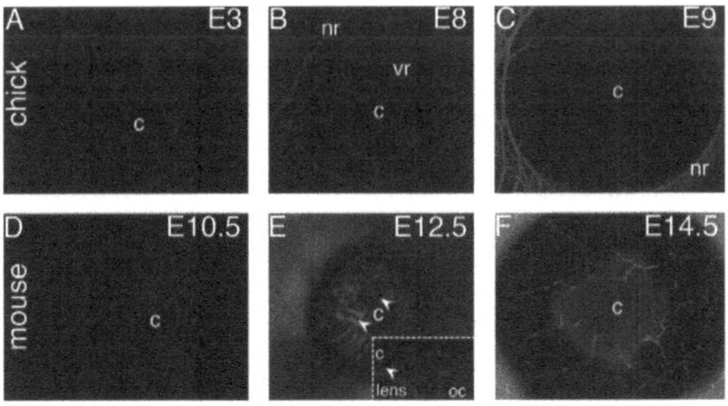

Figure 3 Neurovascular development in the anterior eye. (A and B) Whole-mount Tg(tie1:H2B:eYFP) transgenic quail eyes at E3 and E7 showing that angioblasts and forming vasculature avoid the developing cornea. (B and C) Formation of the pericorneal nerve ring (nr) and innervation of the corneal periphery at E9. (D and E) Endomucin staining of E10.5 and E12.5 mouse embryos showing the vascularization of the anterior eye. Inset in (E) shows that the mouse cornea is not vascularized although it forms adjacent to the pupillary membrane (arrowheads). (F) Projection of sensory nerves into the corneal periphery of an E14.5 mouse embryo. c, cornea; nr, pericorneal nerve ring; vr, vascular ring; oc, optic cup. (See the color plate.)

nerve bundles branch from the pericorneal nerve ring at regular intervals. They give rise to long bundles that innervate the entire corneal radius, as well as short nerve bundles that only innervate the periphery.[66] Corneal nerves populate only the anterior half of the stroma and they bifurcate repeatedly as they project toward the epithelium, which is innervated by smaller terminal nerve bundles (leashes) by E12.[63,66]

Recent studies have shown that unlike chick, mouse corneal nerves do not form a pericorneal nerve ring.[67] Instead, sensory afferents project from four quadrants toward the anterior eye between E12.5 and E13.5. Nerve bundles from all quadrants (except for the dorsal-temporal quadrant) directly innervate the periphery of the presumptive cornea by E13.5 (Fig. 3F). The four major nerve bundles branch repeatedly as they project toward the corneal center, form the stromal plexus, and innervate the epithelium by E16.5. The epithelium is only two-cell layers thick when it is initially innervated and each nerve bundle subsequently ramifies into smaller fascicles that form the subbasal plexus. At approximately 3 weeks after birth, the subbasal nerves orient toward the central cornea, which results in a swirl pattern that persists during postnatal development and adulthood.[68,69] A similar swirl pattern was observed in human.[70] Although studies have shown radial projections of stromal nerves in the adult human cornea similar to chick, their patterning during development remains unclear.

Presumptive corneal nerves are attracted toward the anterior eye by neurotrophic factors secreted by the developing cornea. Nerve growth factor (NGF), brain-derived neurotrophic factor, neurotrophin 3, and glial cell line-derived neurotrophic factor are present in the cornea both during development and in adults.[71–73] Their receptors, the tropomyosin-related kinase (Trk) family of tyrosine kinases (TrkA, TrkB, and TrkC), are spatiotemporally expressed in various patterns in the developing trigeminal ganglion.[74–77] Mice that lack any of the above neurotrophic factors or their receptors show increased apoptosis of sensory neurons during development of the trigeminal ganglion.[77–79] Topical application of NGF on injured corneas has been shown to influence the survival of neuronal and non-neuronal cells.[80–82]

Despite the attractive signals of neurotrophic factors in the developing cornea, tight regulation of nerve projections into the eye is required for proper patterning. This is particularly evident during the formation of the pericorneal nerve ring in birds. Following the observation that the secreted axon guidance molecule, Semaphorin3A (Sema3A) is strongly expressed in the lens vesicle during ocular development,[83,84] we studied its effect on the presumptive corneal nerves. We showed that Sema3A signaling regulates the formation of the pericorneal nerve ring, which is required for proper radial patterning of corneal nerves.[64] The lenticular expression of *Sema3A* and trigeminal expression of *Nrp1* are conserved in the mouse. Disruption of Sema3A/Nrp1 signaling resulted aberrant axon projections into the eye[85] and defective innervation of the cornea.[86] Slit/Robo signaling is also involved in the formation of the pericorneal nerve ring. However, the expression of *Slit2* and *Robo1* are, respectively, downregulated in the cornea and trigeminal ganglion during nerve projection into the corneal periphery.[87,88] It has been suggested that the accumulation of the highly sulfated extracellular keratan sulfate proteoglycan in the anterior cornea blocks the lens-derived Sema3A and Slit signaling, allowing the nerves to project from the midstroma to the corneal epithelium.[89,90] A recent study indicates that polysialic acid on the axon surface plays a role in promoting the fasciculation of sensory nerves from the pericorneal nerve ring during their projection into the corneal periphery.[91]

6. CORNEAL AVASCULARITY DURING DEVELOPMENT

Another important characteristic of the adult cornea is its avascularity despite being surrounded by a dense limbal vessel plexus from which it derives nutrients and oxygen. During development, blood vessel precursors known as angioblasts migrate relatively long distances to form the embryonic vasculature through processes known as vasculogenesis (formation of vasculature directly from endothelial cells) or angiogenesis (formation of vasculature from existing vessels). Our recent study using transgenic Tg(tie1:H2B:eYFP) quail embryos that enable the visualization of fluorescently labeled angioblasts and blood vessels, showed that angioblasts avoid the developing cornea, which is never vascularized despite the presence of proangiogenic factors in the anterior eye (Fig. 3A and B).[92] Proangiogenic factors including vascular endothelial growth factor (Vegf), fibroblast growth factor (Fgf), and platelet-derived growth factor (Pdgf) are expressed in the anterior eye, and the periocular angioblasts express their respective receptors (VegfR1/2, Nrp1/2, FgfR1/2, and PdgfB/RB). This suggests potential roles of the proangiogenic factors in recruiting angioblasts into the anterior eye and initiating the vasculogenesis and angiogenesis processes required for the formation of the limbal vascular plexus. However, antiangiogenic factors including the Semaphorins (Sema3A, Sema3E, Sema3G) and Netrins (Netrin1, Netrin4, Neogenin) are also expressed in the anterior eye, whereas the angioblasts express their respective receptors (Nrp1, PlexinD1, Nrp2, and Unc5B).[92] With the exception of Sema3A, the role of antiangiogenic factors during ocular development is yet to be determined. Our recent study indicates that the presence of Sema3A in the lens plays a role in preventing Vegf-induced angioblast invasion and subsequent vascularization of the developing cornea.[93]

At about E10.5 in the mouse embryo, angioblasts migrate into the optic cup and form the hyaloid vasculature.[94] The hyaloid vasculature and the pupillary membrane give rise to the tunica vasculosa lentis, which forms around the surface of lens in mammals but regresses postnatally.[95,96] The developing mouse cornea remains avascular despite its formation in close proximity to the pupillary membrane (Fig. 3D and E). Analysis of Nrp1$^{(sema-)}$ mutant mice (that lack Nrp1/Sema signaling but have functional Nrp1/Vegf signaling) revealed aberrant angioblast migration into the presumptive cornea and subsequent formation of ectopic vasculature in the corneal stroma.[93] However the ectopic vasculature in the Nrp1$^{(sema-)}$ mutant mice was temporary as these defects were not observed in the adult corneas. These results suggest that Sema3A/Nrp1 signaling prevent vascularization of the cornea during early development, but other antiangiogenic signals are activated later in the cornea, which may prevent vessel stability in the stroma of Nrp1$^{(sema-)}$ mutant mice. An example of such inhibitory signaling is the presence of a soluble form of VegfR1 (also known as sflt1) in the adult corneal epithelium, which has been shown to act as potent an antiangiogenic factor by binding to the Vegf present in the cornea.[97]

In addition to blood vessels, lymphatic vessels also contribute to the limbal vessel plexus. Although numerous studies have examined lymphangiogenesis in adult corneas,[98–101] very little is known about the patterning of the limbal lymphatic vessels during development. Immunostaining with the early marker for lymphatic vessels, Prox1, in the mouse revealed the presence of lymph vasculature in the anterior eye at E17.5.[102] The Prox1-positive vessels later stained for another lymphatic vessel marker Lyve-1 visualized at the corneal limbus at P0 and formed a well-defined limbal lymphatic network in 7-week-old adults. This study also showed that the forkhead transcription factor (Foxc1) plays a role in preventing the sprouting of the limbal blood and lymphatic vessels into the developing cornea.

7. CONCLUSIONS

Improper formation of the cornea is associated with many congenital ocular defects. Historical studies in chick model combined with recent studies in mouse have contributed significantly to our understanding of the cellular and morphogenetic processes that lead to the formation of the cornea. It is increasingly becoming clear that the developmental processes that lead to the formation of the cornea are under elaborate control of multiple families of signaling molecules that are associated with ocular development. The roles of morphogens (retinoic acid), signaling molecules (TGFβ, Wnt, and Fgf), and transcription factors (Pax6, Pitx2, and Fox genes) during corneal development and the human defects associated with their mutations have been reviewed in detail,[19,25,103–108] therefore not included in this review. The current challenge is to understand the mechanisms of how these genes interact at the periocular neural crest cell level to direct their differentiation into corneal endothelium or keratocytes. In addition, studies of the guidance cues that regulate periocular neural crest cell migration, sensory innervation, and/or promote corneal avascularity during development will contribute to better understanding of congenital ocular defects and they might provide insights that may lead to the development of treatments for diseased and injured adult corneas.

ACKNOWLEDGMENTS

This work was supported by NIH EY022158. The author would also like to thank James Spurlin for assisting with the schematics, and members of Lwigale lab for proof reading the manuscript.

REFERENCES

1. Graw J. Eye development. *Curr Top Dev Biol.* 2010;90:343–386. http://dx.doi.org/10.1016/S0070-2153(10)90010-0.
2. Collomb E, Yang Y, Foriel S, et al. The corneal epithelium and lens develop independently from a common pool of precursors. *Dev Dyn.* 2013;242(5):401–413. http://dx.doi.org/10.1002/dvdy.23925.
3. Douarin NML. *The Neural Crest.* Cambridge: Cambridge University Press; 1982.
4. LaBonne C, Bronner-Fraser M. Neural crest induction in Xenopus: evidence for a two-signal model. *Development.* 1998;125(13):2403–2414. http://paperity.org/p/44969369/neural-crest-induction-in-xenopus-evidence-for-a-two-signal-model [Accessed January 29, 2015].
5. García-Castro MI, Marcelle C, Bronner-Fraser M. Ectodermal Wnt function as a neural crest inducer. *Science.* 2002;297(5582):848–851.

6. Patthey C, Edlund T, Gunhaga L. Wnt-regulated temporal control of BMP exposure directs the choice between neural plate border and epidermal fate. *Dev Camb Engl.* 2009;136(1):73–83. http://dx.doi.org/10.1242/dev.025890.
7. Knecht AK, Bronner-Fraser M. Induction of the neural crest: a multigene process. *Nat Rev Genet.* 2002;3(6):453–461. http://dx.doi.org/10.1038/nrg819.
8. Stuhlmiller TJ, García-Castro MI. Current perspectives of the signaling pathways directing neural crest induction. *Cell Mol Life Sci CMLS.* 2012;69(22):3715–3737. http://dx.doi.org/10.1007/s00018-012-0991-8.
9. Johnston MC, Noden DM, Hazelton JL. Origins of avian ocular and periocular tissues. *Exp Eye Res.* 1979;29(1):27–43.
10. Lwigale PY, Cressy PA, Bronner-Fraser M. Corneal keratocytes retain neural crest progenitor cell properties. *Dev Biol.* 2005;288(1):284–293. http://dx.doi.org/10.1016/j.ydbio.2005.09.046.
11. Creuzet S, Vincent C, Couly G. Neural crest derivatives in ocular and periocular structures. *Int J Dev Biol.* 2005;49(2–3):161–171. http://dx.doi.org/10.1387/ijdb.041937sc.
12. Le Douarin NM, Teillet MA. Experimental analysis of the migration and differentiation of neuroblasts of the autonomic nervous system and of neurectodermal mesenchymal derivatives, using a biological cell marking technique. *Dev Biol.* 1974;41(1):162–184.
13. D'Amico-Martel A, Noden DM. An autoradiographic analysis of the development of the chick trigeminal ganglion. *J Embryol Exp Morphol.* 1980;55(1):167–182.
14. Lwigale PY, Conrad GW, Bronner-Fraser M. Graded potential of neural crest to form cornea, sensory neurons and cartilage along the rostrocaudal axis. *Development.* 2004;131:1979–1991.
15. Hay ED, Revel JP. Fine structure of the developing avian cornea. *Monogr Dev Biol.* 1969;1:1–144.
16. Pei YF, Rhodin JA. The prenatal development of the mouse eye. *Anat Rec.* 1970;168(1):105–125. http://dx.doi.org/10.1002/ar.1091680109.
17. Hoar RM. Embryology of the eye. *Environ Health Perspect.* 1982;44:31–34.
18. Sevel D, Isaacs R. A re-evaluation of corneal development. *Trans Am Ophthalmol Soc.* 1988;86:178–207.
19. Sowden JC. Molecular and developmental mechanisms of anterior segment dysgenesis. *Eye Lond Engl.* 2007;21(10):1310–1318. http://dx.doi.org/10.1038/sj.eye.6702852.
20. Hayashi M, Ninomiya Y, Hayashi K, Linsenmayer TF, Olsen BR, Trelstad RL. Secretion of collagen types I and II by epithelial and endothelial cells in the developing chick cornea demonstrated by in situ hybridization and immunohistochemistry. *Development.* 1988;103:27–36.
21. Hay ED. Development of the vertebrate cornea. *Int Rev Cytol.* 1980;63:263–322.
22. Birk DE, Trelstad RL. Extracellular compartments in matrix morphogenesis: collagen fibril, bundle, and lamellar formation by corneal fibroblasts. *J Cell Biol.* 1984;99(6):2024–2033.
23. Cai CX, Fitch JM, Svoboda KK, Birk DE, Linsenmayer TF. Cellular invasion and collagen type IX in the primary corneal stroma *in vitro*. *Dev Dyn.* 1994;201(3):206–215. http://dx.doi.org/10.1002/aja.1002010304.
24. Svoboda KK, Nishimura I, Sugrue SP, Ninomiya Y, Olsen BR. Embryonic chicken cornea and cartilage synthesize type IX collagen molecules with different amino-terminal domains. *Proc Natl Acad Sci USA.* 1988;85(20):7496–7500.
25. Dhouailly D, Pearton DJ, Michon F. The vertebrate corneal epithelium: from early specification to constant renewal. *Dev Dyn.* 2014;243(10):1226–1241. http://dx.doi.org/10.1002/dvdy.24179.
26. Trelstad RL. The Golgi apparatus in chick corneal epithelium: changes in intracellular position during development. *J Cell Biol.* 1970;45(1):34–42.

27. Chaloin-Dufau C, Sun TT, Dhouailly D. Appearance of the keratin pair K3/K12 during embryonic and adult corneal epithelial differentiation in the chick and in the rabbit. *Cell Differ Dev.* 1990;32(2):97–108.
28. Ikeda A, Maisel H, Waggoner D. An immunofluorescent study of cornea development in the chick. *J Embryol Exp Morphol.* 1975;33(2):279–290.
29. Tanifuji-Terai N, Terai K, Hayashi Y, Chikama T, Kao WW-Y. Expression of keratin 12 and maturation of corneal epithelium during development and postnatal growth. *Invest Ophthalmol Vis Sci.* 2006;47(2):545–551. http://dx.doi.org/10.1167/iovs.05-1182.
30. Zieske JD. Corneal development associated with eyelid opening. *Int J Dev Biol.* 2004;48(8–9):903–911. http://dx.doi.org/10.1387/ijdb.041860jz.
31. Rodrigues M, Ben-Zvi A, Krachmer J, Schermer A, Sun TT. Suprabasal expression of a 64-kilodalton keratin (no. 3) in developing human corneal epithelium. *Differ Res Biol Divers.* 1987;34(1):60–67.
32. Wulle KG. Electron microscopy of the fetal development of the corneal endothelium and Descemet's membrane of the human eye. *Invest Ophthalmol.* 1972;11(11):897–904.
33. Cintron C, Covington H, Kublin CL. Morphogenesis of rabbit corneal stroma. *Invest Ophthalmol Vis Sci.* 1983;24(5):543–556.
34. Bronner-Fraser M, Wolf JJ, Murray BA. Effects of antibodies against N-cadherin and N-CAM on the cranial neural crest and neural tube. *Dev Biol.* 1992;153(2):291–301.
35. Nakagawa S, Takeichi M. Neural crest cell-cell adhesion controlled by sequential and subpopulation-specific expression of novel cadherins. *Dev Camb Engl.* 1995;121(5):1321–1332.
36. Rogers CD, Saxena A, Bronner ME. Sip1 mediates an E-cadherin-to-N-cadherin switch during cranial neural crest EMT. *J Cell Biol.* 2013;203(5):835–847. http://dx.doi.org/10.1083/jcb.201305050.
37. Hatta K, Takagi S, Fujisawa H, Takeichi M. Spatial and temporal expression pattern of N-cadherin cell adhesion molecules correlated with morphogenetic processes of chicken embryos. *Dev Biol.* 1987;120(1):215–227.
38. Taneyhill LA, Schiffmacher AT. Cadherin dynamics during neural crest cell ontogeny. *Prog Mol Biol Transl Sci.* 2013;116:291–315. http://dx.doi.org/10.1016/B978-0-12-394311-8.00013-3.
39. Wulle KG, Ruprecht KW, Windrath LC. Electron microscopy of the development of the cell junctions in the embryonic and fetal human corneal endothelium. *Invest Ophthalmol.* 1974;13(12):923–934.
40. Chen Y, Huang K, Nakatsu MN, et al. Identification of novel molecular markers through transcriptomic analysis in human fetal and adult corneal endothelial cells. *Hum Mol Genet.* 2013;22(7):1271–1279. http://dx.doi.org/10.1093/hmg/dds527.
41. Silla ZTV, Naidoo J, Kidson SH, Sommer P. Signals from the lens and Foxc1 regulate the expression of key genes during the onset of corneal endothelial development. *Exp Cell Res.* 2014;322(2):381–388. http://dx.doi.org/10.1016/j.yexcr.2014.01.016.
42. Reneker LW, Silversides DW, Xu L, Overbeek PA. Formation of corneal endothelium is essential for anterior segment development - a transgenic mouse model of anterior segment dysgenesis. *Dev Camb Engl.* 2000;127(3):533–542.
43. Coulombre AJ, Coulombre JL. Corneal development. I. Corneal transparency. *J Cell Physiol.* 1958;51(1):1–11.
44. Fitch JM, Birk DE, Linsenmayer C, Linsenmayer TF. The spatial organization of Descemet's membrane-associated type IV collagen in the avian cornea. *J Cell Biol.* 1990;110(4):1457–1468.

45. Connon CJ, Siegler V, Meek KM, et al. Proteoglycan alterations and collagen reorganisation in the secondary avian cornea during development. *Ophthalmic Res.* 2003;35(4):177–184. http://dx.doi.org/10.1159/000071168.
46. Funderburgh JL, Caterson B, Conrad GW. Keratan sulfate proteoglycan during embryonic development of the chicken cornea. *Dev Biol.* 1986;116(2):267–277.
47. Doane KJ, Ting WH, McLaughlin JS, Birk DE. Spatial and temporal variations in extracellular matrix of periocular and corneal regions during corneal stromal development. *Exp Eye Res.* 1996;62(3):271–283.
48. Dunlevy JR, Beales MP, Berryhill BL, Cornuet PK, Hassell JR. Expression of the keratan sulfate proteoglycans lumican, keratocan and osteoglycin/mimecan during chick corneal development. *Exp Eye Res.* 2000;70(3):349–362. http://dx.doi.org/10.1006/exer.1999.0789.
49. Conrad AH, Conrad GW. The keratocan gene is expressed in both ocular and non-ocular tissues during early chick development. *Matrix Biol J Int Soc Matrix Biol.* 2003;22(4):323–337.
50. Quantock AJ, Boote C, Siegler V, Meek KM. Collagen organization in the secondary chick cornea during development. *Invest Ophthalmol Vis Sci.* 2003;44(1):130–136.
51. Gordon MK, Gerecke DR, Olsen BR. Type XII collagen: distinct extracellular matrix component discovered by cDNA cloning. *Proc Natl Acad Sci USA.* 1987;84(17): 6040–6044.
52. Fitch JM, Linsenmayer CM, Linsenmayer TF. Collagen fibril assembly in the developing avian primary corneal stroma. *Invest Ophthalmol Vis Sci.* 1994;35(3): 862–869.
53. Young BB, Zhang G, Koch M, Birk DE. The roles of types XII and XIV collagen in fibrillogenesis and matrix assembly in the developing cornea. *J Cell Biochem.* 2002;87(2):208–220. http://dx.doi.org/10.1002/jcb.10290.
54. Quantock AJ, Young RD. Development of the corneal stroma, and the collagen-proteoglycan associations that help define its structure and function. *Dev Dyn Off Publ Am Assoc Anat.* 2008;237(10):2607–2621. http://dx.doi.org/10.1002/dvdy.21579.
55. Hart GW. Biosynthesis of glycosaminolgycans during corneal development. *J Biol Chem.* 1976;251(21):6513–6521.
56. Chakravarti S, Petroll WM, Hassell JR, et al. Corneal opacity in lumican-null mice: defects in collagen fibril structure and packing in the posterior stroma. *Invest Ophthalmol Vis Sci.* 2000;41(11):3365–3373.
57. Zhang Y, Conrad AH, Tasheva ES, et al. Detection and quantification of sulfated disaccharides from keratan sulfate and chondroitin/dermatan sulfate during chick corneal development by ESI-MS/MS. *Invest Ophthalmol Vis Sci.* 2005;46(5):1604–1614. http://dx.doi.org/10.1167/iovs.04-1453.
58. Liu C-Y, Birk DE, Hassell JR, Kane B, Kao WW-Y. Keratocan-deficient mice display alterations in corneal structure. *J Biol Chem.* 2003;278(24):21672–21677. http://dx.doi.org/10.1074/jbc.M301169200.
59. Tasheva ES, Koester A, Paulsen AQ, et al. Mimecan/osteoglycin-deficient mice have collagen fibril abnormalities. *Mol Vis.* 2002;8:407–415.
60. Arvidson B. Retrograde axonal transport of horseradish peroxidase from cornea to trigeminal ganglion. *Acta Neuropathol (Berl).* 1977;38(1):49–52.
61. Marfurt CF, Kingsley RE, Echtenkamp SE. Sensory and sympathetic innervation of the mammalian cornea. A retrograde tracing study. *Invest Ophthalmol Vis Sci.* 1989;30(3):461–472.
62. Lwigale PY. Embryonic origin of avian corneal sensory nerves. *Dev Biol.* 2001;239: 323–337.
63. Bee JA, Hay RA, Lamb EM, Devore JJ, Conrad GW. Positional specificity of corneal nerves during development. *Invest Ophthalmol Vis Sci.* 1986;27(1):38–43.

64. Lwigale PY, Bronner-Fraser M. Lens-derived Semaphorin3A regulates sensory innervation of the cornea. *Dev Biol.* 2007;306(2):750–759. http://dx.doi.org/10.1016/j.ydbio.2007.04.012.
65. Kubilus JK, Linsenmayer TF. Developmental guidance of embryonic corneal innervation: roles of Semaphorin3A and Slit2. *Dev Biol.* 2010;344(1):172–184. http://dx.doi.org/10.1016/j.ydbio.2010.04.032.
66. Riley NC, Lwigale PY, Conrad GW. Specificity of corneal nerve positions during embryogenesis. *Mol Vis.* 2001;7:297–304.
67. McKenna CC, Lwigale PY. Innervation of the mouse cornea during development. *Invest Ophthalmol Vis Sci.* 2011;52(1):30–35. http://dx.doi.org/10.1167/iovs.10-5902.
68. Yu CQ, Rosenblatt MI. Transgenic corneal neurofluorescence in mice: a new model for *in vivo* investigation of nerve structure and regeneration. *Invest Ophthalmol Vis Sci.* 2007;48(4):1535–1542. http://dx.doi.org/10.1167/iovs.06-1192.
69. Leiper LJ, Ou J, Walczysko P, et al. Control of patterns of corneal innervation by Pax6. *Invest Ophthalmol Vis Sci.* 2009;50(3):1122–1128. http://dx.doi.org/10.1167/iovs.08-2812.
70. Patel DV, McGhee CNJ. Mapping of the normal human corneal sub-Basal nerve plexus by *in vivo* laser scanning confocal microscopy. *Invest Ophthalmol Vis Sci.* 2005;46(12):4485–4488. http://dx.doi.org/10.1167/iovs.05-0794.
71. Ebendal T, Persson H. Detection of nerve growth factor mRNA in the developing chicken embryo. *Dev Camb Engl.* 1988;102(1):101–106.
72. You L, Kruse FE, Völcker HE. Neurotrophic factors in the human cornea. *Invest Ophthalmol Vis Sci.* 2000;41(3):692–702.
73. Bennett JL, Zeiler SR, Jones KR. Patterned expression of BDNF and NT-3 in the retina and anterior segment of the developing mammalian eye. *Invest Ophthalmol Vis Sci.* 1999;40(12):2996–3005.
74. Arumäe U, Pirvola U, Palgi J, et al. Neurotrophins and their receptors in rat peripheral trigeminal system during maxillary nerve growth. *J Cell Biol.* 1993;122(5):1053–1065.
75. Davies AM. Neurotrophin switching: where does it stand? *Curr Opin Neurobiol.* 1997;7(1):110–118.
76. Ninkina N, Adu J, Fischer A, et al. Expression and function of TrkB variants in developing sensory neurons. *EMBO J.* 1996;15(23):6385–6393.
77. Huang EJ, Wilkinson GA, Fariñas I, et al. Expression of Trk receptors in the developing mouse trigeminal ganglion: *in vivo* evidence for NT-3 activation of TrkA and TrkB in addition to TrkC. *Dev Camb Engl.* 1999;126(10):2191–2203.
78. Piñon LG, Minichiello L, Klein R, Davies AM. Timing of neuronal death in trkA, trkB and trkC mutant embryos reveals developmental changes in sensory neuron dependence on Trk signalling. *Dev Camb Engl.* 1996;122(10):3255–3261.
79. Wilkinson GA, Fariñas I, Backus C, Yoshida CK, Reichardt LF. Neurotrophin-3 is a survival factor *in vivo* for early mouse trigeminal neurons. *J Neurosci Off J Soc Neurosci.* 1996;16(23):7661–7669.
80. Lambiase A, Rama P, Bonini S, Caprioglio G, Aloe L. Topical treatment with nerve growth factor for corneal neurotrophic ulcers. *N Engl J Med.* 1998;338(17):1174–1180. http://dx.doi.org/10.1056/NEJM199804233381702.
81. Lambiase A, Manni L, Rama P, Bonini S. Clinical application of nerve growth factor on human corneal ulcer. *Arch Ital Biol.* 2003;141(2–3):141–148.
82. Aloe L, Rocco ML, Bianchi P, Manni L. Nerve growth factor: from the early discoveries to the potential clinical use. *J Transl Med.* 2012;10:239. http://dx.doi.org/10.1186/1479-5876-10-239.
83. Giger RJ, Wolfer DP, De Wit GM, Verhaagen J. Anatomy of rat semaphorin III/collapsin-1 mRNA expression and relationship to developing nerve tracts during

neuroembryogenesis. *J Comp Neurol.* 1996;375(3):378–392. http://dx.doi.org/10.1002/(SICI)1096-9861(19961118)375:3<378::AID-CNE3>3.0.CO;2-#.
84. Chilton JK, Guthrie S. Cranial expression of class 3 secreted semaphorins and their neuropilin receptors. *Dev Dyn.* 2003;228:726–733.
85. Taniguchi M, Yuasa S, Fujisawa H, et al. Disruption of semaphorin III/D gene causes severe abnormality in peripheral nerve projection. *Neuron.* 1997;19(3):519–530.
86. McKenna CC, Munjaal RP, Lwigale PY. Distinct roles for neuropilin1 and neuropilin2 during mouse corneal innervation. *PLoS One.* 2012;7(5):e37175. http://dx.doi.org/10.1371/journal.pone.0037175.
87. Kubilus JK, Linsenmayer TF. Developmental corneal innervation: interactions between nerves and specialized apical corneal epithelial cells. *Invest Ophthalmol Vis Sci.* 2010;51(2):782–789. http://dx.doi.org/10.1167/iovs.09-3942.
88. Schwend T, Lwigale PY, Conrad GW. Nerve repulsion by the lens and cornea during cornea innervation is dependent on Robo–Slit signaling and diminishes with neuron age. *Dev Biol.* 2012;363(1):115–127. http://dx.doi.org/10.1016/j.ydbio.2011.12.039.
89. Funderburgh JL, Caterson B, Conrad GW. Distribution of proteoglycans antigenically related to corneal keratan sulfate proteoglycan. *J Biol Chem.* 1987;262(24):11634–11640.
90. Zhang Y, Conrad AH, Tasheva ES, et al. Detection and quantification of sulfated disaccharides from keratan sulfate and chondroitin/dermatan sulfate during chick corneal development by ESI-MS/MS. *Invest Ophthalmol Vis Sci.* 2005;46(5):1604–1614. http://dx.doi.org/10.1167/iovs.04-1453.
91. Mao X, Zhang Y, Schwend T, Conrad GW. Effects of polysialic acid on sensory innervation of the cornea. *Dev Biol.* 2014;398:193–205. http://dx.doi.org/10.1016/j.ydbio.2014.11.020.
92. Kwiatkowski S, Munjaal RP, Lee T, Lwigale PY. Expression of pro- and anti-angiogenic factors during the formation of the periocular vasculature and development of the avian cornea. *Dev Dyn Off Publ Am Assoc Anat.* 2013;242(6):738–751. http://dx.doi.org/10.1002/dvdy.23956.
93. McKenna CC, Ojeda AF, Spurlin III J, Kwiatkowski S, Lwigale PY. Sema3A maintains corneal avascularity during development by inhibiting Vegf induced angioblast migration. *Dev Biol.* 2014;391(2):241–250. http://dx.doi.org/10.1016/j.ydbio.2014.04.017.
94. Mitchell CA, Risau W, Drexler HC. Regression of vessels in the tunica vasculosa lentis is initiated by coordinated endothelial apoptosis: a role for vascular endothelial growth factor as a survival factor for endothelium. *Dev Dyn Off Publ Am Assoc Anat.* 1998;213(3):322–333. http://dx.doi.org/10.1002/(SICI)1097-0177(199811)213:3<322::AID-AJA8>3.0.CO;2-E.
95. Ito M, Yoshioka M. Regression of the hyaloid vessels and pupillary membrane of the mouse. *Anat Embryol (Berl).* 1999;200(4):403–411.
96. Poché RA, Larina IV, Scott ML, et al. The Flk1-myr::mCherry mouse as a useful reporter to characterize multiple aspects of ocular blood vessel development and disease. *Dev Dyn.* 2009;238(9):2318–2326. http://dx.doi.org/10.1002/dvdy.21886.
97. Ambati BK, Nozaki M, Singh N, et al. Corneal avascularity is due to soluble VEGF receptor-1. *Nature.* 2006;443(7114):993–997. http://dx.doi.org/10.1038/nature05249.
98. Cursiefen C, Chen L, Dana MR, Streilein JW. Corneal lymphangiogenesis: evidence, mechanisms, and implications for corneal transplant immunology. *Cornea.* 2003;22(3):273–281.
99. Cursiefen C, Maruyama K, Jackson DG, Streilein JW, Kruse FE. Time course of angiogenesis and lymphangiogenesis after brief corneal inflammation. *Cornea.* 2006;25(4):443–447. http://dx.doi.org/10.1097/01.ico.0000183485.85636.ff.

100. Chen L, Cursiefen C, Barabino S, Zhang Q, Dana MR. Novel expression and characterization of lymphatic vessel endothelial hyaluronate receptor 1 (LYVE-1) by conjunctival cells. *Invest Ophthalmol Vis Sci.* 2005;46(12):4536–4540. http://dx.doi.org/10.1167/iovs.05-0975.
101. Nakao S, Hafezi-Moghadam A, Ishibashi T. Lymphatics and Lymphangiogenesis in the Eye. *J Ophthalmol.* 2012;2012:1–11. http://dx.doi.org/10.1155/2012/783163.
102. Seo S, Singh HP, Lacal PM, et al. Forkhead box transcription factor FoxC1 preserves corneal transparency by regulating vascular growth. *Proc Natl Acad Sci USA.* 2012;109(6):2015–2020. http://dx.doi.org/10.1073/pnas.1109540109.
103. Chow RL, Lang RA. Early eye development in vertebrates. *Annu Rev Cell Dev Biol.* 2001;17(1):255–296. http://dx.doi.org/10.1146/annurev.cellbio.17.1.255.
104. Gould DB, Smith RS, John SWM. Anterior segment development relevant to glaucoma. *Int J Dev Biol.* 2004;48(8–9):1015–1029. http://dx.doi.org/10.1387/ijdb.041865dg.
105. Cvekl A, Tamm ER. Anterior eye development and ocular mesenchyme. *BioEssays News Rev Mol Cell Dev Biol.* 2004;26(4):374–386. http://dx.doi.org/10.1002/bies.20009.
106. Gage PJ, Zacharias AL. Signaling "cross-talk" is integrated by transcription factors in the development of the anterior segment in the eye. *Dev Dyn Off Publ Am Assoc Anat.* 2009;238(9):2149–2162. http://dx.doi.org/10.1002/dvdy.22033.
107. Kao WW-Y, Liu H, Zhang J. Wakayama symposium: challenges of future research in ocular surface cell biology. *Ocul Surf.* 2013;11(1):19–24. http://dx.doi.org/10.1016/j.jtos.2012.07.002.
108. Jasrupuria-Agrawal S, Lwigale P. Neural crest cells in eye development. Neural Crest Cells: Evolution, Development and Disease. Ed. Paul Trainor. Elsevier; 2014:189–203.

CHAPTER FIVE

Corneal Epithelial Wound Healing

Chia-Yang Liu[1], Winston Whei-Yang Kao[1]
Edith Crawley Ophthalmic Research Laboratory, Department of Ophthalmology, College of Medicine, University of Cincinnati, Cincinnati, Ohio, USA
[1]Corresponding author: e-mail address: liucg@ucmail.uc.edu; Winston.Kao@UC.Edu

Contents

Abstract

The cornea is important for a clear vision by refracting light onto the lens, which in turn focusing on the retina. To maintain a smooth optical surface, corneal epithelium has to continuously renew itself to function as a barrier so that it protects the eye from various environmental insults. The adult corneal epithelium is maintained homeostatically by an integrated process of cell proliferation, migration, differentiation, stratification, and desquamation/apoptosis. Impairment of this process results in persistent corneal defect, leading to the blindness. Researches throughout the years revealed that appropriate integration and coordination of cell signaling events are responsible for corneal epithelial renewal and wound healing. In this chapter, we will review works done on cell culture, animal models, and human trials to focus on the signaling network during corneal wound healing process which will have potential for the discovery of novel drug to improve corneal wound healing.

1. THREE STAGES OF CORNEAL EPITHELIAL WOUND HEALING

Corneal surface normally remains a dome-shaped and smooth curvature, transparency, and avascularity, so that the incoming light can be uniformly refracted, and that an active repair process resurfaces defects. Upon

Progress in Molecular Biology and Translational Science, Volume 134
ISSN 1877-1173
http://dx.doi.org/10.1016/bs.pmbts.2015.05.002
© 2015 Elsevier Inc.
All rights reserved.

injury, corneal epithelial cells evoke sequential processes striving to quickly seal the wound and to prevent any opportunistic infection that otherwise can result in devastation of the eye and lead to blindness. The corneal epithelial abrasive wound healing can be distinguished anatomically by three continuous stages (Fig. 1): (1) lag phase, a quiescent time to prepare the wound healing, (2) active migration of superficial cell sheet to cover the denuded surface, (3) cell proliferation and stratification, and (4) final reassembly of adhesion structures.

(1) *Lag phase*: The incubation time between epithelial injury and the commencement of reepithelialization. During this phase, the epithelial cells immediately adjacent to the unwounded region flatten and migrate as an intact sheet to seal the wound. It should be noted that cellular proliferation is dispensable in this phase. However, several cytoskeletal proteins,[1] such as talin, vinculin, paxillin, and alpha actinin, and cell surface receptors, such as hyaluronan (HA) receptor (CD44), are actively synthesized. The α6 and β4 integrins located at the basal area of these cells are responsible for the linkage of their cytoskeletal components to the underlying basement membrane. Integrins dissociate from the hemidesmosomes and desmosomes and evenly distribute on the cellular surface. It is also believed that integrins are responsible

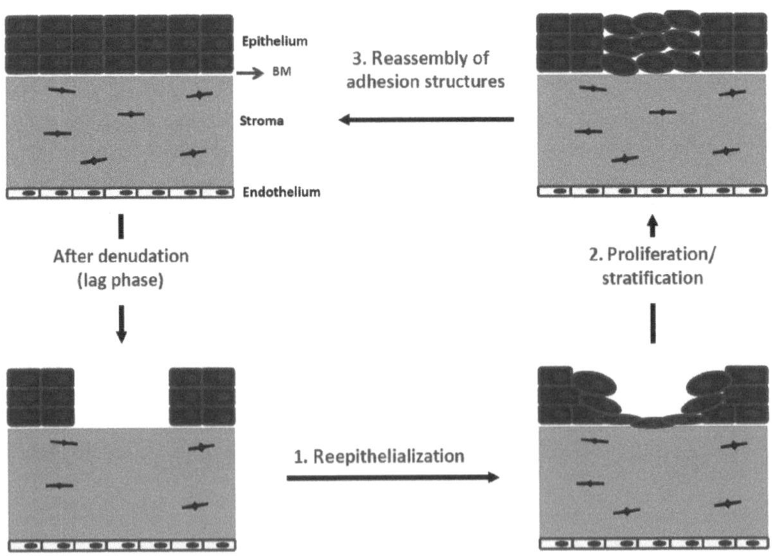

Figure 1 Distinct cellular events in different wound healing stages after corneal epithelial injury.

for a bidirectional signaling between the extracellular matrices and the cytoskeletons.[2]

(2) *Migration/reepithelialization*: Once the lag phase is complete, the multiple layers of cells adjacent to the wound margin extend over the denuded surface to reestablish the integrity of the corneal epithelium.[3] It is believed that both basal and suprabasal (wing) cells participate in the formation of the leading edge.[4] The migratory of cell sheet is highly dependent on the synthesis of cytoplasmic array of actin-rich stress fibers. It has been documented that a topical anesthetic drug, proparacaine, which blocks the synthesis and formation of stress fibers, effectively inhibits the adhesion and migration of the corneal epithelial cells *in vitro*.

(3) *Proliferation, stratification, and differentiation*: Cell proliferation then occurs to repopulate the wound area. The migratory and proliferative responses are compartmentalized in that limbal and peripheral epithelial cells exhibit an enhanced proliferative rate following wounding, while cells at the leading edge of migrating epithelium do not proliferate.[5,6] All the cell proliferation and differentiation even during the wound-healing phase is similar to that which occurs during normal homeostasis termed "X, Y, Z hypothesis" in which "X" stands for the proliferation of basal epithelial cells, "Y" stands for the contribution to the cell mass by centripetal movement of peripheral cells, and "Z" stands for the epithelial cell loss from the surface.[7]

2. EVENTS FOLLOWING CORNEAL EPITHELIAL WOUND

Upon abrasion of the corneal epithelium, there is a regulated disassembly of basement membrane.[8] The wounded basement membrane exposes to the inflammatory cells such as polymorphonuclear neutrophils. They release proteases to cause the degradation of basement membrane components. In addition, proteases in the tears may also play a role in the degradation of the basement membrane, especially after corneal wounding.[9,10] Corneal epithelial cells also can synthesize and release gelatinase B and matrilysin metalloproteinases (MMPs)[11,12] to degrade basement membrane component. Therefore, over a period of time after epithelial abrasion, the basement membrane alters both its structure and function due to the extensively proteolytic degradation. Given enough time after corneal epithelial abrasion, basement membrane loses integrins and other molecules and subsequently compromises the structural integrity and organization. The degraded basement membrane proteins diffuse into the tear film as a result of mechanical friction caused by the blinking of the eyelid.[13]

3. ROLE OF BASEMENT MEMBRANE TO EPITHELIAL WOUND HEALING

Basement membrane disassembly can affect reepithelialization in different ways: (1) epithelial cells at the wound edge expose to underlying stromal extracellular matrix. This induces new integrin expression or activation in migrating epithelial cells.[14] (2) Upon corneal epithelial debridement, cytokines and growth factors such as basic fibroblast growth factor (bFGF) and transforming growth factor-β (TGF-β) are abundantly present in the basement membrane and bind to the extracellular matrix molecules. They modulate intracellular signaling pathways to impact cell proliferation, differentiation, and/or apoptosis in migrating epithelial cells.[15,16] (3) Formation of a stable adhesion complex. The regenerated epithelial cells sit on basement membrane to restore the epithelial barrier function. Therefore, basement membrane plays a central function in the entire process of corneal epithelial wound healing. It is noted that one of the major basement membrane components, laminin-1, regulates a wide variety of cell behaviors including adhesion, proliferation, and differentiation.[17,18] Laminin isoforms 1 and 5 can directly associate with integrin $\alpha6\beta4$ and facilitate cell migration. Migrating epithelial cells synthesize and deposit laminin-1 and -5 in the basement membrane during epithelial wound.[19] Interestingly, approximately at the same time as laminin-1 and -5 syntheses, investigators found an upregulation of connexin and desmoglein-1 and -2. These observations suggested that the basement membrane restoration coincides with the formation of intercellular junctions and the expression of junctional adhesion proteins in corneal epithelial cells.[20]

4. ROLE OF INTEGRINS DURING CORNEAL WOUND HEALING

Integrins are transmembrane receptors that are the bridges for cell–cell and cell–extracellular matrix interactions. They mediate several cellular adhesive functions and trigger signal transduction pathways into the cell. The primary function of integrins is to connect cells to the extracellular matrix. These interactions influence many aspects of cell behavior including cell adhesion, migration, morphology, as well as cellular proliferation and differentiation. They are paramount important in forming a connection between the extracellular matrix on the exterior and the cytoskeleton in the interior.[21] The integrins function in bidirectional transduction of signals across the plasma membrane. Following corneal debridement wound, integrins play an integral part in the healing process. One of the initial changes involves the cells that have migrated beyond the basement membrane because they are exposed to the extracellular matrix. It has been shown that the α2β1 integrin of the migrating epithelial cells interacts with fibrillary type I collagen which dramatically upregulates the synthesis and production of MMP-1 (collagenase-1). MMP-1 then cleaves fibrillar collagen, which in turn dampens the adhesive function but enhances the migration of the epithelial cells *in vitro*.[22] In corneal epithelium, the integrin α6β4 adheres to its basal lamina through which hemidesmosomes are assembled.[23] Following epithelial cell abrasion, the beneath hemidesmosomes need to be disassembled so that epithelial cells can migrate along the substrate. It has been well documented by Gipson *et al.* that the redistribution of α6 and β4 integrins resulted in disassembly of hemidesmosomes which move away from the basal region of the cells during corneal epithelial wound healing.[24] The presence of selective integrin subunits on the cell membrane and the interaction of these α and β subunits with their ligands play a very sensitive role in response to the extracellular environment to restore the cell–matrix adhesion complexes. Thus, a better understanding of these integrin heterodimers bears great potential for us to manipulate the corneal epithelial wound healing process.

5. ROLE OF GROWTH FACTORS DURING CORNEAL WOUND HEALING

Low molecular weight growth factors such as epidermal growth factor (EGF), keratinocyte growth factor 1 (KGF-1, also known as FGF-7; KGF-2, also known as FGF-10), and hepatocyte growth factor (HGF) are very actively involved with corneal tissue remodeling. They are strong mitogens of corneal epithelial cells. EGF existing in various body fluids including tears can increase corneal epithelial cell proliferation both *in vitro* and *in vivo*.[25–27] When epithelial erosions occur, the preexisting EGF in tears can act immediately on the corneal epithelium to promote cell proliferation. KGF-1 and HGF specifically promote proliferation in cells of epithelial origin in a paracrine fashion. KGF-1 is expressed predominantly in corneal stroma, while KGF1 receptor is expressed in corneal epithelial cells.[28] Like EGF,

KGF-1 stimulates corneal epithelial cell proliferation, HGF is also expressed in cultured human stromal fibroblasts, and the HGF receptor in corneal epithelial cells.[29] Both HGF and KGF-1 play important roles in corneal epithelial cell proliferation in a paracrine manner from stroma. It has been known that EGF acts as the basic facilitator for epithelial proliferation, whereas KGF-1 and HGF are additionally upregulated depending on the amount of damage. Interestingly, the differences are apparent between the regional specificity of these growth factors. For example, topical KGF-1 predominantly impacts greatly on the limbal epithelial cells following epithelial wound, whereas EGF acts equally on epithelial cells in the limbus and in the peripheral cornea.[30] Since corneal epithelial stem cells are present in the adult limbal basal region, KGF-1 may act by increasing their proliferation. It is likely that KGF-1 plays a role in the proliferation of limbal and conjunctival epithelial cells, while HGF is involved with the proliferation of central and peripheral corneal epithelial cells. KGF-1 or HGF may depend upon the site of the epithelial defect.

The TGF-β family of proteins comprises three closely related isoforms in mammal: TGF-β1, -β2, and -β3. Various isoforms have been detected in the human tears and in human corneal epithelium. TGF-βs have multifunctional regulatory activity and are known to be intimately associated with the regulation of wound healing. It has been shown that TGF-β antagonizes the actions of EGF on corneal epithelial cells, and that both TGF-β1 and -β2 inhibit the corneal epithelial cell proliferation promoted by KGF-1 and HGF[31] and moderately inhibit the corneal epithelial cell proliferation stimulated by EGF. TGF-β1 or -β2 may be activated at the leading edge of the epithelium, where it might antagonize the proliferative effects of EGF, KGF-1, and/or HGF. Also, it is possible that TGF-β1 and -β2 may play important roles as negative modulators against cell proliferation in the wound-healing process. This is particularly essential for the initial lag phase immediately after epithelial injury.

6. ROLE OF CYTOKINE NETWORKS DURING CORNEAL WOUND HEALING

Cytokines also play multiple roles during corneal epithelial wound healing. Various cytokines are upregulated after abrasive wound to the corneal epithelium. Proinflammatory interleukin 1α (IL-1α) and IL-6 are particularly interesting because they are tremendously upregulated after injury. The levels of IL-1α and IL-6 correlate with the severity of the injury.[32] IL-6 stimulates epithelial migration in the cornea through a fibronectin-dependent mechanism, presumably the increased expression of integrin.[33] IL-1 and EGF synergistically modulate corneal epithelial cell chemotactic migration, suggesting that the combination of IL-1 and EGF offers a strategy to accelerate corneal epithelium wound closure.[34] It has been postulated that IL-1 in the cornea may act in conjunction with EGF in tears to stimulate epithelial wound healing. IL-1 may also induce KGF-1 and HGF expression in corneal fibroblasts causing epithelial cell proliferation.[35,36] It is noteworthy that not all the effects of IL-1 are beneficial because IL-1 also induces the matrix metalloproteinases, leading to corneal stromal melting.[37] Moreover, IL-1 released from injured corneal epithelium mediates keratocyte apoptosis. IL-1 also induces IL-6 production and this has a role in inflammation because it induces lymphocyte differentiation. In addition, IL-1 can induce IL-8 synthesis in both human corneal epithelial cells and human corneal keratocytes. They also suggest that IL-1α may play a more active role in amplifying inflammatory responses in the stroma than in the epithelial layer of the cornea.[38]

7. THE MESENCHYMAL–EPITHELIAL INTERACTIONS DURING WOUND HEALING

As aforementioned, the corneal epithelial wound healing response is brought about by a complex cascade of events involving cytokine (soluble factor)-mediated interactions between the epithelial cells, keratocytes of the corneal stroma, corneal nerves, lacrimal glands, and inflammatory cells of the immune system. The interaction of the various components is critical for the restoration of the epithelial function as a barrier and a refractive surface.[39] Our researches have been focusing on a multifacet protein called lumican (Lum) which belongs to the small leucine-rich proteoglycan (SLRP) family. Lum serves as a constituent of extracellular matrix and is indispensable for the maintenance of corneal transparency. Interestingly, it also serves as a matrikine in which lumican promotes corneal epithelial wound healing and maintains corneal homeostasis by modulating gene expression in normal and diseased tissues.[40–42] Structural–functional analysis has revealed that Lum binds ALK5 (also known as transforming growth factor receptor I, TβRI) via interaction of the C-terminal domain of Lum and the GS (glycine serine-rich) domain of ALK5. Administration of GST-Lum, GST-LumC53 recombinant protein, or a synthetic LumC13 peptide (YECLRVANEVTLN, the very C-terminal end of the human

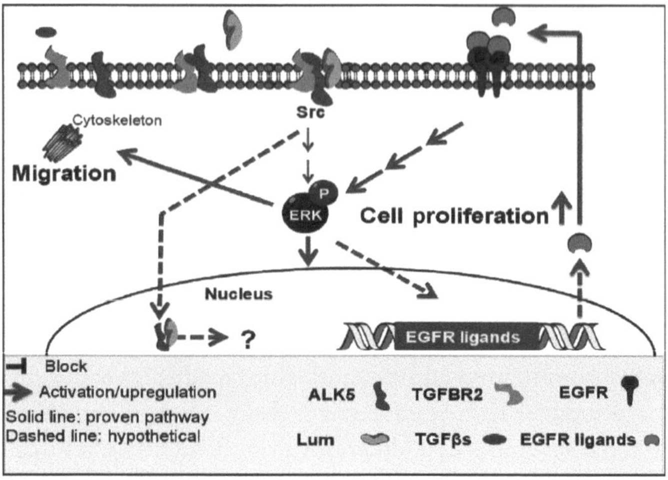

Figure 2 Lumican binds to TβRI (ALK5) to activate EGF ligands in corneal epithelial cells. (See the color plate.)

Lum) is sufficient to enhance wound healing of HTCE (human telomerase-immortalized corneal epithelial cells).[43] The Lum function on promoting HTCE wound healing is abolished by an ALK5 inhibitor and expression of ALK5-shRNA. During wound healing of cultured cells, there is an early activation of pERK1/2 which results from the activation of P2YR upon injury.[44,45] Addition of Lum to scratch-wounded HTCE induced upregulation of EGFR ligands, e.g., epiregulin (EREG), HB-EGF, and TGFα, which provide a feed forward mechanism to sustain pERK1/2 activation for promoting wound healing. Our observations are consistent with reports by other investigators in that ERK regulates EGFR ligand expression. For example, it has been shown that inhibition of pERK1/2 by PD98059 leads to a decrease in EREG and HB-EGF expression, suggesting that ERK has a role in the regulation of EGFR ligand expression.[46,47] The Lum/ALK5 complex is internalized via clathrin-coated vesicles after which ALK5 translocates to nuclei where it may regulate gene expression by collaboration with other nuclear proteins and/or signaling molecules (Fig. 2).

8. CONCLUSIONS

Studies of corneal epithelial wound healing have yielded informative knowledge for our understanding of the roles of growth factors and cytokines in the maintenance of corneal epithelial functions. However, little information is available about the precise roles of individual cytokines in corneal morphogenesis during development and homeostasis in adults. Also, in most cases the information has not been translated successfully to clinical applications. Identification of the signaling components will continue to be an interesting area of corneal wound healing research. Identification of genes that play important roles in maintaining the corneal epithelium remains critical for understanding and treating many devastating corneal diseases. Our observations, for the first time, show that Lum, a member of SLRP, binds ALK5 to promote wound healing. Peptides from the C-terminal domain of Lum can promote wound healing both *in vitro* and *in vivo*, making them promising therapeutic reagents for epithelium wound healing.

REFERENCES

1. Zeiske JD, Gipson IK. Protein synthesis during corneal epithelial wound healing. *Invest Ophthalmol Vis Sci*. 1986;27:1–7.
2. Schoenwaelder SM, Burridge K. Bidirectional signaling between the cytoskeleton and integrins. *Curr Opin Cell Biol*. 1999;11:274–286.
3. Zagon IS, Sassani JW, McLaughlin PJ. Cellular dynamics of corneal wound re-epithelialization in the rat. I. Fate of ocular surface epithelial cells synthesizing DNA prior to wounding. *Brain Res*. 1999;822:149–163.
4. Gipson IK, Sugrue SP. Cell biology of the corneal epithelium. In: Albert DM, Jakobiec FA, eds. *Principles and Practice of Ophthalmology: Basic Sciences*. Philadelphia: Saunders; 1994:3–16.
5. Zeiske JD, Gipson IK. Agents that affect corneal wound healing: modulation of structure and function. In: Albert DM, Jacobiec FA, eds. *Principles and Practice of Ophthalmology*. 2nd ed. Philadelphia: WB Saunders Co.; 2000:364–372.
6. Chung E-H, Hutcheon AEK, Joyce NC, Zieske JD. Synchronization of the G1/S transition in response to corneal debridement. *Invest Ophthalmol Vis Sci*. 1999;40:1952–1958.
7. Thoft RA, Friend J. The X, Y, Z hypothesis of corneal epithelial maintenance. *Invest Ophthalmol Vis Sci*. 1983;24:1442–1443.
8. Wagoner MD, Kenyon KR, Gipson IK, Hanninen LA, Seng WL. Polymorphonuclear neutrophils delay corneal epithelial wound healing in vitro. *Invest Ophthalmol Vis Sci*. 1984;25:1217–1220.
9. Cejkova J. Enzyme histochemistry of corneal wound healing. *Histol Histopathol*. 1998;13:553–564.
10. Sathe S, Sakata M, Beaton AR, Sack RA. Identification, origins and the diurnal role of the principal serine protease inhibitors in human tear fluid. *Curr Eye Res*. 1998;17:348–362.
11. Ye HQ, Azar DT. Expression of gelatinases A and B, and TIMPs-1 and 2 during corneal wound healing. *Invest Ophthalmol Vis Sci*. 1998;39:913–921.
12. Lu PC, Ye H, Maeda M, Azar DT. Immunolocalization and gene expression of matrilysin during corneal wound healing. *Invest Ophthalmol Vis Sci*. 1999;40:20–27.
13. Iglesia DDS, Stepp MA. Disruption of basement after corneal debridement. *Invest Ophthalmol Vis Sci*. 2000;41:1045–1053.
14. Pilcher BK, Dumin JA, Sudbeck BD, Krane SM, Welgus HG, Parkks WC. The activity of collagenase-I is required for keratinocyte migration on a type I collagen matrix. *J Cell Biol*. 1997;137:1445–1457.

15. Dowd CJ, Cooney CL, Nugent MA. Heparan sulfate mediates bFGF transport through basement membrane by diffusion with rapid reversible binding. *J Biol Chem.* 1999;274:5236–5244.
16. Dabin I, Courtois Y. In vitro kinetics of basic fibroblast growth factor diffusion across a reconstituted corneal endothelium. *J Cell Physiol.* 1991;147:396–402.
17. Streuli CH, Schmidhauser C, Bailey N, et al. Laminin mediates tissue-specific gene expression in mammary epithelia. *J Cell Biol.* 1995;129:591–603.
18. Jiang FX, Cram DS, De Aizpurua HJ, Harrison LC. Laminin-1 promotes differentiation of fetal mouse pancreatic beta cells. *Diabetes.* 1999;48:722–730.
19. Suzuki K, Tanaka T, Enoki M, Nishida T. Coordinated reassembly of the basement membrane and junctional proteins during corneal epithelial wound healing. *Invest Ophthalmol Vis Sci.* 2000;41:2495–2500.
20. Torok PG, Mader TH. Corneal abrasions: diagnosis and management. *Am Fam Physician.* 1996;53:3521–3532.
21. Hynes RO. Integrins: a family of cell surface receptors. *Cell.* 1987;48:549–554.
22. Messent AJ, Tuckwell DS, Knaupre V, Humphries MJ, Murphy G, Gavrilovic J. Effects of collagenase cleavage of type I collagen on α2β1 integrin-mediated cell adhesion. *J Cell Sci.* 1998;111:1127–1135.
23. Stepp MA, Spurr-Michaud S, Tisdale A, Elwell J, Gipson IK. α6β4 integrin heterodimer is a component of hemidesmosomes. *Proc Natl Acad Sci USA.* 1990;87:8970–8974.
24. Gipson IK, Spurr-Michaud S, Tisdale A, Elwell J, Stepp MA. Redistribution of the hemidesmosome components α6β4 integrin and bullous pemphigoid antigens during epithelial wound healing. *Exp Cell Res.* 1993;207:86–98.
25. Watanabe H, Ohasi Y, Kinoshita S, Manabe R, Ohshiden K. Distribution of epidermal growth factor in rat ocular and periocular tissues. *Graefes Arch Clin Exp Ophthalmol.* 1993;231:228–232.
26. Wilson SE, Lloyd SA, Kennedy RH. Epidermal growth factor messenger RNA production in human lacrimal gland. *Cornea.* 1991;10:519–524.
27. Wilson SE, He YG, Lloyd SA. EGF, EGF receptor, basic FGF, TGF beta-1, and IL-1alpha mRNA in human corneal epithelial cells and stromal fibroblast. *Invest Ophthalmol Vis Sci.* 1992;33:1756–1765.
28. Sotozono C, Kinoshita S, Kita M, Imanishi J. Paracrine role of keratinocyte growth factor in rabbit corneal epithelial cell growth. *Exp Eye Res.* 1994;59:385–391.
29. Wilson SE, Walker JW, Chwang EL, He YG. Hepatocyte growth factor, keratinocyte growth factor, their receptors, fibroblast growth factor receptor2, and the cells of the cornea. *Invest Ophthalmol Vis Sci.* 1993;35:2544–2561.
30. Kitazawa T, Kinoshita S, Fujita K, et al. The mechanism of accelerated corneal epithelial healing by human epidermal growth factor. *Invest Ophthalmol Vis Sci.* 1990;31:1773–1778.
31. Honma Y, Nishida K, Sotozono C, Kinoshita S. Effect of transforming growth factor-b1 and b2 on in vitro rabbit corneal epithelial cell proliferation promoted by epidermal growth factor, keratinocyte growth factor, or hepatocyte growth factor. *Exp Eye Res.* 1997;65:391–396.
32. Sotozono C, He J, Matsimoto Y, Kita M, Imanishi J, Kinoshita S. Cytokine expression in the alkali-burned cornea. *Curr Eye Res.* 1997;16:670–676.
33. Nishida T, Nakamura M, Mishima H, Otori T. Interleukin 6 promotes epithelial migration by fibronectin-dependent mechanism. *J Cell Physiol.* 1992;153:1–5.
34. Boisjoly HM, Laplante C, Bernatchez SF, Salesse C, Giasson M, Joly MC. Effects of EGF, IL-1 and their combination on in vitro corneal epithelial wound closure and cell chemotaxis. *Exp Eye Res.* 1993;57:293–300.

35. Li DQ, Tseng SCG. Differential regulation of keratinocyte growth factor and hepatocyte growth factor/scatter factor by different cytokines in human corneal and limbal fibroblasts. *J Cell Physiol.* 1997;172:361–372.
36. Malecaze F, Simorre V, Chollet P, et al. Interleukin-6 in tear fluid after photo refractive keratectomy and its effects on keratocytes in culture. *Cornea.* 1997;16:580–587.
37. Girard MT, Matsubara M, Fini ME. Transforming growth factor-beta and interleukin-1 modulate metalloproteinase expression by corneal stromal cells. *Invest Ophthalmol Vis Sci.* 1991;32:2441–2454.
38. Koch AE, Polverini PJ, Kunkel SL, et al. Interleukin-8 as a macrophage-derived mediator of angiogenesis. *Science.* 1992;258:1798–1801.
39. Kao WW, Funderburgh JL, Xia Y, Liu CY, Conrad GW. Focus on molecules: lumican. *Exp Eye Res.* 2006;82:3–4.
40. Kao WW. Ocular surface tissue morphogenesis in normal and disease states revealed by genetically modified mice. *Cornea.* 2006;25(suppl 1):S7–S19.
41. Yeh LK, Chen WL, Li W, et al. Soluble lumican glycoprotein purified from human amniotic membrane promotes corneal epithelial wound healing. *Invest Ophthalmol Vis Sci.* 2005;46:479–486.
42. Robertson DM, Li L, Fisher S, et al. Characterization of growth and differentiation in a telomerase-immortalized human corneal epithelial cell line. *Invest Ophthalmol Vis Sci.* 2005;46:470–478.
43. Yamanaka O, Yuan Y, Coulson-Thomas VJ, et al. Lumican binds ALK5 to promote epithelium wound healing. *PLoS One.* 2013;8:e82730.
44. Klepeis VE, Weinger I, Kaczmarek E, Trinkaus-Randall V. P2Y receptors play a critical role in epithelial cell communication and migration. *J Cell Biochem.* 2004;93:1115–1133.
45. Yang L, Cranson D, Trinkaus-Randall V. Cellular injury induces activation of MAPK via P2Y receptors. *J Cell Biochem.* 2004;91:938–950.
46. Cho MC, Choi HS, Lee S, et al. Epiregulin expression by Ets-1 and ERK signaling pathway in Ki-ras-transformed cells. *Biochem Biophys Res Commun.* 2008;377:832–837.
47. Ingram JL, Rice AB, Santos J, Van Houten B, Bonner JC. Vanadium-induced HB-EGF expression in human lung fibroblasts is oxidant dependent and requires MAP kinases. *Am J Physiol Lung Cell Mol Physiol.* 2003;284:L774–L782.

CHAPTER SIX

Corneal Dystrophies: Overview and Summary

Daniel Schorderet[1]
IRO - Institute for Research in Ophthalmology, Sion, Switzerland
Faculty of Life Sciences, Swiss Federal Institute of Technology, Lausanne, Switzerland
Department of Ophthalmology, University of Lausanne, Lausanne, Switzerland
[1]Corresponding author e-mail address: daniel.schorderet@irovision.ch

Contents

Abstract

In this chapter, I shall discuss the genetics, mode of inheritance and molecular origin of several corneal dystrophies.

1. INTRODUCTION

1.1 Genetics of Corneal Dystrophies (Daniel Schorderet)

Corneal dystrophies represent a heterogeneous group of diseases of the cornea at both genetic and clinical levels. Genetically, autosomal dominant, autosomal recessive, and X-linked recessive modes of inheritance have been described. In this chapter, we shall focus on well-characterized phenotypes. Examples of other corneal dystrophies have been published with tentative modes of inheritance. However, most of them are rare and it is not clear, at this time, whether they represent different entities. Description of the various forms is mostly based on the recent IC3D classification of corneal dystrophies.[1]

1.1.1 Anterior Corneal Dystrophy

1.1.1.1 Meesman Corneal Dystrophy

Initially described by Pameijer in 1935, this disease is characterized by recurrent punctiform epithelial erosions accompanied by pain and lacrimation.[2] The disease can begin early in life, and it usually progresses over time. It is transmitted as an autosomal dominant trait with high penetrance. Irvine *et al.*

Progress in Molecular Biology and Translational Science, Volume 134
ISSN 1877-1173
http://dx.doi.org/10.1016/bs.pmbts.2015.04.004
© 2015 Elsevier Inc.
All rights reserved.

described missense mutations in *KRT3* and *KRT12*, two genes coding for keratin 3 and keratin 12.[3] Keratins are water-insoluble proteins forming the intermediate filaments in epithelial cells. Usually, one basic keratin pairs with an acidic keratin to form tissue-specific heterodimers. In cornea, the two major keratins are KRT3 and KRT12.

1.1.1.2 Epithelial Basement Membrane Dystrophy

First described in 1930 by Vogt, this corneal dystrophy is characterized by the presence of dots, maps, and blebs.[4] This entity has also been named Cogan microcystic epithelial dystrophy, map-dot-fingerprint dystrophy, and epithelial basement membrane dystrophy. Often the disease is asymptomatic and is diagnosed on a routine eye examination. Recurrent erosions may occur.

Most of the cases are sporadic, but if not the disease is transmitted as an autosomal dominant trait. Boutboul *et al.* described mutations in *TGFBI/BIGH3* in two families with autosomal dominant transmission.[5] It is not clear how these mutations affect the structure and function of the protein. No other gene has been associated with this disease.

1.1.1.3 Gelatinous Drop-Like Corneal Dystrophy

First described by Nakaizumi, it is characterized by the presence of subepithelial fluorescein-positive nodules which are visible as small masses with a band-shaped distribution.[6] At histology, the lesions show amyloid deposits and are initially present between the epithelium and Bowman's layer. This disease has an autosomal recessive mode of inheritance and is the consequence of mutations in *TACSTD2*.[7] The role of this protein is still mostly unknown.

1.1.1.4 Thiel–Behnke Corneal Dystrophy

Thiel–Behnke corneal dystrophy, also known as honeycomb, curly fiber, anterior limiting membrane type II, or Bowman's layer type II corneal dystrophy, was first described by Thiel and Behnke.[8] The corneal lesions appear in childhood and are characterized by recurrent erosions. Biomicroscopy shows subepithelial reticular opacities with symmetric honeycomb appearance. This disease is transmitted as an autosomal dominant trait and is due to an R555Q mutation in *TGFBI/BIGH3*.[9]

1.1.1.5 Reis–Bücklers' Corneal Dystrophy

In 1917, Reis reported an inherited form of corneal dystrophy characterized by irregular geographic-like subepithelial opacities.[10] This corneal dystrophy has also been reported under various names, including geographic corneal dystrophy, corneal dystrophy of Bowman's layer type I, and anterior limiting membrane dystrophy type I. Although descriptive names have advantages, it is now accepted that this form should be referred to as Reis–Bücklers' corneal dystrophy. At histology, it is possible to see electron-dense rod-shaped keratoepithelin, the product of *TGFBI/BIGH3*, positive bodies. Reis–Bücklers' corneal dystrophy is inherited as an autosomal dominant disorder and is mostly due to R124L mutations in *TGFBI/BIGH3*.[9] Recently, Piao *et al.* reported a family with Reis–Bücklers' corneal dystrophy and an R555Q mutation.[11]

1.1.2 Stromal

1.1.2.1 Granular Type 1

More than a hundred years ago, Groenouw described a corneal dystrophy that started in early childhood and was progressive.[12] Recurrent erosions are accompanied by pain and photophobia. Clinically, the cornea shows numerous whitish granules sometimes taking the form of "bread crumbs" or rings with a clear center. With age, the number of granules increases and quantitative analysis may be difficult. They are found deeper in the stroma and stain positive with Masson trichrome and with antibody against keratoepithelin. The disease is transmitted as an autosomal dominant trait with complete penetrance. Homozygous affected patients are more heavily affected. Most of the cases are due to an R555W mutation in *TGFBI/BIGH3*.[9] Mutations in other amino acids may also occur but are much less frequent.[13]

1.1.2.2 Granular Type 2

This entity was first described by Folberg *et al.*[14] Clinically atypical granular corneal dystrophy with pathologic features of lattice-like amyloid deposits is seen. This corneal dystrophy is characterized by the presence of superficial white dots that can take the form of stars, snowflakes, or rings. Contrary to the granular type I, the deposits can easily be counted. At histology, the deposits stain with Masson Trichrome or with Congo Red, a staining specific for amyloid, and are keratoepithelin positive. The disease is transmitted as an autosomal dominant trait with mutations in *TGFBI/BIGH3*.[9] The most prevalent mutation is the R124H. This disease is more frequent in Asia where a more severe homozygous form has been described in several families.[15]

1.1.2.3 Lattice Type 1

Biber presented the first description of lattice corneal dystrophy in 1890.[16] Transmission is autosomal dominant and mutations are present in *TGFBI/*

BIGH3.[9] The classical form is accompanied by the R124C mutation. Besides the classical form in which patients experience recurrent erosions during their first decade, appearance of superficial dots together with fine lines starting in the center of the cornea and spreading toward the periphery, several other forms have been described in association with various missense mutations in the fourth FAS1 domain of *TGFBI/BIGH3*.[13]

1.1.2.4 Schnyder

Schnyder or Schnyder crystalline corneal dystrophy is a slowly progressive corneal dystrophy. Diagnosis is usually made in the young adult when a superficial haze and/or subepithelial crystals, hence its name, are observed. Biochemically, there is an abnormal increase in cholesterol and phospholipid deposition in the cornea. Transmission is autosomal dominant and missense mutations are observed in *UBAID1*, a gene involved in vitamin K biosynthesis.[17]

1.1.2.5 Macular Corneal Dytrophy

Also known as Groenouw corneal dystrophy type II, macular corneal dystrophy was first reported by Groenouw.[12] It is characterized by superficial stromal haze appearing between the first and third decade. Initially, the lesions are centrally located but later they extend toward the periphery. Deposits are made of glycosaminoglycans. This disease is transmitted as an autosomal recessive trait and is due to mutations in the carbohydrate sulfotransferase 6 (*CHST6*) gene on chromosome 16q21.[18]

1.1.2.6 Congenital Stromal Dystrophy

As its name implies, this rare corneal dystrophy is present at birth. The patient presents diffuse, bilateral clouding located throughout the stroma. Electron microscopy analysis shows abnormal layers of thin filaments about half the size of normal collagen fibrils. Transmission is autosomal dominant. No gene has been identified as yet.

1.1.2.7 Fleck Corneal Dystrophy

This corneal dystrophy first described by François and Neetens is characterized by the presence of small, ring-shaped white flecks.[19] Discovery is generally fortuitous and the patients are asymptomatic. Transmission is autosomal dominant. Nonsense and missense mutations have been observed in *PIP5K3*, a member of the phosphoinositide 3-kinase family.[20]

1.1.3 Posterior Corneal Dystrophy

1.1.3.1 Congenital Hereditary Endothelial Dystrophy

Posterior hereditary dystrophy has been reported by Maumenee.[21] It is characterized by the appearance, soon after birth, of bilateral diffuse corneal deposits. These deposits can be accompanied by increased corneal thickness, nystagmus, and amblyopia. The disease is transmitted as an autosomal recessive trait. Missense and nonsense mutations in *SLC4A11* have been reported.[22] Type 1 no longer exists.[1]

1.1.3.2 Posterior Polymorphous Corneal Dystrophy

Also known as Schlichting corneal dystrophy, this disease is characterized by the presence of confluent or clustered vesicular and blister-like endothelial lesions. It is rarely congenital and shows a slow or nonprogressive evolution. The disease is transmitted as an autosomal dominant trait with incomplete penetrance. Mutations in at least two genes have been reported: *COL8A2* for PPCD2[23] and *ZEB1* for PPCD3.[2,24] A locus for PPCD1 has been mapped to chromosome 20p11 and mutations have been found in *VSX1*. However, the role of VSX1 has still to be determined.

1.1.3.3 Fuchs Endothelial Corneal Dystrophy

Fuchs corneal dystrophy, named after Fuchs who first described this entity in 1910, has two forms,[25] a late form with symptoms appearing in the fourth decade and an early form starting in the first decade. It is the most frequent form of corneal dystrophy and is characterized by impaired vision in the morning with subsequent improvement during the day. The lesions forming guttae are present in the anterior part of the endothelium. These lesions are initially present in the center of the cornea and then spread toward the periphery. Fuchs' corneal dystrophy is transmitted as an autosomal dominant trait with mutation in *COL8A2* for the early form (FECD1).[23] FECD4 has been associated with mutations in *SLC4A11*[22] and FECD6 with heterozygous mutations in *ZEB1*.[24] It is interesting to note that SLC4A11 and ZEB1 have also been involved in CHED2 and PPCD3, respectively. FECD8 has been associated with mutations in *AGBL1*.[26] Four additional loci have been identified for late-onset Fuchs' corneal dystrophy, so far: FECD2 (13pter–q12.13), FECD3 (18q21.2–q21.3), FECD5 (5q33.1–q35.2), FECD7 (9p24.1–p22.1), and FECD8 (15q25).

REFERENCES

1. Weiss JS, Møller HU, Aldave AJ, et al. IC3D classification of corneal dystrophies—edition 2. *Cornea*. 2015;34:117–159.
2. Pameijer JK. Über eine fremdartige familiare oberflächliche Hornhautveränderung. *Klin Monatsbl Augenheilkd*. 1935;103:516–517.

3. Irvine AD, Corden LD, Swensson O, et al. Mutations in cornea-specific keratin K3 or K12 genes cause Meesmann's corneal dystrophy. *Nat Genet.* 1997;16:184–187.
4. Vogt A. *Lehrbuch und Atlas der Spaltlampenmikroskopie des lebenden Auges.* Berlin: Springer; 1930.
5. Boutboul S, Black GCM, Moore JE, et al. A subset of patients with epithelial basement membrane corneal dystrophy have mutations in TGFBI/BIGH3. *Hum Mutat.* 2006;27:553–557.
6. Nakaizumi K. A rare case of corneal dystrophy. *Acta Soc Ophthalmol Jpn.* 1914; 18:949–950.
7. Tsujikawa M, Kurahashi H, Tanaka T, et al. Identification of the gene responsible for gelatinous drop-like corneal dystrophy. *Nat Genet.* 1999;21:420–423.
8. Thiel HJ, Behnke H. Eine bisher unbekannte subepitheliale hereditäre Hornhautdystrophie. *Klin Monbl Augenheilkd.* 1967;150:862–874.
9. Munier FL, Korvatska E, Djemai A, et al. Kerato-epithelin mutations in four 5q31-linked corneal dystrophies. *Nat Genet.* 1997;15:247–251.
10. Reis W. Familiäre, fleckige Hornhautentartung. *Dtsch Med Wochenschr.* 1917;43:575.
11. Piao MZ, Zhou XT, Chu RY. Arg555Gln mutation of *TGFBI* gene in geographical-type Reis-Bücklers corneal dystrophy in a Chinese family. *J Int Med Res.* 2012; 40:1149–1155.
12. Groenouw A. Knötchenförmige Hornhauttrübungen (Noduli corneae). *Arch Augenheilkd.* 1890;21:281–289.
13. Munier FL, Frueh BE, Othenin-Girard P, et al. BIGH3 mutation spectrum in corneal dystrophies. *Invest Ophthalmol Vis Sci.* 2002;43:949–954.
14. Folberg R, Alfonso E, Croxatto JO, et al. Clinically atypical granular corneal dystrophy with pathologic features of lattice-like amyloid deposits: a study of three families. *Ophthalmology.* 1988;95:46–51.
15. Mashima Y, Konishi M, Nakamura Y, et al. Severe form of juvenile corneal stromal dystrophy with homozygous R124H mutation in the keratoepithelin gene in five Japanese patients. *Br J Ophthalmol.* 1998;82:1280–1284.
16. Biber H. *Über einige seltene Hornhauterkrankungen: die oberflächliche gittrige Keratitis.* Zürich: Inaugural Dissertation; 1890.
17. Orr A, Dube MP, Marcadier J, et al. Mutations in the UBIAD1 gene, encoding a potential prenyltransferase, are causal for Schnyder crystalline corneal dystrophy. *PLoS One.* 2007;8:e685.
18. Akama TO, Nishida K, Nakayama J, et al. Macular corneal dystrophy type I and type II are caused by distinct mutations in a new sulphotransferase gene. *Nat Genet.* 2000; 26:237–241.
19. François J, Neetens A. Nouvelle dystrophie hérédofamiliale du parenchyme cornéen (hérédo-dystrophie mouchetée). *Bull Soc Belge Ophtalmol.* 1957;114:641–646.
20. Li S, Tiab L, Jiao X, et al. Mutations in PIP5K3 are associated with Francois-Neetens mouchetee fleck corneal dystrophy. *Am J Hum Genet.* 2005;77:54–63.
21. Maumenee AE. Congenital hereditary corneal dystrophy. *Am J Ophthalmol.* 1960; 50:1114–1124.
22. Vithana EN, Morgan P, Sundaresan P, et al. Mutations in sodium-borate cotransporter SLC4A11 cause recessive congenital hereditary endothelial dystrophy (CHED2). *Nat Genet.* 2006;38:755–757.
23. Biswas S, Munier FL, Yardley J, et al. Missense mutations in COL8A2, the gene encoding the alpha-2 chain of type VIII collagen, cause two forms of corneal endothelial dystrophy. *Hum Mol Genet.* 2001;10:2415–2423.
24. Krafcha CM, Pawar H, Moroi SE, et al. Mutations in TCF8 cause posterior polymorphous corneal dystrophy and ectopic expression of COL4A3 by corneal endothelial cells. *Am J Hum Genet.* 2005;77:694–708.
25. Fuchs E. Dystrophia epithelialis corneae. *Arch Ophthalmol.* 1910;76:478–508.
26. Riazuddin SA, Vasanth S, Katsanis N, Gottsch JD. Mutations in AGBL1 cause dominant late-onset Fuchs corneal dystrophy and alter protein-protein interaction with TCF4. *Am J Hum Genet.* 2013;93:758–764.

CHAPTER SEVEN

Fuchs Corneal Dystrophy

Allen O. Eghrari, S. Amer Riazuddin, John D. Gottsch[1]
Wilmer Eye Institute, Johns Hopkins University School of Medicine, Baltimore, Maryland, USA
[1]Corresponding author e-mail address: jgottsch@jhmi.edu

Contents

Abstract

Fuchs corneal dystrophy (FCD) is a hereditary, progressive disease of the posterior cornea which results in excrescences of Descemet membrane, endothelial cell loss, corneal edema, and, in late stages, bullous keratopathy. Structural changes are noted principally in Descemet membrane and the endothelium, with thickening of Descemet membrane, loss of barrier function, and increased corneal hydration, although secondary effects occur throughout all layers. Multiple chromosomal loci and, more recently, causal genetic mutations have been identified for this complex disorder, including in *TCF8*, *SLC4A11*, *LOXHD1*, and *AGBL1*. A trinucleotide repeat in *TCF4* correlates strongly with disease status and interacts in common pathways with previously identified genes. Dysregulation of pathways involving oxidative stress and apoptosis, epithelial-to-mesenchymal transition, microRNA, mitochondrial genes, and unfolded protein response has been implicated in FCD pathogenesis.

Progress in Molecular Biology and Translational Science, Volume 134
ISSN 1877-1173
http://dx.doi.org/10.1016/bs.pmbts.2015.04.005
© 2015 Elsevier Inc.
All rights reserved.

1. INTRODUCTION

Fuchs corneal dystrophy (FCD) is a progressive, hereditary, bilateral corneal condition first described a century ago by the Austrian ophthalmologist Ernst Fuchs.[1] Patients present in the fifth to sixth decade of life with blurry morning vision that increases in duration as the disease progresses. Primarily a condition of the posterior cornea, characteristic features include the formation of focal excrescences of Descemet membrane termed "guttae," a progressive increase in corneal edema, loss of endothelial cell density over years, and end-stage disease manifested by formation of painful epithelial bullae as the cornea decompensates in its ability to maintain stromal dehydration.

Recent advances in our understanding of the genetic and pathophysiological mechanisms of the disease as well as the application of new imaging modalities provide unique insight into the development of and potential therapeutic options for FCD.

2. STRUCTURAL CHANGES IN THE FCD CORNEA

The characteristic clinical feature of FCD is the presence of guttae, focal nodules in Descemet membrane which are anvil shaped and protrude posteriorly into the endothelial plane, and are several microns in diameter. Guttae are visible clinically with slit-lamp biomicroscopy, and in early stages help to determine the clinical phenotype (Fig. 1).

Unlike peripheral, age-associated nodules or pigment deposition on the endothelium from the anterior chamber, these excrescences begin in the center and over the course of decades, begin to coalesce and present peripherally. The grading scale commonly utilized and described by Krachmer 35 years ago reflects this phenomenon; a severity score between one and five is given based on the number and distribution of guttae. A score of one, reflecting minimal and generally asymptomatic disease, is defined as more than 12 central guttae, whereas a cluster of central confluent guttae are graded as two, and higher scores represent increasing diameters of distribution of confluent guttae.[2]

Although the traditional model of FCD suggested that severe disease is marked by corneal decompensation and diffuse stromal and epithelial edema in the presence of diffuse guttae, assessment of a large number of patients with FCD across various levels of severity suggests that the progressive development of edema is a gradual process that occurs over time,[3] with symptoms developing in later stages over the course of decades.

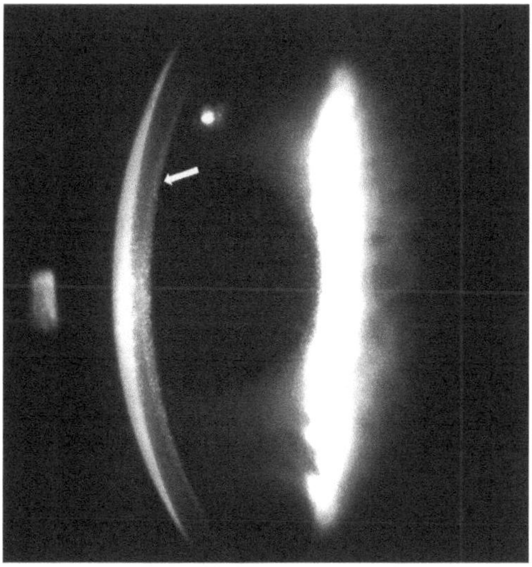

Figure 1 Direct illumination of guttae with slit-lamp biomicroscopy. In the slit of light seen passing through the cornea from left (anterior surface) to right (posterior surface), the beaten-metal appearance of guttae is appreciated posteriorly in light reflected from Descemet membrane (arrow).

We will subsequently explore pathological structural changes in the FCD-affected cornea.

2.1 Epithelium

Prior to the advent of slit-lamp biomicroscopy, FCD was identified in its severe stages, and Fuchs initially described the disease as *dystrophia epithelialis corneae.*[1] The epithelium is generally intact in early stages of classic Fuchs dystrophy, a phenomenon evidenced by a lack of change in anterior keratometry in patients undergoing endothelial keratoplasty, which would be expected if distortion of the epithelium were to occur in the disease process.

However, in severe FCD, loss of pump function in the edematous cornea results in anterior migration of fluid through the corneal stroma and the formation of painful epithelial bullae.

2.2 Bowman's Layer

In vivo confocal microscopy of patients with FCD reveals an abnormal Bowman's layer in approximately half of patients with FCD, with diffuse bright reflection and a paucity or absence of nerves.[4]

2.3 Stroma

Anterior stromal changes occur during periods of corneal edema with formation of stromal haze and visual higher order aberrations, a phenomenon that may persist after endothelial keratoplasty.[5] Confocal microscopy suggests that after endothelial transplantation, subepithelial haze rather than interface haze is the predominant source of forward light scatter.[6]

Collagen fibers in the anterior stroma are tightly packed relative to the posterior stroma, in which fibrils have less connections. As a result, corneal edema results in increased hydration of the posterior cornea and bulging of the posterior stroma into the anterior chamber. This focal thickening in the central corneal relative to the periphery can be appreciated through anterior segment optical coherence tomography or Scheimpflug photography; the latter has been utilized to demonstrate an increased center-to-peripheral thickness ratio in FCD-affected corneas compared to controls.[7]

2.4 Descemet Membrane

Descemet membrane is normally composed of two layers, an anterior banded layer approximately 3 μm in thickness which develops *in utero* and is relatively constant in thickness over time, and a posterior nonbanded layer which is deposited by the endothelium and which thickens over an individual lifetime to approximately 10 μm in an adult. Descemet membrane in FCD is marked by the presence of a thick posterior banded layer, a posterior-most layer in which collagen fibrils approximately 10–20 nm in diameter and with 110 nm banding are deposited by the endothelium. This posterior banded layer measures approximately 16 μm in thickness and correlates directly with clinical severity of FCD.[8] It is in this posterior banded layer that focal excrescences of Descemet membrane produce posteriorly and tent up the endothelium.

In some cases, particularly those in which the cornea has suffered severe edema, a fibrillar layer is present, without fibroblasts or cellular elements. This has also been described as a posterior collagenous layer.[9,10]

Guttae develop in the central cornea in early disease and are formed peripherally in more advanced disease. However, a propensity for the development of guttae in the inferotemporal quadrant has been documented in multiple families using retroillumination photography[11,12] and specular microscopy.[13] Although the causes for this are unclear, the inferotemporal cornea is associated with a thinner cornea and potentially increased light exposure (Fig. 2).

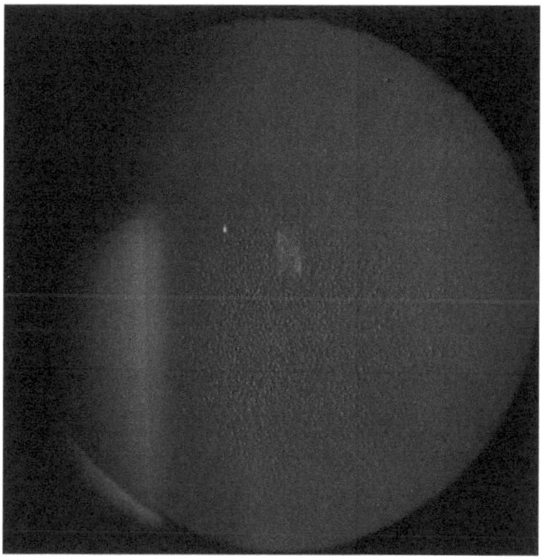

Figure 2 Retroillumination photography reveals the distribution of guttae throughout the cornea, a process which is largely central but increases in the periphery over time.

2.5 Endothelium

The endothelium in both normal and FCD-affected corneas is a single monolayer of cells which is responsible for maintaining stromal dehydration, and measures approximately 5 μm in thickness. As endothelial cell density decreases, *in vivo* confocal microscopy reveals increased pleomorphism and polymegethism.[14]

Polymerase chain reaction array analysis of human Descemet membrane–endothelial complexes reveals upregulation of numerous extracellular matrix-related genes in FCD, including collagen and fibronectin, suggesting significant matrix alterations in FCD pathophysiology.[15]

Transmission electron microscopy of FCD corneas demonstrates some endothelial cells with cytoplasmic filaments, increased rough endoplasmic reticulum, and cytoplasmic processes, bearing similarity to fibroblasts; intercellular vacuoles remain in areas associated with loss of cells.[9]

Decreased aquaporin expression has been detected in FCD corneal endothelium and may contribute to corneal edema.[16]

3. GENETIC BASIS OF FCD

Through the passage of a century after Fuchs' initial description of FCD, an understanding of the genetic basis of disease has largely correlated with the knowledge and application of human genetics, with rapid advances made in the twenty-first century.

Although the possibility of a hereditary component was proposed within a few years after Fuchs' publication of his findings, pedigrees presented in the 1970s assisted to establish its autosomal dominant pattern of inheritance.[2,17,18]

3.1 Genetic Linkage Analysis

To date, large family studies of FCD have resulted in the identification of several chromosomal loci associated with FCD.

The *FCD1* locus, the first to be associated with classic, late-onset FCD, was localized to a 26.4-Mb interval between 13pTel and 13q12.13 through recruitment and analysis of a large Caucasian family.[19] A total of 13 affected and 3 unaffected individuals across three generations were recruited for genotyping, and linkage using 399 STR markers from the MD10 genotyping panel revealed significant two-point logarithm of odds scores of 3.91 at D13S1236 and 3.80 at D13S1304, assuming fully penetrant dominant inheritance. Notably, two children in this family whose parents were both affected demonstrated clinical signs of late-onset FCD at a young age.

The *FCD2* locus was identified in three large families and localized to 18q21.2–18q21.32.[20] This was the first locus associated with multiple families, suggesting that this area may represent a common locus. Linkage analyses were performed using 1107 STR markers, and maximum two-point LOD scores among the three families were 3.41, 2.89, and 2.45. Differences in haplotypes among families raised the possibility of heterogeneity of the locus. Using retroillumination photography, progression in *FCD2*-associated families was compared with that in *FCD1*-related families and found to demonstrate a less marked rate of development of guttae.[12]

The *FCD3* locus was identified through a large family with 17 examined individuals, 10 of which manifested clinical signs of disease. This interval was localized to chromosome 5 using an SNP array, followed by refinement of the region with an STR marker panel to 5q33.1–5q35.2.[21]

The *FCD4* locus was identified in the region 9p22.1–9p24.1 and was the first to demonstrate interaction between two pathologic alleles in this disease. After mutations in TCF8 were identified among some affected members of this family, linkage analyses were performed conditioned to the presence of the *TCF8* mutation, and refined using STR markers, with maximum LOD scores of 3.09 at D9S168 and 3.20 at D9S256, and a linkage interval spanning 14.3 Mb between recombinations at D91681 and D9S1684.[22] While either the pathologic mutation in TCF8 or the causal *FCD4* haplotype was associated with clinically significant disease, those with the presence of both genetic inputs demonstrated markedly severe disease, generally progressing to corneal transplantation.

A number of additional loci have been identified with large numbers of small pedigrees which, although below the typical LOD threshold of 3.0, may serve as promising targets in the future. Using SNP linkage panels and DNA from 92 individuals from 22 families, four loci were identified on chromosome 1 (rs760594), chromosome 7 (rs257376), chromosome 15 (rs352476), and chromosome 17 (rs938350).[23] A follow-up study using 215 individuals from 64 families identified six linkage peaks with maximum HLOD scores over 2, and including two over 3: one in chromosome 10 (rs1889974) with HLOD score of 3.37 and one in chromosome 15 (rs235512) with HLOD score of 3.53.[68]

3.2 Causal Genetic Mutations

Coding mutations in four genes have been identified to be causal for the FCD phenotype: *TCF8*, *SLC4A11*, *LOXHD1*, and *AGBL1*.

Mutations in *TCF8* had been previously implicated in posterior polymorphous corneal dystrophy, which similar to FCD primarily affects the posterior cornea. To investigate whether this gene may be associated with the FCD phenotype, an initial screen of 74 Chinese individuals with FCD for *TCF8* mutations identified a novel variant, p.N696S, present in 1 sporadic case and absent in 93 controls.[24] In a large family also linked to the *FCD4* locus, a causal missense mutation in *TCF8* was identified, p.Q840P.[22] Four additional pathogenic mutations (p.N78T, p.P649A, p.Q810P, and p.A905T) were discovered through sequencing of exonic regions of TCF8 in 384 unrelated FCD-affected individuals. Of these five mutations, three occurred at sites that are highly evolutionarily conserved in vertebrates (p.Q810P, p.Q840P, and p.A905T) and two at moderately conserved sites. In contrast to frameshift, nonsense, or lost-start-codon mutations associated with PPCD, the causal alleles in TCF8 for FCD contain missense mutations. Functional assessment using zebrafish embryos revealed an inability of human mRNA containing each of the FCD-linked mutations to fully rescue developmental abnormalities caused by morpholino oligonucleotide knockdown of endogenous TCF8. While the variants p.N78T and p.Q810P partially rescued the phenotype, the other three variants demonstrated phenotypes identical to morpholino injection alone, with successful rescue limited to wild-type mRNA.

The second gene for which causal mutations have been associated with FCD, *SLC4A11*, codes for a sodium-borate cotransporter that can also stimulate cell growth and proliferation through the MAPK pathway.[25] In the corneal endothelium, SLC4A11 demonstrates robust Na^+-coupled OH^- transport, but does not transport $B(OH)_4^-$ or HCO_3^-.[26] Expression of this gene was shown to be downregulated in FCD through serial analysis of gene expression.[27] Mutations in this gene have been associated with the autosomal recessive form of congenital hereditary endothelial dystrophy[28] and systemically in Harboyan syndrome, a condition which includes both corneal dystrophy and deafness. A screen of 89 FCD-affected individuals, 64 of Chinese and 25 of Indian descent, revealed four previously unreported mutations (p.S33SfsX18, p.E399K, p.G709E, and p.T754M) which were not present in 354 ethnically matched controls, and acquired with autosomal dominant inheritance.[29] These missense mutations were shown in a biochemical assay of cell surface localization and confocal immunolocalization to demonstrate defective localization to the cell surface. A large study sequencing all coding regions of SLC4A11 in 192 FCD-affected and 192 unaffected individuals resulted in the identification of an additional seven missense mutations (p.E167D, p.R282P, p.Y526C, p.V575M, p.G583D, p.G742R, and p.G834S) among affected individuals, with no such mutations among the control samples.[30] Sorting Intolerant from Tolerant (SIFT) and PolyPhen predicted that five of these mutations were deleterious. Loss of SLC4A11 was explored utilizing short hairpin RNA against SLC4A11 in an immortalized human corneal endothelial cell (HCEC) line to downregulate SLC4A11 gene expression; knockdown suppressed growth and reduced viability among HCECs, associated with increased apoptosis in SLC4A11-depleted cells.[31]

The third gene with FCD-associated causal mutations, *LOXHD1*, was identified in a family previously linked the *FCD2* locus in chromosome 18. *LOXHD1* codes for a protein composed of polycystin/lipoxygenase/alpha-toxin domains, which modulate plasma membrane targeting. Sequencing of 36 Mb on chromosome 18 using a custom exon-capture array in one affected and one unaffected member of three families linked to the *FCD2* locus identified a p.R547C mutation in LOXHD1 in one affected individual, a variant not present in either the 1000 Genomes project or 1500 Exomes from the University of Seattle Human Exome Variant database. PolyPhen 2 predicted this variant to be pathogenic and it was absent in

384 unaffected control individuals. Immunohistochemistry was utilized to identify LOXHD1 in mouse corneas, both in the epithelial and endothelial layers. Immunostaining of the affected proband cornea, one from a FCD-affected individual known not to bear the identified mutation in LOXHD1, and a control with keratoconus revealed increased LOXHD1 staining in Descemet membrane and corneal endothelium in the LOXHD1-associated Fuchs cornea. Three-dimensional reconstruction of the LOXHD1 protein predicted that most mutations resided on the protein surface, likely affecting interaction with other proteins.[32]

The fourth gene with an identified causal mutation, AGBL1, codes for ATP/GTP-binding protein-like 1. The cytosolic carboxypeptidase family, of which AGBL1 is a member, catalyzes deglutamylation of polyglutamylated proteins. Polyglutamylation is a reversible posttranslational protein modification which generates glutamate side chains on tubulins and other proteins, a process which has been shown to be necessary for neuronal survival.[33] In a three-generation family with 12 affected individuals, a genome-wide linkage analysis with a multilocus model was conducted, conditioned on severe disease resulting from two disease alleles, mild disease from a single allele, and offspring of two affected parents receiving the disease allele from the parent otherwise unrelated to the pedigree. Two positive signals were noted on 3p and 15q, which were investigated further with next-generation sequencing. A C-to-T transition, c.3082C > T in AGBL1, was identified, which results in a premature stop codon; this nonsense allele segregated with the FCD haplotype under the multilocus model. Sequencing of FCD-affected individuals resulted in the identification of two other individuals with the same nonsense mutation and the identification of a second allele, a heterozygous missense variant, c.2969G > C, neither of which was present among 384 ethnically matched controls. To investigate the role of AGBL1, immunoprecipitated lysates were probed with anti-Myc antibody and variant forms of the protein demonstrated reduced binding affinity with TCF4. The results overall indicated that AGBL1 interacts specifically with TCF4, but not with TCF8.[34]

3.3 Association Studies

A genome-wide association study by Baratz and colleagues among 130 cases and 260 controls matched by age and sex interrogated 338,727 SNPs, of which one reached threshold for genome-wide significance: rs613872, an intronic variant in *TCF4*. The minor allele (*G*) frequency was significantly enriched in FCD cases (G=0.37) compared to controls (0.14), a finding replicated in a second group of 150 cases and 150 controls (0.43 among cases and 0.15 among controls). The odds ratio associated with the minor allele was notable in heterozygotes (5.5) and particularly homozygotes (30).[35] In the United States, these findings were replicated in a large cohort of 450 cases and 360 controls with odds ratio of 4.5,[36] and in a Midwestern population of 82 patients with FCD and 163 control subjects.[37] The population of Tangier, an island in the Chesapeake Bay with a high prevalence of FCD, was found to also demonstrate a high minor allele frequency of 0.37, comparable to affected cohorts in the aforementioned studies.[38] Globally, findings were replicated in an Australian cohort of 105 cases and 275 controls with odds ratio of 4.05.[39] In a Chinese cohort of 57 affected and 121 unaffected individuals, two other SNPs in TCF4 were significantly associated with FCD affectation with an odds ratio over 2[40]; however, the rs613872 minor allele was absent among the affected cohort.

Together, these findings indicated strong association but raised the possibility of a yet undetermined causal allele in TCF4. Subsequently, testing of 66 affected and 63 unaffected individuals for an intronic TGC trinucleotide repeat in TCF4 demonstrated, in cases with over 50 TGC repeats, sensitivity of 79% and specificity of 96%, even greater specificity than had been previously determined with the rs613872 single nucleotide polymorphism.[41] This association with the CTG 18.1 trinucleotide repeat expansion was replicated in 120 affected Caucasian individuals and 100 controls ($p=6.5\times10^{-25}$) and the two polymorphisms were found to be in linkage disequilibrium. In this cohort, the odds ratio of each copy of the CTG18.1 expanded allele was 32.3, and in 29 families, the expanded allele cosegregated with the trait in 15 families (52%) with complete penetrance.[42] Trans-ethnic replication of the CTG18.1 repeat expansion has been described in the Chinese population with 57 affected and 121 control individuals; the expanded CTG18.1 allele was associated ($p=4.7\times10^{-14}$) with FCD with odds ratio of 66.5.[43] Replication was also performed in 44 affected and 108 control patients from an Indian population sample, with significance achieved for the CTG18.1 expanded allele ($p=2\times10^{-4}$); 34% of affected and 5% of control subjects were found to harbor over 50 CTG repeats.[44] A comprehensive assessment of 1866 genetic variants, with sequencing of the TCF4 coding region, introns, and flanking sequence surrounding the two variants, revealed that no single causative allele correlated with all cases of FCD, but the trinucleotide repeat is the most strongly associated of the two TCF4 variants.[45]

Clusterin is a molecular chaperone which plays a role in extracellular matrix interaction and is overexpressed in FCD relative to pseudophakic bullous keratopathy and normal control corneas[46] and has been shown to colocalize in guttae.[47] In the aforementioned Australian cohort exploring association with TCF4, additional SNPs were interrogated and one found to be significant in CLU (rs17466684, $p=0.003$) with odds ratio of 1.85.[39] Immunohistochemistry revealed anti-CLU labeling in FCD epithelium but absent in control epithelium, while labeling in Descemet membrane was limited to its anterior portion in FCD and more diffusely distributed in controls. Both affected and control endothelium expressed CLU, although endothelial cell density was decreased in affected individuals.[39] Polymerase chain reaction array analyses reveal upregulation of clusterin expression in human FCD-affected Descemet membrane–endothelium complexes, which demonstrate increased immunohistochemistry labeling among 40 samples when compared to controls.[15]

TGF-β1 protein has also been found to be overexpressed in FCD and in addition to clusterin, colocalizes in guttae.[47] An association with a single haplotype of TGF-β1 SNPs and FCD affectation reached statistical significance among Caucasian Australians ($p=0.011$), with odds ratio of 2.29.[39]

4. FUNCTIONAL MECHANISMS IN FCD

4.1 Oxidative Damage and Apoptosis

Loss of corneal endothelial cell density is a characteristic feature of FCD and is associated with increased apoptotic processes. A study of 47 FCD-affected corneal buttons revealed a significantly higher percentage of apoptotic endothelial cells by nucleus labeling assay relative to controls, a phenomenon also supported by TUNEL assay.[48] Subsequent studies with TUNEL assay of entire corneas revealed significantly higher apoptotic cell numbers not only in the corneal endothelium but also in the stroma and epithelium in patients with FCD compared to controls.[49]

Serial analysis of gene expression reveals decreased transcription levels of antioxidant molecules in FCD-affected corneal endothelium, including glutathione *S*-transferase-pi.[27]

A number of apoptosis-related factors are found more frequently in FECD-affected corneas. Immunohistochemical study of FCD corneas reveals a lower number of p27- and survivin-positive epithelial cells, and higher cathepsin-positive epithelial cells in FCD corneas relative to controls.[50] Descemet membrane extracted from FCD corneas undergoing endothelial keratoplasty reveals abundant advanced glycation end products in Descemet membrane, molecules which are associated with oxidative stress, inflammation, and aging.[51] A comparison of immortalized FCD and control corneal endothelial cell lines in which oxidative stress was induced with *tert*-butyl hydroperoxide (*t*BHP) reveals significantly increased activation of p53 in FCD cells relative to controls (Fig. 3).[52]

Peroxiredoxins are expressed in human corneal endothelium and protect against oxidative stress by removing cellular hydrogen peroxide. Proteomic analysis reveals significantly decreased expression of Prx-2, -3, and -5 in FCD cells compared to controls, a finding supported by decreased Prx-2 mRNA detected by real-time PCR analysis.[53]

Evidence suggests that the oxidant–antioxidant balance in FCD-affected endothelial cells shifts to a pro-oxidant state. In addition to downregulation of peroxiredoxins, PCR array also demonstrates transcriptional downregulation of other antioxidants, including SOD2, MT3, and TXNDR1 in HCECs, without a compensatory increase in other antioxidants such as catalase or glutathione peroxidases and/or transferases.[54] These downregulated antioxidants share a common promoter region, the antioxidant response element, to which the transcription factors nuclear factor erythroid 2-related factor-1 (Nrf1) and -2 (Nrf2) bind and induce upregulation of antioxidant enzyme genes in settings of oxidative stress. The identification of a

Figure 3 Descemet membrane removed from a patient at the time of endothelial keratoplasty reveals classic anvil-shaped excrescences which protrude posteriorly, termed guttae.

significant decrease in Nrf2 protein in FCD endothelial cells compared to controls is further supported by depleted expression of a major Nrf2-regulated antioxidant gene, HO-1,[54] and significant decrease in the Nrf2 protein stabilizer DJ-1.[55]

Identification of dysregulated oxidative stress pathways offers potential avenues for future therapeutic targets. Sulforaphane is a naturally occurring glucosinolate found in green leafy vegetables which modifies sulfhydryl residues of Keap1, causing the release and activation of Nrf2. In human unaffected and FCD-affected immortalized corneal endothelial cell lines subjected to oxidative stress with *t*BHP, pretreatment with sulforaphane decreased intracellular production of reactive oxygen species, enhanced nuclear translocation of Nrf2, and decreased p53 staining.[56]

4.2 Mitochondrial Dysregulation

FCD is principally a disease of the central cornea, and early research suggested regional differences in mitochondrial cytochrome oxidase (CO) activity in FCD-affected corneal endothelium, with decreased CO centrally and increased CO in the periphery.[57]

Serial analysis of gene expression in FCD and normal corneal endothelium revealed significant underexpression in FCD of multiple mitochondrial genes involved in electron transport and oxidative phosphorylation, including NADH dehydrogenase subunits 1, 2, and 4, cytochrome *b*, cytochrome *c* oxidase subunit III, and ATP synthase F0 subunit 6.[27]

Comparison of oxidative DNA damage in FCD-affected and unaffected corneal endothelium was conducted with high-sensitivity ELISA to detect the concentration of 8-OHdG (an oxidative DNA damage marker). This was significantly higher in FCD-affected corneal endothelium than in controls, and immunolabeling revealed that this marker colocalized to mitochondria.[54]

To test if FCD was associated with a systemic deficiency in mitochondrial enzymes, a separate study challenged mitochondria in peripheral blood lymphocytes with hydrogen peroxide in 35 FCD patients and 32 controls; increased mitochondrial DNA (mtDNA) damage and a higher ratio of a 4977-bp deletion were found in FCD patients compared to controls.[58]

In a study of mtDNA variants on FCD susceptibility, a total of 530 cases and 498 controls of European descent were assessed for 10 mtDNA variants defining European haplogroups, and replication analyses conducted in cohorts of 3200 and 3043 individuals. The mtDNA variant A10398G was significantly associated with FCD with odds ratio of 0.72, suggesting that this variant confers a protective effect; this persisted even after adjusting for smoking status. Haplogroup I was significantly associated with FCD even after adjusting for smoking or the TCF4 intronic variant rs613872.[59]

4.3 Epithelial–Mesenchymal Transition

Although the corneal endothelium is often considered to be in a state of arrested growth, with endothelial cells migrating to cover defects in the endothelial layer, evidence exists of stem cell markers in the posterior limbus and corneal endothelium, and markers of proliferation are present after wounding of the endothelium.[60] Fibroblastic features of some endothelial cells in FCD[9] may be suggestive of alterations in the process of maturation and differentiation, such as epithelial–mesenchymal transition (EMT).

Several genes associated with FCD such as TCF8 and TCF4 are implicated directly or indirectly in the EMT pathway. In breast cancer cells, TCF8 serves as a transcriptional repressor and downregulates E-cadherin, inducing a transition to a mobile, fibroblastoid appearance. TCF4 modulates multiple pathways, including TGF-β, and upregulates TCF8 expression.[61] TGF-βI induces EMT in colon cancer cells[62] and silencing of clusterin in lung adenocarcinoma cells appears to induce a mesenchymal-to-epithelial transition through the ERK/Slug pathway.[63]

4.4 Unfolded Protein Response

Several genes implicated in FCD pathogenesis modify protein folding mechanisms, supporting a possible role for the unfolded protein response. CLU is involved with protein folding and colocalizes to guttae, and pathogenic LOXHD1 alleles induce misfolded aggregates in corneal endothelial cells.[32]

In a comparison of FCD corneas with keratoconus and normal controls, misfolded proteins were found to be increased in FCD.[64] Of 10 corneas with FCD and 9 corneas with non-FCD corneal dystrophy, all FCD corneas exhibited prominent rough endoplasmic reticulum, in contrast to only a third of non-FCD corneas. FCD-associated corneas also demonstrated increased immunofluorescent labeling for GRP78, phospho-eIF2a, and CHOP, markers for unfolded protein response relative to both normal and keratoconus corneas. Similarly, the apoptosis markers caspase 3 and caspase 9 were both significantly increased in FCD corneas, suggesting that increased misfolded protein in FCD may contribute to an increased unfolded protein response and caspase-dependent apoptosis.[64]

4.5 MicroRNA

MicroRNA (miRNA) dysregulation is implicated in numerous pathophysiological pathways, including those associated with FCD and other corneal dystrophies with similar phenotypes. A mutation in miR-184 has been shown to be causative for EDICT syndrome, which includes a beaten-metal appearance to the corneal endothelium.[65]

The miR-200 family and miR-205 have been shown to regulate EMT by targeting of TCF8 and SIP1,[66] although the nature of miRNA interaction with dysfunctional TCF8 transcripts is unclear at this time.

Study of miRNA expression in HCECs reveals significant downregulation of at least 87 miRNAs in FCD compared with normal endothelium, and a significant decrease in DICER1, which encodes an endoribonuclease critical to miRNA biogenesis.[67] Evidence of downregulation of three miR-29 family members (miR-29a-3p, miR-29b-2-5p, and miR-29c-5p) is further supported by upregulation of collagen I and collagen IV, mRNA targets which are associated with extracellular matrix production, and increased endothelial laminin protein expression FCD.[67]

REFERENCES

1. Fuchs E. Dystrophia epithelialis corneae. *Graefes Arch Clin Exp Ophthalmol.* 1910;76(3):478–508.
2. Krachmer JH, Purcell Jr JJ, Young CW, Bucher KD. Corneal endothelial dystrophy. A study of 64 families. *Arch Ophthalmol.* 1978;96(11):2036–2039.
3. Kopplin LJ, Przepyszny K, Schmotzer B, et al. Relationship of Fuchs endothelial corneal dystrophy severity to central corneal thickness. *Arch Ophthalmol.* 2012;130(4):433–439.
4. Mustonen RK, McDonald MB, Srivannaboon S, Tan AL, Doubrava MW, Kim CK. In vivo confocal microscopy of Fuchs' endothelial dystrophy. *Cornea.* 1998;17(5):493–503.
5. Patel SV, Baratz KH, Maguire LJ, Hodge DO, McLaren JW. Anterior corneal aberrations after Descemet's stripping endothelial keratoplasty for Fuchs' endothelial dystrophy. *Ophthalmology.* 2012;119(8):1522–1529.
6. Baratz KH, McLaren JW, Maguire LJ, Patel SV. Corneal haze determined by confocal microscopy 2 years after Descemet stripping with endothelial keratoplasty for Fuchs corneal dystrophy. *Arch Ophthalmol.* 2012;130(7):868–874.
7. Repp DJ, Hodge DO, Baratz KH, McLaren JW, Patel SV. Fuchs' endothelial corneal dystrophy: subjective grading versus objective grading based on the central-to-peripheral thickness ratio. *Ophthalmology.* 2013;120(4):687–694.
8. Bourne WM, Johnson DH, Campbell RJ. The ultrastructure of Descemet's membrane III. Fuchs' dystrophy. *Arch Ophthalmol.* 1982;100(12):1952–1955.
9. Iwamoto T, DeVoe AG. Electron microscopic studies on Fuchs'combined dystrophy. I. Posterior portion of the cornea. *Invest Ophthalmol.* 1971;10(1):9–28.

10. Waring 3rd GO. Posterior collagenous layer of the cornea. Ultrastructural classification of abnormal collagenous tissue posterior to Descemet's membrane in 30 cases. *Arch Ophthalmol.* 1982;100(1):122–134.
11. Meadows DN, Eghrari AO, Riazuddin SA, Emmert DG, Katsanis N, Gottsch JD. Progression of Fuchs corneal dystrophy in a family linked to the FCD1 locus. *Invest Ophthalmol Vis Sci.* 2009;50(12):5662–5666.
12. McGlumphy EJ, Yeo WS, Riazuddin SA, et al. Age-severity relationships in families linked to FCD2 with retroillumination photography. *Invest Ophthalmol Vis Sci.* 2010;51(12):6298–6302.
13. Fujimoto H, Maeda N, Soma T, et al. Quantitative regional differences in corneal endothelial abnormalities in the central and peripheral zones in Fuchs' endothelial corneal dystrophy. *Invest Ophthalmol Vis Sci.* 2014;55(8):5090–5098.
14. Rokita-Wala I, Mrukwa-Kominek E, Gierek-Ciaciura S. Changes in corneal structure observed with confocal microscopy during Fuchs endothelial dystrophy. *Klin Oczna.* 2000;102(5):339–344.
15. Weller JM, Zenkel M, Schlotzer-Schrehardt U, Bachmann BO, Tourtas T, Kruse FE. Extracellular matrix alterations in late-onset Fuchs' corneal dystrophy. *Invest Ophthalmol Vis Sci.* 2014;55(6):3700–3708.
16. Kenney MC, Atilano SR, Zorapapel N, Holguin B, Gaster RN, Ljubimov AV. Altered expression of aquaporins in bullous keratopathy and Fuchs' dystrophy corneas. *J Histochem Cytochem.* 2004;52(10):1341–1350.
17. Cross HE, Maumenee AE, Cantolino SJ. Inheritance of Fuchs' endothelial dystrophy. *Arch Ophthalmol.* 1971;85(3):268–272.
18. Rosenblum P, Stark WJ, Maumenee IH, Hirst LW, Maumenee AE. Hereditary Fuchs' dystrophy. *Am J Ophthalmol.* 1980;90(4):455–462.
19. Sundin OH, Jun AS, Broman KW, et al. Linkage of late-onset Fuchs corneal dystrophy to a novel locus at 13pTel-13q12.13. *Invest Ophthalmol Vis Sci.* 2006;47(1):140–145.
20. Sundin OH, Broman KW, Chang HH, Vito EC, Stark WJ, Gottsch JD. A common locus for late-onset Fuchs corneal dystrophy maps to 18q21.2-q21.32. *Invest Ophthalmol Vis Sci.* 2006;47(9):3919–3926.
21. Riazuddin SA, Eghrari AO, Al-Saif A, et al. Linkage of a mild late-onset phenotype of Fuchs corneal dystrophy to a novel locus at 5q33.1-q35.2. *Invest Ophthalmol Vis Sci.* 2009;50(12):5667–5671.
22. Riazuddin SA, Zaghloul NA, Al-Saif A, et al. Missense mutations in TCF8 cause late-onset Fuchs corneal dystrophy and interact with FCD4 on chromosome 9p. *Am J Hum Genet.* 2010;86(1):45–53.
23. Afshari NA, Li YJ, Pericak-Vance MA, Gregory S, Klintworth GK. Genome-wide linkage scan in Fuchs endothelial corneal dystrophy. *Invest Ophthalmol Vis Sci.* 2009;50(3):1093–1097.
24. Mehta JS, Vithana EN, Tan DT, et al. Analysis of the posterior polymorphous corneal dystrophy 3 gene, TCF8, in late-onset Fuchs endothelial corneal dystrophy. *Invest Ophthalmol Vis Sci.* 2008;49(1):184–188.
25. Park M, Li Q, Shcheynikov N, Zeng W, Muallem S. NaBC1 is a ubiquitous electrogenic Na+-coupled borate transporter essential for cellular boron homeostasis and cell growth and proliferation. *Mol Cell.* 2004;16(3):331–341.
26. Jalimarada SS, Ogando DG, Vithana EN, Bonanno JA. Ion transport function of SLC4A11 in corneal endothelium. *Invest Ophthalmol Vis Sci.* 2013;54(6):4330–4340.
27. Gottsch JD, Bowers AL, Margulies EH, et al. Serial analysis of gene expression in the corneal endothelium of Fuchs' dystrophy. *Invest Ophthalmol Vis Sci.* 2003;44(2):594–599.

28. Vithana EN, Morgan P, Sundaresan P, et al. Mutations in sodium-borate cotransporter SLC4A11 cause recessive congenital hereditary endothelial dystrophy (CHED2). *Nat Genet.* 2006;38(7):755–757.
29. Vithana EN, Morgan PE, Ramprasad V, et al. SLC4A11 mutations in Fuchs endothelial corneal dystrophy. *Hum Mol Genet.* 2008;17(5):656–666.
30. Riazuddin SA, Vithana EN, Seet LF, et al. Missense mutations in the sodium borate cotransporter SLC4A11 cause late-onset Fuchs corneal dystrophy. *Hum Mutat.* 2010;31(11):1261–1268.
31. Liu J, Seet LF, Koh LW, et al. Depletion of SLC4A11 causes cell death by apoptosis in an immortalized human corneal endothelial cell line. *Invest Ophthalmol Vis Sci.* 2012;53(7):3270–3279.
32. Riazuddin SA, Parker DS, McGlumphy EJ, et al. Mutations in LOXHD1, a recessive-deafness locus, cause dominant late-onset Fuchs corneal dystrophy. *Am J Hum Genet.* 2012;90(3):533–539.
33. Rogowski K, van Dijk J, Magiera MM, et al. A family of protein-deglutamylating enzymes associated with neurodegeneration. *Cell.* 2010;143(4):564–578.
34. Riazuddin SA, Vasanth S, Katsanis N, Gottsch JD. Mutations in AGBL1 cause dominant late-onset Fuchs corneal dystrophy and alter protein-protein interaction with TCF4. *Am J Hum Genet.* 2013;93(4):758–764.
35. Baratz KH, Tosakulwong N, Ryu E, et al. E2-2 protein and Fuchs's corneal dystrophy. *N Engl J Med.* 2010;363(11):1016–1024.
36. Li YJ, Minear MA, Rimmler J, et al. Replication of TCF4 through association and linkage studies in late-onset Fuchs endothelial corneal dystrophy. *PLoS One.* 2011;6(4):e18044.
37. Stamler JF, Roos BR, Wagoner MD, et al. Confirmation of the association between the TCF4 risk allele and Fuchs endothelial corneal dystrophy in patients from the Midwestern United States. *Ophthalmic Genet.* 2013;34(1–2):32–34.
38. Eghrari AO, McGlumphy EJ, Iliff BW, et al. Prevalence and severity of Fuchs corneal dystrophy in Tangier Island. *Am J Ophthalmol.* 2012;153(6):1067–1072.
39. Kuot A, Hewitt AW, Griggs K, et al. Association of TCF4 and CLU polymorphisms with Fuchs' endothelial dystrophy and implication of CLU and TGFBI proteins in the disease process. *Eur J Hum Genet.* 2012;20(6):632–638.
40. Thalamuthu A, Khor CC, Venkataraman D, et al. Association of TCF4 gene polymorphisms with Fuchs' corneal dystrophy in the Chinese. *Invest Ophthalmol Vis Sci.* 2011;52(8):5573–5578.
41. Wieben ED, Aleff RA, Tosakulwong N, et al. A common trinucleotide repeat expansion within the transcription factor 4 (TCF4, E2-2) gene predicts Fuchs corneal dystrophy. *PLoS One.* 2012;7(11):e49083.
42. Mootha VV, Gong X, Ku HC, Xing C. Association and familial segregation of CTG18.1 trinucleotide repeat expansion of TCF4 gene in Fuchs' endothelial corneal dystrophy. *Invest Ophthalmol Vis Sci.* 2014;55(1):33–42.
43. Xing C, Gong X, Hussain I, et al. Transethnic replication of association of CTG18.1 repeat expansion of TCF4 gene with Fuchs' corneal dystrophy in Chinese implies common causal variant. *Invest Ophthalmol Vis Sci.* 2014;55(11):7073–7078.
44. Nanda GG, Padhy B, Samal S, Das S, Alone DP. Genetic association of TCF4 intronic polymorphisms, CTG18.1 and rs17089887 with Fuchs' endothelial corneal dystrophy in Indian population. *Invest Ophthalmol Vis Sci.* 2014;55(11):7674–7680.
45. Wieben ED, Aleff RA, Eckloff BW, et al. Comprehensive assessment of genetic variants within TCF4 in Fuchs' endothelial corneal dystrophy. *Invest Ophthalmol Vis Sci.* 2014;55(9):6101–6107.

46. Jurkunas UV, Bitar MS, Rawe I, Harris DL, Colby K, Joyce NC. Increased clusterin expression in Fuchs' endothelial dystrophy. *Invest Ophthalmol Vis Sci*. 2008;49(7): 2946–2955.
47. Jurkunas UV, Bitar M, Rawe I. Colocalization of increased transforming growth factor-beta-induced protein (TGFBIp) and clusterin in Fuchs endothelial corneal dystrophy. *Invest Ophthalmol Vis Sci*. 2009;50(3):1129–1136.
48. Borderie VM, Baudrimont M, Vallee A, Ereau TL, Gray F, Laroche L. Corneal endothelial cell apoptosis in patients with Fuchs' dystrophy. *Invest Ophthalmol Vis Sci*. 2000;41(9):2501–2505.
49. Szentmary N, Szende B, Suveges I. Epithelial cell, keratocyte, and endothelial cell apoptosis in Fuchs' dystrophy and in pseudophakic bullous keratopathy. *Eur J Ophthalmol*. 2005;15(1):17–22.
50. Szentmary N, Szende B, Suveges I. P53, CD95, cathepsin and survivin pathways in Fuchs' dystrophy and pseudophakic bullous keratopathy corneas. *Histol Histopathol*. 2008;23(8):911–916.
51. Wang Z, Handa JT, Green WR, Stark WJ, Weinberg RS, Jun AS. Advanced glycation end products and receptors in Fuchs' dystrophy corneas undergoing Descemet's stripping with endothelial keratoplasty. *Ophthalmology*. 2007;114(8):1453–1460.
52. Azizi B, Ziaei A, Fuchsluger T, Schmedt T, Chen Y, Jurkunas UV. p53-regulated increase in oxidative-stress-induced apoptosis in Fuchs endothelial corneal dystrophy: a native tissue model. *Invest Ophthalmol Vis Sci*. 2011;52(13):9291–9297.
53. Jurkunas UV, Rawe I, Bitar MS, et al. Decreased expression of peroxiredoxins in Fuchs' endothelial dystrophy. *Invest Ophthalmol Vis Sci*. 2008;49(7):2956–2963.
54. Jurkunas UV, Bitar MS, Funaki T, Azizi B. Evidence of oxidative stress in the pathogenesis of Fuchs endothelial corneal dystrophy. *Am J Pathol*. 2010;177(5):2278–2289.
55. Bitar MS, Liu C, Ziaei A, Chen Y, Schmedt T, Jurkunas UV. Decline in DJ-1 and decreased nuclear translocation of Nrf2 in Fuchs endothelial corneal dystrophy. *Invest Ophthalmol Vis Sci*. 2012;53(9):5806–5813.
56. Ziaei A, Schmedt T, Chen Y, Jurkunas UV. Sulforaphane decreases endothelial cell apoptosis in Fuchs endothelial corneal dystrophy: a novel treatment. *Invest Ophthalmol Vis Sci*. 2013;54(10):6724–6734.
57. Tuberville AW, Wood TO, McLaughlin BJ. Cytochrome oxidase activity of Fuchs' endothelial dystrophy. *Curr Eye Res*. 1986;5(12):939–947.
58. Czarny P, Seda A, Wielgorski M, et al. Mutagenesis of mitochondrial DNA in Fuchs endothelial corneal dystrophy. *Mutat Res Fundam Mol Mech Mutagen*. 2014;760:42–47.
59. Li YJ, Minear MA, Qin X, et al. Mitochondrial polymorphism A10398G and haplogroup I are associated with Fuchs' endothelial corneal dystrophy. *Invest Ophthalmol Vis Sci*. 2014;55(7):4577–4584.
60. McGowan SL, Edelhauser HF, Pfister RR, Whikehart DR. Stem cell markers in the human posterior limbus and corneal endothelium of unwounded and wounded corneas. *Mol Vis*. 2007;13:1984–2000.
61. Sobrado VR, Moreno-Bueno G, Cubillo E, et al. The class I bHLH factors E2-2A and E2-2B regulate EMT. *J Cell Sci*. 2009;122(pt 7):1014–1024.
62. Ma C, Rong Y, Radiloff DR, et al. Extracellular matrix protein betaig-h3/TGFBI promotes metastasis of colon cancer by enhancing cell extravasation. *Genes Dev*. 2008;22(3):308–321.
63. Chou TY, Chen WC, Lee AC, Hung SM, Shih NY, Chen MY. Clusterin silencing in human lung adenocarcinoma cells induces a mesenchymal-to-epithelial transition through modulating the ERK/slug pathway. *Cell Signal*. 2009;21(5):704–711.
64. Engler C, Kelliher C, Spitze AR, Speck CL, Eberhart CG, Jun AS. Unfolded protein response in Fuchs endothelial corneal dystrophy: a unifying pathogenic pathway? *Am J Ophthalmol*. 2010;149(2), 194–202.e2.

65. Iliff BW, Riazuddin SA, Gottsch JD. A single-base substitution in the seed region of miR-184 causes EDICT syndrome. *Invest Ophthalmol Vis Sci.* 2012;53(1):348–353.
66. Gregory PA, Bert AG, Paterson EL, et al. The miR-200 family and miR-205 regulate epithelial to mesenchymal transition by targeting ZEB1 and SIP1. *Nat Cell Biol.* 2008;10(5):593–601.
67. Matthaei M, Hu J, Kallay L, et al. Endothelial cell microRNA expression in human late-onset Fuchs' dystrophy. *Invest Ophthalmol Vis Sci.* 2014;55(1):216–225.
68. Li YJ, Minear MA, Rimmler J, et al. Replication of TCF4 through association and linkage studies in late-onset Fuchs endothelial corneal dystrophy. *PLoS One.* 2011;6(4):e18044.

CHAPTER EIGHT

Molecular Pathogenesis of Corneal Dystrophies: Schnyder Dystrophy and Granular Corneal Dystrophy type 2

Eung Kweon Kim[*,†,‡,1], **Hun Lee**[*,†], **Seung-il Choi**[†]
[*]Department of Ophthalmology, Vision Research Institute, Severance Hospital, Yonsei University College of Medicine, Seoul, Republic of Korea
[†]Corneal Dystrophy Research Institute, Yonsei University College of Medicine, Seoul, Republic of Korea
[‡]BK21 Plus Project for Medical Science and Severance Biomedical Science Institute, Yonsei University College of Medicine, Seoul, Republic of Korea
[1]Corresponding author: e-mail address: eungkkim@yuhs.ac

Contents

Abstract

The International Committee for Classification of Corneal Dystrophies (IC3D) provides updated data to ophthalmologists by incorporating traditional definitions of corneal dystrophies with new genetic, clinical, and pathologic information. Recent advances in the genetics of corneal dystrophies facilitate more precise classifications and elucidate each classification's molecular mechanisms. Unfortunately, the molecular mechanisms and underlying pathogenic mechanisms have remained obscure, with the exception of Schnyder corneal dystrophy (CD), granular CD type 2 (GCD2), and Fuch's endothelial CD. Here, we review the pathogenesis of Schnyder CD and GCD2.

1. PATHOGENESIS

1.1 Schnyder Corneal Dystrophy

Schnyder corneal dystrophy (CD) is characterized by an abnormal deposition of cholesterol and phospholipids in the corneal epithelium and stroma,

Progress in Molecular Biology and Translational Science, Volume 134
ISSN 1877-1173
http://dx.doi.org/10.1016/bs.pmbts.2015.05.003
© 2015 Elsevier Inc.
All rights reserved.

which leads to progressive corneal opacification and loss of visual acuity.[1–3] Mutations in the UbiA prenyltransferase domain-containing 1 (*UBIAD1*) gene on chromosome 1p36 were recently shown to be the main cause of Schnyder CD.[4–6] Additionally, Weiss *et al.*[5,6] reported that N102S, a missense mutation in *UBIAD1*, was the cause of most Schnyder CD cases. Even though the exact pathogenic mechanism of Schnyder CD remains unknown, several theories on the development of this rare, autosomal dominant, and stromal dystrophy have been suggested.

1.1.1 Abnormal Lipid Metabolism

Chemical comparison between normal corneas and those with Schnyder CD has shown a 10-fold increase in cholesterol levels and a fivefold increase in phospholipid levels in corneas with Schnyder CD.[7] Furthermore, immunohistochemical analyses have shown that high-density lipoprotein (HDL) deposition is more remarkable than low-density lipoprotein deposition. However, corneal changes in eyes with Schnyder CD have not been shown to be associated with systemic cholesterol levels. Changing systemic cholesterol levels did not alter the progression of corneal opacification, and dyslipidemia severity was not correlated with crystalline formation.[8–10] Thus, local corneal abnormalities in HDL metabolism are suspected to play a crucial role in the development of Schnyder CD.

Other evidence also suggests that local factors affect lipid and cholesterol processing and/or transport, which lead to corneal cholesterol deposition in tissue. Battisti *et al.*[11] found that lipid material was present in the cornea and in the skin fibroblasts of patients with Schnyder CD and normal serum cholesterol levels. Interestingly, Mehta *et al.*[12] analyzed lipids in conjunctival and skin fibroblasts from an Italian patient (proband) with Schnyder CD. They found no evidence of lipid deposition in either conjunctiva or skin. This result may have been attributable to ethnic or genetic mutation differences in the examined Chinese family than in the previously examined Italian family.

The UBIAD1 protein is the product of the most frequent causative genetic mutation of Schnyder CD. The protein interacts with the COOH-terminal portion of apolipoprotein E, which helps mediate cholesterol removal from cells.[5,13,14] Additionally, UBIAD1 protein is thought to be an integral membrane protein that contains a prenyltransferase domain, which is associated with cholesterol synthesis. Therefore, defects in UBIAD1 protein could result in both insufficient cholesterol removal from the cornea and local cholesterol overproduction.[4,5] Further studies on the relationship between UBIAD1 protein and cholesterol metabolism are needed. Cholesterol transport out of the cornea and cholesterol production within the cornea should be examined and will be critical to better understanding underlying pathogenic mechanisms of Schnyder CD.

1.1.2 Mitochondrial Changes

Nickerson *et al.*[15] demonstrated that the mitochondrial UBIAD1 protein is involved in cholesterol metabolism in eyes with Schnyder CD. Immunohistochemistry with antibodies specific for UBIAD1 protein was performed on keratocytes from a healthy donor and from a Schnyder CD proband (N102S mutation). These analyses showed that both wild-type and N102S mutant UBIAD1 protein colocalized with a subunit of the oxidative phosphorylation (OXPHOS) complex I (NADH dehydrogenase, a mitochondrial marker). It was concluded that both wild-type and N102S mutant UBIAD1 protein molecules were localized to the mitochondria on a subcellular level. This suggests that mislocalization of the N102S mutant protein in the mitochondria was not a factor in Schnyder CD development.[15] Although the mitochondrial UBIAD1 protein is expected to play a role in the pathogenesis of Schnyder CD, the interaction between UBIAD1 protein and apolipoprotein E outside of the mitochondria also needs to be considered. The known interaction between UBIAD1 protein and apolipoprotein E, combined with the absence of apolipoprotein E in mitochondria,[13,16] suggests an effect of the *UBIAD1* mutation outside of the mitochondria.

1.1.3 Oxidative Stress

Possible contributions of nitric oxide oxidation and lipid peroxidation to the pathogenesis of Schnyder CD have been investigated.[17] Immunohistochemical examinations with primary antibodies against inducible nitric oxide synthase (i-NOS), constitutive nitric oxide synthase (e-NOS), and nitrotyrosine were conducted on cryosectioned corneas from three patients with Schnyder CD who had undergone penetrating keratoplasty. Compared with normal corneas, corneas with Schnyder CD had increased signals for both i-NOS and nitrotyrosine in the corneal epithelium. However, immune-reactivity for e-NOS was detected in the corneal epithelium and anterior stroma in both normal corneas and those with Schnyder CD. Additionally, significantly increased levels of malondialdehyde

(MDA), the end product of free radical initiated lipid peroxidation, were found in the aqueous humor of patients with Schnyder CD compared to normal patients.[17] Based upon this evidence, further investigations are needed to better understand how oxidative stress is involved in Schnyder CD pathogenic mechanisms.

1.2 Granular corneal dystrophy type 2 and *TGFBI*-Linked Corneal Dystrophies

Granular corneal dystrophy type 2 (GCD2) is an autosomal-dominant disorder caused by an arginine-to-histidine substitution at codon 124 (R124H) in the transforming growth factor β-induced gene (*TGFBI*) on chromosome 5q31.[18] Hallmarks of GCD2 are an age-dependent progressive accumulation of hyaline and amyloid and the production of TGFBIp in the corneal epithelia and stroma, which interfere with corneal transparency.[19,20] The TGFBIp, an extracellular matrix protein, deposits in corneal tissue in the three major forms of dominant *TGFBI*-linked CDs, which are granular CD (types 1 and 2), lattice CD (types 1, 3, and 4), and diffuse Bowman's layer deposits CD (Reis Buehler CD and Thiel Behnke CD).[19] Many histological features of GCD2 are also shared with those of both granular and lattice CD.[21]

1.2.1 Corneal Fibroblasts in Eyes with Granular corneal dystrophy type 2

1.2.1.1 Morphological Properties

Corneal fibroblasts, also known as keratocytes, are the major cell type in the stroma and are estimated to occupy between 9% and 17% of tissue cellular volume.[22] It has been shown that the human cornea has a cell density of approximately 23,000 cells/mm^3.[22,23] Corneal fibroblasts are mainly involved in maintaining corneal transparency and the extracellular matrix environment.[24]

In eyes with GCD2, corneal fibroblasts are increased in size, have intracellular deposits, and appear with a senescence-like morphology.[25,26] Additionally, GCD2 corneal fibroblasts have a much more extensive pattern of colocalized mutant TGFBIp with cathepsin D (a lysosomal enzyme) than wild-type cells.[25]

1.2.2 The Role of Oxidative Stress in Granular corneal dystrophy type 2

In mammalian cells, aerobic metabolic processes lead to the production of reactive oxygen species (ROS) in mitochondria and peroxisomes. However, excessive ROS can cause oxidative damage to proteins, lipids, and DNA.

Compared to other tissues, the cornea is particularly vulnerable to oxidative stress for several reasons. The cornea must be transparent to allow light transmission to the retina. Therefore, it is avascular and constantly exposed to a wide spectrum of light, including ultraviolet (UV) light. It is well known that UV exposure causes tissue stress and ROS generation, and the cornea absorbs most of the UV light entering the eye.[27] It is rich in antioxidant enzyme systems (Fig. 1) that aid in the removal of ROS generated by UV light absorption.[28] A fair amount of evidence, presented below, suggests that oxidative stress plays a role in the pathogenesis of GCD2.

1.2.2.1 Oxidative Stress and ROS-Scavenging Mechanisms

In mammalian cells, mitochondria produce the majority of ROS.[29] During mitochondrial oxidative phosphorylation reactions, oxygen is reduced to water (H_2O) by the passage of electrons through electron transport chain redox reactions coupled with proton transfer across the inner mitochondrial membrane. As a side effect of this process, superoxide ($O_2^{\bullet-}$), hydroxyl radicals ($^{\bullet}OH$), and hydrogen peroxide (H_2O_2) are formed.[30] Excessive production of O_2^- and H_2O_2 can result in oxidative injury to the tissue.

The ROS can react with membrane lipids, proteins, and nucleic acids, and elevated ROS concentrations have been implicated in the pathogenesis

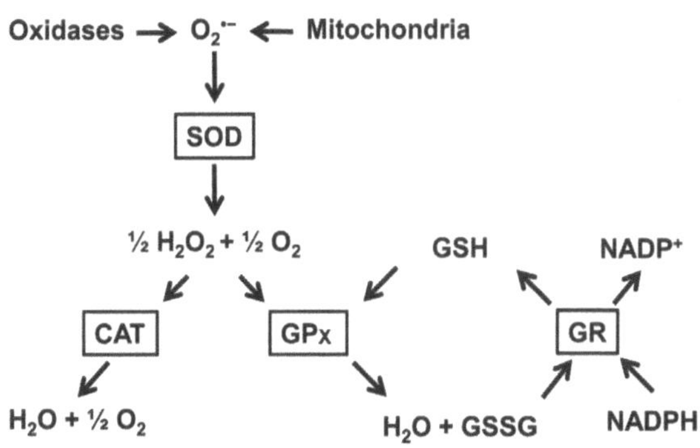

Figure 1 Enzymatic and nonenzymatic antioxidant systems in mammalian cells. Superoxide dismutase (SOD) converts a superoxide radical (O_2^-) into hydrogen peroxide (H_2O_2). Then, catalase (CAT) and glutathione peroxidase (GPx) convert H_2O_2 into water. This way, two toxic species (O_2^- and H_2O_2) are converted into water, a harmless substance. The GPx enzyme requires several secondary enzymes (glutathione reductase [GR]) and cofactors (reduced glutathione [GSH], reduced nicotinamide adenine dinucleotide phosphate [NADPH]) to function. In this reaction, GR is considered a secondary antioxidant enzymes because it enables GPx to function, but does not directly act on reactive oxygen species (ROS). GSSG, oxidized glutathione.

of many human diseases.[31] Because of this, many types of mammalian cells have developed both enzymatic and nonenzymatic ROS-scavenging mechanisms to prevent ROS accumulation (Fig. 1). Imbalances in the redox-active species metabolism promote oxidizing agent formation and can result in oxidative stress. Increased amounts of MDA,[32,33] 4-hydroxy-2-nonenal (HNE),[32,33] protein carbonyl groups,[34,35] and 8-hydroxy-2′-deoxyguanosine (8-OHdG)[36,37] have been frequently associated with oxidative damage (Fig. 2).

1.2.2.2 Mitochondrial Oxidative Damage

When mitochondrial features were examined by electron microscopy in heterozygous GCD2 corneas, keratocytes were found to be enlarged. These images clearly demonstrated that GCD2 corneas have many mitochondria that are dilated, degenerative, and/or contain vesicles with amorphous material. In homozygous corneal fibroblasts cultured for four to eight

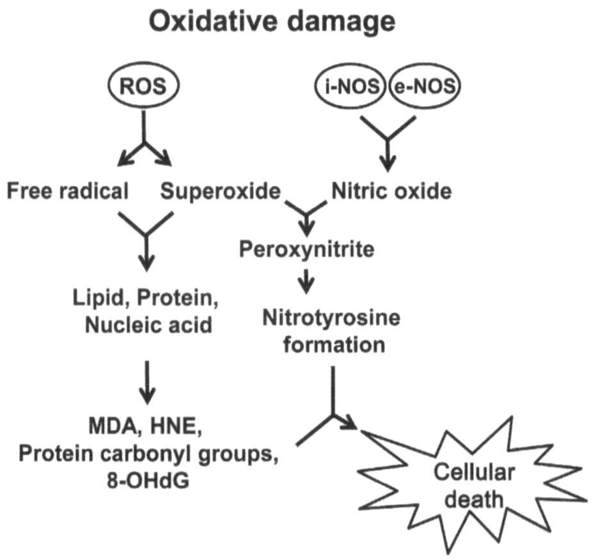

Figure 2 Mechanism by which cellular oxidative damage occurs. Oxygen free radicals and superoxide are produced during pathological conditions. These damage proteins, lipids, and nucleic acids, leading to formation of malondialdehyde (MDA), 4-hydroxynonenal (HNE), protein carbonyl groups, and 8-hydroxy-2-deoxyguanosine (8-OHdG). Additionally, in the presence of nitric oxide (NO), originating from inducible NO synthase (i-NOS) and constitutive NO synthase (e-NOS), superoxide radicals (O_2^-) lead to peroxynitrite ($ONOO^-$) formation. Both of these changes lead to increased oxidative stress, which damages macromolecules and, unless counteracted, ultimately causes cellular death. ROS, reactive oxygen species.

passages, MitoTracker and cytochrome *c* staining showed an increased mitochondrial activity in mutated cells in early passages. Decreases in depolarized mitochondria, cellular proliferation, and complexes I through V expression were observed in late-passage mutant cells. Furthermore, treatment with the antioxidant, butylated hydroxyanisole, has been shown to decrease intracellular ROS, increase mutant cell proliferation, and reduce mitochondrial oxidative damage.[38]

1.2.2.3 Oxidative Damage in Corneal Fibroblasts and Tissue

Choi *et al.*[39] demonstrated that oxidative damage is involved in GCD2 pathogenesis by showing that levels of the MDA, 4-HNE, and protein carbonyl groups were significantly elevated in GCD2 corneal fibroblasts compared to unaffected cells. This study also examined MDA distribution in the corneal tissue from GCD2 patients using immunohistochemical staining. Although MDA immunoreactivity was detectable in corneal epithelial cells from age-matched normal controls, MDA immunoreactivity was significantly higher in the corneal stroma and epithelia from GCD2 patients.[39]

1.2.2.4 Altered Antioxidant Enzyme Systems and ROS Production in Corneal Fibroblasts

It has been well documented that antioxidant enzyme expression increases in response to increases in ROS, H_2O_2, and other oxidative stress products. Expression of Cu/Zn-superoxide dismutase (SOD), Mn-SOD, glutathione peroxidase (GPx), and glutathione reductase (GR) proteins is significantly higher in GCD2 corneal fibroblasts than in normal cells. Interestingly, the level of catalase (*CAT*) mRNA was significantly higher in GCD2 corneal fibroblasts than in wild-type corneal fibroblasts. In contrast, the level of catalase protein was dramatically lower in GCD2 than in wild-type corneal fibroblasts. Catalase immunoreactivity was also examined using confocal immunostaining and was found to be weaker in GCD2 corneal fibroblasts than in wild-type corneal fibroblasts. Catalase activity was also significantly lower in GCD2 corneal fibroblasts than in wild-type corneal fibroblasts. Intracellular ROS and H_2O_2 levels were significantly elevated in GCD2 corneal fibroblasts compared to wild-type corneal fibroblasts.[39]

1.2.2.5 Oxidative Stress-Induced Cell Death

The effect of oxidative stress on GCD2 corneal fibroblasts was examined by exposing cells to H_2O_2. Early morphological events of apoptosis were observed in H_2O_2-treated cells. Additionally, viability of GCD2 corneal

Table 1 Evidence That Oxidative Stress Affects Fibroblasts in Eyes with Granular corneal dystrophy type 2

Factors	Increase	Decrease	References
Oxygen free radical production	ROS, H_2O_2		39
Antioxidant enzymes	Cu/Zn-SOD, Mn-SOD, GPx, GR	Catalase	39
Oxidative damage markers	MDA, HNE, protein carbonyl groups		39
Mitochondrial function	Fragmentations, activity		25,38
Apoptotic proteins	Bax, Box	Bcl-2, Bcl-xL	39
Cell death to oxidative	Susceptibility		39

Bax and Box, proapoptotic proteins; Bcl-2 and Bcl-xL, antiapoptotic proteins; ROS, reactive oxygen species; H_2O_2, hydrogen peroxide; SOD, superoxide dismutase; GPx, glutathione peroxidase; GR, glutathione reductase; MDA, malondialdehyde; HNE, 4-hydroxynonenal protein carbonyl groups; 8-OHdG, 8-hydroxy-2-deoxyguanosine.

fibroblasts, especially in homozygous corneal fibroblasts, was significantly lower than that of wild-type cells after H_2O_2 exposure.[39] These results (Table 1) were the first evidence that oxidative stress is involved in GCD2 pathogenesis.

1.2.3 Accumulation and Degradation of Transforming Growth Factor β-Induced Protein in Corneal Fibroblasts

Mutant TGFBIp has been shown to accumulate in the autophagolysosomal compartment of corneal fibroblasts.[25,26] Developing a method to induce degradation of intracellular mutant TGFBIp may be a potential therapeutic strategy for the treatment of the GCD2.

1.2.3.1 Protein Degradation Systems in Mammalian Cells

Many degenerative disorders are characterized by the accumulation of intracellular or extracellular protein aggregates, which eventually leads to cellular death. Cells have special surveillance systems to control the quality of intracellular protein and organelles. Two major proteolytic systems, the autophagy and the ubiquitin/proteasome system (UPS), are responsible for removing abnormal proteins.[40] Several studies have shown that the accumulation of disease-related proteins is highly dependent on both the autophagy system and the UPS.[41,42] Degradation of proteins by the UPS

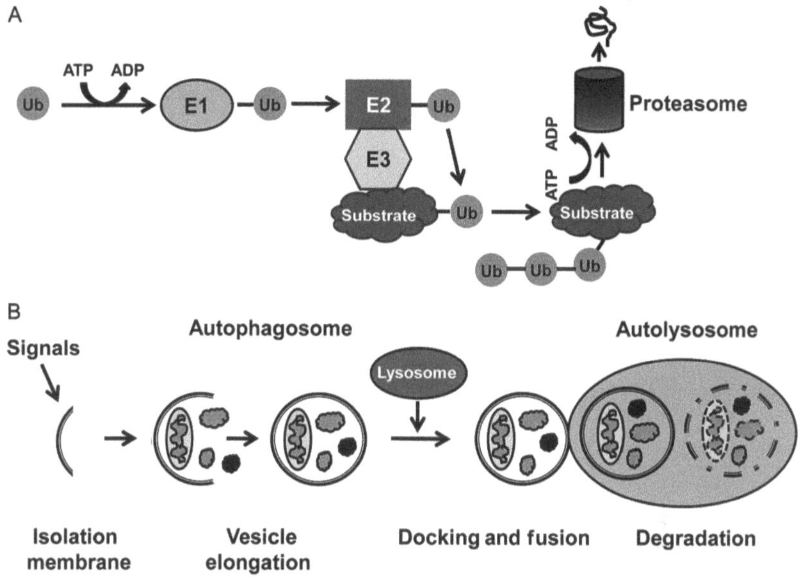

Figure 3 The ubiquitin–proteasome and autophagy–lysosome systems. (A) Degradation of a target substrate by the ubiquitin (Ub)/proteasome system occurs in two successive steps. The conjugation of Ub to the protein substrate, and the degradation of the Ub-tagged substrate by the 26S proteasome. Ubiquitin is activated by the Ub-activating enzyme, E1,[46] in a process that requires adenosine triphosphate (ATP). Ubiquitin is transferred to the Ub-conjugating enzyme, E2.[47] The E2 enzyme transfers the activated Ub moiety to the protein substrate that is specifically bound to a unique Ub ligase, E3.[48] Successive conjugation of Ub generates a polyubiquitin chain.[49] The polyubiquitin chain serves as a binding and degradation signal for the 26S proteasome,[46] and the substrate is degraded into short peptides.[47] (B) The autophagy system begins with isolation membrane formation. An autophagosome (vesicle) is formed through membrane elongation, and the autophagosome engulfs bulk cytoplasm, including the target cargo. The autophagosome outer membrane docks and directly fuses with a lysosome to form an autolysosome. The sequestered material is degraded inside the autolysosome and recycled later.

is also known to be a tightly regulated and highly specific process.[43] Degradation of a target protein by the UPS occurs in two successive steps.[44,45] The first step involves conjugation of several ubiquitin molecules to the target protein, a process that is catalyzed by ubiquitin-activating enzyme (E1), ubiquitin-conjugating enzyme (E2), and a substrate-specific ubiquitin-protein ligase (E3). The second step is the degradation of the tagged protein into small peptides by the 26S proteasome complex, which recognizes only ubiquitin-conjugated proteins (Fig. 3A).

The autophagy functions by sequestering molecules and organelles, such as depolarized mitochondria and aberrant protein aggregates, for degradation by lysosomes. It is considered a major player in cellular quality control. Autophagy can be classified into the following three types: microautophagy, chaperone-mediated autophagy, and macroautophagy.[50] Of these, macroautophagy, hereafter referred to as autophagy, has been studied the most. In autophagy, bulky cytoplasmic materials and organelles are packaged into double-membrane-bound vesicles, known as autophagosomes, which then fuse with lysosomes. After this, damaged material is degraded with lysosomal enzymes (Fig. 3B).[51]

1.2.3.2 Insufficient Autophagy in Granular corneal dystrophy type 2 Corneal Fibroblasts

Abnormal intracellular accumulation of mutant TGFBIp was demonstrated by comparing levels of *TGFBI* mRNA (real-time PCR) and TGFBIp (Western blotting) in wild-type and GCD2 cells. The higher levels in the GCD2 cells suggest that mutant TGFBIp is insufficiently broken down in GCD2 corneal fibroblasts.[26] When an adenovirus-mediated gene delivery system was introduced to normal corneal fibroblasts to induce TGFBIp overexpression, high-molecular-weight bands corresponding to dimeric and oligomeric TGFBIp molecules were detected, along with major immunoreactive bands at ~70 kDa.[26] These TGFBIp immunoreactive bands from cells that overexpressed TGFBIp were similar to those seen in corneas from homozygous GCD2 patients.[26]

Mutant TGFBIp colocalized extensively with microtubule-associated protein 1 light chain 3b (LC3)-enriched cytosolic vesicles, as well as cathepsin D and lysosomal enzymes in GCD2 corneal fibroblasts.[25,26] This indicates that TGFBIp is degraded by autophagy and not by the UPS.[25] Autophagy in GCD2 is known to be induced via mTOR signaling.[26]

1.2.3.3 Autophagy Suppression Increases Susceptibility to Cell Death in Granular corneal dystrophy type 2 Cells

A growing amount of evidence indicates that an autophagy can protect a cell,[52,53] but, paradoxically, may also contribute to its death.[54,55] Inhibiting autophagy resulted in more of a decrease in cell proliferation in cultured GCD2 corneal fibroblasts than in wild-type corneal fibroblasts.[25] Furthermore, treatment with bafilomycin A1, an inhibitor of autophagy fusion, caused activation of caspase-3 and poly (ADP-ribose) polymerase 1 (PARP1) in GCD2 corneal fibroblasts.[25] These results suggest that defective autophagy may contribute to GCD2 corneal fibroblast dysfunction.

1.2.3.4 The Effect of Activated Autophagy on Mutant TGFBIp Clearance

Autophagy induction has been suggested as a therapeutic treatment for age-related diseases.[56] Additionally, aggregation-prone proteins can be cleared by autophagy induction.[57] These findings suggest that autophagy induction might be a treatment option for GCD2 because autophagy is insufficient to remove mutant TGFBIp from GCD2 corneal fibroblasts.[26] When rapamycin, an autophagy inducer, was applied to cultured GCD2 corneal fibroblasts, mutant TGFBIp levels were reduced in GCD2 corneal fibroblasts, but normal TGFBIp levels in wild-type corneal fibroblasts did not change.[25] This suggests that rapamycin, or its related analogs, may be suitable therapeutic agents for eyes with GCD2. Rapamycin is especially promising because it is currently used in the clinical setting to treat several diseases.[25,56]

Melatonin has also been suggested as a possible autophagy activator and therapeutic agent for GCD2. Choi *et al.*[58] demonstrated that melatonin induces autophagy, via the mammalian target of rapamycin (mTOR)-dependent pathway, and consequently eliminates mutant TGFBIp from cells of GCD2 patients (Table 2). In addition, the application of melatonin

Table 2 Effect of Melatonin on the Transforming Growth Factor β-Induced Protein (TGFBIp), Autophagy, and Mammalian Target of Rapamycin (mTOR) Signaling Pathway

Effects	WT	GCD2
LC3 I	↑	↑
LC3 II	↑	↑
Autophagosome	↑	n.t.
ATG5	↑	n.t.
mTOR	n.c.	n.t.
p-mTOR (2481)	n.c.	n.t.
p-mTOR (2448)	↑	↑
GβL	n.c.	n.t.
Raptor	n.c.	n.t.
Rictor	n.c.	n.t.
TGFBIp	↓	↓

All results originally reported in Choi *et al.*[58] n.t. indicates not tested. n.c. indicates not changed significantly. LC3, light chain 3; ATG5, autophagy gene 5; p-mTOR, phosphor-mTOR; GβL, G-protein β-subunit-like protein.

counteracted autophagy inhibition that was induced with bafilomycin A1.[58] Interestingly, combination treatment with melatonin and rapamycin had a greater effect on mutant TGFBIp clearance than either drug alone.[58] Therefore, given its ability to activate autophagy and to remove mutant TGFBIp from cells, melatonin may be a possible therapy for GCD2, as shown in the literature (Table 2).

1.2.3.5 Effect of Mitomycin C on Corneas from Eyes with Granular corneal dystrophy type 2

When cultured keratocytes from normal, heterozygote GCD2, and homozygote GCD2 corneas were incubated with mitomycin C (MMC), cell viability was reduced in all three cell types in a dose- and time-dependent manner. Application of MMC caused apoptosis in GCD2 corneal cells, with homozygous cells being most vulnerable. Reverse transcription polymerase chain reaction (RT-PCR) and immunoblotting analyses showed that MMC reduced *TGFBI* mRNA levels and TGFBIp levels in the cell and in culture media for all cell types. These data suggest that MMC can induce keratocyte apoptosis, by reducing keratocyte TGFBIp production.[59] Ha *et al.*[60] showed that MMC application during photorefractive keratectomy to a cornea with GCD2 did not inhibit GCD2 exacerbation after laser ablation. These data suggest that GCD2 exacerbation after central corneal laser ablation may result from factors not related to TGFBIp production. Because TGFBIp can be absorbed from outside of the cell, the role of keratocyte function in GCD2 pathogenesis should be further evaluated.[61]

1.2.4 TGFBIp Regulation via the Transforming Growth Factor β Signaling Pathway: Therapeutic Applications

Strategies for treating *TGFBI*-linked CD have focused on reducing TGFBIp expression using short-interfering RNA.[62,63] Because TGFBIp expression is induced by transforming growth factor β (TGF-β), reducing intracellular TGFBIp levels by controlling TGF-β signaling may also be a useful approach for targeted drug intervention in *TGFBI*-linked CDs.[64]

1.2.4.1 The TGF-β Signaling Pathway

The TGF-β molecule is a multifunctional cytokine that regulates diverse cellular and physiological processes, including proliferation, apoptosis, differentiation, and extracellular matrix homeostasis.[65] This pathway is initiated by TGF-β binding to type I (TβRI) and type II (TβRII) TGF-β

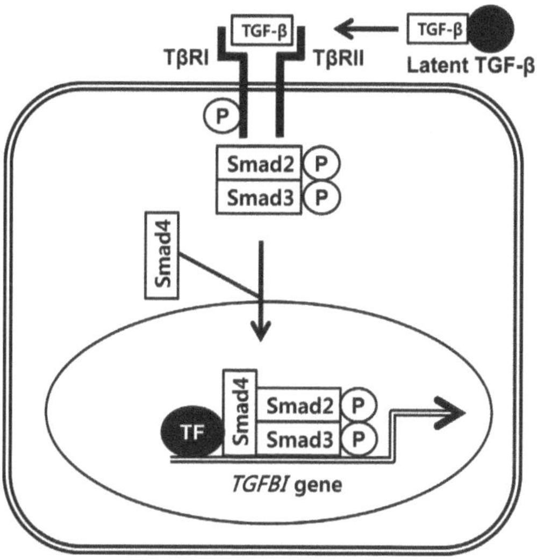

Figure 4 The transforming growth factor β (TGF-β) signaling pathway. The TGF-β molecule binds to its type II receptor (TβRII) on the cellular membrane. This induces type I (TβRI) and type II (TβRII) receptor complex formation. Once TGF-β1 binds to this receptor complex, intracellular Smad2 and Smad3 are recruited and phosphorylated at their C-termini by TβRI. Phosphorylated Smad2 and Smad3 subsequently bind to Smad4 and translocate to the nucleus to initiate target gene expression in collaboration with cofactors and other transcription factors (TFs).

receptors, both of which are transmembrane receptors that can activate multiple downstream signaling pathways to alter gene transcription (Fig. 4). The activated receptor complex phosphorylates the downstream transcription factors, Smad2 and Smad3. This leads to their association with Smad4. The Smad complex then translocates to the nucleus to regulate transcription of target genes (Fig. 4).[66]

1.2.4.2 Inhibition of TGF-β Signaling by Lithium

The TGFBIp was first identified in an adenocarcinoma cell line, where it was shown to be upregulated with TGF-β treatment.[67] Choi *et al.*[64] also demonstrated that TGF-β treatment induces TGFBIp expression in corneal fibroblasts. They also showed that lithium treatment reduces pSmad3 (S423/425) levels and the expression of normal and mutant TGFBIp (Table 3). Therefore, lithium might be a potential therapeutic agent for *TGFBI*-linked CD by TGF-β1 signaling inhibition (Table 3).

Table 3 Effect of Lithium on Transforming Growth Factor β-Induced Protein (TGFBIp), Autophagy, Cellular Proliferation, and Transforming Growth Factor β Signaling

Effects	WT	GCD2
TGFBIp	↓	↓
LC3 I	n.c.	n.t.
LC3 II	↑	n.t.
Smad3	↓	n.t.
p-Smad3 (S423/425)	n.c.	n.t.
GSK3α/β	n.c.	n.t.
GSK3α/β (S21/9)	↑	n.t.
Cell proliferation	n.c.	n.c.

All results originally reported in Choi *et al.*[64] n.t. indicates not tested. n.c. indicates not changed significantly. WT, wild type; GCD2, granular corneal dystrophy type 2; LC3, light chain 3; GSK3, glycogen synthase kinase 3.

2. CONCLUSION

Through a review of the recent literature, we summarize previously reported potential pathogenic mechanisms of Schnyder CD and GCD2. Several possible agents have been proposed for treating GCD2. However, medical treatment of both dystrophies is still needed and requires further basic research before clinical testing and final application to human eyes.

ACKNOWLEDGMENTS

This work was supported by the National Research Foundation of Korea (NRF) grant funded by the Korea government (MEST) (No. 2011-0028699).

REFERENCES

1. Weiss JS. Schnyder crystalline dystrophy sine crystals. Recommendation for a revision of nomenclature. *Ophthalmology*. 1996;103:465–473.
2. Weiss JS. Visual morbidity in thirty-four families with Schnyder crystalline corneal dystrophy (an American Ophthalmological Society thesis). *Trans Am Ophthalmol Soc*. 2007;105:616–648.
3. Weiss JS. Schnyder's dystrophy of the cornea. A Swede-Finn connection. *Cornea*. 1992;11:93–101.
4. Orr A, Dube MP, Marcadier J, et al. Mutations in the UBIAD1 gene, encoding a potential prenyltransferase, are causal for Schnyder crystalline corneal dystrophy. *PLoS One*. 2007;2:e685.

5. Weiss JS, Kruth HS, Kuivaniemi H, et al. Mutations in the UBIAD1 gene on chromosome short arm 1, region 36, cause Schnyder crystalline corneal dystrophy. *Invest Ophthalmol Vis Sci.* 2007;48:5007–5012.
6. Weiss JS, Kruth HS, Kuivaniemi H, et al. Genetic analysis of 14 families with Schnyder crystalline corneal dystrophy reveals clues to UBIAD1 protein function. *Am J Med Genet.* 2008;146A:271–283.
7. Gaynor PM, Zhang WY, Weiss JS, et al. Accumulation of HDL apolipoproteins accompanies abnormal cholesterol accumulation in Schnyder's corneal dystrophy. *Arterioscler Thromb Vasc Biol.* 1996;16:992–999.
8. Lisch W, Weidle EG, Lisch C, et al. Schnyder's dystrophy. Progression and metabolism. *Ophthalmic Paediatr Genet.* 1986;7:45–56.
9. Sysi R. Xanthoma corneae as hereditary dystrophy. *Br J Ophthalmol.* 1950;34:369–374.
10. McCarthy M, Innis S, Dubord P, et al. Panstromal Schnyder corneal dystrophy. A clinical pathologic report with quantitative analysis of corneal lipid composition. *Ophthalmology.* 1994;101:895–901.
11. Battisti C, Dotti MT, Malandrini A, et al. Schnyder corneal crystalline dystrophy: description of a new family with evidence of abnormal lipid storage in skin fibroblasts. *Am J Med Genet.* 1998;75:35–39.
12. Mehta JS, Vithana EN, Venkataraman D, et al. Analysis of conjunctival fibroblasts from a proband with Schnyder corneal dystrophy. *Mol Vis.* 2008;14:1277–1281.
13. McGarvey TW, Nguyen TB, Malkowicz SB. An interaction between apolipoprotein E and TERE1 with a possible association with bladder tumor formation. *J Cell Biochem.* 2005;95:419–428.
14. Zhang WY, Gaynor PM, Kruth HS. Apolipoprotein E produced by human monocyte-derived macrophages mediates cholesterol efflux that occurs in the absence of added cholesterol acceptors. *J Biol Chem.* 1996;271:28641–28646.
15. Nickerson ML, Kostiha BN, Brandt W, et al. UBIAD1 mutation alters a mitochondrial prenyltransferase to cause Schnyder corneal dystrophy. *PLoS One.* 2010;5:e10760.
16. McGarvey TW, Nguyen T, Tomaszewski JE, et al. Isolation and characterization of the TERE1 gene, a gene down-regulated in transitional cell carcinoma of the bladder. *Oncogene.* 2001;20:1042–1051.
17. Gatzioufas Z, Charalambous P, Loew U, et al. Evidence of oxidative stress in Schnyder corneal dystrophy. *Br J Ophthalmol.* 2010;94:1262–1264.
18. Munier FL, Korvatska E, Djemai A, et al. Kerato-epithelin mutations in four 5q31-linked corneal dystrophies. *Nat Genet.* 1997;15:247–251.
19. Klintworth GK. Advances in the molecular genetics of corneal dystrophies. *Am J Ophthalmol.* 1999;128:747–754.
20. Korvatska E, Henry H, Mashima Y, et al. Amyloid and non-amyloid forms of 5q31-linked corneal dystrophy resulting from kerato-epithelin mutations at Arg-124 are associated with abnormal turnover of the protein. *J Biol Chem.* 2000;275:11465–11469.
21. Kennedy SM, McNamara M, Hillery M, Hurley C, Collum LM, Giles S. Combined granular lattice dystrophy (Avellino corneal dystrophy). *Br J Ophthalmol.* 1996;80:489–490.
22. Hahnel C, Somodi S, Weiss DG, Guthoff RF. The keratocyte network of human cornea: a three-dimensional study using confocal laser scanning fluorescence microscopy. *Cornea.* 2000;19:185–193.
23. McLaren JW, Nau CB, Kitzmann AS, Bourne WM. Keratocyte density: comparison of two confocal microscopes. *Eye Contact Lens.* 2005;31:28–33.
24. Jester JV, Moller-Pedersen T, Huang J, et al. The cellular basis of corneal transparency: evidence for 'corneal crystallins'. *J Cell Sci.* 1999;112(pt 5):613–622.
25. Choi SI, Kim BY, Dadakhujaev S, et al. Impaired autophagy and delayed autophagic clearance of transforming growth factor beta-induced protein (TGFBI) in granular corneal dystrophy type 2. *Autophagy.* 2012;8:1782–1797.

26. Choi SI, Maeng YS, Kim KS, Kim TI, Kim EK. Autophagy is induced by raptor degradation via the ubiquitin/proteasome system in granular corneal dystrophy type 2. *Biochem Biophys Res Commun*. 2014;450:1505–1511.
27. Wenk J, Brenneisen P, Meewes C, et al. UV-induced oxidative stress and photoaging. *Curr Probl Dermatol*. 2001;29:83–94.
28. Rao NA, Romero JL, Fernandez MA, Sevanian A, Marak Jr GE. Role of free radicals in uveitis. *Surv Ophthalmol*. 1987;32:209–213.
29. Halliwell B, Gutteridge JMC. *Free Radicals in Biology and Medicine*. 2nd ed. Oxford: Clarendon; 1989.
30. Mates JM, Perez-Gomez C, Nunez de Castro I. Antioxidant enzymes and human diseases. *Clin Biochem*. 1999;32:595–603.
31. Pham-Huy LA, He H, Pham-Huy C. Free radicals, antioxidants in disease and health. *Int J Biomed Sci*. 2008;4:89–96.
32. Esterbauer H, Schaur RJ, Zollner H. Chemistry and biochemistry of 4-hydroxynonenal, malonaldehyde and related aldehydes. *Free Radic Biol Med*. 1991;11:81–128.
33. Gutteridge JM, Halliwell B. The measurement and mechanism of lipid peroxidation in biological systems. *Trends Biochem Sci*. 1990;15:129–135.
34. Berlett BS, Stadtman ER. Protein oxidation in aging, disease, and oxidative stress. *J Biol Chem*. 1997;272:20313–20316.
35. Dalle-Donne I, Giustarini D, Colombo R, Rossi R, Milzani A. Protein carbonylation in human diseases. *Trends Mol Med*. 2003;9:169–176.
36. Kasai H, Hayami H, Yamaizumi Z, Saito H, Nishimura S. Detection and identification of mutagens and carcinogens as their adducts with guanosine derivatives. *Nucleic Acids Res*. 1984;12:2127–2136.
37. Kasai H, Nishimura S. Hydroxylation of deoxyguanosine at the C-8 position by ascorbic acid and other reducing agents. *Nucleic Acids Res*. 1984;12:2137–2145.
38. Kim TI, Kim H, Lee DJ, Choi SI, Kang SW, Kim EK. Altered mitochondrial function in type 2 granular corneal dystrophy. *Am J Pathol*. 2011;179:684–692.
39. Choi SI, Kim TI, Kim KS, et al. Decreased catalase expression and increased susceptibility to oxidative stress in primary cultured corneal fibroblasts from patients with granular corneal dystrophy type II. *Am J Pathol*. 2009;175:248–261.
40. Ciechanover A. Proteolysis: from the lysosome to ubiquitin and the proteasome. *Nat Rev Mol Cell Biol*. 2005;6:79–87.
41. Levine B, Kroemer G. Autophagy in the pathogenesis of disease. *Cell*. 2008;132:27–42.
42. Maynard CJ, Bottcher C, Ortega Z, et al. Accumulation of ubiquitin conjugates in a polyglutamine disease model occurs without global ubiquitin/proteasome system impairment. *Proc Natl Acad Sci USA*. 2009;106:13986–13991.
43. Glickman MH, Ciechanover A. The ubiquitin-proteasome proteolytic pathway: destruction for the sake of construction. *Physiol Rev*. 2002;82:373–428.
44. Dantuma NP, Masucci MG. Stabilization signals: a novel regulatory mechanism in the ubiquitin/proteasome system. *FEBS Lett*. 2002;529:22–26.
45. Reinstein E, Ciechanover A. Narrative review: protein degradation and human diseases: the ubiquitin connection. *Ann Intern Med*. 2006;145:676–684.
46. Goldberg AL. Nobel committee tags ubiquitin for distinction. *Neuron*. 2005;45:339–344.
47. Wilkinson KD. Ubiquitin: a Nobel protein. *Cell*. 2004;119:741–745.
48. Jemal A, Siegel R, Ward E, et al. Cancer statistics, 2006. *CA Cancer J Clin*. 2006;56:106–130.
49. Pickart CM, Eddins MJ. Ubiquitin: structures, functions, mechanisms. *Biochim Biophys Acta*. 2004;1695:55–72.
50. Cuervo AM. Autophagy: in sickness and in health. *Trends Cell Biol*. 2004;14:70–77.
51. Mizushima N. Autophagy: process and function. *Genes Dev*. 2007;21:2861–2873.

52. Qu X, Yu J, Bhagat G, et al. Promotion of tumorigenesis by heterozygous disruption of the beclin 1 autophagy gene. *J Clin Invest.* 2003;112:1809–1820.
53. Yue Z, Jin S, Yang C, Levine AJ, Heintz N. Beclin 1, an autophagy gene essential for early embryonic development, is a haploinsufficient tumor suppressor. *Proc Natl Acad Sci USA.* 2003;100:15077–15082.
54. Daido S, Kanzawa T, Yamamoto A, Takeuchi H, Kondo Y, Kondo S. Pivotal role of the cell death factor BNIP3 in ceramide-induced autophagic cell death in malignant glioma cells. *Cancer Res.* 2004;64:4286–4293.
55. Xue L, Fletcher GC, Tolkovsky AM. Autophagy is activated by apoptotic signalling in sympathetic neurons: an alternative mechanism of death execution. *Mol Cell Neurosci.* 1999;14:180–198.
56. Rubinsztein DC. The roles of intracellular protein-degradation pathways in neurodegeneration. *Nature.* 2006;443:780–786.
57. Ravikumar B, Duden R, Rubinsztein DC. Aggregate-prone proteins with polyglutamine and polyalanine expansions are degraded by autophagy. *Hum Mol Genet.* 2002;11:1107–1117.
58. Choi SI, Kim KS, Oh JY, Jin JY, Lee GH, Kim EK. Melatonin induces autophagy via an mTOR-dependent pathway and enhances clearance of mutant-TGFBIp. *J Pineal Res.* 2013;54:361–372.
59. Kim TI, Choi SI, Lee HK, Cho YJ, Kim EK. Mitomycin C induces apoptosis in cultured corneal fibroblasts derived from type II granular corneal dystrophy corneas. *Mol Vis.* 2008;14:1222–1228.
60. Ha BJ, Kim TI, Choi SI, et al. Mitomycin C does not inhibit exacerbation of granular corneal dystrophy type II induced by refractive surface ablation. *Cornea.* 2010;29:490–496.
61. Choi SI, Maeng YS, Kim TM, Lee Y, Kim YS, Kim EK. Lysosomal trafficking of TGFBIp via caveolae-mediated endocytosis. *PLoS One.* 2015;10:e0119561.
62. Courtney DG, Atkinson SD, Moore JE, et al. Development of allele-specific gene-silencing siRNAs for TGFBI Arg124Cys in lattice corneal dystrophy type I. *Invest Ophthalmol Vis Sci.* 2014;55:977–985.
63. Yellore VS, Rayner SA, Aldave AJ. TGFB1-induced extracellular expression of TGFBIp and inhibition of TGFBIp expression by RNA interference in a human corneal epithelial cell line. *Invest Ophthalmol Vis Sci.* 2011;52:757–763.
64. Choi SI, Kim BY, Dadakhujaev S, et al. Inhibition of TGFBIp expression by lithium: implications for TGFBI-linked corneal dystrophy therapy. *Invest Ophthalmol Vis Sci.* 2011;52:3293–3300.
65. Hedgepeth CM, Conrad LJ, Zhang J, et al. Activation of the Wnt signaling pathway: a molecular mechanism for lithium action. *Dev Biol.* 1997;185:82–91.
66. Shi Y, Massague J. Mechanisms of TGF-beta signaling from cell membrane to the nucleus. *Cell.* 2003;113:685–700.
67. Skonier J, Neubauer M, Madisen L, Bennett K, Plowman GD, Purchio AF. cDNA cloning and sequence analysis of beta ig-h3, a novel gene induced in a human adenocarcinoma cell line after treatment with transforming growth factor-beta. *DNA Cell Biol.* 1992;11:511–522.

SECTION II

Lens

CHAPTER NINE

Overview of the Lens

J. Fielding Hejtmancik*, Alan Shiels[†,1]
*Ophthalmic Genetics and Visual Function Branch, National Eye Institute, National Institutes of Health, Bethesda, Maryland, USA
[†]Department of Ophthalmology and Visual Sciences, Washington University School of Medicine, St. Louis, Missouri, USA
[1]Corresponding author e-mail address: Shiels@vision.wustl.edu

Contents

Abstract

In order to accomplish its function of transmitting and focusing light, the crystalline lens of the vertebrate eye has evolved a unique cellular structure and protein complement. These distinct adaptations have provided a rich source of scientific discovery ranging from biochemistry and genetics to optics and physics. In addition, because of these adaptations, lens cells persist for the lifetime of an organism, providing an excellent model of the aging process. The chapters dealing with the lens will demonstrate how the different aspects of lens biology and biochemistry combine in this singular refractive organ to accomplish its critical role in the visual system.

1. INTRODUCTION

Like the lens in a camera, the basic function of the eye lens is to transmit and focus light onto the retina. To facilitate this, it contains one of the highest concentrations of proteins of any tissue. The lens has been studied scientifically for over a century, beginning in 1833 when Sir David Brewster deduced the fine structure of the cod lens using only a candle and a finely ruled steel bar.[1] In 1894, Mörner first described high concentrations of soluble structural proteins we now call crystallins,[2] and Spemann developed the concept of inductive interactions in development by studying the lens in 1901.[3] Renwick mapped a cataract locus, one of the first autosomal loci to be localized,[4] and chicken lens δ-crystallins were among the first mRNAs

Progress in Molecular Biology and Translational Science, Volume 134
ISSN 1877-1173
http://dx.doi.org/10.1016/bs.pmbts.2015.04.006
© 2015 Elsevier Inc.
All rights reserved.

to be isolated and cloned.[5] Thus, in addition to being important in the study of inherited diseases, the lens has also been a model system invaluable for developmental and structural biology.

2. STRUCTURE AND CELLS OF THE LENS

Weighing about 65 mg at birth, the human lens increases in weight to about 160 mg by the age of 10 at which time growth slows substantially so that it weighs about 250 mg by the age of 90.[6,7] As much as 60% of the total mass of the lens can be made up of proteins, much higher than almost any other tissue.[8] The lens is surrounded by a collagenous capsule, on which the anterior-facing basal poles of the epithelial cells rest, as do the basal poles of the fiber cells facing posteriorly (Fig. 1).[9,10] The capsule acts as a barrier to diffusion and contributes to shaping the lens during accommodation.[11,12] Its main components are type IV collagen, laminin, entactin, perlecan, type XVIII collagen, heparin sulfate proteoglycan, and fibronectin,[13,14] of which the first four are major structural molecules that self-assemble to form a matrix. The capsular filaments, of uniform size and aligned in a parallel fashion, are thinnest at the posterior pole and thicken to a maximum at the equator, where the lens zonules insert.[14] Fibrillin and elastin fibers also integrate in the equatorial region, especially in the outer zonular.[15] The lens capsule is first detectable at 5–6 weeks of gestation in humans[16] and is produced continually throughout life[12] anteriorly by the cuboidal epithelium and more slowly posteriorly by the fiber cells.

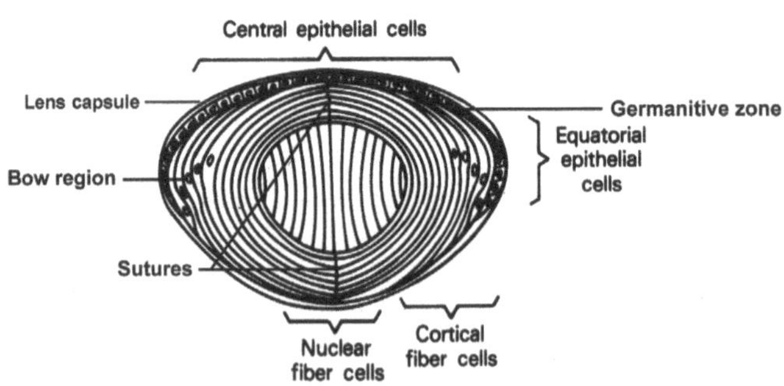

Figure 1 Human lens structure. Anterior epithelial cells divide at the 10 and 2 o'clock positions. Cells then move laterally, eventually inverting in the bow region, at which time they elongate and begin degrading their organelles to form cortical fiber cells. Central nuclear fiber cells elongate from the posterior epithelia early in development. The ends of the more peripheral secondary fiber cells abut at the sutures, which are shown here as vertical lines but are seen clinically as the anterior and posterior Y structures.

Mitotic division in the lens occurs in the germinative zone of the anterior epithelium located just anterior to the equator. The anterior epithelial cells of the lens are connected by gap junctions,[17] allowing exchange of low-molecular-weight metabolites and ions. They have few or no tight junctions that would make the extracellular spaces impermeable to these molecules.[18,19] Anterior cuboidal epithelial cells also are rich in organelles and contain large amounts of cytoskeletal proteins such as microtubules, spectrin, α-actinin, actin, myosin, and vimentin, presumably to help stabilize the cell structures during accommodation.[20–22] Both lens epithelial and especially fiber cells contain large amounts of crystallins.

Fiber cells make up the lens nucleus. Layers of nucleated cortical fiber cells form highly ordered concentric shells around the nonnucleated central fiber cells which make up the fetal nucleus, with the ends of the peripheral fiber cells abutting in sutures anteriorly and posteriorly. Both the ordered arrangement of the fiber cells and their sutures as well as their intracellular structure are important for light transmission and lens transparency.[23–25] Also contributing to transparency is the presence of only minimal extracellular space between fiber cells, which have many interdigitations.[9,26] Junctional complexes between adjacent fiber cells allow for exchange of metabolites.[21,22] Lens crystallins, which make up about 90% of the water-soluble protein, are the main soluble components of fiber cells, along with cytoskeletal components, including actin, myosin, vimentin, α-actinin, and microtubules.[27]

The lens is composed of a single cell type that follows a developmental pattern, beginning as a member of the germinative zone in the single layer of anterior epithelial cells overlaying the fiber cell mass.[26] Epithelial cells then migrate laterally toward the equator, where they begin to elongate and invert to form secondary fibers. In order to increase light transmission, organelles such as mitochondria, Golgi bodies, and both rough and smooth ER are degraded in the differentiating lens fiber cells so that they are absent from nuclear fiber cells. The density of their cell membranes increases, approaching that of the cytoplasm, which also decreases light scattering.[28] As the cells elongate newer cortical fiber cells are layered over them so that they are moved toward the lens nucleus, stretching anteriorly from the cuboidal epithelial cells posteriorly to the posterior capsule. Transcriptional control plays a significant role in the differential synthesis of lens crystallins (see Ref. 29). The distribution of β-crystallin mRNAs in chickens[30] and the β- and γ-crystallin proteins and mRNAs in rats[31,32] provides examples of the spatial and temporal control of crystallin gene expression during lens development.

3. TRANSPARENCY

The main optical function of the lens is to transmit light, focusing it on the retina. The cornea contributes about 80% of total refraction, while the lens fine-tunes the focusing of light onto the retina. Although the human lens is colorless at birth, there is a gradual increase in yellowish pigmentation with age[33] probably due to the production of 3-hydroxykynurenine and other metabolites of tryptophan that filter UV light.[34] The lens transmits light with wavelengths up to 1200 nm efficiently, but transmits very little light below 390 nm. 1200 nm is well above the limit of visual perception, about 720 nm. As discussed previously, the architecture and cellular contents of the lens are critical for its transparency. The transparency and high-refractive index of cells in the lens result from tight packing of their proteins, providing a constant refractive index over distances approximating the wavelength of the transmitted light.[24,25] In fact, as lens proteins are diluted to concentrations below that found in the lens, about 450 mg/ml, light scattering actually increases,[35,36] because dilution decreases the weak interactions between unlike proteins that occur at high concentrations and help to maintain lens transparency.[37,38] Finally, there is a gradual increase in the refractive index of the human lens from 1.38 (73–80% H_2O) in the cortex to 1.42 (68% H_2O) in the nucleus, in part due to an enrichment of tightly packed γ-crystallins.[39]

4. AGING

The inability of cells to be replaced in the encapsulated lens combined with the inability of lens cell proteins to turn over in the nonnucleated fiber cells makes the lens particularly susceptible to damage with aging and environmental insults such as UV light and other oxidative stresses.[40] This results in a decrease in transmission of light and focusing even in the normal aged lens so that the intensity of light reaching the retina is reduced by about 10-fold by 80 years of age.[41] It also increases susceptibility to senescent cataract and presbyopia, especially in individuals exposed to environmental insults or having a genetic proclivity.[42] With increasing age, vacuoles and multilamellar bodies develop between lens fiber cells, occasionally disrupting the fiber plasma membrane.[43] In addition, most of the elaborate cytoskeletal structure found in lens cells disappears with age,[44] so that by the fifth decade presbyopia develops with loss of the ability to accommodate.[45,46]

Enzymatic activity in the lens decreases with age, especially in the central cells of the lens nucleus where the cells are older than those in the cortical nucleus and especially the anterior epithelial cells.[47] This compromised intracellular homeostasis might be exacerbated by the decreased metabolic coupling of the active cortex and the inactive nucleus that occurs in older lenses, in part associated with decreased gap junction coupling.[48,49] This is particularly relevant for the enzymes that produce a reducing environment by maintaining high levels of reduced glutathione, such as glutathione reductase and glucose-6-phosphate dehydrogenase.[50] Decreases in the activity of these and other reducing enzymes decrease defenses against oxidative damage in the lens and exacerbate damage to crystallins and other metabolic support systems.[51] Finally, as the lens ages intracellular Na^+ and Ca^{2+} concentrations rise, probably due to an increase in lens permeability or decrease in ion channel pumping efficiency.[52]

Lens crystallins also show age-related changes that might interfere with lens transparency.[37] Between 10 and 50 years of age crystallin modification increases,[53] as does the level of high-molecular-weight aggregates and water-insoluble protein.[54] Because of their chaperone activity, this is especially notable in the α-crystallins, but is also seen in the β- and γ-crystallins.[55,56] Crystallins, membranes, and enzymes are also cleaved and partially degraded, including the nonenzymatic cleavage of αA-crystallin at the bond between Asn101 and Glu102.[57] In what might be a positive feedback effect, cleavage products of βA3-crystallin appear to inhibit the chaperone activity of α-crystallin chaperone.[58] γ-Crystallins, and particularly γS-crystallin, are often subject to proteolysis, degradation, and modification in age-dependent cataracts, being broken down to low-molecular-weight peptides.[59–61]

As the lens ages both the amino- and carboxyl-terminal arms of up to half of the intrinsic membrane protein AQP0 (MP26) molecules undergo proteolysis, forming MP22.[62] Other posttranslational modifications of AQP0 also occur with aging including C-terminal phosphorylation, possibly involved in intercellular trafficking, and glycation, which influences AQP0 interaction with calmodulin. However, the precise functional significance of these remains unclear.[63,64] The lens contains proteasomes, which preferentially degrade oxidized proteins[65] tagged with the protease cofactor ubiquitin,[66] whose activity is increased by oxidative stress.[67] These proteinases are balanced during aging by inhibitors including the chaperones HSP90 and α-crystallin.[59]

Covalent modifications of crystallins and other lens proteins also increase with aging, with increases in oxidation of methionine, deamidation of asparagine and glutamine residues, disulfide bridges, backbone cleavage, and racemization of aspartic acid residues.[59,68,69] Deamidation can destabilize βA3-crystallin, causing it to aggregate,[70] while deamidation of glutamines at the interface of γD-crystallin can also destabilize it.[71] Asp151 in αA-crystallin is especially susceptible to racemization because it forms a succinimide intermediate easily.[72] Racemization at both Asp58 and Asp151 can lead to increased aggregation and decreased chaperone activity and is enhanced by mutations of nearby residues.[73] Finally, phosphorylation and nonenzymatic glycosylation (glycation) also occur, especially affecting the ε-amino groups of lysine residues.[57,74,75] These can participate in the Maillard reaction, resulting in nondisulfide covalent cross-links, increased pigmentation, and nontryptophan fluorescence.[76] Glycation of α-crystallin can also decrease its chaperone function, eventually resulting in aggregation.[77] Lens proteins can also undergo carbamylation with aging or other insult, and this can induce cataract,[78] which has been proposed to be the mechanism of cataract associated with chronic diarrhea and its resultant uremia.[79] Thus, the development and biology of the lens is directed at establishing transparency and focusing of light, and then defending this highly specialized system against damage by age and environmental insults.

REFERENCES

1. Brewster D. On the anatomical and optical structure of the crystalline lens of animals, particularly that of cod. *Philos Trans R Soc Lond.* 1833;123:323–332.
2. Morner CT. Untersuchungen der protein-substanzen in den lichtbrechenden Medien des Auges. *Hoppe Seylers Z Physiol Chem.* 1894;18:61–106.
3. Spemann H. Uber Korrelationen in der Entwicklung des Auges. *Verh Anat Ges.* 1901;15:61–79.
4. Renwick JH, Lawler SD. Probable linkage between a congenital cataract locus and the Duffy blood group locus. *Ann Hum Genet.* 1963;27:67–84.
5. Zelenka PS, Piatigorsky J. Isolation and in vitro translation of delta-crystallin mRNA from embryonic chick lens fibers. *Proc Natl Acad Sci USA.* 1974;71:1896–1900.
6. Harding JJ, Rixon KC, Marriott FHC. Men have heavier lenses than women of the same age. *Exp Eye Res.* 1977;25:651.
7. Augusteyn RC. Growth of the human eye lens. *Mol Vis.* 2007;13:252–257.
8. Wistow GJ, Piatigorsky J. Lens crystallins: the evolution and expression of proteins for a highly specialized tissue. *Annu Rev Biochem.* 1988;57:479–504.

9. Kuszak JR. Embryology and anatomy of the lens. In: Tasman W, Jaeger EA, eds. *Duane's Clinical Ophthalmology*. Philadelphia: J.B. Lippincott; 1990:1–9.
10. Danysh BP, Duncan MK. The lens capsule. *Exp Eye Res*. 2009;88(2):151–164.
11. Koretz JF, Handelman GH. How the human eye focuses. *Sci Am*. 1988;256:92–99.
12. Fisher RF, Pettet BE. The postnatal growth of the capsule of the human crystalline lens. *J Anat*. 1972;112:207–214.
13. Parmigiani C, McAvoy J. Localisation of laminin and fibronectin during rat lens morphogenesis. *Differentiation*. 1986;28:53–61.
14. Cammarata PR, Cantu-Crouch D, Oakford L, Morrill A. Macromolecular organization of the bovine lens capsule. *Tissue Cell*. 1986;18:83–97.
15. Mir S, Wheatley HM, Hussels IE, Whittum-Hudson JA, Traboulsi EI. A comparative histologic study of the fibrillin microfibrillar system in the lens capsule of normal subjects and subjects with Marfan syndrome. *Invest Ophthalmol Vis Sci*. 1998;39(1):84–93.
16. Mann I. *The Development of the Human Eye*. New York: Grune and Stratton; 1964.
17. Goodenough DA, Dick JSB, Lyons JE. Lens metabolic cooperation: a study of mouse lens transport and permeability visualized with freeze-substitution autoradiography and electron microscopy. *J Cell Biol*. 1980;86:576–589.
18. Gorthy WC, Snavely MR, Berrong ND. Some aspects of transport and digestion in the lens of the normal young adult rat. *Exp Eye Res*. 1971;12:112–119.
19. Rae JL, Stacey T. Lanthanum and procion yellow as extracellular markers in the crystalline lens of the rat. *Exp Eye Res*. 1979;28:1–21.
20. Ramaekers FCS, Bloemendal H. Cytoskeletal and contractile structures in lens cell differentiation. In: Bloemendal H, ed. *Molecular and Cellular Biology of the Eye Lens*. New York: John Wiley & Sons; 1981:85–136.
21. Benedetti L, Dunia I, Ramaekers FCS, Kibbelaar MA. Lenticular plasma membranes and cytoskeleton. In: Bloemendal H, ed. *Molecular and Cellular Biology of the Eye Lens*. New York: John Wiley & Sons; 1981:137–188.
22. Alcala H, Maisel H. Biochemistry of lens plasma membranes and cytoskeleton. In: Maisel H, ed. *The Ocular Lens*. New York: Marcel Dekker Inc.; 1985:169–222.
23. Kuszak JR, Zoltoski RK, Sivertson C. Fibre cell organization in crystalline lenses. *Exp Eye Res*. 2004;78(3):673–687.
24. Benedek GB. Theory of transparency of the eye. *Appl Opt*. 1971;10:459–473.
25. Delaye M, Tardieu A. Short-range order of crystallin proteins accounts for eye lens transparency. *Nature*. 1983;302:415–417.
26. Rafferty NS. Lens morphology. In: Maisel H, ed. *The Ocular Lens*. New York: Marcel Dekker Inc.; 1985:1–60.
27. Ireland M, Maisel H. A family of lens fiber cell specific proteins. *Lens Eye Toxic Res*. 1989;6:623–638.
28. Michael R, van Marle J, Vrensen GF, van den Berg TJ. Changes in the refractive index of lens fibre membranes during maturation—impact on lens transparency. *Exp Eye Res*. 2003;77(1):93–99.
29. Piatigorsky J. Gene expression and genetic engineering in the lens. Friedenwald lecture. *Invest Ophthalmol Vis Sci*. 1987;28:9–28.
30. Hejtmancik JF, Beebe DC, Ostrer H, Piatigorsky J. delta- and beta-Crystallin mRNA levels in the embryonic and posthatched chicken lens: temporal and spatial changes during development. *Dev Biol*. 1985;109:72–81.
31. van Leen RW, van Roozendaal KEP, Lubsen NH, Schoenmakers JG. Differential expression of crystallin genes during development of the rat eye lens. *Dev Biol*. 1987;120:457–464.
32. Aarts HJ, Lubsen NH, Schoenmakers JG. Crystallin gene expression during rat lens development. *Eur J Biochem*. 1989;183:31–36.
33. Lerman S. *Radiant Energy and the Eye*. New York: MacMillan; 1980.

34. Korlimbinis A, Truscott RJ. Identification of 3-hydroxykynurenine bound to proteins in the human lens. A possible role in age-related nuclear cataract. *Biochemistry*. 2006; 45(6):1950–1960.
35. Bettelheim FA, Siew EL. Effect of changes in concentration upon lens turbidity as predicted by the random fluctuation theory. *Biophys J*. 1983;41:29–33.
36. Delaye M, Gromiec A. Mutual diffusion of crystallin proteins at finite concentrations: a light scattering study. *Biopolymers*. 1983;22:1203–1221.
37. Takemoto L, Sorensen CM. Protein-protein interactions and lens transparency. *Exp Eye Res*. 2008;87(6):496–501.
38. Stradner A, Foffi G, Dorsaz N, Thurston G, Schurtenberger P. New insight into cataract formation: enhanced stability through mutual attraction. *Phys Rev Lett*. 2007;99(19): 198103.
39. Uhlhorn SR, Borja D, Manns F, Parel JM. Refractive index measurement of the isolated crystalline lens using optical coherence tomography. *Vision Res*. 2008;48(27):2732–2738.
40. Lynnerup N, Kjeldsen H, Heegaard S, Jacobsen C, Heinemeier J. Radiocarbon dating of the human eye lens crystallines reveal proteins without carbon turnover throughout life. *PLoS One*. 2008;3(1):e1529.
41. Sample PA, Esterson FD, Weinreb RN, Boynton RM. The aging lens: in vivo assessment of light absorption in 84 human eyes. *Invest Ophthalmol Vis Sci*. 1988;8:1306–1311.
42. Bron AJ, Vrensen GF, Koretz J, Maraini G, Harding JJ. The ageing lens. *Ophthalmologica*. 2000;214(1):86–104.
43. Vrensen G, Kappelhof J, Willikens B. Aging of the human lens. *Lens Eye Toxic Res*. 1990;7:1–30.
44. Kuszak JR, Deutsch TA, Brown HG. Anatomy of aged and senile cataractous lenses. In: Albert D, Jacobiec F, eds. *Principles and Practice of Ophthalmology: Basic Sciences*. Philadelphia: W.B. Saunders; 1994:82–97.
45. Koretz JF, Kaufman PL, Neider MW, Goeckner PA. Accommodation and presbyopia in the human eye—aging of the anterior segment. *Vision Res*. 1989;29:1685–1692.
46. Davson H. *Physiology of the Eye*. New York: Pergamon Press; 1990.
47. Hockwin O, Ohrloff C. The eye in the elderly: lens. In: Platt D, ed. *Geriatrics*. Berlin: Springer-Verlag; 1984:373–424.
48. Truscott RJ. Age-related nuclear cataract: oxidation is the key. *Exp Eye Res*. 2005;80(5):709–725.
49. Gao J, Wang H, Sun X, et al. The effects of age on lens transport. *Invest Ophthalmol Vis Sci*. 2013;54(12):7174–7187.
50. Ganea E, Harding JJ. Glutathione-related enzymes and the eye. *Curr Eye Res*. 2006;31(1):1–11.
51. Wei M, Xing KY, Fan YC, Libondi T, Lou MF. Loss of thiol repair systems in human cataractous lenses. *Invest Ophthalmol Vis Sci*. 2015;56(1):598–605.
52. Duncan G, Hightower KR, Gandolfi SA, Tomlinson J, Maraini G. Human lens membrane cation permeability increases with age. *Invest Ophthalmol Vis Sci*. 1989;30(8):1855–1859.
53. Lampi KJ, Ma Z, Hanson SR, et al. Age-related changes in human lens crystallins identified by two-dimensional electrophoresis and mass spectrometry. *Exp Eye Res*. 1998;67(1):31–43.
54. Datiles III MB, Ansari RR, Suh KI, et al. Clinical detection of precataractous lens protein changes using dynamic light scattering. *Arch Ophthalmol*. 2008;126(12):1687–1693.
55. Roy D, Spector A. Absence of low-molecular weight alpha-crystallin in nuclear region of old human lens. *Proc Natl Acad Sci USA*. 1976;73:3484–3487.
56. McFall-Ngai MJ, Ding LL, Takemoto LJ, Horwitz J. Spatial and temporal mapping of the age-related changes in human lens crystallins. *Exp Eye Res*. 1985;41:745–758.
57. Voorter CE, De Haard-Hoekman WA, Roersma ES, Meyer HE, Bloemendal H, de Jong WW. The in vivo phosphorylation sites of bovine alpha B-crystallin. *FEBS Lett*. 1989;259:50–52.

58. Rao G, Santhoshkumar P, Sharma KK. Anti-chaperone betaA3/A1(102-117) peptide interacting sites in human alphaB-crystallin. *Mol Vis*. 2008;14:666–674.
59. Harding JJ, Crabbe MJC. The lens: development, proteins, metabolism and cataract. In: Davson H, ed. 3rd ed. Orlando: Academic Press; 1984:207–492. The Eye; vol IB.
60. Straatsma BR, Horwitz J, Takemoto LJ, Lightfoot DO, Ding LL. Clinicobiochemical correlations in aging-related human cataract. *Am J Ophthalmol*. 1984;97:457–469.
61. David LL, Shearer TR. Role of proteolysis in lenses: a review. *Lens Eye Toxic Res*. 1989;6:725–747.
62. Horwitz J, Wong MM. Peptide mapping by limited proteolysis in sodium dodecyl sulfate of the main intrinsic polypeptides isolated from human and bovine lens plasma membranes. *Biochim Biophys Acta*. 1980;622:134–143.
63. Ball LE, Garland DL, Crouch RK, Schey KL. Post-translational modifications of aquaporin 0 (AQP0) in the normal human lens: spatial and temporal occurrence. *Biochemistry*. 2004;43(30):9856–9865.
64. Swamy-Mruthinti S. Glycation decreases calmodulin binding to lens transmembrane protein, MIP. *Biochim Biophys Acta*. 2001;1536(1):64–72.
65. Wagner BJ, Margolis JW, Garland D, Roseman JE. Bovine lens neutral proteinase preferentially hydrolyses oxidatively modified glutamine synthetase. *Exp Eye Res*. 1986;43:1141–1143.
66. Jahngen JH, Lipman RD, Eisenhauer DA, Jahngen Jr EG, Taylor A. Aging and cellular maturation cause changes in ubiquitin-eye lens protein conjugates. *Arch Biochem Biophys*. 1990;276:32–37.
67. Shang F, Gong X, Taylor A. Activity of ubiquitin-dependent pathway in response to oxidative stress. Ubiquitin-activating enzyme is transiently up-regulated. *J Biol Chem*. 1997;272(37):23086–23093.
68. Masters PM, Bada JL, Zigler Jr JS. Aspartic acid racemisation in the human lens during ageing and in cataract formation. *Nature*. 1977;268:71–73.
69. Hanson SR, Hasan A, Smith DL, Smith JB. The major in vivo modifications of the human water-insoluble lens crystallins are disulfide bonds, deamidation, methionine oxidation and backbone cleavage. *Exp Eye Res*. 2000;71(2):195–207.
70. Takata T, Oxford JT, Demeler B, Lampi KJ. Deamidation destabilizes and triggers aggregation of a lens protein, betaA3-crystallin. *Protein Sci*. 2008;17(9):1565–1575.
71. Flaugh SL, Mills IA, King J. Glutamine deamidation destabilizes human gammaD-crystallin and lowers the kinetic barrier to unfolding. *J Biol Chem*. 2006;281(41):30782–30793.
72. Groenen PJ, van den Ijssel PR, Voorter CE, Bloemendal H, de Jong WW. Site-specific racemization in aging alpha A-crystallin. *FEBS Lett*. 1990;269:109–112.
73. Nakamura T, Sakai M, Sadakane Y, et al. Differential rate constants of racemization of aspartyl and asparaginyl residues in human alpha A-crystallin mutants. *Biochim Biophys Acta*. 2008;1784(9):1192–1199.
74. Spector A, Chiesa R, Sredy J, Garner W. cAMP-dependent phosphorylation of bovine lens alpha-crystallin. *Proc Natl Acad Sci USA*. 1985;82:4712–4716.
75. Garlick RL, Mazer JS, Chylack Jr LT, Tung WH, Bunn HF. Nonenzymatic glycation of human lens crystallin. Effect of aging and diabetes mellitus. *J Clin Investig*. 1984;74:1742–1749.
76. Augusteyn RC. Distribution of fluorescence in the human cataractous lens. *Ophthalmic Res*. 1975;7:217–224.
77. Kumar PA, Kumar MS, Reddy GB. Effect of glycation on alpha-crystallin structure and chaperone-like function. *Biochem J*. 2007;408(2):251–258.
78. Harding JJ. Possible causes of the unfolding of proteins in cataract and a new hypothesis to explain the high prevalence of cataract in some countries. In: Regnault F, Hockwin O, Courtois Y, eds. *Aging of the Lens*. Amsterdam: Elsevier; 1980:71–80.
79. Harding JJ, Rixon KC. Carbamylation of lens proteins: a possible factor in cataractogenesis in some tropical countries. *Exp Eye Res*. 1980;31:567–571.

CHAPTER TEN

Lens Development and Crystallin Gene Expression

Ales Cvekl[1], Rebecca McGreal, Wei Liu
Departments of Genetics and Ophthalmology and Visual Sciences, Albert Einstein College of Medicine, Bronx, New York, USA
[1]Corresponding Author: e-mail address: ales.cvekl@einstein.yu.edu

Contents

Abstract

The eye and lens represent excellent models to understand embryonic development at cellular and molecular levels. Initial 3D formation of the eye depends on a reciprocal invagination of the lens placode/optic vesicle to form the eye primordium, i.e., the optic cup partially surrounding the lens vesicle. Subsequently, the anterior part of the lens vesicle gives rise to the lens epithelium, while the posterior cells of the lens vesicle differentiate into highly elongated lens fibers. Lens fiber differentiation involves cytoskeletal rearrangements, cellular elongation, accumulation of crystallin proteins, production of extracellular matrix for the lens capsule, and degradation of organelles. This chapter summarizes recent advances in lens development and provides insights into the regulatory mechanisms and differentiation at the level of chromatin structure and dynamics, the emerging field of noncoding RNAs, and novel strategies to fill the gaps in our understanding of lens development.

Progress in Molecular Biology and Translational Science, Volume 134
ISSN 1877-1173
http://dx.doi.org/10.1016/bs.pmbts.2015.05.001
© 2015 Elsevier Inc.
All rights reserved.

ABBREVIATIONS

3D three-dimensional
BMP bone morphogenetic protein
ECM extracellular matrix
ES embryonic stem
FGF fibroblast growth factor
GRN gene regulatory network
RA retinoic acid

1. INTRODUCTION

Embryonic lens development is a critical step during eye organogenesis, by which the optic cup and lens vesicle are formed through a reciprocal invagination of the lens placode and optic vesicle.[1] Building on this structural foundation, eye morphogenesis proceeds through the formation of the eye anterior segment using the surface ectoderm, migratory neural crest cells, and undifferentiated cells located at the outer margins of the optic cup. These "late" processes of eye morphogenesis employ the lens as a local "tissue organizer,"[2] as these processes are driven by growth factors generated by the lens,[3–5] as well as by other tissues, including the periocular mesenchyme[6] and iris.[7] Thus, it is important to consider that studies of lens development are not only important for understanding its own formation but are also inseparable from studies of the entire eye morphogenesis.

For embryologists, lens formation has represented an advantageous model system to understand the process of embryological induction for over 100 years. In the past 25 years, a number of developmental geneticists and molecular biologists joined the field to seek better understanding of lens development at the cellular and molecular levels. These studies generated a complex yet incomplete picture of cellular and molecular processes that govern lens formation and differentiation. The unique properties of the ocular lens that are required for its function include transparency, light refraction, and elasticity. These properties originate from a unique internal cellular organization of the lens as well as its status as an avascular, noninnervated, and encapsulated tissue. At the cellular level, the lens comprises a sheet of cuboidal anterior epithelium connected with a mass of highly elongated lens fibers. A hallmark property of these fibers is expression and accumulation of high levels of a relatively small group of water-soluble proteins called lens crystallins. The crystallins are essential for lens transparency and refractive power, and multiple recent studies have shown that they play additional "nonrefractive" roles such as negative regulators of cell death and modulators of autophagy and tissue remodeling. This chapter describes the basic principles of vertebrate lens development in the mouse model with a particular focus on crystallin gene function and expression.

2. EARLY LENS DEVELOPMENT

2.1 Embryonic Lens Development: Overview

The initial morphological process indicating lens formation is a thickening of the head surface ectoderm to generate lens placodes (Fig. 1), which are comprised of groups of cells with palisade-like morphology. Through a series of morphological movements including the lens placode and the underlying optic vesicle, a three-dimensional (3D) lens primordium is generated in the form of the lens vesicle. Its posterior cells exit the cell cycle and undergo terminal differentiation to an elongated morphology, filling up the lumen of the lens vesicle. In contrast, anterior cells of the lens vesicle differentiate into the lens epithelium while retaining their proliferative ability. Cell division within the lens epithelium generates new cells, and this process enables cells located at the lens equator to differentiate into secondary lens fibers. These fibers form outer shells of cells, and this process of very slow lens growth continues throughout its lifespan. Taken together, lens anatomy and morphology provide a unique system in which individual cell location coupled with its shape (i.e., state of elongation) genuinely reflects cell differentiation status and a cross-section through the lens simultaneously displays cells of the same embryological/progenitor cell origin at different stages of their differentiation.[1]

From a historical perspective, amphibian lens formation has been established as a leading experimental model to study the inductive processes during embryogenesis that specify novel cell types for over 100 years, starting with the pioneering studies of Spemman and Mencl published between 1901 and 1912.[8,9] In subsequent years, this experimental system was further employed by a number of prominent scientists, including Lewis,[10] Waddington,[11] Coulombre and Coulombre,[12,13] and Jacobson,[14–16] and reexamined with a set of new molecular tools, not available for earlier studies, by Grainger.[17] The phenomenon of "embryonic induction" involves specification of a new cell type that requires directed exchange of signals between the "inducing" and "induced" cells. An important intrinsic factor is the "competence" of individual cells to "read" inductive signals, in other words, to which cell fates the progenitor cells were preprogrammed.

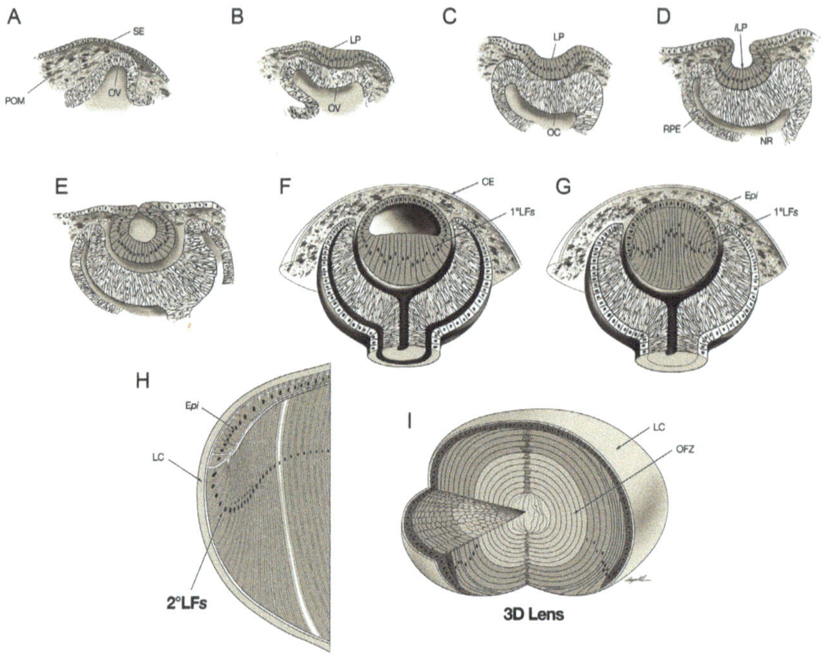

Figure 1 Embryonic lens development (mouse). (A) E9.0 (prospective lens ectoderm–surface ectoderm). (B) E9.5 (lens placode). (C) E10 (lens placode at the beginning of invagination). (D) E10.5 (transition from invaginating lens placode to lens pit). (E) E11 (open lens vesicle). (F) E12.5 (onset of primary lens fiber cell differentiation). (G) E13.5–E14.5 (completion of primary lens fiber cell elongation–secondary lens fiber cell formation). (H) Secondary lens fiber cell differentiation as a driving force of continuous lens growth. Apical–basal polarity of lens epithelial and lens fiber cells is indicated. Anterior lens epithelium, ALE; corneal epithelium, CE; invaginating lens placode, iLP; invaginating lens placode/pit, iLP/p; lens capsule, LC; lens epithelium, Epi; lens placode, LP; neuroretina, NR; optic vesicle, OV; periocular mesenchyme, POM; primary lens fibers, 1° LFs; prospective lens ectoderm, PLE; retinal pigmented epithelium, RPE; secondary lens fibers, 2° LFs; surface ectoderm, SE. *From Cvekl and Ashery-Padan.*[1] (See the color plate.)

Molecular cell biology studies conducted during the past 20 years have revealed that lens inductive signals are represented by specific growth factors produced by "inducing" cells and their receptors, which are localized on the cell surface of the "induced" cell. The cell-fate decisions are then executed by specific transcriptional machinery localized within the cell nuclei of the "induced" cells. The "intrinsic" factors include specific combinations of lineage-specific DNA-binding transcription factors, which control expression of additional regulatory as well as structural genes specific for the tissue of interest. Thus, understanding lens development requires identification of a minimal set of "core" lens regulatory transcription factors and their connectivity with extracellular signaling.

Lens development and differentiation have been popular subjects of many detailed reviews. In this paragraph, we provide a comprehensive list of references organized into individual groups. For general reviews on lens development, see Refs. 1,18–28; for general reviews on eye development, see Refs. 29–41; for reviews related to early stages of lens development culminating with lens placode formation, see Refs. 27,42–47; and for reviews on extracellular signaling and lens development, see Refs. 22,24,48–52. The following reviews discuss in detail several aspects of lens development and differentiation, including anatomy and lens morphology,[53,54] biochemistry and biology of crystallins,[55–57] cell cycle exit,[58,59] connexins and other lens membrane proteins,[60–62] degradation of subcellular organelles in lens fibers,[63,64] biology of the lens extracellular matrix (ECM)/lens capsule,[65] biology of lens mitochondria,[66] mouse[67–69] and human lens models and cataract,[70–73] and lens transdifferentiation and regeneration.[74–76] Through the evo-devo approach in conjunction with comparative genomic studies, lens development also represents an exciting model to understand cellular and molecular principles of eye evolution.[77–83]

Lens morphogenesis can be divided into at least four specific phases, each characterized by the use of novel molecular tools, and the redeployment of molecular tools that were used at an earlier stage of lens development, as well as the use of modified molecular pathways employed by other cells ranging from neurons to erythrocytes. These four phases include (1) lens placode formation; (2) invagination of the lens placode to form the lens vesicle; (3) cell cycle exit, primary lens fiber cell differentiation, and lens epithelium differentiation; and (4) lens growth and secondary lens fiber cell formation (Fig. 1). Given the large body of work, namely through genetic studies including loss-of- and gain-of-function experiments, mouse lens development is presented here as a leading experimental model. Nevertheless, some critical insights into lens formation were established using other model vertebrates, namely frog, chick, and zebrafish. The main differences between these models include lens fiber cell differentiation through delamination of the lens placode in zebrafish,[28] and variable expression domains of DNA-binding transcription factors and growth factors within all model organisms examined.[47,84,85] The four phases of lens formation pertinent to the mouse model are discussed in subsequent sections.

2.2 Lens Placode Formation

A series of embryological processes following neural plate formation culminate with the formation of the neural tube and lens placode that is considered as the initial, i.e., morphologically observable, phase of lens formation. Our knowledge about the "preplacodal" stages of lens development is based on a cohesive set of relatively recent embryological studies using chick, frog, and zebrafish models. These studies approached lens placode formation as one of many parallel embryonic processes that result in formation of distinct neurogenic (e.g., olfactory, otic, trigeminal, and epibranchial placode) and nonneurogenic (adenohypophyseal and lens placode) ectodermal sensory placodes. The current model postulates that a border region between the neuroectoderm and nonneural ectoderm (i.e., prospective epidermis) gives rise in the anterior part to the anterior preplacodal ectoderm, while in the more posterior part of the border a distinct population of neural crest cells are being formed.[46] The first experimental evidence of the preplacodal region was obtained through embryological manipulations involving rotation of a band of ectoderm from the anterior region of the frog neurula, followed by morphological analysis of individual placode formation.[86] A series of studies conducted in the last decade revealed that the anterior preplacodal ectoderm is a mixed population of cells that are specified at a single-cell level within a field of cells, not from a discrete region of the surface ectoderm implicated by "classical" embryological studies.[19,42,87] According to the current model, the most anterior adenohypophyseal, olfactory, and lens placodes are formed from a common progenitor, the anterior preplacodal cell. Following their cell-fate specification, the individual cells migrate and ultimately converge along the anterio-posterior axis as cell-fate determined groups of cells that give rise to the specific placodes.

At the molecular level, signaling processes that promote lens placode formation combined with a group of DNA-binding transcription factors which are expressed in the anterior preplacodal ectoderm as well as in the prospective lens ectoderm and lens placode have been identified.[27] Genetic studies in mice have assigned the most prominent roles for the lens placode formation to bone morphogenetic protein (BMP) and retinoic acid (RA) signaling, in conjunction with activities of two homeodomain-containing proteins, Pax6 and Six3 (Fig. 2). In addition, expression of Sox2, an HMG group member and "core" embryonic stem (ES) cell circuitry regulatory protein, is established within the Pax6/Six3 expression domain; however, Sox2 plays roles following lens placode formation. Similarly, expression of transcription factor AP-2α commences early in the surface ectoderm, though its role in mouse lens morphogenesis is also manifested later.

The optic vesicle plays at least two major roles in lens placode formation. The first role is "instructive," as the optic vesicle generates BMP4 and depletion of BMP4 in *Bmp4*$^{-/-}$ mouse embryos is not compatible with lens placode formation, although expression of Pax6 is maintained in the surface

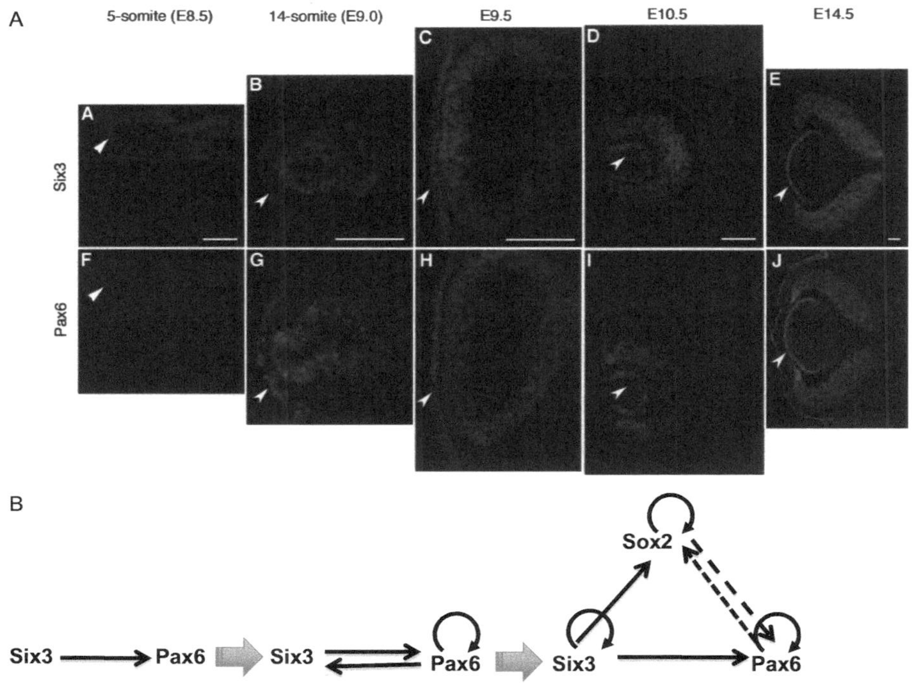

Figure 2 Expression of Pax6 and Six3 during early stages of lens formation and regulatory mechanisms involving Pax6, Six3, and Sox2. (A) Six3 is expressed first at the surface ectoderm (E8.5), followed by Pax6 (∼E8.75). (B) Onset of Pax6 expression in the presumptive lens ectoderm requires Six3. At E9.0, Pax6 and Six3 cross-regulate each other and autoregulation of Pax6 is being established. At E9.5, expression of Sox2 in the invaginating lens placode is dependent on Six3 and transitionally also on Pax6. *Images from Liu* et al.[88] (See the color plate.)

ectoderm.[89] Expression of BMP4 is controlled by LIM domain and homeodomain factor Lhx2, expressed in the optic vesicle but not in the presumptive lens ectoderm.[90] In contrast to mouse *Bmp4*$^{-/-}$ embryos, *Lhx2*$^{-/-}$ embryos lose expression of Pax6 in the prospective lens ectoderm.[91] The role of BMP7 in lens placode formation remains to be clarified as lens defects are dependent on genetic background of the mice used.[92–94] Consistent with studies of BMP4, conditional inactivation of type I BMP receptors expressed in the prospective lens ectoderm, *Bmpr1a* and *Acvr1*, disrupts lens placode formation and further shows that BMP receptor signaling also regulates both cell survival and proliferation during lens placode formation.[95]

Another component of the lens-inducing system is RA signaling.[48,49] RA is generated by multiple enzymes at different places during mouse eye development. At the beginning of eye morphogenesis, RA is generated by Raldh2/Aldh1a2 enzymes within the temporal periocular mesenchyme that is in contact with the optic vesicle (Fig. 1). Consequently, RA signaling becomes active in the optic vesicle at E8.5, but not in the surface ectoderm. In the surface ectoderm, RA signaling activity commences at E8.75 and continues throughout lens placode formation and invagination. In *Raldh2*$^{-/-}$ and compound *Raldh1*$^{-/-}$;*Raldh2*$^{-/-}$ murine embryos, the prospective lens ectoderm and optic cup fail to invaginate to form the lens vesicle and optic cup.[96] As no lens placode-specific markers were analyzed in these embryos, the transition from the prospective lens ectoderm to lens placode remains to be characterized. Taken together, paracrine RA signaling is required for the reciprocal invagination of the lens pit/optic cup.[96,97]

2.3 Invagination of the Lens Placode to Form the Lens Vesicle

A reciprocal invagination between the lens placode and the underlying optic vesicle gives rise to the lens vesicle and optic cup (Fig. 1). This process involves multiple cellular components, including cytoskeletal reorganization, change of cell shape from cylindrical to conical by apical constriction, filopodia formation which originate in the invaginating lens cells and make contact with the basal lamina of the optic vesicle neuroepithelium, as well as the ECM between the lens ectoderm and the optic vesicle.[98–101] A transitional structure, the lens pit, is formed next, and the lens vesicle is formed by pinching off from the surface ectoderm, followed by its restoration. Cell adhesion molecules N- and E-cadherin, as well as programmed cell death, play a major role in the formation of the lens vesicle.[102,103] Thus, within approximately 48 h of embryonic development, the sheet of columnar epithelium is converted into a truly 3D structure, the lens vesicle.

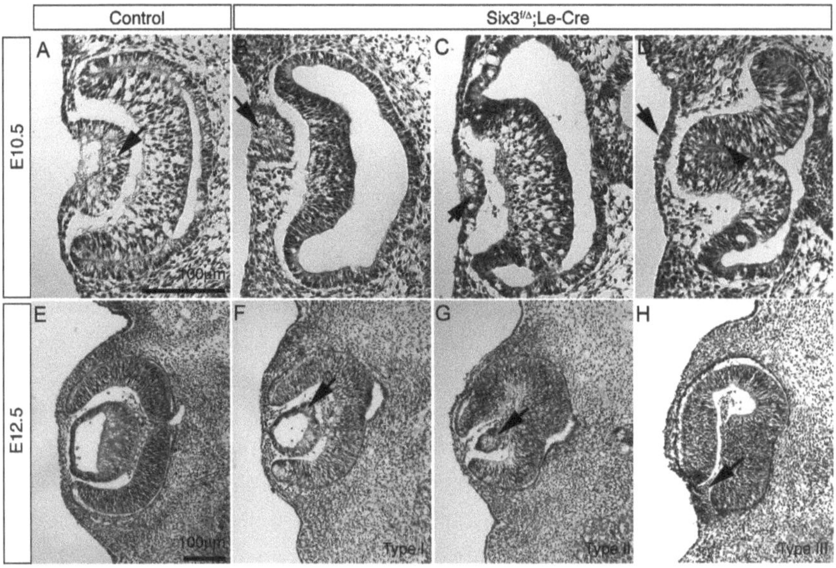

Figure 3 Depletion of Six3 in the presumptive lens ectoderm disrupts lens morphogenesis. Conditional inactivation of Six3 with Le-Cre causes a range of lens defects (A–H); in the most severe case the lens does not form (D, H).[88] (See the color plate.)

At the molecular level, the invaginating lens placode continues expression of Pax6, Six3, and Sox2, while expression of c-Maf, Foxe3, Pitx3, and Prox1 is initiated as the lens placode increases its size and changes its shape (Fig. 1). As illustrated in subsequent sections on cell cycle exit and crystallin gene expression, Pitx3 and Prox1 play roles in cell cycle exit of the posterior cells of the lens vesicle. In addition, c-Maf, Pitx3, and Prox1 subsequently play major roles in crystallin gene expression. Inactivation of Six3,[88] as well as of evolutionarily conserved Mab21l1 protein[104] implicated in BMP signaling,[105] resulted in dramatically increased apoptosis within the lens placode (Fig. 3). Collectively, lens placode invagination is an excellent model for understanding the cellular basis of lens 3D organization, as well as for cellular and molecular understanding of the morphogenetic movements of sheets of epithelial cells during embryonic development.

3. LENS DIFFERENTIATION

3.1 Cell Cycle Exit, Primary Lens Fiber Cell Differentiation, and Lens Epithelium Differentiation

The lens vesicle is a 3D primordium of the fully formed lens. It has to be precisely positioned within the eyeball to enable proper formation of the optical axis. The lens vesicle is located within the optic cup and behind the nascent anterior segment. Within this arrangement, different parts of the lens vesicle are exposed to different combinations of growth factors that polarize its structure into the anterior or posterior cell compartments. Under the influence of BMPs and fibroblast growth factors (FGFs) generated by the neuroretina, the cells forming the posterior part of the lens vesicle exit the cell cycle and undergo their terminal differentiation to generate the fiber cell lens compartment, comprising only from postmitotic cells. In contrast, the anterior cells of the lens vesicle differentiate into the lens epithelium, which is comprised of a mixture of cells characterized by different proliferative indexes.[106] Most recent studies identified a population of Sox2-expressing cells ($Sox2^{+}$) within the lens epithelium that exhibit properties of adult stem-like cells.[107] Although cell division within the epithelium is a relatively rare event, generation of novel cells has a major impact on lens growth throughout its entire lifespan. The epithelial cells localized in the equatorial region of the lens differentiate into newly formed shells of fiber cells and lens grows in size throughout its entire lifespan.

A flawless execution of cell cycle exit and lens fiber terminal differentiation is a prerequisite for lens transparency and refraction.[27,54] The current model of cell cycle exit in the lens vesicle postulates that extracellular signals (BMPs and FGFs) result in activation of expression and/or posttranslational modifications of a specific group of DNA-binding transcription factors in the posterior part of the lens vesicle. More recent studies have shown a contribution of Notch signaling in the control of lens cell cycle exit.[108,109] The integrative function of these factors converges into a spatiotemporal control of the expression of proteins that inhibit cell cycle progression, the cyclin kinase inhibitors. A large body of genetic studies have identified key regulatory roles for canonical cell cycle regulatory proteins, including pRb, E2Fs, p53, Cdk2, Cdk4, Cdkn1b/$p27^{Kip1}$, and Cdkn1c/$p57^{Kip2}$.[58,59] The lack of Cdk6 expression in lens may be due to the repressor role of Pax6 as shown in mouse cortical progenitor cells.[110] Notable differences exist between expression of E2Fs, pocket, and other proteins between lens fibers and lens epithelium (see Table 1) to assure tight control of the balance between proliferation and differentiation.[58] A range of lens cellular and morphological defects are linked to disrupted regulation of primary lens fiber formation from the posterior cells of the lens vesicle. These include proliferation/limited differentiation of these cells instead of their terminal differentiation that can be accompanied by apoptosis in the presumptive lens fiber cell compartment,[121,129] disrupted lens polarity (i.e., anterior or posterior shifts of the lens equator), which reduces either the epithelial or fiber cell

Table 1 Common and Different Expression Domains of Cell Cycle Regulatory Proteins in Mouse Lens

Proteins	Epithelium	Fibers	References
Cyclins	Cyclin A, B, D, E (low)	Cyclin B, D	111–115
Cyclin-dependent kinases (Cdks)	Cdk2, Cdk4, Cdc2	Cdk4, Cdc2	112,113,116
Cyclin kinase inhibitors (CKIs)	Cdkn2a/p16-INK4a/ARF, Cdkn1c/p57^{Kip2} (low)	Cdkn1b/p27^{Kip1}, Cdkn1c/p57^{Kip2}	108,114, 117–120
Pocket proteins	pRb, p107, p130	pRb, p107	121–123
E2Fs	E2F1, E2F2, E2F3, E2F4, E2F5	E2F1, E2F3, E2F5	123–128

compartment concomitant with expansion of the other compartment,[109,130,131] and depletion of lens epithelium due to premature differentiation of all cells into abnormal lens fibers.[132]

Studies of chicken embryonic explants have focused on understanding upstream BMP and FGF signaling during primary lens differentiation. A balance of BMP and FGF signals, evaluated through the use of specific agonists of these pathways (noggin to inhibit BMP signaling and a small drug SU5402 to inhibit function of FGF receptors), is critical for proper cell cycle exit.[133] BMP activity promotes FGF-dependent cell cycle exit, while FGF signaling alone is not sufficient to elicit cell cycle exit.[133] In addition, multiple studies revealed that expression of tumor suppressor protein Ccdc80/equarin in lens transitional zone is a novel marker for this stage of lens differentiation in a chick model.[133–137] FGF signaling in primary lens fiber cell differentiation was dissected in the mouse using multiple genetic approaches, including somatic and conditional knockouts and transgenic models. The principal conclusion from these studies is that mouse embryos that lack three FGF receptor genes, Fgfr1, Fgfr2, and Fgfr3, generate abnormal lens vesicles that do not produce any elongated lens fibers.[138] At the molecular level, expression of c-Maf, Prox1, Etv1/ER81, and Etv5/ERM is strongly reduced in compound FGF receptor mutants[138] as well as in other mouse models, including studies on the *Ndst1*,[139,140] *Frs2α*,[141,142] and *Mapk1/Erk2* genes.[143] Finally, disrupted Notch signaling impairs primary lens fiber cell differentiation, as indicated by an abnormal lens transitional zone, shifted anteriorly in conditionally inactivated *Rbpj*$^{-/-}$ E11.5–E12.5 mouse embryos[108,109]; however, potential cross talk of Notch with BMP and FGF signaling pathways in the lens remains to be established.

Somatic and conditional inactivation of multiple DNA-binding transcription factors, including Gata3,[117] Pax6,[103] Pitx3,[144] and Prox1,[118] results in disrupted cell cycle exit during primary lens differentiation. Despite more than 20 years of studies of cell cycle coupled lens fiber cell differentiation, a number of important questions remain unanswered. For example, the link between BMP and FGF signaling and the functions of Gata3, Pax6, Pitx3, Prox1, and other possible regulators of cell cycle exit of lens precursor cells remain to be identified. The current data show binding of Notch signaling regulated DNA-binding transcription factor Hey1/Herp2 to the Cdkn1c/p57^{Kip2} promoter.[108] Our ongoing studies are aimed to establish binding of general signal-regulated transcription factors (e.g., c-Jun, Etv5/ERM, and Smad1/5/8), in conjunction with lens-preferred local activators (e.g., Gata3 and Prox1), to both the Cdkn1b/p27^{Kip1} and Cdkn1c/p57^{Kip2} promoters and their distal enhancers in lens chromatin.

The anterior cells of the lens vesicle differentiate into the lens epithelium (Fig. 1),[145] a monolayer of lens epithelial cells overlying the apical tips of elongated lens fibers.[146] Lens epithelial integrity is mediated by E- and N-cadherins and β-catenin (gene names: Cdh1, Cdh2, and Ctnnb1, respectively).[102,147] Depletion of transcription factor AP-2α following formation of the lens vesicle generates a multilayered abnormal lens epithelium.[148] Interestingly, our knowledge about differential gene and protein expression between lens epithelium and lens fibers is still in its infancy. Recently, RNA-seq studies of microdissected embryonic chick lenses revealed the first molecular signatures of distinct lens transcriptomes at a quantitative level.[149,150] Upregulation of batteries of genes that function in lens differentiation and organelle clearance has been noted. Follow-up studies using other model organisms are needed to fill this important gap in our knowledge of distinct lens compartments.

The lens capsule, an acellular structure formed by ECM proteins such as collagen IV, laminin, nidogen, and perlecan,[65] is an important environment that plays multiple roles in extracellular signaling and lens differentiation. It has been shown that the lens capsule contains FGF2 and that it is released by matrix metalloproteinase 2 to facilitate lens epithelial cell viability.[151,152] Heparan sulfate proteoglycans are also present in the lens capsule and one of their functions is to bind FGFs.[153] Heparan sulfate biogenesis in the lens is controlled by the enzyme Ndst1[139] working in conjunction with FGF signaling. Another secreted proteoglycan protein in the lens capsule is Ccdc80/equarin (see above), a syndecan 3 binding partner.[136] Given the role of Pax6 in regulation of expression of fibronectin 1 (Fn1), versican (Vcan/Cspg2), and collagen Col3a1,[99,154] it is not surprising that human aniridic lenses, carrying mutations in *PAX6*, have compromised capsules that negatively impact surgical removal of the abnormal lenses.[155]

3.2 Lens Growth and Secondary Lens Fiber Cell Formation

Lens growth is a consequence of cell proliferation within the epithelium. Cell divisions in the epithelium impact alignment of cells at the lens equator. The cells that pass the equator exit the cell cycle and differentiate as secondary lens fibers. Thus, lens growth consists of three tightly linked processes: terminal differentiation of primary lens fibers to form the initial lens structure, proliferation within the lens epithelium to generate cells from which the secondary lens fiber are formed, and secondary lens fiber cell terminal differentiation to add new outer cell layers to the lens. Consequently, cross-sectional analysis of the lens reveals cells at different stages of their differentiation and their status can be inferred from their location within the individual lens compartment. Lens growth and secondary lens fiber cell differentiation have been reviewed in detail elsewhere.[1,18,22,24,48,50–52,63] Here, we focus on the concept that signal transduction pathways employed during the early stages of lens development are reused (BMP, FGF, Notch, and RA signaling), while additional signaling is mediated by cadherins, integrins, insulin-like growth factor receptor-1 (IGF-1R)/NF-κB, phosphatidylinositol 3-kinase (PI3K), the mitochondrial cell death pathway, and Wnts, which participate in both primary and secondary lens fiber cell differentiation. The evidence that secondary lens fiber cell differentiation is regulated by a combination of BMP and FGF signaling is supported by studies in chick system,[156] coupled with recent studies showing that BMP signaling is activated in FGF2-treated rat lens explants.[157] Earlier studies have shown that bFGF/FGF2 at concentrations 40–50 ng/ml induces many aspects of lens fiber cell differentiation including cellular elongation and accumulation of crystallins.[22,52] Disrupted lens fiber cell differentiation was found in RXRα$^{-/-}$;RARγ$^{-/-}$ compound mutant embryonic lenses,[158] suggesting an additional role of RA signaling in lens development, probably via αB-[159] and γ-crystallin gene expression.[160–162] Attenuated Notch signaling also impairs secondary lens fiber cell differentiation as evidenced by the presence of a disrupted lens transitional zone in conditionally inactivated *Rbpj*$^{-/-}$ E18.5 mouse embryos.[109] Taken together, these results suggest that secondary lens fiber cell differentiation also employs BMP, FGF, Notch, and RA signaling; however, the individual molecular components of each pathway during primary and secondary lens fiber cell differentiation may differ.

Proper lens morphogenesis requires participation of additional signal transduction pathways, including Wnt/planar cell polarity (Wnt/PCP) signaling,[163] cadherins,[164] integrins,[165–167] IGF-1R/NF-κB,[167] PI3K,[168] mitochondrial cell death pathway,[167,168] and MAPK/JNK-mTORC1 signaling.[169] The employment of these pathways reflects specific needs during lens fiber cell elongation, including inhibition of apoptosis by the IGF-1R/NF-κB pathway[170] and β1-integrin signaling[165]; formation of cilium at the apical surface of the lens fiber cell (see below); regulation of lens cell cytoskeleton reorganization by cadherins, integrins, Rho-GTP proteins, and tropomodulin[164,167,171–174]; establishment of lens-specific beaded filaments comprised of filensin and Bfsp2[175,176]; and degradation of subcellular organelles,[63] such the endoplasmic reticulum, Golgi apparatus, mitochondria, and nuclei by autophagy and its related processes.[149,169,177] A group of recent studies demonstrated a key role for the Wnt/PCP pathway in directing lens fiber morphology[51] as each lens fiber cell has a primary cilium on its hexagonal apical surface. The cilia are polarized toward the side that faces the anterior pole. Depletion of multiple protein components of PCP pathway (e.g., *Rac1*, *Vangl2*, and *Celsr1*) disrupts cilium orientation and lens fiber morphology.[51,163] Collectively, studies of lens fiber cell differentiation through identification and molecular dissection of specific components represent a fertile soil for complete understanding of the internal logic of this critical process in establishing a fully functional 3D lens.

3.3 Crystallin Gene Expression and Its "Core" Gene Regulatory Network in the Lens

Crystallins are the most prevalent proteins in the lens. Comprising two families, α- and βγ-crystallins, they make up 90% of water-soluble proteins of the mammalian lens.[178] They are highly organized providing a refractive index gradient, which allows for transparency of the lens. Once thought to be solely lens proteins, crystallins have since been identified outside of the lens, providing several important functions (see Table 2).

Multiple crystallins are expressed as early as in the nascent lens placode and during its invagination, while their expression levels are typically increased by a factor of one to three orders of magnitude in differentiating lens fibers. During lens fiber cell differentiation, expression of all crystallins is upregulated at the transcriptional level, while their expression in lens epithelium is generally lower compared to the fiber cells, as shown in chick

Table 2 Location and Known Functions of Mouse Crystallins

Crystallin	Location	Function	References
αA (Cryaa)	Eye: lens, retina, cornea Other tissues: brain, liver, lung, spleen, skin, small intestine	Apoptosis regulator, autokinase, chaperone differentiation, structural protein	179–191
αB (Cryab)	Eye: lens, retina Other tissues: brain, heart, liver, kidney, peripheral nerves, placenta, skeletal muscle, skin, spinal cord	Angiogenesis, apoptosis regulator, autokinase, chaperone, chaperone-mediated autophagy (CMA), heat-shock protein, lens structural protein	179,182,184,187, 189,192–202
βB1 (Crybb1)	Eye: lens	Lens structural protein	203,204
βB2 (Crybb2)	Eye: lens, retina Other tissues: brain, testes	Lens structural protein	205–207
βB3 (Crybb3)	Eye: lens, retina	Lens structural protein	208,209
βA1/A3 (Cryba1)	Eye: lens, retina, cornea	Autophagy, lens structural protein	191,207,210–212
βA2 (Cryba2)	Eye: lens, retina	Lens structural protein	207,213
βA4 (Cryba4)	Eye: lens, retina	Lens structural protein	207,214
γA (Cryga)	Eye: lens, retina	Lens structural protein	68,208,215
γB (Crugb)	Eye: lens, retina	Lens structural protein	215,216
γC (Crygc)	Eye: lens, retina, cornea	Lens structural protein	191,217,218
γD (Crygd)	Eye: lens, retina	Lens structural protein	216,219,220
γE (Cryge)	Eye: lens, retina	Lens structural protein	68,218
γF (Crygf)	Eye: lens, retina, cornea	Lens structural protein	68,188,191
γS (Crygs)	Eye: lens, retina, cornea	Lens structural protein	191,216,221

model.[149] The precise spatial and temporal expression of individual crystallins in the lens remains to be fully established. Complete data at the mRNA and protein levels are only available for mouse αA- and αB-crystallins[222–224] and rat βA3/A1-crystallin.[225] αB-crystallin is found in the E9.5 lens placode and its expression continues throughout lens morphogenesis in both lens epithelium and fibers.[222,223] Expression of mouse αA-crystallin commences later in the posterior part of the invaginating lens pit and is also expressed in both compartments of the lens.[223,224] Expression of rat βA3/A1 also commences in the posterior part of the invaginating lens vesicle and later in the differentiating lens fibers.[225]

Elucidation of the molecular mechanisms that account for lens-specific/lens-preferred crystallin gene expression began with the identification of the mouse αA-crystallin promoter,[226] which was actually the first mammalian promoter identified with a remarkable tissue specificity. A number of studies conducted around the world have led to the identification of individual *cis*-acting promoter and enhancer elements followed by identification of a relatively small group of *trans*-acting DNA-binding transcription factors (Table 3) that regulate one or more crystallin genes. Crystallin gene regulation is mediated by two "core" factors, c-Maf and Pax6, in combination with Hsf4, Prox1, RARβ/RXRβ, Six3, and Sox1 (Fig. 4). Multiple combinations of these factors, such as Pax6/c-Maf, Pax6/Sox2, c-Maf/Sox1, and Pax6/RARβ/RXRβ, generate specific synergistic effects in crystallin gene regulation that explain their high levels of expression in the lens.[162,227,240,241] In addition, regulation of c-Maf by p53 through binding to its promoter has been established[242] and it has been shown that p53 also directly binds distal and proximal regulatory regions of αA- and βA3/A1-crystallins.[232] In addition, c-Maf, Pax6, and Sox1/2 proteins recruit CBP and p300 histone acetyltransferases and regulate crystallin gene expression.[228] Several signal-regulated transcription factors were also implicated in crystallin gene expression, including ArH, CREB, c-Jun, Etv5/ERM, LEF/TCF, RARβ/RXRβ, and Sp1 (see Table 4). These signal-regulated transcription factors include canonical targets of extracellular signaling, such as RA-activated nuclear receptors RARβ/RXRβ and Wnt-regulated LEF/TCF DNA-binding factors.

The involvement of Maf proteins in eye development requires additional comments. Mafs are bZIP proteins that form homo- and heterodimers with themselves or with other bZIP proteins such as members of AP-1 and ATF/CREB families of transcription factors[251] (Table 3). Mafs are divided into two classes: large (MafA, MafB, c-Maf, and NRL) and small Mafs (MafF, MafG, and MafK). Interestingly, the expression domains of Maf genes differ between vertebrate model organisms examined (chicken, mouse, frog, and zebrafish) and this phenomenon was called "expression domain shuffling."[84] In mouse, the roles of individual Mafs in eye development have been rigorously examined using gene targeting. Lens fiber cell

Table 3 DNA-Binding Transcription Factors That Regulate Crystallin Gene Expression

Name	Expression Domain	Structure and Function	References
c-Maf	iLP, LV, upregulated in 1° and 2° LFs Other tissues: bone, blood, ear, kidney, skin, etc.	bZIP as homo- or heterodimer Activates α- and β/γ-crystallin gene expression	162,227–230
Gata3	Posterior part of LV, upregulated in 1° and 2° LFs Other tissues: blood, breast, ear, skin, etc.	ZnF monomer Regulates γ-crystallins but direct or indirect mechanism unknown	117
Hsf4	Upregulated in 1° and 2° LFs Other tissues: liver	HTH, trimer or more units Activates γ-crystallin gene expression	231
p53	Ubiquitous expression	Tetramer with a palindrome Regulates Cryaa, Cryba1	232
Pax6	PLE, iLP, LV, 1° and 2° LFs, ALE Other tissues: retina, cornea, iris, lacrimal gland, parts of forebrain and cerebellum, pancreas, adenohypophyseal, and olfactory placodes	PD and HD monomer Activates α- and β/γ-crystallin gene expression	161,162, 233,234
Pitx3	iLP, LV, 1° and 2° LFs Other tissues: parts of forebrain	HD monomer Activates α- and β/γ-crystallin gene expression	144,235
Prox1	iLP, LV, upregulated in 1° and 2° LFs Other tissues: retina, liver, vascular system	HD monomer Activates Crygb, Crygd, and Crybb1 gene expression	118,236,237
Six3	iLP, LV, ALE Other tissues: anterior forebrain, retina	HD monomer Inhibits γ-crystallin gene expression	162,238
Sox1	Upregulated in 1° and 2° LFs Other tissues: neuroectoderm	HMG-box monomer Activates Crygb, Crygd, Cryge, and Crygf	227,239
Sox2	PLE, iLP, LV, lens stem-like cells within ALE Other tissues: ES cells, neuroectoderm, retina	HMG-box monomer Activates chicken δ1-crystallin (Asl1/Crygd1) gene expression	240

Prospective lens ectoderm, PLE; lens placode, LP; invaginating lens placode, iLP; lens vesicle, LV; primary lens fibers, 1° LFs; secondary lens fibers, 2° LFs; anterior lens epithelium, ALE; basic leucine zipper, bZIP; homedomain, HD; helix–turn–helix domain, HTH; paired domain, PD; high-mobility group, HMG box; Zn-finger, ZnF.

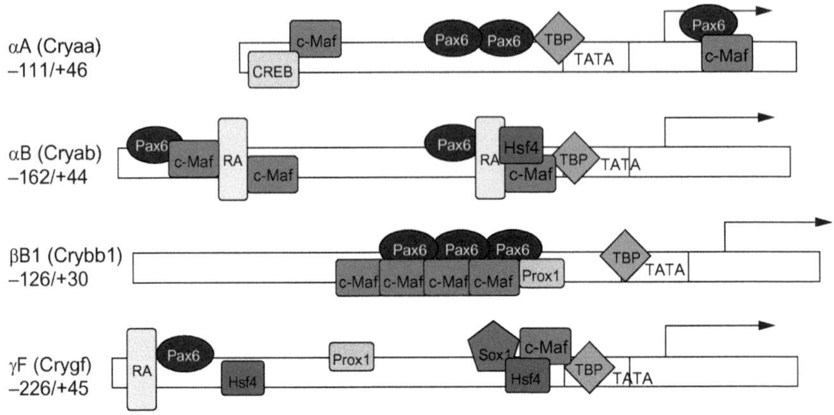

Figure 4 Interaction of DNA-binding transcription factors with a representative panel of vertebrate crystallin promoters and enhancers. Four crystallin promoters are shown: mouse αA-, mouse αB-, chicken βB1-, and mouse γF-crystallins. Transcription factors c-Maf, CREB, Hsf4, Pax6, Prox1, RA-responsive nuclear factors (e.g., RARβ/RXRβ), Sox1, and TBP are indicated by different shapes and colors. (See the color plate.)

differentiation is severely abrogated in *c-Maf*-null embryos.[229,252,253] In contrast, $MafA^{-/-}$, $MafB^{-/-}$, and $MafA^{-/-}$;$MafB^{-/-}$ lenses and eyes appear normal.[254] Inactivation of NRL, expressed in the retina and lens, results in transformation of rods into S cones, while lens formation was not disrupted.[255] In the chick model, MafA/L-Maf has been widely used as a molecular marker for analysis of various steps of lens formation[133,256,257] and was implicated as a cell-fate regulatory protein and direct regulator of αA-crystallin gene expression.[258] However, to date, no conclusive data exist to show that MafA/L-Maf and/or c-Maf directly regulate αA-crystallin gene expression in chick lens chromatin. Taken together, the large number of homo- and heterodimers can be formed between individual Maf, AP-1, and ATF/CREB family members,[259] coupled with identification of human and mouse mutations in c-Maf/MAF proteins causing cataracts,[260,261] raises the possibility that novel crystallin gene regulatory mechanisms mediated by these factors as well as novel cataract-causing mutations may be identified within these gene families.[262]

Control of c-Maf expression by Pax6[233] and reciprocal regulation of Six3 and Pax6[88,263] have been established in the early embryonic lens. While c-Maf acts as an activator of many, if not all, crystallin genes in lens fibers,[229,252,253] expression of Six3 in cultured lens epithelial cells attenuates expression of γ-crystallins.[227,238] Together, Pax6, c-Maf, and crystallins are organized as a "core" gene regulatory network (GRN) that is assisted by

Table 4 Signal-Regulated DNA-Binding Transcription Factors Implicated in Crystallin Gene Regulation

	Lens/Eye Expression Domain	Structure[a] and Function	References
ArH	TBD	bHLH/PAS Maybe regulated by cAMP, mTOR, and Wnt/β-catenin signaling	243
CREB	LV, 1° LFs, other stages: TBD	bZIP, homo- and heterodimers cAMP-regulated phosphorylation at Ser133	234
c-Jun	Posterior part of LV, upregulated in 1° and 2° LFs	bZIP, homo- and heterodimers regulated by FGF/MAPK signaling; activates αA-crystallin	244–246
Etv5/ERM	iLP, LV, 1° and 2° LFs, ALE	Winged HTH Maybe regulated by FGF/MAPK signaling at the level of phosphorylation; activates αA-crystallin	244
LEF/TCF		HMG-box Wnt/β-catenin signaling and regulation of γ-crystallins	247,248
RAR/RXR	RARβ/RXRβ are in whole embryonic lens	ZnF Retinoic acid signaling; activates αB- and γ-crystallins	159–162, 227,249
Sp1	TBD	ZnF Regulated by sumoylation; targets β-crystallins	250

[a]Basic helix–loop–helix/Per-Arnt-Sim, bHLH/PAS; to be determined, TBD; for other abbreviations, see Table 1.

Hsf4, Prox1, Sox1, and other DNA-binding transcription factors (Fig. 5). Importantly, evidence exists that expression of c-Maf in the lens is directly regulated by FGF signaling via AP-1 (c-Jun) and Ets (Etv5/ERM)-regulated transcription factors (see Table 3) that recognize the −272/−70 promoter region of c-Maf.[244] In these studies, it has also been shown that the mouse αA-crystallin promoter and its 5′-distal enhancer (DCR1) bind c-Jun and Etv5/ERM (Fig. 5). FGF2 strongly activates expression of the αA-crystallin gene promoter coupled with the DCR1 enhancer compared to the promoter alone.[234] Interestingly, FGF signaling also regulates expression of a

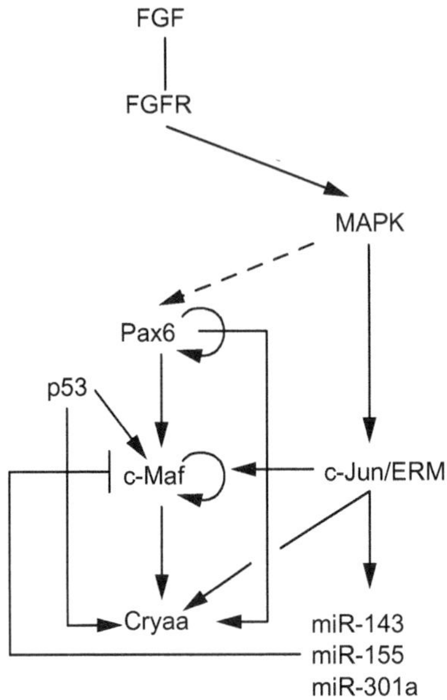

Figure 5 GRNs that control αA-crystallin gene expression. The "core" GRN for crystallin gene expression is comprised of Pax6, c-Maf, and individual crystallins and functions through a feed-forward loop that involves Pax6 and c-Maf autoregulation.[233] Recent studies have also established a link between FGF signaling and crystallin gene expression mediated by the AP-1 (c-Jun) and Ets (Etv5/ERM) transcription factors.[244] FGF signaling controls the transcription of αA-crystallin via the 5′-enhancer (DCR1).[234] MAPK/ERKs regulate expression of AP-1 and Ets factors at the level of transcription and/or by direct phosphorylation of the proteins. It is unclear if ERKs directly phosphorylate Pax6 proteins (dotted arrow) in the lens as suggested by multiple studies.[264,265] Thus, it will be necessary to evaluate the potential role of protein phosphatase 1 in this process.[266] Similarly, it remains to be determined if other crystallin genes are regulated by AP-1/Ets factors. The FGF-induced microRNAs (miR-143, miR-155, and miR-301a) negatively regulate expression of c-Maf. The regulation of c-Maf and αA-crystallin by p53 also requires additional studies, as the p53-binding sites in the Cryaa locus are outside of the functionally tested promoter. *Updated diagram originally from Cvekl and Ashery-Padan.*[1]

large set of miRNAs that modulate gene expression during FGF2-induced lens fiber cell differentiation.[157] At least three miRNAs, including miR-143, miR-155, and miR-301a, target the 3′-untranslated region of c-Maf.[157] Modulation of c-Maf expression by these miRNAs is also shown in Fig. 5. Thus, the current model of crystallin gene expression takes into account DNA-binding transcriptional factors, extracellular signaling, and ncRNAs, a significantly more detailed and nuanced compared to the earlier models summarized elsewhere.[20,25]

Other factors that are likely regulated by FGF signaling in the lens include Hsf4 and Prox1. It has been shown that an Hsf4b splice variant is phosphorylated by ERK at threonine 472, and this modification promotes its nuclear translocation and increases its DNA-binding activity.[267–269] Attenuation of FGF signaling via compound inactivation of Fgf1, Fgfr2, and Fgfr3 reduced expression of both c-Maf and Prox1 in the posterior parts of the lens vesicle.[138] Phosphorylation of Pax6 has been documented outside of the lens.[264,265] The homeodomain-interacting protein kinase 2 (Hipk2) phosphorylates Pax6 to enhance its interaction with p300 histone acetyltransferase,[270] an important regulator of crystallin gene expression[228] and lens fiber cell terminal differentiation.[157] Pax6 proteins are also targets of protein phosphatase-1 to reverse their phosphorylation.[266] Finally, both c-Maf[271] and p32/33 Pax6 protein that lacks the paired domain[272] can be sumoylated; however, the role of these posttranslational modifications in crystallin gene expression in the lens remains to be tested. Taken together, crystallin gene expression in the lens is regulated by lineage-specific DNA binding in combination with signal-regulated transcription factors, although some of these factors may belong to both of these categories (e.g., c-Maf, Hsf4, and Prox1).

Given an expanding role of crystallins outside of the lens, especially in the retina and optic nerve, it is surprising that very little is known about crystallin gene regulation outside of the lens, with a notable exception being αB-crystallin. In addition to the lens (see above), transcriptional control of the αB-crystallin gene has been dissected in the cornea, cardiac and skeletal muscle, and lung.[222,273–277] Expression analysis of a 148-kb BAC clone containing the mouse αA-crystallin locus with an integrated reported gene encoding the green fluorescent protein did not reveal any notable retinal expression of the reporter gene, raising the possibility that retinal-specific enhancers are located in more distal regions or even on different chromosomes.[224] Thus, future studies of crystallin gene expression in the lens and other ocular tissues will be necessary to establish molecular mechanisms of gene control of this important family of genes (see below).

4. CONCLUSIONS AND FUTURE DIRECTIONS

The long-term goals in the field of embryonic lens development and crystallin regulatory biology include the completion of understanding lens placode formation, as well as subsequent stages of lens morphogenesis within the 3D-space of the embryo. At the cellular level, it will be necessary to comprehend extracellular signaling between the prospective lens cells and other placodal progenitors, and the lens-inducing tissues, such as the "early" (cardiac mesoderm, foregut endoderm, etc.) and the "late" (optic vesicle) inducers. While the periocular mesenchyme (E8.5–E9.5) is the source of RA for lens placode invagination and optic cup formation, these mesenchymal cells also generate lens inhibitory signals through expression of TGF-β factors that trigger expression of Smad3 and Wnt/β-catenin activity to inhibit Pax6 expression in the nonlens ectoderm.[278,279] At the molecular level, the goal is to identify a full complement of regulatory genes, such as DNA-binding transcription factors and their links with signal transduction, chromatin-remodeling proteins, and posttranslational regulatory mechanisms involved in activation and/or repression of the regulatory transcription factors that control them. Studies of the lens crystallins are thought to be pivotal in understanding the physiochemical environment of the lens fibers and molecular basis of lens transparency, refraction, and elasticity. It is reasonable to expect additional studies on the multifunctional nature of crystallins in a range of processes, including tissue remodeling, inhibition of cell death, regulation of autophagy, and organelle degradation in order to elucidate their "non-crystallin" roles in the lens and nonlens tissues such as the retinal and central nervous system neurons.[280–282] The role of Pax6 in regulation of αA-crystallin and βB2-crystallin in neuronal cells[281] suggests that Pax6 regulates crystallin gene expression in retinal ganglion cells and other ocular cells; however, experimental proofs of this concept are still in progress. Studies focused on crystallin gene expression are not only leading models to understand gene regulation in the eye but also more general principles of gene control during embryonic development and organogenesis.

The following specific examples of interesting problems in lens biology directly related to the topics of this chapter are critically important next steps for understanding lens development. First, while it is clear that BMP4 is a critical signaling molecule for lens placode formation, expression of Pax6 and Six3 in the presumptive lens ectoderm is retained in *Bmp4*$^{-/-}$ embryos.[89] Thus, it is likely that additional DNA-binding transcription factors with key roles in this process remain to be identified and/or that posttranslational modification of Pax6 and Six3 may have critical roles in the ability of these proteins to recruit chromatin-remodeling enzymes that are downstream of BMP4-controlled signaling. Second, lens progenitor cells, like many other cells in the embryo, are prone to apoptosis and require both BMP and FGF signaling for their continued survival.[95,283] There are additional regulatory proteins that mediate prosurvival functions, including

Mab21l1,[104] Pitx3,[144] and Six3[88]; however, their molecular roles as inhibitors of apoptosis in the invaginating lens placode remain to be elucidated and integrated into the better understood BMP/FGF-controlled pathways. Third, RA signaling remains poorly understood during the early stages of lens placode formation and separation of the lens vesicle from the surface ectoderm. Pharmacological studies have shown a range of ocular abnormalities, including in the lens, in a vitamin A-deficiency model.[284] Early studies outside of the lens have established a link between AP-2 transcription factors and RA signaling.[285] AP-2α is a proven regulator of lens formation[286,287] and both AP-2β and AP-2γ proteins are also expressed in the eye[288]; nevertheless, the link between RA signaling and AP-2 factors in the lens remains to be experimentally tested. Fourth, the emerging field of noncoding RNAs, including miRNAs and lncRNAs, as well as the biology of RNA-binding proteins, in lens development and maintenance also requires special attention. The role of miRNAs in lens development is best illustrated via miR-204 that is embedded in intron 6 of the *Trpm3* gene.[289] Pax6 directly regulates expression of miR-204,[290,291] while eye regulatory genes, such as *Meis2* and *Sox11*, are among the predicted and/or validated miR-204 target genes.[290] Similarly, the best-characterized RNA-binding protein in lens fiber cells is Tdrd7, which regulates posttranscriptional control of multiple lens proteins.[292] Advances in massively parallel "next-gen" DNA sequencing facilitate studies of these ncRNAs and RNA-binding proteins and should provide increased insight into the RNA biology of the lens.

The crystallin GRN represents an advantageous system to understand the molecular and cellular principles of coordinated gene expression. It has been shown in other systems that coregulated genes are spatially organized within the nucleus in order to form specific 3D chromosomal contacts to establish "transcriptional factories." These factories are made from loops of chromosomal regions that encode proteins that work together, such as globin genes in red blood cells.[293] Furthermore, distal (i.e., enhancers) and proximal (i.e., promoters) regulatory regions of transcription are spatially connected via intra- or even interchromosomal loops to facilitate retention and recycling of proteins and enzymes involved in gene expression. Given the high input of the lens transcriptional apparatus to produce large quantities of crystallin mRNAs coupled with the "narrow" window between the completion of primary lens fiber cell differentiation (E14.5 in mouse) and degradation of nuclei and cessation of transcription (E18.5 in mouse), the lens transcriptional apparatus encounters lens-specific challenges that are not found in other postmitotic cells. Thus, it will be important to identify 3D structures of crystallin loci and whether individual crystallin genes assemble into "transcriptional" factories and which genes participate in these "collective" decisions. The 4C assay and similar technologies now are available to answer these important questions.

The "additional" roles of crystallins need to be explored in detail both in the lens and outside in the eye and other organs and tissues. As many of these "additional" roles probably evolved prior to lens formation, their better understanding may bring unexpected functions both in normal and in disease-affected tissues. Much evidence suggests that crystallins function as complexes, and the deletion of one may be compensated by another family member. Unexpected findings of crystallin roles are likely as supported by recent studies demonstrating binding of αB-crystallin to the argonaute protein 2 (Ago2), a catalytic subunit of the RNA-induced silencing complex.[294] Therefore, future studies should involve combinatorial deletions of two or more crystallins as described elsewhere.[1]

Novel approaches to study normal and abnormal lens development, i.e., cataractogenesis, are based on differentiation of ES and patient-specific iPS cells toward lens precursor cells and lentoid bodies.[295–297] Lentoid bodies are biconcave 3D structures resembling the primitive lens. However, the majority of cells in the lentoid bodies do not elongate and degrade their nuclei. Thus, additional experimentation is necessary to engineer fully differentiated lentoids with epithelial/fiber mass cellular composition. On the other hand, the current procedures are sufficient to generate lentoids from human iPS cells that carry specific mutations of interest and to use these cells for a range of studies. For example, derivation of iPS cells from patients carrying a panel of PAX6 mutations can provide materials "at large-scale" that can lead to identification of those processes disrupted in individual patients using genomic and proteomic approaches. Similarly, lentoid bodies that carry specific crystallin mutations can be used as sources of proteins for various biochemical and physicochemical tests to elucidate the molecular basis of lens structural abnormalities caused by the individual mutations of particular interest.

In conclusion, over 110 years of embryological, cellular, and molecular lens research have provided detailed insights into the vertebrate lens development. Nevertheless, a large number of outstanding questions remain to be addressed. Recent advances in genome science, mapping of protein-coding and noncoding transcriptomes, epigenome, and proteomes ("omics") allow for the first time in history to find definitive answers to the questions asked. As in the past, many lessons first discovered in lens should be widely applicable for a better understanding of related processes in other organs and tissues.

ACKNOWLEDGMENTS

This work was supported by NIH Grants R01 EY012200 (A.C.), EY014237 (A.C.), and EY022645 (W.L.) and an unrestricted grant from Research to Prevent Blindness to the Department of Ophthalmology and Visual Sciences.

REFERENCES

1. Cvekl A, Ashery-Padan R. The cellular and molecular mechanisms of vertebrate lens development. *Development*. 2014;00:1–16.
2. Beebe DC, Coats JM. The lens organizes the anterior segment: specification of neural crest cell differentiation in the avian eye. *Dev Biol*. 2000;220(2):424–431.
3. Ittner LM, Wurdak H, Schwerdtfeger K, et al. Compound developmental eye disorders following inactivation of TGFbeta signaling in neural-crest stem cells. *J Biol*. 2005;4(3):11.
4. Chang B, Smith RS, Peters M, et al. Haploinsufficient Bmp4 ocular phenotypes include anterior segment dysgenesis with elevated intraocular pressure. *BMC Genet*. 2001;2:18.
5. Lwigale PY, Bronner-Fraser M. Lens-derived Semaphorin3A regulates sensory innervation of the cornea. *Dev Biol*. 2007;306(2):750–759.
6. Gage PJ, Zacharias AL. Signaling "cross-talk" is integrated by transcription factors in the development of the anterior segment in the eye. *Dev Dyn*. 2009;238(9):2149–2162.
7. Davis-Silberman N, Ashery-Padan R. Iris development in vertebrates; genetic and molecular considerations. *Brain Res*. 2008;1192:17–28.
8. Spemann H. Über korrelationen in der Entwicklung des Auges. *Verh Anat Ges*. 1901;15:61–79.
9. Mencl E. Ein Fall von beiderseitiger Augenlinsenausbildung während der Abwesenheit von Augenblasen. *Arch EntwMech Org*. 1903;16(2):328–339.
10. Lewis WH. The experimental production of cyclopia in the fish embryo (Fundulus heteroclitus). *Anat Rec*. 1909;3(4):175–181.
11. Waddington CH. The origin of competence for lens formation in the amphibia. *J Exp Biol*. 1936;13:86–91.
12. Coulombre JL, Coulombre AJ. Lens development: fiber elongation and lens orientation. *Science*. 1963;142(3598):1489–1490.
13. Coulombre AJ. Experimental embryology of the vertebrate eye. *Investig Ophthalmol*. 1965;4:411–419.
14. Jacobson AG. The determination and positioning of the nose, lens and ear. I. Interactions within the ectoderm, and between the ectoderm and underlying tissues. *J Exp Zool*. 1963;154:273–283.
15. Jacobson AG. The determination and positioning of the nose, lens and ear. II The role of the endoderm. *J Exp Zool*. 1963;154:285–291.
16. Jacobson AG. The determination and positioning of the nose, lens and ear. III. Effects of reversing the antero-posterior axis of epidermis, neural plate and neural fold. *J Exp Zool*. 1963;154:293–303.
17. Saha MS, Spann CL, Grainger RM. Embryonic lens induction: more than meets the optic vesicle. *Cell Differ Dev*. 1989;28(3):153–171.
18. Piatigorsky J. Lens differentiation in vertebrates. A review of cellular and molecular features. *Differentiation*. 1981;19(3):134–153.
19. Grainger RM. Embryonic lens induction: shedding light on vertebrate tissue determination. *Trends Genet*. 1992;8(10):349–355.
20. Cvekl A, Piatigorsky J. Lens development and crystallin gene expression: many roles for Pax-6. *BioEssays*. 1996;18(8):621–630.

21. Duncan MK, Xie L, David LL, et al. Ectopic Pax6 expression disturbs lens fiber cell differentiation. *Invest Ophthalmol Vis Sci.* 2004;45(10):3589–3598.
22. Lovicu FJ, McAvoy JW. Growth factor regulation of lens development. *Dev Biol.* 2005;280(1):1–14.
23. Donner AL, Lachke SA, Maas RL. Lens induction in vertebrates: variations on a conserved theme of signaling events. *Semin Cell Dev Biol.* 2006;17(6):676–685.
24. Robinson ML. An essential role for FGF receptor signaling in lens development. *Semin Cell Dev Biol.* 2006;17(6):726–740.
25. Cvekl A, Duncan MK. Genetic and epigenetic mechanisms of gene regulation during lens development. *Prog Retin Eye Res.* 2007;26(6):555–597.
26. Medina-Martinez O, Jamrich M. Foxe view of lens development and disease. *Development.* 2007;134(8):1455–1463.
27. Gunhaga L. The lens: a classical model of embryonic induction providing new insights into cell determination in early development. *Philos Trans R Soc Lond B Biol Sci.* 2011;366(1568):1193–1203.
28. Greiling TM, Clark JI. New insights into the mechanism of lens development using zebra fish. *Int Rev Cell Mol Biol.* 2012;296:1–61.
29. Oliver G, Gruss P. Current views on eye development. *Trends Neurosci.* 1997;20(9):415–421.
30. Ashery-Padan R, Gruss P. Pax6 lights-up the way for eye development. *Curr Opin Cell Biol.* 2001;13(6):706–714.
31. Chow RL, Lang RA. Early eye development in vertebrates. *Annu Rev Cell Dev Biol.* 2001;17:255–296.
32. Livesey FJ, Cepko CL. Vertebrate neural cell-fate determination: lessons from the retina. *Nat Rev Neurosci.* 2001;2(2):109–118.
33. Graw J, Loster J. Developmental genetics in ophthalmology. *Ophthalmic Genet.* 2003;24(1):1–33.
34. Hatakeyama J, Kageyama R. Retinal cell fate determination and bHLH factors. *Semin Cell Dev Biol.* 2004;15(1):83–89.
35. Bailey TJ, El-Hodiri H, Zhang L, Shah R, Mathers PH, Jamrich M. Regulation of vertebrate eye development by Rx genes. *Int J Dev Biol.* 2004;48(8-9):761–770.
36. Moore K, Vetter M. Retinal development. In: Moody S, ed. *Principles of Developmental Genetics.* New York: Elsevier, Inc.; 2007.
37. Cvekl A, Mitton KP. Epigenetic regulatory mechanisms in vertebrate eye development and disease. *Heredity (Edinb).* 2010;105(1):135–151.
38. Fuhrmann S. Eye morphogenesis and patterning of the optic vesicle. *Curr Top Dev Biol.* 2010;93:61–84.
39. Heavner W, Pevny L. Eye development and retinogenesis. *Cold Spring Harb Perspect Biol.* 2012;4(12):a008391.
40. Shaham O, Menuchin Y, Farhy C, Ashery-Padan R. Pax6: a multi-level regulator of ocular development. *Prog Retin Eye Res.* 2012;31(5):351–376.
41. Sinn R, Wittbrodt J. An eye on eye development. *Mech Dev.* 2013;130(6-8):347–358.
42. Baker CV, Bronner-Fraser M. Vertebrate cranial placodes I. Embryonic induction. *Dev Biol.* 2001;232(1):1–61.
43. Lang RA. Pathways regulating lens induction in the mouse. *Int J Dev Biol.* 2004;48(8–9):783–791.
44. Streit A. Early development of the cranial sensory nervous system: from a common field to individual placodes. *Dev Biol.* 2004;276(1):1–15.
45. McCabe KL, Bronner-Fraser M. Molecular and tissue interactions governing induction of cranial ectodermal placodes. *Dev Biol.* 2009;332(2):189–195.
46. Patthey C, Gunhaga L. Signaling pathways regulating ectodermal cell fate choices. *Exp Cell Res.* 2014;321(1):11–16.

47. Saint-Jeannet JP, Moody SA. Establishing the pre-placodal region and breaking it into placodes with distinct identities. *Dev Biol.* 2014;389(1):13–27.
48. Cvekl A, Wang WL. Retinoic acid signaling in mammalian eye development. *Exp Eye Res.* 2009;89(3):280–291.
49. Duester G. Keeping an eye on retinoic acid signaling during eye development. *Chem Biol Interact.* 2009;178(1–3):178–181.
50. Walker J, Menko AS. Integrins in lens development and disease. *Exp Eye Res.* 2009;88(2):216–225.
51. Sugiyama Y, Stump RJ, Nguyen A, et al. Secreted frizzled-related protein disrupts PCP in eye lens fiber cells that have polarised primary cilia. *Dev Biol.* 2010;338(2):193–201.
52. West-Mays JA, Pino G, Lovicu FJ. Development and use of the lens epithelial explant system to study lens differentiation and cataractogenesis. *Prog Retin Eye Res.* 2010;29(2):135–143.
53. Kuszak JR, Zoltoski RK, Tiedemann CE. Development of lens sutures. *Int J Dev Biol.* 2004;48(8–9):889–902.
54. Bassnett S, Shi Y, Vrensen GF. Biological glass: structural determinants of eye lens transparency. *Philos Trans R Soc Lond B Biol Sci.* 2011;366(1568):1250–1264.
55. Andley UP. Crystallins in the eye: function and pathology. *Prog Retin Eye Res.* 2007;26(1):78–98.
56. Horwitz J. Alpha crystallin: the quest for a homogeneous quaternary structure. *Exp Eye Res.* 2009;88(2):190–194.
57. Slingsby C, Wistow GJ, Clark AR. Evolution of crystallins for a role in the vertebrate eye lens. *Protein Sci.* 2013;22(4):367–380.
58. Griep AE, Zhang P. *Development of the Ocular Lens.* Cambridge, UK: Cambridge University Press; 2004.
59. Griep AE. Cell cycle regulation in the developing lens. *Semin Cell Dev Biol.* 2006;17(6):686–697.
60. Verkman AS. Role of aquaporin water channels in eye function. *Exp Eye Res.* 2003;76(2):137–143.
61. Gong X, Cheng C, Xia CH. Connexins in lens development and cataractogenesis. *J Membr Biol.* 2007;218(1–3):9–12.
62. Schey KL, Wang Z, Wenke J L, Qi Y. Aquaporins in the eye: expression, function, and roles in ocular disease. *Biochim Biophys Acta.* 2014;1840(5):1513–1523.
63. Bassnett S. On the mechanism of organelle degradation in the vertebrate lens. *Exp Eye Res.* 2009;88(2):133–139.
64. Wride MA. Lens fibre cell differentiation and organelle loss: many paths lead to clarity. *Philos Trans R Soc Lond B Biol Sci.* 2011;366(1568):1219–1233.
65. Danysh BP, Duncan MK. The lens capsule. *Exp Eye Res.* 2009;88(2):151–164.
66. Brennan LA, Kantorow M. Mitochondrial function and redox control in the aging eye: role of MsrA and other repair systems in cataract and macular degenerations. *Exp Eye Res.* 2009;88(2):195–203.
67. Collinson JM, Hill RE, West JD. Analysis of mouse eye development with chimeras and mosaics. *Int J Dev Biol.* 2004;48(8–9):793–804.
68. Graw J, Neuhauser-Klaus A, Klopp N, Selby PB, Loster J, Favor J. Genetic and allelic heterogeneity of Cryg mutations in eight distinct forms of dominant cataract in the mouse. *Invest Ophthalmol Vis Sci.* 2004;45(4):1202–1213.
69. Graw J. Genetics of crystallins: cataract and beyond. *Exp Eye Res.* 2009;88(2):173–189.
70. Churchill A, Graw J. Clinical and experimental advances in congenital and paediatric cataracts. *Philos Trans R Soc Lond B Biol Sci.* 2011;366(1568):1234–1249.
71. Hingorani M, Hanson I, van Heyningen V. Aniridia. *Eur J Hum Genet.* 2012;20(10):1011–1017.
72. Wistow G. The human crystallin gene families. *Hum Genomics.* 2012;6:26.

73. Shiels A, Hejtmancik JF. Genetics of human cataract. *Clin Genet.* 2013;84(2):120–127.
74. Tsonis PA, Del Rio-Tsonis K. Lens and retina regeneration: transdifferentiation, stem cells and clinical applications. *Exp Eye Res.* 2004;78(2):161–172.
75. Tsonis PA, Madhavan M, Tancous EE, Del Rio-Tsonis K. A newt's eye view of lens regeneration. *Int J Dev Biol.* 2004;48(8–9):975–980.
76. Tsonis PA. Regeneration via transdifferentiation: the lens and hair cells. *Hear Res.* 2007;227(1–2):28–31.
77. Gehring WJ, Ikeo K. Pax 6: mastering eye morphogenesis and eye evolution. *Trends Genet.* 1999;15(9):371–377.
78. Lamb TD, Collin SP, Pugh Jr EN. Evolution of the vertebrate eye: opsins, photoreceptors, retina and eye cup. *Nat Rev Neurosci.* 2007;8(12):960–976.
79. Jonasova K, Kozmik Z. Eye evolution: lens and cornea as an upgrade of animal visual system. *Semin Cell Dev Biol.* 2008;19(2):71–81.
80. Nilsson DE. The evolution of eyes and visually guided behaviour. *Philos Trans R Soc Lond B Biol Sci.* 2009;364(1531):2833–2847.
81. Vopalensky P, Kozmik Z. Eye evolution: common use and independent recruitment of genetic components. *Philos Trans R Soc Lond B Biol Sci.* 2009;364(1531):2819–2832.
82. Lamb TD. Evolution of phototransduction, vertebrate photoreceptors and retina. *Prog Retin Eye Res.* 2013;36:52–119.
83. Nilsson DE. Eye evolution and its functional basis. *Vis Neurosci.* 2013;30(1–2):5–20.
84. Coolen M, Sii-Felice K, Bronchain O, et al. Phylogenomic analysis and expression patterns of large Maf genes in Xenopus tropicalis provide new insights into the functional evolution of the gene family in osteichthyans. *Dev Genes Evol.* 2005;215(7):327–339.
85. Ogino H, Yasuda K. Sequential activation of transcription factors in lens induction. *Develop Growth Differ.* 2000;42(5):437–448.
86. Jacobson AG. Inductive processes in embryonic development. *Science.* 1966;152(3718): 25–34.
87. Streit A. The preplacodal region: an ectodermal domain with multipotential progenitors that contribute to sense organs and cranial sensory ganglia. *Int J Dev Biol.* 2007;51(6–7):447–461.
88. Liu W, Lagutin OV, Mende M, Streit A, Oliver G. Six3 activation of Pax6 expression is essential for mammalian lens induction and specification. *EMBO J.* 2006;25(22):5383–5395.
89. Furuta Y, Hogan BL. BMP4 is essential for lens induction in the mouse embryo. *Genes Dev.* 1998;12(23):3764–3775.
90. Yun S, Saijoh Y, Hirokawa KE, et al. Lhx2 links the intrinsic and extrinsic factors that control optic cup formation. *Development.* 2009;136(23):3895–3906.
91. Porter FD, Drago J, Xu Y, et al. Lhx2, a LIM homeobox gene, is required for eye, forebrain, and definitive erythrocyte development. *Development.* 1997;124(15):2935–2944.
92. Luo G, Hofmann C, Bronckers AL, Sohocki M, Bradley A, Karsenty G. BMP-7 is an inducer of nephrogenesis, and is also required for eye development and skeletal patterning. *Genes Dev.* 1995;9(22):2808–2820.
93. Wawersik S, Purcell P, Rauchman M, Dudley AT, Robertson EJ, Maas R. BMP7 acts in murine lens placode development. *Dev Biol.* 1999;207(1):176–188.
94. Morcillo J, Martinez-Morales JR, Trousse F, Fermin Y, Sowden JC, Bovolenta P. Proper patterning of the optic fissure requires the sequential activity of BMP7 and SHH. *Development.* 2006;133(16):3179–3190.
95. Rajagopal R, Huang J, Dattilo LK, et al. The type I BMP receptors, Bmpr1a and Acvr1, activate multiple signaling pathways to regulate lens formation. *Dev Biol.* 2009;335(2):305–316.

96. Mic FA, Molotkov A, Molotkova N, Duester G. Raldh2 expression in optic vesicle generates a retinoic acid signal needed for invagination of retina during optic cup formation. *Dev Dyn.* 2004;231(2):270–277.
97. Molotkov A, Molotkova N, Duester G. Retinoic acid guides eye morphogenetic movements via paracrine signaling but is unnecessary for retinal dorsoventral patterning. *Development.* 2006;133(10):1901–1910.
98. Chauhan BK, Disanza A, Choi SY, et al. Cdc42- and IRSp53-dependent contractile filopodia tether presumptive lens and retina to coordinate epithelial invagination. *Development.* 2009;136(21):3657–3667.
99. Huang J, Rajagopal R, Liu Y, et al. The mechanism of lens placode formation: a case of matrix-mediated morphogenesis. *Dev Biol.* 2011;355(1):32–42.
100. Plageman Jr TF, Chauhan BK, Yang C, et al. A Trio-RhoA-Shroom3 pathway is required for apical constriction and epithelial invagination. *Development.* 2011;138(23):5177–5188.
101. Lang RA, Herman K, Reynolds AB, Hildebrand JD, Plageman Jr TF. p120-catenin-dependent junctional recruitment of Shroom3 is required for apical constriction during lens pit morphogenesis. *Development.* 2014;141(16):3177–3187.
102. Pontoriero GF, Smith AN, Miller LA, Radice GL, West-Mays JA, Lang RA. Co-operative roles for E-cadherin and N-cadherin during lens vesicle separation and lens epithelial cell survival. *Dev Biol.* 2009;326(2):403–417.
103. Shaham O, Smith AN, Robinson ML, Taketo MM, Lang RA, Ashery-Padan R. Pax6 is essential for lens fiber cell differentiation. *Development.* 2009;136(15):2567–2578.
104. Yamada R, Mizutani-Koseki Y, Hasegawa T, Osumi N, Koseki H, Takahashi N. Cell-autonomous involvement of Mab21l1 is essential for lens placode development. *Development.* 2003;130(9):1759–1770.
105. Baldessari D, Badaloni A, Longhi R, Zappavigna V, Consalez GG. MAB21L2, a vertebrate member of the Male-abnormal 21 family, modulates BMP signaling and interacts with SMAD1. *BMC Cell Biol.* 2004;5(1):48.
106. Zhou M, Leiberman J, Xu J, Lavker RM. A hierarchy of proliferative cells exists in mouse lens epithelium: implications for lens maintenance. *Invest Ophthalmol Vis Sci.* 2006;47(7):2997–3003.
107. Arnold K, Sarkar A, Yram MA, et al. Sox2(+) adult stem and progenitor cells are important for tissue regeneration and survival of mice. *Cell Stem Cell.* 2011;9(4):317–329.
108. Jia J, Lin M, Zhang L, York JP, Zhang P. The Notch signaling pathway controls the size of the ocular lens by directly suppressing p57Kip2 expression. *Mol Cell Biol.* 2007;27(20):7236–7247.
109. Rowan S, Conley KW, Le TT, Donner AL, Maas RL, Brown NL. Notch signaling regulates growth and differentiation in the mammalian lens. *Dev Biol.* 2008;321(1):111–122.
110. Mi D, Carr CB, Georgala PA, et al. Pax6 exerts regional control of cortical progenitor proliferation via direct repression of Cdk6 and hypophosphorylation of pRb. *Neuron.* 2013;78(2):269–284.
111. Gao CY, Bassnett S, Zelenka PS. Cyclin B, p34cdc2, and H1-kinase activity in terminally differentiating lens fiber cells. *Dev Biol.* 1995;169(1):185–194.
112. Fromm L, Overbeek PA. Regulation of cyclin and cyclin-dependent kinase gene expression during lens differentiation requires the retinoblastoma protein. *Oncogene.* 1996;12(1):69–75.
113. He HY, Gao C, Vrensen G, Zelenka P. Transient activation of cyclin B/Cdc2 during terminal differentiation of lens fiber cells. *Dev Dyn.* 1998;211(1):26–34.

114. Zhang P, Wong C, DePinho RA, Harper JW, Elledge SJ. Cooperation between the Cdk inhibitors p27(KIP1) and p57(KIP2) in the control of tissue growth and development. *Genes Dev.* 1998;12(20):3162–3167.
115. Hyde RK, Griep AE. Unique roles for E2F1 in the mouse lens in the absence of functional pRB proteins. *Invest Ophthalmol Vis Sci.* 2002;43(5):1509–1516.
116. Gomez Lahoz E, Liegeois NJ, Zhang P, et al. Cyclin D- and E-dependent kinases and the p57(KIP2) inhibitor: cooperative interactions in vivo. *Mol Cell Biol.* 1999;19(1):353–363.
117. Maeda A, Moriguchi T, Hamada M, et al. Transcription factor GATA-3 is essential for lens development. *Dev Dyn.* 2009;238(9):2280–2291.
118. Wigle JT, Chowdhury K, Gruss P, Oliver G. Prox1 function is crucial for mouse lens-fibre elongation. *Nat Genet.* 1999;21(3):318–322.
119. Zhang P, Liegeois NJ, Wong C, et al. Altered cell differentiation and proliferation in mice lacking p57KIP2 indicates a role in Beckwith-Wiedemann syndrome. *Nature.* 1997;387(6629):151–158.
120. Govindarajan V, Overbeek PA. Secreted FGFR3, but not FGFR1, inhibits lens fiber differentiation. *Development.* 2001;128(9):1617–1627.
121. Morgenbesser SD, Williams BO, Jacks T, DePinho RA. p53-dependent apoptosis produced by Rb-deficiency in the developing mouse lens. *Nature.* 1994;371(6492):72–74.
122. Jiang Z, Zacksenhaus E, Gallie BL, Phillips RA. The retinoblastoma gene family is differentially expressed during embryogenesis. *Oncogene.* 1997;14(15):1789–1797.
123. Rampalli AM, Gao CY, Chauthaiwale VM, Zelenka PS. pRb and p107 regulate E2F activity during lens fiber cell differentiation. *Oncogene.* 1998;16(3):399–408.
124. Chen Q, Hung FC, Fromm L, Overbeek PA. Induction of cell cycle entry and cell death in postmitotic lens fiber cells by overexpression of E2F1 or E2F2. *Invest Ophthalmol Vis Sci.* 2000;41(13):4223–4231.
125. Chen Q, Liang D, Yang T, Leone G, Overbeek PA. Distinct capacities of individual E2Fs to induce cell cycle re-entry in postmitotic lens fiber cells of transgenic mice. *Dev Neurosci.* 2004;26(5-6):435–445.
126. Liu Y, Zacksenhaus E. E2F1 mediates ectopic proliferation and stage-specific p53-dependent apoptosis but not aberrant differentiation in the ocular lens of Rb deficient fetuses. *Oncogene.* 2000;19(52):6065–6073.
127. Saavedra HI, Wu L, de Bruin A, et al. Specificity of E2F1, E2F2, and E2F3 in mediating phenotypes induced by loss of Rb. *Cell Growth Differ.* 2002;13(5):215–225.
128. Wenzel PL, Chong JL, Saenz-Robles MT, et al. Cell proliferation in the absence of E2F1-3. *Dev Biol.* 2011;351(1):35–45.
129. Zacksenhaus E, Jiang Z, Chung D, Marth JD, Phillips RA, Gallie BL. pRb controls proliferation, differentiation, and death of skeletal muscle cells and other lineages during embryogenesis. *Genes Dev.* 1996;10(23):3051–3064.
130. Robinson ML, MacMillan-Crow LA, Thompson JA, Overbeek PA. Expression of a truncated FGF receptor results in defective lens development in transgenic mice. *Development.* 1995;121(12):3959–3967.
131. Robinson ML, Overbeek PA, Verran DJ, et al. Extracellular FGF-1 acts as a lens differentiation factor in transgenic mice. *Development.* 1995;121(2):505–514.
132. He S, Xie Q, Limi S, *et al.* Chromatin remodeling enzyme Smarca5/Snf2h regulates embryonic lens differentiation and denucleation. *Development.* 2015; in submission.
133. Jarrin M, Pandit T, Gunhaga L. A balance of FGF and BMP signals regulates cell cycle exit and Equarin expression in lens cells. *Mol Biol Cell.* 2012;23(16):3266–3274.
134. Mu H, Ohta K, Kuriyama S, et al. Equarin, a novel soluble molecule expressed with polarity at chick embryonic lens equator, is involved in eye formation. *Mech Dev.* 2003;120(2):143–155.

135. Song X, Sato Y, Felemban A, et al. Equarin is involved as an FGF signaling modulator in chick lens differentiation. *Dev Biol.* 2012;368(1):109–117.
136. Song X, Sato Y, Sekiguchi K, Tanaka H, Ohta K. Equarin is involved in cell adhesion by means of heparan sulfate proteoglycan during lens development. *Dev Dyn.* 2013;242(1):23–29.
137. Song X, Tanaka H, Ohta K. Multiple roles of Equarin during lens development. *Develop Growth Differ.* 2014;56(3):199–205.
138. Zhao H, Yang T, Madakashira BP, et al. Fibroblast growth factor receptor signaling is essential for lens fiber cell differentiation. *Dev Biol.* 2008;318(2):276–288.
139. Pan Y, Woodbury A, Esko JD, Grobe K, Zhang X. Heparan sulfate biosynthetic gene Ndst1 is required for FGF signaling in early lens development. *Development.* 2006;133(24):4933–4944.
140. Qu X, Hertzler K, Pan Y, Grobe K, Robinson ML, Zhang X. Genetic epistasis between heparan sulfate and FGF-Ras signaling controls lens development. *Dev Biol.* 2011;355(1):12–20.
141. Gotoh N, Ito M, Yamamoto S, et al. Tyrosine phosphorylation sites on FRS2alpha responsible for Shp2 recruitment are critical for induction of lens and retina. *Proc Natl Acad Sci USA.* 2004;101(49):17144–17149.
142. Li H, Tao C, Cai Z, et al. Frs2alpha and Shp2 signal independently of Gab to mediate FGF signaling in lens development. *J Cell Sci.* 2014;127(pt 3):571–582.
143. Madakashira BP, Kobrinski DA, Hancher AD, et al. Frs2alpha enhances fibroblast growth factor-mediated survival and differentiation in lens development. *Development.* 2012;139(24):4601–4612.
144. Medina-Martinez O, Shah R, Jamrich M. Pitx3 controls multiple aspects of lens development. *Dev Dyn.* 2009;238(9):2193–2201.
145. Martinez G, de Iongh RU. The lens epithelium in ocular health and disease. *Int J Biochem Cell Biol.* 2010;42(12):1945–1963.
146. Dawes LJ, Sugiyama Y, Lovicu FJ, Harris CG, Shelley EJ, McAvoy JW. Interactions between lens epithelial and fiber cells reveal an intrinsic self-assembly mechanism. *Dev Biol.* 2014;385(2):291–303.
147. Cain S, Martinez G, Kokkinos MI, et al. Differential requirement for beta-catenin in epithelial and fiber cells during lens development. *Dev Biol.* 2008;321(2):420–433.
148. Kerr CL, Zaveri MA, Robinson ML, Williams T, West-Mays JA. AP-2alpha is required after lens vesicle formation to maintain lens integrity. *Dev Dyn.* 2014;243(10):1298–1309.
149. Chauss D, Basu S, Rajakaruna S, et al. Differentiation state-specific mitochondrial dynamic regulatory networks are revealed by global transcriptional analysis of the developing chicken lens. *G3.* 2014;4(8):1515–1527.
150. Hoang TV, Kumar P, Sutharzan S, Tsonis PA, Liang C, Robinson ML. Comparative transcriptome analysis of epithelial and fiber cells in newborn mouse lenses with RNA sequencing. *Mol Vis.* 2014;20:1491–1517.
151. Tholozan FM, Gribbon C, Li Z, et al. FGF-2 release from the lens capsule by MMP-2 maintains lens epithelial cell viability. *Mol Biol Cell.* 2007;18(11):4222–4231.
152. Wu W, Tholozan FM, Goldberg MW, Bowen L, Wu J, Quinlan RA. A gradient of matrix-bound FGF-2 and perlecan is available to lens epithelial cells. *Exp Eye Res.* 2014;120:10–14.
153. Schulz MW, Chamberlain CG, McAvoy JW. Binding of FGF-1 and FGF-2 to heparan sulphate proteoglycans of the mammalian lens capsule. *Growth Factors.* 1997;14(1):1–13.
154. Wolf LV, Yang Y, Wang J, et al. Identification of pax6-dependent gene regulatory networks in the mouse lens. *PLoS One.* 2009;4(1):e4159.
155. Neuhann IM, Neuhann TF. Cataract surgery and aniridia. *Curr Opin Ophthalmol.* 2010;21(1):60–64.

156. Boswell BA, Overbeek PA, Musil LS. Essential role of BMPs in FGF-induced secondary lens fiber differentiation. *Dev Biol.* 2008;324(2):202–212.
157. Wolf L, Gao CS, Gueta K, et al. Identification and characterization of FGF2-dependent mRNA: microRNA networks during lens fiber, cell differentiation. *G3.* 2013;3(12):2239–2255.
158. Lohnes D, Mark M, Mendelsohn C, et al. Function of the retinoic acid receptors (RARs) during development (I). Craniofacial and skeletal abnormalities in RAR double mutants. *Development.* 1994;120(10):2723–2748.
159. Gopal-Srivastava R, Cvekl A, Piatigorsky J. Involvement of retinoic acid/retinoid receptors in the regulation of murine alphaB-crystallin/small heat shock protein gene expression in the lens. *J Biol Chem.* 1998;273(28):17954–17961.
160. Tini M, Otulakowski G, Breitman ML, Tsui LC, Giguere V. An everted repeat mediates retinoic acid induction of the gamma F-crystallin gene: evidence of a direct role for retinoids in lens development. *Genes Dev.* 1993;7(2):295–307.
161. Kralova J, Czerny T, Spanielova H, Ratajova V, Kozmik Z. Complex regulatory element within the gammaE- and gammaF-crystallin enhancers mediates Pax6 regulation and is required for induction by retinoic acid. *Gene.* 2002;286(2):271–282.
162. Chauhan BK, Yang Y, Cveklova K, Cvekl A. Functional interactions between alternatively spliced forms of Pax6 in crystallin gene regulation and in haploinsufficiency. *Nucleic Acids Res.* 2004;32(5):1696–1709.
163. Sugiyama Y, Lovicu FJ, McAvoy JW. Planar cell polarity in the mammalian eye lens. *Organogenesis.* 2011;7(3):191–201.
164. Leonard M, Zhang L, Bleaken BM, Menko AS. Distinct roles for N-Cadherin linked c-Src and fyn kinases in lens development. *Dev Dyn.* 2013;242(5):469–484.
165. Samuelsson AR, Belvindrah R, Wu C, Muller U, Halfter W. Beta1-integrin signaling is essential for lens fiber survival. *Gene Regul Syst Bio.* 2007;1:177–189.
166. Hayes JM, Hartsock A, Clark BS, Napier HR, Link BA, Gross JM. Integrin alpha5/fibronectin1 and focal adhesion kinase are required for lens fiber morphogenesis in zebrafish. *Mol Biol Cell.* 2012;23(24):4725–4738.
167. Basu S, Rajakaruna S, De Arcangelis A, Zhang L, Georges-Labouesse E, Menko AS. alpha6 integrin transactivates insulin-like growth factor receptor-1 (IGF-1R) to regulate caspase-3-mediated lens epithelial cell differentiation initiation. *J Biol Chem.* 2014;289(7):3842–3855.
168. Weber GF, Menko AS. Phosphatidylinositol 3-kinase is necessary for lens fiber cell differentiation and survival. *Invest Ophthalmol Vis Sci.* 2006;47(10):4490–4499.
169. Basu S, Rajakaruna S, Reyes B, Van Bockstaele E, Menko AS. Suppression of MAPK/JNK-MTORC1 signaling leads to premature loss of organelles and nuclei by autophagy during terminal differentiation of lens fiber cells. *Autophagy.* 2014;10(7):1193–1211.
170. Basu S, Rajakaruna S, Menko AS. Insulin-like growth factor receptor-1 and nuclear factor kappaB are crucial survival signals that regulate caspase-3-mediated lens epithelial cell differentiation initiation. *J Biol Chem.* 2012;287(11):8384–8397.
171. Maddala R, Reneker LW, Pendurthi B, Rao PV. Rho GDP dissociation inhibitor-mediated disruption of Rho GTPase activity impairs lens fiber cell migration, elongation and survival. *Dev Biol.* 2008;315(1):217–231.
172. Nowak RB, Fischer RS, Zoltoski RK, Kuszak JR, Fowler VM. Tropomodulin1 is required for membrane skeleton organization and hexagonal geometry of fiber cells in the mouse lens. *J Cell Biol.* 2009;186(6):915–928.
173. Scheiblin DA, Gao J, Caplan JL, et al. Beta-1 integrin is important for the structural maintenance and homeostasis of differentiating fiber cells. *Int J Biochem Cell Biol.* 2014;50:132–145.

174. Tripathi BK, Lowy DR, Zelenka PS. The Cdk5 activator P39 specifically links muskelin to myosin II and regulates stress fiber formation and actin organization in lens. *Exp Cell Res*. 2014;330:186–198.
175. FitzGerald PG. Lens intermediate filaments. *Exp Eye Res*. 2009;88(2):165–172.
176. Song S, Landsbury A, Dahm R, Liu Y, Zhang Q, Quinlan RA. Functions of the intermediate filament cytoskeleton in the eye lens. *J Clin Invest*. 2009;119(7):1837–1848.
177. Costello MJ, Brennan LA, Basu S, et al. Autophagy and mitophagy participate in ocular lens organelle degradation. *Exp Eye Res*. 2013;116:141–150.
178. Bloemendal H. The vertebrate eye lens. *Science*. 1977;197(4299):127–138.
179. Horwitz J. Alpha-crystallin can function as a molecular chaperone. *Proc Natl Acad Sci USA*. 1992;89(21):10449–10453.
180. Andley UP, Song Z, Wawrousek EF, Fleming TP, Bassnett S. Differential protective activity of alphaA- and alphaB-crystallin in lens epithelial cells. *J Biol Chem*. 2000;275(47):36823–36831.
181. Boyle DL, Takemoto L. EM immunolocalization of alpha-crystallins: association with the plasma membrane from normal and cataractous human lenses. *Curr Eye Res*. 1996;15(5):577–582.
182. Kantorow M, Piatigorsky J. Alpha-crystallin/small heat shock protein has autokinase activity. *Proc Natl Acad Sci USA*. 1994;91(8):3112–3116.
183. Kato K, Shinohara H, Kurobe N, Goto S, Inaguma Y, Ohshima K. Immunoreactive alpha A crystallin in rat non-lenticular tissues detected with a sensitive immunoassay method. *Biochim Biophys Acta*. 1991;1080(2):173–180.
184. Plater ML, Goode D, Crabbe MJ. Effects of site-directed mutations on the chaperone-like activity of alphaB-crystallin. *J Biol Chem*. 1996;271(45):28558–28566.
185. Rao NA, Saraswathy S, Wu GS, Katselis GS, Wawrousek EF, Bhat S. Elevated retina-specific expression of the small heat shock protein, alphaA-crystallin, is associated with photoreceptor protection in experimental uveitis. *Invest Ophthalmol Vis Sci*. 2008;49(3):1161–1171.
186. Srinivasan AN, Nagineni CN. Bhat SP alpha A-crystallin is expressed in non-ocular tissues. *J Biol Chem*. 1992;267(32):23337–23341.
187. van den IPR, Overkamp P, Knauf U, Gaestel M, de Jong WW. Alpha A-crystallin confers cellular thermoresistance. *FEBS Lett*. 1994;355(1):54–56.
188. Xi J, Farjo R, Yoshida S, Kern TS, Swaroop A, Andley UP. A comprehensive analysis of the expression of crystallins in mouse retina. *Mol Vis*. 2003;9:410–419.
189. Yaung J, Jin M, Barron E, et al. alpha-Crystallin distribution in retinal pigment epithelium and effect of gene knockouts on sensitivity to oxidative stress. *Mol Vis*. 2007;13:566–577.
190. Yaung J, Kannan R, Wawrousek EF, Spee C, Sreekumar PG, Hinton DR. Exacerbation of retinal degeneration in the absence of alpha crystallins in an in vivo model of chemically induced hypoxia. *Exp Eye Res*. 2008;86(2):355–365.
191. Wu F, Lee S, Schumacher M, Jun A, Chakravarti S. Differential gene expression patterns of the developing and adult mouse cornea compared to the lens and tendon. *Exp Eye Res*. 2008;87(3):214–225.
192. Mao YW, Liu JP, Xiang H, Li DW. Human alphaA- and alphaB-crystallins bind to Bax and Bcl-X(S) to sequester their translocation during staurosporine-induced apoptosis. *Cell Death Differ*. 2004;11(5):512–526.
193. McGreal RS, Kantorow WL, Chauss DC, Wei J, Brennan LA, Kantorow M. alphaB-crystallin/sHSP protects cytochrome c and mitochondrial function against oxidative stress in lens and retinal cells. *Biochim Biophys Acta*. 2012;1820(7):921–930.
194. Horwitz J, Bova MP, Ding LL, Haley DA, Stewart PL. Lens alpha-crystallin: function and structure. *Eye*. 1999;13(pt 3b):403–408.

195. Lowe J, McDermott H, Pike I, Spendlove I, Landon M, Mayer RJ. alpha B crystallin expression in non-lenticular tissues and selective presence in ubiquitinated inclusion bodies in human disease. *J Pathol.* 1992;166(1):61–68.
196. Nagineni CN, Bhat SP. Alpha B-crystallin is expressed in kidney epithelial cell lines and not in fibroblasts. *FEBS Lett.* 1989;249(1):89–94.
197. Dimberg A, Rylova S, Dieterich LC, et al. alphaB-crystallin promotes tumor angiogenesis by increasing vascular survival during tube morphogenesis. *Blood.* 2008;111(4): 2015–2023.
198. Inaguma Y, Hasegawa K, Goto S, Ito H, Kato K. Induction of the synthesis of hsp27 and alpha B crystallin in tissues of heat-stressed rats and its suppression by ethanol or an alpha 1-adrenergic antagonist. *J Biochem.* 1995;117(6):1238–1243.
199. Iwaki T, Kume-Iwaki A, Goldman JE. Cellular distribution of alpha B-crystallin in non-lenticular tissues. *J Histochem Cytochem.* 1990;38(1):31–39.
200. Iwaki T, Kume-Iwaki A, Liem RK, Goldman JE. Alpha B-crystallin is expressed in non-lenticular tissues and accumulates in Alexander's disease brain. *Cell.* 1989;57(1): 71–78.
201. Kase S, He S, Sonoda S, et al. alphaB-crystallin regulation of angiogenesis by modulation of VEGF. *Blood.* 2010;115(16):3398–3406.
202. Klemenz R, Andres AC, Frohli E, Schafer R, Aoyama A. Expression of the murine small heat shock proteins hsp 25 and alpha B crystallin in the absence of stress. *J Cell Biol.* 1993;120(3):639–645.
203. Chen WV, Fielding Hejtmancik J, Piatigorsky J, Duncan MK. The mouse beta B1-crystallin promoter: strict regulation of lens fiber cell specificity. *Biochim Biophys Acta.* 2001;1519(1-2):30–38.
204. Mackay DS, Boskovska OB, Knopf HL, Lampi KJ, Shiels A. A nonsense mutation in CRYBB1 associated with autosomal dominant cataract linked to human chromosome 22q. *Am J Hum Genet.* 2002;71(5):1216–1221.
205. Magabo KS, Horwitz J, Piatigorsky J, Kantorow M. Expression of betaB(2)-crystallin mRNA and protein in retina, brain, and testis. *Invest Ophthalmol Vis Sci.* 2000;41(10):3056–3060.
206. Litt M, Carrero-Valenzuela R, LaMorticella DM, et al. Autosomal dominant cerulean cataract is associated with a chain termination mutation in the human beta-crystallin gene CRYBB2. *Hum Mol Genet.* 1997;6(5):665–668.
207. Piri N, Song M, Kwong JM, Caprioli J. Modulation of alpha and beta crystallin expression in rat retinas with ocular hypertension-induced ganglion cell degeneration. *Brain Res.* 2007;1141:1–9.
208. Templeton JP, Wang X, Freeman NE, et al. A crystallin gene network in the mouse retina. *Exp Eye Res.* 2013;116:129–140.
209. Riazuddin SA, Yasmeen A, Yao W, et al. Mutations in betaB3-crystallin associated with autosomal recessive cataract in two Pakistani families. *Invest Ophthalmol Vis Sci.* 2005;46(6):2100–2106.
210. Valapala M, Wilson C, Hose S, et al. Lysosomal-mediated waste clearance in retinal pigment epithelial cells is regulated by CRYBA1/betaA3/A1-crystallin via V-ATPase-MTORC1 signaling. *Autophagy.* 2014;10(3):480–496.
211. Ma B, Sen T, Asnaghi L, et al. betaA3/A1-Crystallin controls anoikis-mediated cell death in astrocytes by modulating PI3K/AKT/mTOR and ERK survival pathways through the PKD/Bit1-signaling axis. *Cell Death Dis.* 2011;2:e217.
212. Srivastava OP, Srivastava K, Harrington V. Age-related degradation of betaA3/A1-crystallin in human lenses. *Biochem Biophys Res Commun.* 1999;258(3):632–638.
213. Puk O, Ahmad N, Wagner S, Hrabe de Angelis M, Graw J. First mutation in the betaA2-crystallin encoding gene is associated with small lenses and age-related cataracts. *Invest Ophthalmol Vis Sci.* 2011;52(5):2571–2576.

214. Zhou G, Zhou N, Hu S, Zhao L, Zhang C, Qi Y. A missense mutation in CRYBA4 associated with congenital cataract and microcornea. *Mol Vis*. 2010;16:1019–1024.
215. Mehra S, Kapur S, Vasavada AR. Polymorphisms of the gamma crystallin A and B genes among Indian patients with pediatric cataract. *J Postgrad Med*. 2011;57(3):201–205.
216. Fort PE, Freeman WM, Losiewicz MK, Singh RS, Gardner TW. The retinal proteome in experimental diabetic retinopathy: up-regulation of crystallins and reversal by systemic and periocular insulin. *Mol Cell Proteomics*. 2009;8(4):767–779.
217. Guo Y, Su D, Li Q, et al. A nonsense mutation of CRYGC associated with autosomal dominant congenital nuclear cataracts and microcornea in a Chinese pedigree. *Mol Vis*. 2012;18:1874–1880.
218. Vazquez-Chona F, Song BK, Geisert Jr EE. Temporal changes in gene expression after injury in the rat retina. *Invest Ophthalmol Vis Sci*. 2004;45(8):2737–2746.
219. Hansen L, Yao W, Eiberg H, et al. Genetic heterogeneity in microcornea-cataract: five novel mutations in CRYAA, CRYGD, and GJA8. *Invest Ophthalmol Vis Sci*. 2007;48(9):3937–3944.
220. Smith RS, Hawes NL, Chang B, et al. Lop12, a mutation in mouse Crygd causing lens opacity similar to human Coppock cataract. *Genomics*. 2000;63(3):314–320.
221. Sinha D, Esumi N, Jaworski C, Kozak CA, Pierce E, Wistow G. Cloning and mapping the mouse Crygs gene and non-lens expression of [gamma]S-crystallin. *Mol Vis*. 1998;4:8.
222. Haynes 2nd JI, Duncan MK, Piatigorsky J. Spatial and temporal activity of the alpha B-crystallin/small heat shock protein gene promoter in transgenic mice. *Dev Dyn*. 1996;207(1):75–88.
223. Robinson ML, Overbeek PA. Differential expression of alpha A- and alpha B-crystallin during murine ocular development. *Invest Ophthalmol Vis Sci*. 1996;37(11):2276–2284.
224. Wolf L, Yang Y, Wawrousek E, Cvekl A. Transcriptional regulation of mouse alpha A-crystallin gene in a 148 kb Cryaa BAC and its derivates. *BMC Dev Biol*. 2008;8:88.
225. Parthasarathy G, Ma B, Zhang C, et al. Expression of betaA3/A1-crystallin in the developing and adult rat eye. *J Mol Histol*. 2011;42(1):59–69.
226. Overbeek PA, Chepelinsky AB, Khillan JS, Piatigorsky J, Westphal H. Lens-specific expression and developmental regulation of the bacterial chloramphenicol acetyltransferase gene driven by the murine alpha A-crystallin promoter in transgenic mice. *Proc Natl Acad Sci USA*. 1985;82(23):7815–7819.
227. Yang Y, Chauhan BK, Cveklova K, Cvekl A. Transcriptional regulation of mouse alphaB- and gammaF-crystallin genes in lens: opposite promoter-specific interactions between Pax6 and large Maf transcription factors. *J Mol Biol*. 2004;344(2):351–368.
228. Chen Q, Dowhan DH, Liang D, Moore DD, Overbeek PA. CREB-binding protein/p300 co-activation of crystallin gene expression. *J Biol Chem*. 2002;277(27):24081–24089.
229. Ring BZ, Cordes SP, Overbeek PA, Barsh GS. Regulation of mouse lens fiber cell development and differentiation by the Maf gene. *Development*. 2000;127(2):307–317.
230. Civil A, van Genesen ST, Lubsen NH. c-Maf, the gammaD-crystallin Maf-responsive element and growth factor regulation. *Nucleic Acids Res*. 2002;30(4):975–982.
231. Fujimoto M, Izu H, Seki K, et al. HSF4 is required for normal cell growth and differentiation during mouse lens development. *EMBO J*. 2004;23(21):4297–4306.
232. Ji WK, Tang XC, Yi M, et al. p53 directly regulates alphaA- and betaA3/A1-crystallin genes to modulate lens differentiation. *Curr Mol Med*. 2013;13(6):968–978.
233. Xie Q, Cvekl A. The orchestration of mammalian tissue morphogenesis through a series of coherent feed-forward loops. *J Biol Chem*. 2011;286(50):43259–43271.
234. Yang Y, Stopka T, Golestaneh N, et al. Regulation of alphaA-crystallin via Pax6, c-Maf, CREB and a broad domain of lens-specific chromatin. *EMBO J*. 2006;25(10):2107–2118.

235. Wada K, Matsushima Y, Tada T, et al. Expression of truncated PITX3 in the developing lens leads to microphthalmia and aphakia in mice. *PLoS One*. 2014;9(10):e111432.
236. Cui W, Tomarev SI, Piatigorsky J, Chepelinsky AB, Duncan MK. Mafs, Prox1, and Pax6 can regulate chicken betaB1-crystallin gene expression. *J Biol Chem*. 2004;279(12):11088–11095.
237. Chen X, Taube JR, Simirskii VI, Patel TP, Duncan MK. Dual roles for Prox1 in the regulation of the chicken betaB1-crystallin promoter. *Invest Ophthalmol Vis Sci*. 2008;49(4):1542–1552.
238. Lengler J, Krausz E, Tomarev S, Prescott A, Quinlan RA, Graw J. Antagonistic action of Six3 and Prox1 at the gamma-crystallin promoter. *Nucleic Acids Res*. 2001;29(2):515–526.
239. Nishiguchi S, Wood H, Kondoh H, Lovell-Badge R, Episkopou V. Sox1 directly regulates the gamma-crystallin genes and is essential for lens development in mice. *Genes Dev*. 1998;12(6):776–781.
240. Kamachi Y, Uchikawa M, Tanouchi A, Sekido R, Kondoh H. Pax6 and SOX2 form a co-DNA-binding partner complex that regulates initiation of lens development. *Genes Dev*. 2001;15(10):1272–1286.
241. Rajaram N, Kerppola TK. Synergistic transcription activation by Maf and Sox and their subnuclear localization are disrupted by a mutation in Maf that causes cataract. *Mol Cell Biol*. 2004;24(13):5694–5709.
242. Liu FY, Tang XC, Deng M, et al. The tumor suppressor p53 regulates c-Maf and Prox-1 to control lens differentiation. *Curr Mol Med*. 2012;12(8):917–928.
243. Liu S, Piatigorsky J. Regulation of mouse small heat shock protein alphab-crystallin gene by aryl hydrocarbon receptor. *PLoS One*. 2011;6(4):e17904.
244. Xie Q, McGreal RS, Reneker LW, Musil LS, Cvekl A. Regulation of c-Maf and αA-crystallin in Ocular Lens by FGF Signaling. *J Biol Chem*. 2015, in revision.
245. Cvekl A, Sax CM, Bresnick EH, Piatigorsky J. A complex array of positive and negative elements regulates the chicken alpha A-crystallin gene: involvement of Pax-6, USF, CREB and/or CREM, and AP-1 proteins. *Mol Cell Biol*. 1994;14(11):7363–7376.
246. Ilagan JG, Cvekl A, Kantorow M, Piatigorsky J, Sax CM. Regulation of alphaA-crystallin gene expression. Lens specificity achieved through the differential placement of similar transcriptional control elements in mouse and chicken. *J Biol Chem*. 1999;274(28):19973–19978.
247. Lyu J, Joo CK. Wnt signaling enhances FGF2-triggered lens fiber cell differentiation. *Development*. 2004;131(8):1813–1824.
248. Antosova B, Smolikova J, Borkovcova R, et al. Ectopic activation of Wnt/beta-catenin signaling in lens fiber cells results in cataract formation and aberrant fiber cell differentiation. *PLoS One*. 2013;8(10):e78279.
249. Mori M, Ghyselinck NB, Chambon P, Mark M. Systematic immunolocalization of retinoid receptors in developing and adult mouse eyes. *Invest Ophthalmol Vis Sci*. 2001;42(6):1312–1318.
250. Gong L, Ji WK, Hu XH, et al. Sumoylation differentially regulates Sp1 to control cell differentiation. *Proc Natl Acad Sci USA*. 2014;111(15):5574–5579.
251. Yang Y, Cvekl A. Large Maf transcription factors: cousins of AP-1 proteins and important regulators of cellular differentiation. *Einstein J Biol Med*. 2007;23(1):2–11.
252. Kawauchi S, Takahashi S, Nakajima O, et al. Regulation of lens fiber cell differentiation by transcription factor c-Maf. *J Biol Chem*. 1999;274(27):19254–19260.
253. Kim JI, Li T, Ho IC, Grusby MJ, Glimcher LH. Requirement for the c-Maf transcription factor in crystallin gene regulation and lens development. *Proc Natl Acad Sci USA*. 1999;96(7):3781–3785.
254. Takeuchi T, Kudo T, Ogata K, et al. Neither MafA/L-Maf nor MafB is essential for lens development in mice. *Genes Cells*. 2009;14(8):941–947.

255. Mears AJ, Kondo M, Swain PK, et al. Nrl is required for rod photoreceptor development. *Nat Genet.* 2001;29(4):447–452.
256. Ochi H, Ogino H, Kageyama Y, Yasuda K. The stability of the lens-specific Maf protein is regulated by fibroblast growth factor (FGF)/ERK signaling in lens fiber differentiation. *J Biol Chem.* 2003;278(1):537–544.
257. Reza HM, Urano A, Shimada N, Yasuda K. Sequential and combinatorial roles of maf family genes define proper lens development. *Mol Vis.* 2007;13:18–30.
258. Ogino H, Yasuda K. Induction of lens differentiation by activation of a bZIP transcription factor. L-Maf. *Science.* 1998;280(5360):115–118.
259. Vinson C, Myakishev M, Acharya A, Mir AA, Moll JR, Bonovich M. Classification of human B-ZIP proteins based on dimerization properties. *Mol Cell Biol.* 2002;22(18):6321–6335.
260. Jamieson RV, Perveen R, Kerr B, et al. Domain disruption and mutation of the bZIP transcription factor, MAF, associated with cataract, ocular anterior segment dysgenesis and coloboma. *Hum Mol Genet.* 2002;11(1):33–42.
261. Perveen R, Favor J, Jamieson RV, Ray DW, Black GC. A heterozygous c-Maf transactivation domain mutation causes congenital cataract and enhances target gene activation. *Hum Mol Genet.* 2007;16(9):1030–1038.
262. Lachke SA, Ho JW, Kryukov GV, et al. iSyTE: integrated Systems Tool for eye gene discovery. *Invest Ophthalmol Vis Sci.* 2012;53(3):1617–1627.
263. Goudreau G, Petrou P, Reneker LW, Graw J, Loster J, Gruss P. Mutually regulated expression of Pax6 and Six3 and its implications for the Pax6 haploinsufficient lens phenotype. *Proc Natl Acad Sci USA.* 2002;99(13):8719–8724.
264. Mikkola I, Bruun JA, Bjorkoy G, Holm T, Johansen T. Phosphorylation of the transactivation domain of Pax6 by extracellular signal-regulated kinase and p38 mitogen-activated protein kinase. *J Biol Chem.* 1999;274(21):15115–15126.
265. Yoo YD, Huang CT, Zhang X, Lavaute TM, Zhang SC. Fibroblast growth factor regulates human neuroectoderm specification through ERK1/2-PARP-1 pathway. *Stem Cells.* 2011;29(12):1975–1982.
266. Yan Q, Liu WB, Qin J, et al. Protein phosphatase-1 modulates the function of Pax-6, a transcription factor controlling brain and eye development. *J Biol Chem.* 2007;282(19): 13954–13965.
267. Hu Y, Mivechi NF. Association and regulation of heat shock transcription factor 4b with both extracellular signal-regulated kinase mitogen-activated protein kinase and dual-specificity tyrosine phosphatase DUSP26. *Mol Cell Biol.* 2006;26(8): 3282–3294.
268. Hu YZ, Zhang J, Li S, et al. The transcription activity of heat shock factor 4b is regulated by FGF2. *Int J Biochem Cell Biol.* 2013;45(2):317–325.
269. Zhang J, Ma Z, Wang J, et al. Regulation of Hsf4b nuclear translocation and transcription activity by phosphorylation at threonine 472. *Biochim Biophys Acta.* 2014;1843(3): 580–589.
270. Kim EA, Noh YT, Ryu MJ, et al. Phosphorylation and transactivation of Pax6 by homeodomain-interacting protein kinase 2. *J Biol Chem.* 2006;281(11):7489–7497.
271. Leavenworth JW, Ma X, Mo YY, Pauza ME. SUMO conjugation contributes to immune deviation in nonobese diabetic mice by suppressing c-Maf transactivation of IL-4. *J Immunol.* 2009;183(2):1110–1119.
272. Yan Q, Gong L, Deng M, et al. Sumoylation activates the transcriptional activity of Pax-6, an important transcription factor for eye and brain development. *Proc Natl Acad Sci USA.* 2010;107(49):21034–21039.
273. Dubin RA, Gopal-Srivastava R, Wawrousek EF, Piatigorsky J. Expression of the murine alpha B-crystallin gene in lens and skeletal muscle: identification of a muscle-preferred enhancer. *Mol Cell Biol.* 1991;11(9):4340–4349.

274. Gopal-Srivastava R, Piatigorsky J. The murine alpha B-crystallin/small heat shock protein enhancer: identification of alpha BE-1, alpha BE-2, alpha BE-3, and MRF control elements. *Mol Cell Biol.* 1993;13(11):7144–7152.
275. Gopal-Srivastava R, Haynes 2nd JI, Piatigorsky J. Regulation of the murine alpha B-crystallin/small heat shock protein gene in cardiac muscle. *Mol Cell Biol.* 1995;15(12):7081–7090.
276. Gopal-Srivastava R, Kays WT, Piatigorsky J. Enhancer-independent promoter activity of the mouse alphaB-crystallin/small heat shock protein gene in the lens and cornea of transgenic mice. *Mech Dev.* 2000;92(2):125–134.
277. Haynes 2nd JI, Gopal-Srivastava R, Frederikse PH, Piatigorsky J. Differential use of the regulatory elements of the alpha B-crystallin enhancer in cultured murine lung (MLg), lens (alpha TN4-1) and muscle (C2C12) cells. *Gene.* 1995;155(2):151–158.
278. Grocott T, Frost V, Maillard M, et al. The MH1 domain of Smad3 interacts with Pax6 and represses autoregulation of the Pax6 P1 promoter. *Nucleic Acids Res.* 2007;35(3): 890–901.
279. Grocott T, Johnson S, Bailey AP, Streit A. Neural crest cells organize the eye via TGF-beta and canonical Wnt signalling. *Nat Commun.* 2011;2:265.
280. Ganguly K, Favor J, Neuhauser-Klaus A, et al. Novel allele of crybb2 in the mouse and its expression in the brain. *Invest Ophthalmol Vis Sci.* 2008;49(4):1533–1541.
281. Ninkovic J, Pinto L, Petricca S, et al. The transcription factor Pax6 regulates survival of dopaminergic olfactory bulb neurons via crystallin alphaA. *Neuron.* 2010;68(4): 682–694.
282. Sun M, Holter SM, Stepan J, et al. Crybb2 coding for betaB2-crystallin affects sensorimotor gating and hippocampal function. *Mamm Genome.* 2013;24(9-10):333–348.
283. Garcia CM, Huang J, Madakashira BP, et al. The function of FGF signaling in the lens placode. *Dev Biol.* 2011;351(1):176–185.
284. Zhou J, Kochhar DM. Regulation of AP-2 and apoptosis in developing eye in a vitamin A-deficiency model. *Birth Defects Res A Clin Mol Teratol.* 2003;67(1):41–53.
285. Luscher B, Mitchell PJ, Williams T, Tjian R. Regulation of transcription factor AP-2 by the morphogen retinoic acid and by second messengers. *Genes Dev.* 1989;3(10): 1507–1517.
286. West-Mays JA, Zhang J, Nottoli T, et al. AP-2alpha transcription factor is required for early morphogenesis of the lens vesicle. *Dev Biol.* 1999;206(1):46–62.
287. Pontoriero GF, Deschamps P, Ashery-Padan R, et al. Cell autonomous roles for AP-2alpha in lens vesicle separation and maintenance of the lens epithelial cell phenotype. *Dev Dyn.* 2008;237(3):602–617.
288. Bassett EA, Korol A, Deschamps PA, et al. Overlapping expression patterns and redundant roles for AP-2 transcription factors in the developing mammalian retina. *Dev Dyn.* 2012;241(4):814–829.
289. Conte I, Carrella S, Avellino R, et al. miR-204 is required for lens and retinal development via Meis2 targeting. *Proc Natl Acad Sci USA.* 2010;107(35):15491–15496.
290. Shaham O, Gueta K, Mor E, et al. Pax6 regulates gene expression in the vertebrate lens through miR-204. *PLoS Genet.* 2013;9(3):e1003357.
291. Xie Q, Ung D, Khafizov K, Fiser A, Cvekl A. Gene regulation by PAX6: structural-functional correlations of missense mutants and transcriptional control of Trpm3/ miR-204. *Mol Vis.* 2014;20:270–282.
292. Lachke SA, Alkuraya FS, Kneeland SC, et al. Mutations in the RNA granule component TDRD7 cause cataract and glaucoma. *Science.* 2011;331(6024):1571–1576.
293. Ragoczy T, Bender MA, Telling A, Byron R, Groudine M. The locus control region is required for association of the murine beta-globin locus with engaged transcription factories during erythroid maturation. *Genes Dev.* 2006;20(11):1447–1457.

294. Neppl RL, Kataoka M, Wang DZ. Crystallin-alphaB regulates skeletal muscle homeostasis via modulation of argonaute2 activity. *J Biol Chem*. 2014;289(24):17240–17248.
295. Yang C, Yang Y, Brennan L, Bouhassira EE, Kantorow M, Cvekl A. Efficient generation of lens progenitor cells and lentoid bodies from human embryonic stem cells in chemically defined conditions. *FASEB J*. 2010;24(9):3274–3283.
296. Qiu X, Yang J, Liu T, Jiang Y, Le Q, Lu Y. Efficient generation of lens progenitor cells from cataract patient-specific induced pluripotent stem cells. *PLoS One*. 2012;7(3):e32612.
297. Anchan RM, Lachke SA, Gerami-Naini B, *et al.* Pax6- and Six3-mediated induction of lens fate in mouse and human ES cells. *PLoS One*; 2014;9(12):e115106.

CHAPTER ELEVEN

Lens Biology and Biochemistry

J. Fielding Hejtmancik*, S. Amer Riazuddin†, Rebecca McGreal‡,§, Wei Liu‡,§, Ales Cvekl‡,§, Alan Shiels¶,1

*Ophthalmic Genetics and Visual Function Branch, National Eye Institute, National Institutes of Health, Bethesda, Maryland, USA
†The Wilmer Eye Institute, Johns Hopkins University School of Medicine, Baltimore, Maryland, USA
‡Department of Genetics and Ophthalmology, Albert Einstein College of Medicine, Bronx, New York, USA
§Department of Ophthalmology and Visual Sciences, Albert Einstein College of Medicine, Bronx, New York, USA
¶Department of Ophthalmology and Visual Sciences, Washington University School of Medicine, St. Louis, Missouri, USA
[1]Corresponding author: e-mail address: Shiels@vision.wustl.edu

Contents

Abstract

The primary function of the lens resides in its transparency and ability to focus light on the retina. These require both that the lens cells contain high concentrations of densely packed lens crystallins to maintain a refractive index constant over distances approximating the wavelength of the light to be transmitted, and a specific arrangement of anterior epithelial cells and arcuate fiber cells lacking organelles in the nucleus to avoid blocking transmission of light. Because cells in the lens nucleus have shed their organelles, lens crystallins have to last for the lifetime of the organism, and are specifically adapted to this function. The lens crystallins comprise two major families: the βγ-crystallins are among the most stable proteins known and the α-crystallins, which have a chaperone-like function. Other proteins and metabolic activities of the lens are primarily organized to protect the crystallins from damage over time and to maintain homeostasis of the lens cells. Membrane protein channels maintain osmotic and ionic balance across the lens, while the lens cytoskeleton provides for the specific shape of the lens cells, especially the fiber cells of the nucleus. Perhaps most importantly, a large part of

Progress in Molecular Biology and Translational Science, Volume 134
ISSN 1877-1173
http://dx.doi.org/10.1016/bs.pmbts.2015.04.007
© 2015 Elsevier Inc.
All rights reserved.

the metabolic activity in the lens is directed toward maintaining a reduced state, which shelters the lens crystallins and other cellular components from damage from UV light and oxidative stress. Finally, the energy requirements of the lens are met largely by glycolysis and the pentose phosphate pathway, perhaps in response to the avascular nature of the lens. Together, all these systems cooperate to maintain lens transparency over time.

ABBREVIATIONS

AIM absent in melanoma
AQP0 aquaporin 0, or MIP, major intrinsic protein
ATP adenosine triphosphate
ATPase adenosine triphosphatase
BFSP1 beaded filament specific protein 1, or CP-115, or filensin
BFSP2 beaded filament specific protein 2, or CP-49, or phakinin
cAMP cyclic adenosine monophosphate
cGMP cyclic guanosine monophosphate
ERK extracellular signal-regulated kinase, or MAPK, mitogen-activated protein kinase
FGF fibroblast growth factor
GFAP glial fibrillary acidic protein
GJA1 gap junction protein A1, or connexin-43, Cx43
GJA3 gap junction protein A3, or connexin-46, Cx46
GJA8 gap junction protein A8, or connexin-50, Cx50
GLUT1 glucose transporter 1
GLUT3 glucose transporter 3
GSH glutathione
IP_3 inositol trisphosphate
MAPK mitogen-activated protein kinase, or ERK, extracellular signal-regulated kinase
MEK mitogen-activated protein kinase kinase
MIP major intrinsic protein, or AQP0, aquaporin 0
NCAM neural-cell adhesion molecule
PEDF pigment epithelium-derived factor
Raf Raf-1 protooncogene
RPE retinal pigment epithelium
TGF-β transforming growth factor beta
UVB ultraviolet B

1. LENS BIOLOGY: OVERVIEW

This overview of lens biology is necessarily limited and will be based primarily on those functions that support transparency and focusing of light, the primary function of the lens. That function is accomplished largely by the combination of the microarchitecture of the lens, comprising anterior epithelial cells and arcuate fiber cells lacking organelles in the nucleus, and on a molecular level the densely packed lens crystallins. Other biological activities of the lens function primarily to protect these two complementary systems from disruption and damage by age and external insults, especially as related to UV light and oxidative stress. While this is a very simplistic view of the lens, it does provide a perspective for consideration of the various biological pathways required for lens function.

2. CRYSTALLINS AND LENS BIOLOGY

Crystallins are the most prevalent proteins in the lens. Comprising of two families, α- and βγ-crystallins, they make up 90% of water-soluble proteins of the mammalian lens.[1] They are highly organized and provide a refractive index gradient, which allows for transparency of the lens. Once thought to be solely lens proteins, crystallins have since been identified outside of the lens, providing several important functions (see Table 1). Organelle degradation, including endoplasmic reticulum, Golgi apparatus, mitochondria, and nuclei, which occurs during late stages of lens fiber cell differentiation, is also imperative for lens transparency and any disruption of this process will result in light scattering and ultimately cataract.[2,3] Since differentiated lens cells lack the entire apparatus to produce new proteins, crystallins are not turned over and those in the center of the lens are thus among the oldest proteins in the body. Maintenance of lens proteins is therefore crucial for ocular health and for the prevention of lens opacities.

2.1 α-Crystallins

α-Crystallin monomers include an α-crystallin domain that is highly conserved in all members of the small heat-shock protein family. In the lens, α-crystallin proteins exist as globular aggregates roughly 600–900 kDa in mass,[4] whose quaternary structure has been suggested to behave as a protein micelle.[5,6] αA- and αB-Crystallins have been shown to occupy equivalent and dynamic positions in the aggregate, with subunit exchange occurring easily,[6] and by a synchrotron radiation X-ray solution scattering study of a truncation mutant[7–9] retaining chaperone activity. This suggests that αB-crystallin is composed of flexible monomers with an extended surface area.[10] Cryo-electron microscopy has shown that recombinant α-crystallin has variable monomer packing consisting of a hollow central core with a surrounding protein shell.[11] α-Crystallin monomers appear to dimerize and then associate into a large complex, with amino and

Table 1 Summary of Lens Crystallin Gene Families

Distribution	Crystallin	(Related) or Identical
Represented in all vertebrates	α	(Small heat-shock proteins)
		(*Schistosoma mansoni* antigen p40)
	β	(*Myxococcus xanthus* protein S)
		(*Physarum polycephalum* spherulin 3a)
	γ	(EDSP)
		(A1M1, AIMIL, CRYBG3)
Birds and reptiles	δ1	(Argininocuccinate lyase)
	δ2	Argininosuccinate lyase
	ε	Lactate dehydrogenase B
Some mammals	ζ	(NADPH:quinone reductase), CRYZ and CRYZL1 in humans
	π	Glyceraldehyde 3-phosphate dehydrogenase
	ι	(Cellular retinol-binding protein)
	η	Cytoplasmic aldehyde dehydrogenase
	λ	(Hydroxyacyl CoA dehydrogenase)
	μ	(Ornithine cyclodeaminase) CRYM in humans
	BHMT	Betaine–homocysteine methyltransferase (BHMT2)
Many species	τ	α-Enolase
Frogs	ρ	(NADPH-dependent reductases)
Cephalopods	SL11/ LOPS4	Glutathione *S*-transferase
	S	(Glutathione *S*-transferase)
	Ω/L	(Aldehyde dehydrogenase)
Jellyfish	J	(Novel proteins)

carboxyl regions of the αB-crystallin monomers interacting with residues from the corresponding region of their partners.[12] Although αA- and αB-crystallins seem to occupy equivalent positions in the α-crystallin complex, they show different tissue-specific expression patterns, show different phenotypes when knocked out in mice,[13,13] differ in their phosphorylation patterns,[14,15] structural properties, and chaperone activity.[16] In addition, αB-crystallin fractionates with rough endoplasmic reticulum while αA-crystallin fractionates with smooth membranes,[17] and only αB is stress inducible[18,19] suggesting that each fulfills a unique role in the lens.

α-Crystallins are phosphorylated in both cyclic adenosine monophosphate (cAMP)-dependent[20,21] and -independent[14,15] manner. They can be dephosphorylated by a calcineurin-related enzyme found primarily in the lens epithelial cells and to a lesser extent in fiber cells.[22] α-Crystallin phosphorylation is of unknown significance as it does not seem to affect chaperone activity[23] or inhibition of glial fibrillary acidic protein (GFAP) and vimentin assembly.[24] However, it does stabilize actin filaments and decrease cytochalasin-dependent depolymerization.[25] The phosphorylation of αB-crystallin is stimulated by stress, including hydrogen peroxide treatment of cultured lenses[23] and heat treatment of heart and diaphragm muscles.[26]

Both αA- and αB-crystallin can protect β- and γ-crystallins as well as other lens proteins from thermal aggregation, although not recycling these proteins.[27,28] α-Crystallin can improve revival of glutathione reductase activity in human cortical extracts of both cataractous and clear lens by thioredoxin and thioredoxin reductase[29] and regenerate sorbitol dehydrogenase activity.[30] This does not require adenosine triphosphate (ATP) hydrolysis, although it is enhanced in the presence of ATP, and αB-crystallin chaperone activity is increased by the products of common metabolic pathways such as glutathione and pantethine.[31] α-Crystallin chaperone activity requires its C-terminal domain,[32] whose IXI/V motif is important in inter-subunit interactions and chaperone site accessibility.[33] During this process, the target protein apparently lodges in the fenestrated outer shell of the α-crystallin complex,[34] and its structure transitions into a multimeric molten globular state resulting in the appropriate placement of its hydrophobic surfaces.[35] Binding of target proteins by α-crystallin is cooperative, and mutations of βB1-crystallin increasing binding by αA- but not αB-crystallin.[36] Zinc binds to α-crystallin and appears to increase its stability and chaperone activity.[37] αA-Crystallin peptides inhibit amyloid fibril formation by amyloid beta protein,[38,39] and α-crystallin suppresses microtubule assembly, maintaining a pool of unassembled tubulin.[40] All of these characteristics probably protect against cataractogenesis by reducing the aggregation of partially denatured proteins that accumulate within the lens during aging, holding them in high-molecular-weight complexes that do not form a nidus for precipitation and light scattering. Thus, the chaperone function of α-crystallins serves to protect the lens from toxicity caused by damaged proteins denaturing with age and it probably acts similarly in nonlenticular tissues. In addition, peptides from α-crystallin can inhibit apoptosis of lens epithelia and lens opacification.[41]

α-Crystallin expression is not limited to the lens. αB-crystallin is widely expressed, including in the heart, skeletal muscle, kidney, lung, brain, and retina, although at lower levels than in the lens.[42,43] In skeletal muscle, αB-crystallin expression increases by stretching,[44] and in the heart it forms aggregates of 400–650 kDa.[45] In contrast, αA-crystallin is expressed constitutively at low levels in many tissues and somewhat higher in the spleen and thymus.[46] αB-Crystallin has been shown to block p53-dependent apoptosis by inhibiting RAS activation,[47] and mitochondria in stressed cardiac muscle.[48] α-Crystallin enhances survival of optic nerve axons after a crush injury[49] and is upregulated in regenerating neuron axons in monkeys.[50] As mentioned above, in cultured cells αB- but not αA-crystallin is inducible by stress, including heat[19] and osmotic shock.[18] Possibly related to this, αB-crystallin is found at high levels in the brains of patients with Alexander disease,[51] brains of scrapie-infected hamsters,[52] and fibroblasts from patients with Werner syndrome.[53]

The ability of the α-crystallins to inhibit apoptosis is well documented.[54,55] α-Crystallins have been shown to prevent apoptosis induced by a number of factors, including etoposide and staurosporine,[56] UV radiation,[54] sorbitol,[56] TNFα,[57] hydrogen peroxide,[58] and okadaic acid.[59] Kamradt and colleagues demonstrated that overexpressing αB-crystallin slows caspase-3 maturation, thus inhibiting apoptosis.[60] Similarly, overexpression of αA- or αB-crystallin in lens epithelial cells confers resistance against several forms of stress, including thermal and photochemical.[57] Andley and colleagues demonstrated that αA-crystallin shows two- to threefold higher levels of antiapoptotic activity than equivalent amounts of αB-crystallin in cultured lens epithelial cells.[57] Mutations in α-crystallin, that were shown to reduce chaperone function, also diminished the antiapoptotic activity of αA-crystallin.[61] Similarly, activation of apoptosis was seen in αB-R120G-induced desmin-related myopathy.[62] αA-/αB-Crystallin double-knockout mice exhibit caspase-dependent secondary lens fiber cell disintegration.[63]

The mechanism of apoptotic inhibition by α-crystallin has been shown to be different for the two subunits. αB-crystallin is able to prevent apoptosis through repression of Raf/MEK/ERK signaling, whereas αA-crystallin activates the Akt surviving pathway to inhibit apoptosis.[54] αA-Crystallin has also been shown to prevent apoptosis by enhancing phosphoinositide 3-kinase activity.[64] Moreover, αB-crystallin was shown to prevent caspase 3 activation and therefore apoptosis in both astrocytes[65] and cardiomyocytes.[66]

Preservation of mitochondrial function is essential for viability of both lens and retinal cells.[67–69] It has been demonstrated that αB-crystallin translocates to the mitochondria under oxidative stress conditions, protecting its function and preventing oxidation of cytochrome *c*. However, the chaperone function of α-crystallin was not required for this mitochondrial protective function.[70]

2.2 βγ-Crystallins

The β- and γ-crystallins show sequence similarities of 30% in the aligned regions of their central globular domains, which have a similar tertiary structure, and together form a βγ-crystallin superfamily.[71–73] βγ-Crystallins have a common two-domain structure comprising four repeated "Greek key" motifs because they resemble the classical pattern found on the borders of ancient Greek pottery. Each Greek key motif consists of an extremely stable, torqued β-pleated sheet.[74] The first and second Greek key motifs make up the N-terminal domain, while the third and fourth form the C-terminal domain of the protein. The two domains are connected by a linker or connecting peptide (Fig. 1). Because hydrogen bonding occurs between amino acids of different motifs in different domains, all four motifs need to be present for the high stability seen in the βγ-crystallins.[75] In addition, interactions between the interface of the amino- and carboxyl-terminal domains help to stabilize the structure.[76]

While each of the four Greek key motifs is similar to the others, the first motif shows higher homology to the third, and the second to the fourth, consistent with an evolution of the βγ-crystallin family starting from a single-motif through a two motif intermediate that is reduplicated, creating the present βγ-crystallin core structure (see Fig. 1).[77] This intermediary evolutionary step is represented by spherulin 3a, *Ciona* βγ-crystallin, and nitrollin, while absent in melanoma 1 (AIM1) has expanded to six domains.[78,79] Many evolutionary βγ-crystallin homologs bind Ca^{2+}, which increases their stability,[80] as has also been suggested for mammalian β-crystallins.[81] In addition to being extremely stable, the Greek key motif structure can also quench UV irradiation, helping to prevent photooxidation of the lens as it ages.[82]

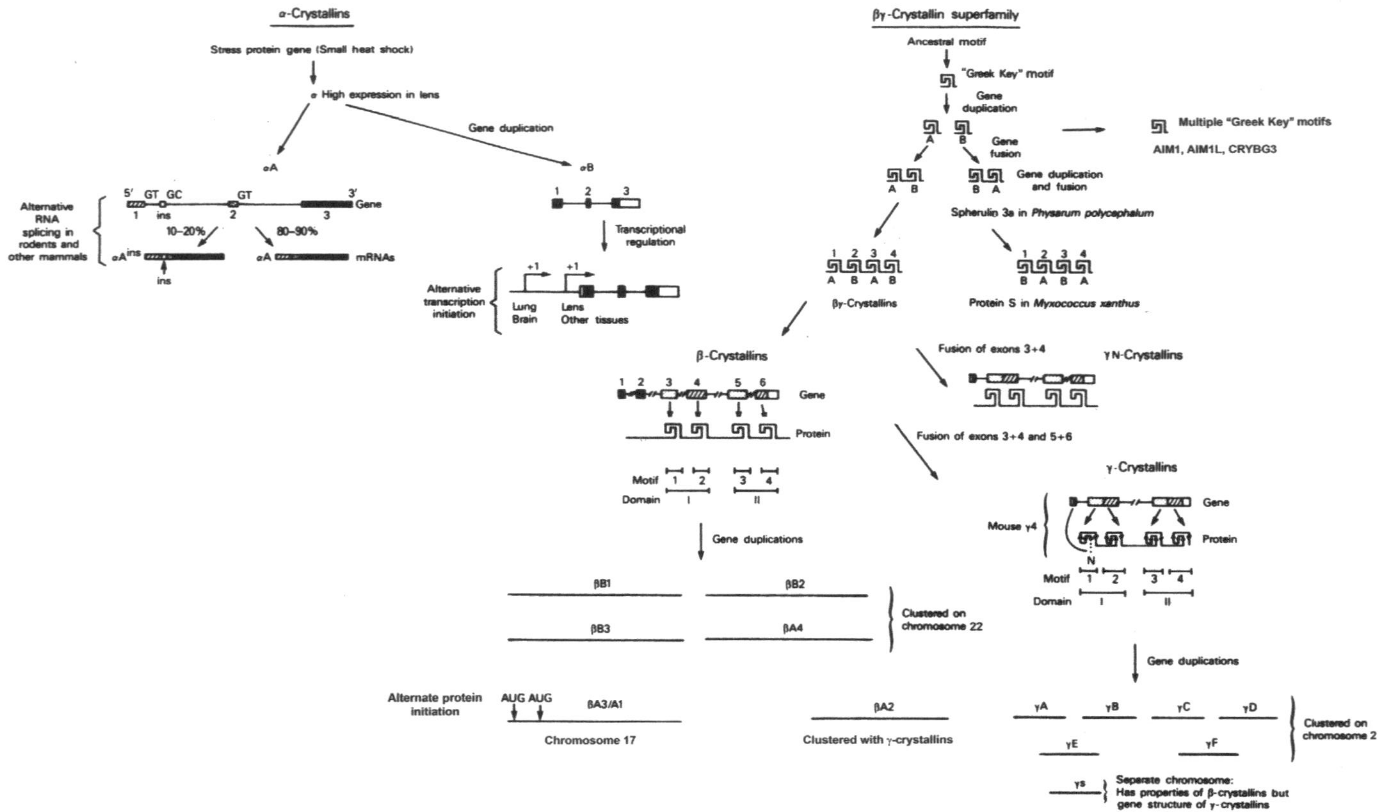

Figure 1 Evolution of the genomic and protein structures of the ubiquitous α-, β-, and γ-crystallins. The α-crystallins are chaperones related to heat-shock proteins, while the βγ-crystallins belong to a superfamily related to stress proteins. During evolution, multiple events have increased divergence of the crystallin families, including gene duplication, fusion of exons, and the use of multiple transcription and protein synthesis initiation sites.

β-Crystallin monomers associate initially into homo- and heterodimers of about 50 kDa, which then further associate in a more complicated fashion into complexes of 150–200 kDa, especially in the lens.[83] The crystal structure of βB2-crystallin dimers shows an extended linker connecting the domains. The widely separated amino and carboxyl domains of one β-crystallin polypeptide pair through "domain swapping" in which the amino domain of the first pairs with the carboxyl domain of the second, while the carboxyl domain of the first pairs with the amino domain of the second. In contrast, in the monomeric γ-crystallins, this connecting peptide folds back on itself so that the two domains of a single molecule pair with each other.[84] The domain swapping pairing of the β-crystallins allows higher oligomerization through further association of the dimers.[85,86] βB1-Crystallin has a stronger tendency than βB2-crystallin to form higher order oligomers both *in vitro*[87] and *in vivo*.[88] Under physiological conditions, β-crystallin complexes are able to exchange monomers rapidly, indicating that they are in a reversible equilibrium rather than being static structures.[89] Formation of β-crystallin dimers is entropically driven, with the tightness of association as the temperature increases.[90] Substitution of the γ2-crystallin connecting peptide into murine βA3-crystallin and, conversely, part of the βB2-crystallin connecting peptide into γ2-crystallin by site-specific mutagenesis has no effect on association, suggesting that the connecting peptide itself does not explain dimerization of β-crystallins while the γ-crystallins remain monomers.[91,92]

While it is unclear whether cleavage of β-crystallin arms represents development or aging, loss of the terminal extensions definitely alters their association and stability. Truncation of the amino-terminal extensions increases affinity of association of monomers of βA3-crystallin into dimers[93] but decreases that of βB2-crystallin.[90] Truncation of the terminal arms of βA3-crystallin increases its susceptibility to photooxidation by ultraviolet B (UVB).[94] Truncation of βB1-crystallin increases its tendency to oligomerize and causes it to undergo two phase transitions, resembling those for cold cataract in γ-crystallins.[95] However, removal of amino- and carboxyl-terminal arms of rat βB2-crystallin[96] did not alter its association properties nor did changing the sequence or deleting part of the amino-terminal arms of chick βB1-crystallin, or conversion of rat βA3- from βA1-crystallin by deletion of the 17 amino acids by which they differ.[97] As with the bonding of γ-crystallins at their interdomain interface, dimerization of β-crystallins stabilizes them and increases their solubility.[98]

The highly symmetrical and correspondingly extremely stable γ-crystallins have a molecular mass of about 21 kDa.[84] The γ-crystallins are expressed specifically in the lens fibers and are thus the major crystallins in the lens nucleus, contributing to it having the highest protein concentration and being the most dehydrated and hence the hardest section of the lens. The γ-crystallins are adapted for high-density molecular packing.[99] The γ-crystallins are abundant in almost all mammals including humans, but not in birds and reptiles, which use other proteins as their major nuclear lens crystallins. Species, such as fish and rodents, with high concentrations of γ-crystallins, have hard lenses that lack the accommodative powers of the softer lenses found in birds and reptiles. γ-Crystallins, and especially γDEF, can form "cold cataracts", a reversible opacity which occurs on cooling of the lens.[100]

γS-Crystallin, which was previously termed βs-crystallin, represents an intermediate between the β- and γ-crystallins.[101,102] γS-Crystallin shares many physical and chemical properties with the β-crystallins. The γS-crystallin protein is slightly larger than most γ-crystallins, having 177 residues, and includes an amino-terminal arm like the β-crystallins.[101] Its isoelectric point is lower than that of most γ-crystallins, being closer to those of the β-crystallins. Finally, as opposed to other γ-crystallins, the amino-terminus of γS-crystallin is blocked, similar to the β-crystallins. The human γS-crystallin gene is located on chromosome 3, while the remaining γ-crystallin genes are located in a cluster on chromosome 2q33-35. Developmentally, γS-crystallin is expressed later than other γ-crystallins, and continues into adulthood, when expression of other γ-crystallins is low or has ceased.[103] Unlike other γ-crystallins, γS-crystallin is found in birds and reptiles, as are the β-crystallins.[104,105] However, unlike β-crystallins, γS-crystallin exists in solution as a monomeric protein as do the remaining γ-crystallins. One important criterion for assigning γS-crystallin to the γ-crystallins is that its gene structure consists of three exons, as do the other γ-crystallins but not the β-crystallins, which consist of six exons.[6,106]

In contrast to the ubiquitous α- and βγ-crystallins, taxon-specific crystallins are expressed at high concentrations in the lens (usually at least 10% of the soluble protein), but tend only to be present in one or a few species.[107] Many taxon-specific crystallins are likely to have arisen by "gene-sharing," in which a single gene product usually retaining its original function in nonlens tissues, acquires an additional function in the lens.[13,108,109] Use of a single gene product for multiple separate functions subjects it to additional evolutionary selection. In addition, a mutation in a regulatory sequence altering gene expression may result in an additional function for the encoded protein before or without gene duplication or loss of its original function. Gene duplication and divergent evolution and specialization of function for one of the two resulting proteins may occur later, as has been suggested to have happened for the α- and δ-crystallins.[110] While it is expected that taxon-specific crystallins might have enzymatic activity in addition to their role in the lens, similar evolutionary and functional principles appear to apply to the ubiquitous crystallins as well, as the extralenticular roles of the crystallins have been examined.

Novel roles of crystallins in autophagy and tissue remodeling have recently come to light. For example, loss of Cryba1 was shown to lead to lysosomal dysregulation and the impairment of both autophagy and phagocytosis in retinal pigment epithelium (RPE) cells.[111] Also, mutation of the rat βA3/A1-crystallin gene (*Nuc1* allele) impairs phagosome degradation in rat RPE.[112] In addition, autophagy was inhibited in the αB-crystallin R120G mutant lenses leading to larger autophagosomes compared to wild-type lenses, possibly due to a defect in protein degradation after autophagosome formation.[113] α-Crystallin has also been localized to the leading edges of migrating lens epithelial cells and possibly plays a role in actin dynamics during cell migration,[114] suggesting a possible function for α-crystallin in tissue remodeling. A potential role for β- and γ-crystallins in the vascular remodeling of the eye has also been proposed.[115] A possible role has also been suggested for γS-crystallin in fiber cell maturation due to its functional role in the stabilization of actin and "shepherding" of filaments.[116] Finally, it has been recently suggested that αB-crystallin acts as a molecular guard in mouse decidualization, playing a role during early pregnancy.[117]

A number of mouse models have been generated to help analyze the functions of crystallins *in vivo* further. Knockout models for αA- and αB-crystallin and double αA/αB knockouts have been made. The lenses of αB-knockout mice appeared normal compared to wild-type mice, although skeletal muscle degeneration was observed.[118] However, the lenses of αA-crystallin-deficient mice were smaller and exhibited progressive opacification several weeks after birth.[119] The double-knockout mice were born with small opaque lenses, much more severe than the individual subunit knockouts, suggesting a compensatory activity by the two α-crystallin subunits.[63] Knocking out the mouse γS-crystallin gene appears to result in disorganized actin filaments and interferes with fiber cell maturation,[116] and knockout of the βB2-crystallin gene results in age-related cataract at several months of age.[120] The "noncrystallin" functions of crystallins include a range of activities, including inhibition of apoptosis, chaperone-like functions, cytoskeletal remodeling, kinase activity, regulation of autophagy, cytochrome *c* redox state, and denucleation, although which of these specific functions are responsible for the phenotypes in knockout mice is at present unclear.

3. MEMBRANE PROTEINS

Membrane-associated proteins account for approximately 2% of lens proteins. They have molecular masses ranging from about 10 to over 250 kDa. Some, such as N-cadherin, a 135-kDa intrinsic membrane protein which may be involved in cell–cell adhesion, are components of the cytoskeletal structure.[121] Similarly, the calpactins, extrinsic membrane proteins attached to the membrane through calcium, are probably involved in membrane–cytoskeleton interactions and fiber cell elongation.[122,123] Neural-cell adhesion molecule 2 (NCAM 2) appears to play a role in cell adhesion, contributing to the proper arrangement of gap junctions in the developing lens fiber cells.[124] Other membrane proteins include enzymes such as glyceraldehyde 3-phosphate dehydrogenase, other glycolytic enzymes and channels on the endoplasmic reticulum,[125] and a variety of other enzymes such as adenosine triphosphatases (ATPases).

In addition, there are two additional intrinsic membrane proteins that are also highly expressed in lens fiber cells. Major intrinsic protein (MIP), a hydrophobic 28 kDa, 263-amino-acid protein is the most abundant. It is part of the aquaporin family of water channels, and is also known as aquaporin 0 (AQP0). AQP0 is highly homologous to AQP1, and like it has six transmembrane domains, three extracellular loops, two intracellular loops around a central channel, and cytoplasmic amino- and carboxyl-terminal domains.[126] The MIP monomer sequence also shows a structural symmetry of two tandem repeats, each containing a "hemipore" composed of three transmembrane helices with a hydrophobic loop containing a conserved Asn-Pro-Ala (NPA) motif. These fold into the membrane to form a functional water pore. The presence of two evolutionarily conserved NPA-core motifs, directly involved in water transport, is consistent with a gene duplication event in the aquaporin gene family.[127]

X-ray and electron crystallography has largely confirmed the originally predicted hourglass structure of MIP monomers, and further confirmed that MIP forms homotetramers, consistent with the square arrays observed by ultrastructural analyses of native plasma membranes from the lens core. Not only can MIP purified from the lens core form single-layered 2D crystals of tetramers, but it also can form head-to-head tetramers (double-layered 2D crystals of junctional octamers) of similar size and thickness to the 11–13 nm thin junctions seen in ultrastructural analysis of lens plasma membranes.[126,128,129] Conversion of nonjunctional MIP tetramers to junctional MIP octamers is associated with posttranslational proteolytic truncation of the cytoplasmic N- and C-termini, and stabilized by proline-mediated hydrophobic interactions between the extracellular loops of apposed tetramers.[126,130,131] In addition, atomic force microscopy shows that AQP0 arrays are surrounded by connexins to form junctional microdomains important for both adhesion and channels for intercellular communication and transport.[132]

Physiological studies have shown that MIP functions as a relatively slow water channel, being less than 10-fold as active as AQP1 in both the classical heterologous expression system in *Xenopus* oocytes as well as membrane vesicles isolated from lens fiber cells.[133] However, in spite of its low conductivity, experiments in lens membrane vesicles isolated from MIP-deficient mice show that because of its abundancy in the lens, aquaporin provides about 80% of water transport in wild-type mouse lenses.[133,134] In addition, the water channel activity of MIP in lens vesicles increases as the pH is lowered to 6.5 and is raised by increasing Ca^{2+} ion concentration in a calmodulin-dependent manner.[135,136]

Recently, molecular dynamic simulations have indicated that both junctional and nonjunctional forms of MIP can transport water at similar rates, with both showing an average channel occupancy of about five water molecules.[137] In addition, extracellular water appears to access the pores of junctional MIP,[138] suggesting that junctional MIP could carry out both the water transport and cell-to-cell adhesion functions needed for lens homeostasis. In addition to the self-association into tetramers shown in water transport and junction formation, MIP is suggested to interact with several other proteins in the lens, including connexins,[139] filensin and phakinin (beaded filament proteins),[140] and crystallins[141] through its C-terminus, perhaps providing a means of regulation or for additional functions.

4. GAP JUNCTION PROTEINS

In the mature lens, which is avascular, a network of gap junction channels facilitates intercellular communication and metabolic cooperation.[142,143] This is accomplished by exchange of ions such as Na^+, K^+,

Ca^{2+}, and Cl^-, second messengers such as cAMP, cGMP, and inositol trisphosphate (IP_3), and small metabolites including glucose and amino acids. There are at least three gap junction proteins, called connexins, in the lens, having overlapping but distinct expression patterns. Gap junction protein A1 (GJA1, connexin-43, Cx43), with a molecular mass of about 43 kDa, is expressed predominantly in the anterior epithelia.[144] Gap junction protein A8 (GJA8, connexin 50, Cx50), with a molecular mass of about 50 kDa, is expressed in anterior epithelial cells along with GJA1, but its expression continues in the elongating fiber cells along with gap junction protein A3 (GJA3, connexin 46, Cx46), with a molecular mass of about 46 kDa, to make them the most highly expressed gap junction proteins in the lens nucleus.[145,146] Each gene consists of a single coding exon and encodes a protein with four transmembrane domains separated by two extracellular loops and a cytoplasmic loop with cytoplasmic amino- and carboxyl-termini.[35] Connexin monomers associate to form hexamers, also known as hemichannels or connexons, and gap junction channels are formed when two connexons in the membranes of adjacent cells associate in the extracellular space. Gap junction plaques are formed in lens fiber cell membranes when up to hundreds of gap junction channels cluster to form, and appear as 16–18 nm thick junctions seen in electron micrographs.[128] In addition, as discussed above, connexins appear to associate with AQP0 to form junctional microdomains.[132]

Gja3 and Gja8 function in the internal microcirculation system of the lens, which delivers glucose, amino acids, and antioxidants such as glutathione to the central fiber cells.[143] In this model, it is proposed that Na^+ ions carry the current with metabolites into the lens primarily via the extracellular spaces at the anterior and posterior poles, and also carry the current out of the lens through cytoplasmic diffusion gradients through gap junctions to be transported out of the lens by the Na^+–K^+ pumps in the anterior epithelial cells. In mice with both Gja3 and Gja8 knocked out, the lenses develop swelling and degeneration of mature fiber cells in the inner lens, resulting in cataract formation. This suggests that these two connexins participate in the intercellular transport of proteins, metabolites, and ions within the central lens nucleus.[147] In this light, Gja3 and Gja8 have been implicated in helping to establish the uniform distribution of protein seen among cells of the lens core, and GJA3 gap junctions have been shown to provide a pathway for glutathione (GSH) diffusion from the outer cortex to the lens nucleus.[148,149] In lens fiber cells, both connexins GJA3 and GJA8 undergo posttranslational phosphorylation. This is hypothesized to regulate proteolytic cleavage and turnover as well as downregulate cell–cell communication and increase protein stability.[150–152]

5. CYTOSKELETAL PROTEINS

Many cytoskeletal proteins common to other tissues are found in the lens, such as actin, α-actinin, ankyrin, myosin, spectrin, and vimentin. An extended network of proteins including spectrin and actin is found immediately below the cell membrane, similar to that seen in erythrocytes,[153] and through tropomodulin 1 contributes both to early lens development to the maintenance of hexagonal cell shape of differentiating fiber cells of the lens cortex.[154–156] Microtubules, which are rare in epithelial cells but are found arrayed lengthwise in the peripheral cytoplasm of cortical fiber cells and decrease again in nuclear fiber cells, contain α- and β-tubulins.[157] Along with other cytoskeletal components, microtubules may contribute to establishing and maintaining the elongated shape of fiber cells as well as being involved in the migration of chromosomes in dividing lens epithelial cells.[158] Actin filaments interact closely with cell membranes of lens cells[159,160] through adherens junctions as well as proteins such as ezrin, radixin, and protein 4.1.[161–163] It has been suggested that they might facilitate accommodation.[164–166] These components of the basal membrane complex undergo rearrangement as the fiber cells elongate and migrate from the capsule to the lens sutures,[167] forming a terminal web to stabilize the ends of lens fibers at the sutures.[168] Nonmuscle isoforms of β- and γ-actin also occur in lens microfilaments (also called thin filaments).[169] Tropomodulin and α-actinin have been shown to associate with actin in lens microfilaments, especially in elongating cortical fiber cells.[170,171]

Vimentin is usually found in cells of mesenchymal origin, but it is part of intermediate filaments in lens cells.[155] These 10 nm filaments can occur as extrinsic membrane proteins, but are more commonly found in the cytoplasm.[172] Vimentin-based intermediate filaments occur primarily in anterior epithelial cells, where they are required for repair of damage to the lens epithelium.[173] While some vimentin is expressed in superficial cortical cells, vimentin-containing filaments are replaced by BFSP2 filaments deeper in the cortex.[155] Vimentin can be phosphorylated[174] and defects in its phosphorylation can cause aneuploidy of lens epithelia resulting in microphthalmia and cataract.[175] Vimentin expression is requisite on pigment epithelium-derived factor (PEDF) expression in epithelia cells,[176] and triples in lens development in embryonic chickens, decreasing after hatching.[177]

Transfection of vimentin promoters into cultured cells has identified a complex set of positive and negative *cis*-regulatory 5′-flanking sequences that control vimentin expression.[177,178] An intermediate filament protein GFAP, usually found in cells of neurectodermal origin, shows a similar expression pattern: high in anterior epithelial cells and decreasing upon differentiation to fiber cells.[106,179]

As vimentin-containing intermediate filaments disappear during fiber cell differentiation, they are replaced by beaded filaments,[180] cytoskeletal proteins that appear to be uniquely expressed in the lens.[77,181] Beaded filaments consist of a central backbone filament 7–9 nm wide containing BFSP1 (CP-115, filensin), along which 12–15 nm globular protein particles containing BFSP1 as well as BFSP2 (CP-49, phakinin) are spaced.[155] Both BFSP1 and BFSP2 belong to the intermediate filament family, although they are highly divergent.[182] They appear to be critical in maintaining the shape, size, and stiffness of the lens,[183] probably through regulating fiber cell geometry in coordination with the spectrin–actin membrane skeleton mediated through Tmod1.[184] This is critical for lens transparency and coincides with the rearranging and remodeling of fiber cells as they transit from the cortical lens to the nucleus.[185]

The α-crystallins appear to be critical for the assembly, maintenance, and remodeling of the lens cytoskeleton. BFSP1 and BFSP2 copolymerize *in vitro*, forming 10 nm fibers similar in appearance to intermediate filaments seen in nonlens tissues.[186,187] These proteins fail to assemble with vimentin, and substitution of the vimentin rod or tail domain for that of BFSP1 will inhibite fiber assembly.[188] However, addition of α-crystallin during the assembly process results in the formation of a structure similar to a beaded chain.[186] In addition, not only will addition of α-crystallins result in the correct beaded filament-type structure, but it also inhibits assembly of both GFAP and vimentin into filaments in an ATP-dependent manner,[24] resulting in their transition from formed filaments to a soluble state.

6. LENS METABOLISM

6.1 Energy Metabolism

The intermediary metabolism of the lens is generally similar to those of other tissues. The major differences arise from the lens being avascular, so that it has to obtain most of its nutrients through the aqueous humor.[189] In addition, the loss of intracellular organelles, including mitochondria, during fiber cell differentiation places additional requirements on lens metabolism. Thus, utilization of various metabolic pathways for energy production in the whole lens is quite different from most tissues, with most glucose in the lens metabolized through anaerobic glycolysis. Enzymes composing the citric acid cycle are only found in the anterior epithelia, which retain their mitochondria. Because of its greater efficiency, the citric acid cycle provides 20–30% of the total lens while consuming only roughly 3% of the glucose.[190,191] Although the lens can control its ion balance, maintain high-energy phosphate levels, and carry out protein synthesis in the absence of oxygen, blocking the Embden–Meyerhof pathway by exposing the lens to iodoacetate, which inhibits 3-phosphoglyceraldehyde dehydrogenase, causes ionic changes with accompanying swelling and cataracts.[192,193] Conversely, exposure of cultured lenses to oxidative stress induced by exposure to hydrogen peroxide[194] or hyperbaric oxygen[195] increases hexokinase activity and thus stimulates the pentose phosphate pathway.

The osmotic hypothesis suggests that aldose reductase activity might increase sorbitol and thus protect the lens from daily diet and disease-related changes in osmolality of the aqueous humor,[196] similar to the action of sorbitol in the renal medulla.[197] This proposes that cataracts secondary to diabetes mellitus and galactosemia share a common pathogenic mechanism in which aldose reductase reduces glucose to sorbitol and galactose (more readily) to galactitol.[198] As opposed to sorbitol, which is further metabolized by sorbitol dehydrogenase, galactitol remains in the lens, potentially damaging the lens cells by increasing its intracellular fluid in response to the increased osmotic pressure. This can cause swelling of the lens cells, increasing their membrane permeability with secondary electrolyte abnormalities, and finally resulting in metabolic dysfunction.[199] Cataracts occurring secondary to polyol accumulation occur in transgenic mice expressing the aldose reductase gene in their lenses, which normally lack this enzyme.[200] While human lenses have reduced aldose reductase compared to the rat, human lenses from diabetic patients cultured *in vitro* accumulate more polyols compared to lenses from nondiabetic controls, and aldose reductase inhibitors inhibit this polyol accumulation.[201] Osmotic stress can also induce a variety of FGF and TGF-β signals and increased signaling through the mitogen-activated protein kinase (MAPK) pathway.[202] However, the mechanism of sugar cataract pathogenesis remains controversial, and the possibility that inhibition of aldose reductase might reduce accumulation of polyols and the accompanying osmotic damage and cataract in diabetics is still being investigated.[203]

6.2 Maintenance of a Reduced State

As discussed in Section 4 of the chapter Overview of the Lens, oxidative damage to lens constituents, especially that from photooxidation, is a major threat to lens transparency. This accumulates over the lifetime of an individual and can include not only proteins but also lipids and even nucleic acids. It is particularly problematic when it affects those very systems that provide for lens homeostasis. Loss of the reducing environment in lens cells can result in a variety of oxidative modifications to lens proteins, including ascorbylation. When this impacts the lens crystallins, it can alter their short range interactions, causing aggregation and cataract.[204] Hyperbaric oxygen treatment of guinea pigs results in disulfide bond formation and cross-linking of lens crystallins with formation of aggregates and resultant light scattering.[205] When βB3-crystallin is oxidized, it interacts with γ-crystallins to increase their sensitivity to thermal aggregation and cause light scattering.[206] One major source of oxidative stress in the lens is photooxidation from ultraviolet light, and it is possible that other types of insults such as osmotic stress might also tax the reducing environment of the lens.[207,208] In the normal human eye, peroxides are generated in both lens fiber and anterior epithelial cells, maintaining the concentration of H_2O_2 at around 30 mM, although it can be significantly elevated with cataracts.[209] In addition, it has been suggested that vitreal degeneration can increase oxidative stress on the posterior lens.[210]

Under optimal conditions, the lens can respond to chromic oxidative stress by increasing production of reducing agents, and has developed a comprehensive and versatile system to defend against oxidative stress, and repair oxidative damage when the reducing environment is overwhelmed.[211] The most notable of these defenses is high levels of glutathione, the most abundant low-molecular-weight thiol in the human lens, which serves to maintain a strong reducing environment.[212,213] The high levels of reduced glutathione are maintained by a variety of intracellular enzymes including glutathione reductase as well as reducing power generated through the pentose phosphate pathway. A gradient of reduced glutathione exists in the lens, from the peak concentration in the metabolically active anterior epithelium followed by the cortical regions and finally lowest in the nucleus.[214] While glutathione levels in the anterior epithelium and cortex are not age dependent in the normal lens, they are decreased markedly in most types of cataractous lenses.[215] Lens glutathione serves as a sulfhydryl buffer that maintains protein thiol groups in a reduced state as well as protecting other residues against oxidative damage. However, under oxidative stress, glutathione is found in mixed disulfides with various lens proteins, including crystallins such as γS-crystallin, possibly altering its association behavior.[216] However, thioredoxin and thioredoxin reductase can rescue enzymatic function of oxidized glyceraldehyde 3-phosphate dehydrogenase, and this activity is enhanced by α-crystallin.[29] Similarly, methionine sulfoxide reductase can reverse oxidation of methionine, conferring additional resistance to oxidative stress.[67]

Detoxification of hydrogen peroxide in the lens can be carried out by both the glutathione redox cycle and catalase, although glutathione reductase activity appears to be more critical for this function.[217] Catalase is found in peroxisomes and acts on higher levels of H_2O_2.[218] In contrast, enzymes such as glutathione reductase and peroxidase, that are part of the glutathione redox cycle, occur throughout the cytoplasm. Although the mercapturic acid pathway, including glutathione *S*-transferase, also helps to protect lens proteins from oxidative damage,[219] glutathione reductase has a primary role in maintaining lens glutathione in a reduced state.[213] That glutathione *S*-transferases have been important in lens biology for a long period during evolution is suggested by their close similarity to S-crystallins, the major crystallins in cephalopods.[220] Because of the effectiveness of these combined reducing pathways, only about 2–5% of glutathione in the lens is normally in its oxidized state. The oxidative defense systems including catalase, glutathione reductase, and glutathione peroxidase cooperate and show considerable functional overlap, so that defects in one pathway can be partially compensated by others.[221] Efforts are underway to utilize peroxidases as anticataract agents by protecting the lens for oxidative stress.[504] Additional protection against oxidative stress is provided by heat-shock proteins, especially α-crystallins, which can also act through protection of mitochondria in a manner independent of their chaperone activity.[70]

6.3 Osmoregulation

Osmoregulation in the lens occurs through active transport, in which Na^+/K^+-dependent ATPase exchanges sodium for potassium into the lens, and is associated with the lens microcirculation mentioned earlier.[222] Most of the lens Na^+/K^+-dependent ATPase is found in the apicolateral membranes of the anterior epithelial cells,[223,224] but there is also some activity in the anterior cortex and the fiber membranes abutting the sutures.[225,226] The actively exchanged cations are followed passively by diffusion of chloride and water, and transport might be purinergic control.[227] Osmotic and ionic homeostasis of the lens is a complex process. There is ion transport through the intercellular spaces of epithelia and fiber cells, followed by water and solutes. These are then actively transported into the fiber cells, and access the central fiber cells through coupling by gap junctions. Finally, metabolic end products exit the fiber cell mass following through the lens equator, probably coupled to Na^+/K^+ transport.[228] This microcirculation occurs largely in the superficial cortex. There is also transport of macromolecules and cholesterol by caveolae and coated vesicles.[222] Similarly, calcium homeostasis is also complex, involving not only Ca^{2+}-ATPase but also calcium channels and a variety of means to sequester calcium intracellularly including binding proteins and active transport across the endoplasmic reticulum. Ca^{2+}-ATPase shows the highest specific activity in the anterior epithelium, so that the lens has a lower Ca^{2+} concentration than the aqueous or vitreous humor.[229–231] Among other problems, disruption of the Ca^{2+} circulation can lead to opacity through increased calcium levels resulting in the activation of proteinases and then cleavage and aggregation of crystallins.[232]

The lens controls movement of various macromolecules in different fashions. It has been proposed that metabolites might pass from the aqueous chamber into the lens, and this has been demonstrated for albumin.[233] The first barrier a molecule must pass to reach the avascular lens is the capsule. It provides a limit for diffusion: horseradish peroxidase, with a molecular weight of 40,000 kDa, can diffuse across the lens capsule, but ferritin, with a molecular weight of 500,000, cannot,[234] and experiments with dextran show an exclusion limit of about 150 kDa.[235] Thus, low-molecular-weight crystallins can penetrate the lens capsule but not the higher molecular weight α-crystallin.[236] A second barrier is the layer of anterior epithelial cells, but these can also be penetrated by low-molecular-weight proteins such as horseradish peroxidase and various dyes.[237] However, passage of metabolites between the epithelial cells and fiber cells is probably through transcytotic processes mediated by caveolae and clathrin-coated vesicles rather than, or perhaps in addition to, the relatively small number of gap junctions connecting them.[238,239] Lens fiber cells are connected by an extensive network of communicating channels.[143] Sugar transport in the lens occurs with glucose transporters 1 and 3 (GLUT1 and GLUT3), similarly to that seen in many other tissues, including myocytes and blood cells.[240–242]

This brief overview cannot provide a detailed examination of all the protein families, metabolic pathways, and homeostatic systems that are present in the lens. A number of additional systems have not been covered at all, including RNA granules during lens, and especially fiber cell development,[243] autophagy in lens development and maintenance of transparency,[244] and the underappreciated role of epithelial cell mitochondria in preserving lens fiber cell function.[245]

REFERENCES

1. Wistow G. The human crystallin gene families. *Hum Genomics*. 2012;6:26.
2. Bassnett S. On the mechanism of organelle degradation in the vertebrate lens. *Exp Eye Res*. 2009;88(2):133–139.
3. Costello MJ, Brennan LA, Basu S, et al. Autophagy and mitophagy participate in ocular lens organelle degradation. *Exp Eye Res*. 2013;116:141–150.
4. Horwitz J. Alpha-crystallin. *Exp Eye Res*. 2003;76(2):145–153.
5. Augusteyn RC, Koretz JF. A possible structure for alpha-crystallin. *FEBS Lett*. 1987;222:1–5.
6. Thomson JA, Augusteyn RC. On the structure of alpha-crystallin: construction of hybrid molecules and homopolymers. *Biochim Biophys Acta*. 1989;994:246–252.
7. Hendriks W, Weetink H, Voorter CE, Sanders J, Bloemendal H, de Jong WW. The alternative splicing product alpha Ains-crystallin is structurally equivalent to alpha A and alpha B subunits in the rat alpha-crystallin aggregate. *Biochim Biophys Acta*. 1990;1037:58–65.
8. Bova MP, Ding LL, Horwitz J, Fung BK. Subunit exchange of alphaA-crystallin. *J Biol Chem*. 1997;272(47):29511–29517.
9. Sun TX, Liang JN. Intermolecular exchange and stabilization of recombinant human alphaA- and alphaB-crystallin. *J Biol Chem*. 1998;273(1):286–290.
10. Feil IK, Malfois M, Hendle J, van Der ZH, Svergun DI. A novel quaternary structure of the dimeric alpha-crystallin domain with chaperone-like activity. *J Biol Chem*. 2001;276(15):12024–12029.
11. Haley DA, Horwitz J, Stewart PL. The small heat-shock protein, alphaB-crystallin, has a variable quaternary structure. *J Mol Biol*. 1998;277(1):27–35.
12. Murugesan R, Santhoshkumar P, Sharma KK. Role of alphaBI5 and alphaBT162 residues in subunit interaction during oligomerization of alphaB-crystallin. *Mol Vis*. 2008;14:1835–1844.
13. de Jong WW, Hendriks W, Mulders JW, Bloemendal H. Evolution of eye lens crystallins: the stress connection. *Trends Biochem Sci*. 1989;14:365–368.
14. Kantorow M, Piatigorsky J. Alpha-crystallin/small heat shock protein has autokinase activity. *Proc Natl Acad Sci USA*. 1994;91:3112–3116.
15. Kantorow M, Horwitz J, van Boekel MA, de Jong WW, Piatigorsky J. Conversion from oligomers to tetramers enhances autophosphorylation by lens alpha A-crystallin. Specificity between alpha A- and alpha B-crystallin subunits. *J Biol Chem* 1995;270:17215–17220.
16. Sun TX, Das BK, Liang JJ. Conformational and functional differences between recombinant human lens alphaA- and alphaB-crystallin. *J Biol Chem*. 1997;272:6220–6225.
17. Gangalum RK, Horwitz J, Kohan SA, Bhat SP. alphaA-crystallin and alphaB-crystallin reside in separate subcellular compartments in the developing ocular lens. *J Biol Chem*. 2012;287(50):42407–42416.
18. Dasgupta S, Hohman TC, Carper D. Hypertonic stress induces alpha B-crystallin expression. *Exp Eye Res*. 1992;54:461–470.
19. Klemenz R, Frohli E, Steiger RH, Schafer R, Aoyama A. Alpha B-crystallin is a small heat shock protein. *Proc Natl Acad Sci USA*. 1991;88:3652–3656.

20. Spector A, Chiesa R, Sredy J, Garner W. cAMP-dependent phosphorylation of bovine lens alpha-crystallin. *Proc Natl Acad Sci USA*. 1985;82:4712–4716.
21. Voorter CE, Mulders JW, Bloemendal H, de Jong WW. Some aspects of the phosphorylation of alpha-crystallin A. *Eur J Biochem*. 1986;160:203–210.
22. Chiesa R, Spector A. The dephosphorylation of lens alpha-crystallin A chain. *Biochem Biophys Res Commun*. 1989;162:1494–1501.
23. Wang K, Ma W, Spector A. Phosphorylation of alpha-crystallin in rat lenses is stimulated by H2O2 but phosphorylation has no effect on chaperone activity. *Exp Eye Res*. 1995;61(1):115–124.
24. Nicholl ID, Quinlan RA. Chaperone activity of alpha-crystallins modulates intermediate filament assembly. *EMBO J*. 1994;13:945–953.
25. Wang K, Spector A. alpha-Crystallin stabilizes actin filaments and prevents cytochalasin-induced depolymerization in a phosphorylation-dependent manner. *Eur J Biochem*. 1996;242:56–66.
26. Ito H, Okamoto K, Nakayama H, Isobe T, Kato K. Phosphorylation of alphaB-crystallin in response to various types of stress. *J Biol Chem*. 1997;272(47):29934–29941.
27. Horwitz J, Huang QL, Ding L, Bova MP. Lens alpha-crystallin: chaperone-like properties. *Methods Enzymol*. 1998;290:365–383.
28. Horwitz J. Alpha-crystallin can function as a molecular chaperone. *Proc Natl Acad Sci USA*. 1992;89:10449–10453.
29. Yan H, Lou MF, Fernando MR, Harding JJ. Thioredoxin, thioredoxin reductase, and alpha-crystallin revive inactivated glyceraldehyde 3-phosphate dehydrogenase in human aged and cataract lens extracts. *Mol Vis*. 2006;12:1153–1159.
30. Marini I, Moschini R, Del CA, Mura U. Alpha-crystallin: an ATP-independent complete molecular chaperone toward sorbitol dehydrogenase. *Cell Mol Life Sci*. 2005;62(5):599–605.
31. Clark JI, Huang QL. Modulation of the chaperone-like activity of bovine alpha-crystallin. *Proc Natl Acad Sci USA*. 1996;93(26):15185–15189.
32. Muchowski PJ, Bassuk JA, Lubsen NH, Clark JI. Human alphaB-crystallin. Small heat shock protein and molecular chaperone. *J Biol Chem*. 1997;272(4):2578–2582.
33. Pasta SY, Raman B, Ramakrishna T, Rao C. The IXI/V motif in the C-terminal extension of alpha-crystallins: alternative interactions and oligomeric assemblies. *Mol Vis*. 2004;10:655–662.
34. Regini JW, Grossmann JG, Timmins P, et al. X-ray- and neutron-scattering studies of alpha-crystallin and evidence that the target protein sits in the fenestrations of the alpha-crystallin shell. *Invest Ophthalmol Vis Sci*. 2007;48(6):2695–2700.
35. Kumar NM, Gilula NB. The gap junction communication channel. *Cell*. 1996; 84(3):381–388.
36. McHaourab HS, Kumar MS, Koteiche HA. Specificity of alphaA-crystallin binding to destabilized mutants of betaB1-crystallin. *FEBS Lett*. 2007;581(10):1939–1943.
37. Biswas A, Das KP. Zn2+ enhances the molecular chaperone function and stability of alpha-crystallin. *Biochemistry*. 2008;47(2):804–816.
38. Santhoshkumar P, Sharma KK. Inhibition of amyloid fibrillogenesis and toxicity by a peptide chaperone. *Mol Cell Biochem*. 2004;267(1–2):147–155.
39. Tanaka N, Tanaka R, Tokuhara M, Kunugi S, Lee YF, Hamada D. Amyloid fibril formation and chaperone-like activity of peptides from alphaA-crystallin. *Biochemistry*. 2008;47(9):2961–2967.
40. Xi JH, Bai F, McGaha R, Andley UP. Alpha-crystallin expression affects microtubule assembly and prevents their aggregation. *FASEB J*. 2006;20(7):846–857.
41. Nahomi RB, Wang B, Raghavan CT, et al. Chaperone peptides of alpha-crystallin inhibit epithelial cell apoptosis, protein insolubilization, and opacification in experimental cataracts. *J Biol Chem*. 2013;288(18):13022–13035.

42. Bhat SP, Nagineni CN. Alpha B subunit of lens-specific protein alpha-crystallin is in other ocular and non-ocular tissues. *Biochem Biophys Res Commun.* 1989;158: 319–325.
43. Dubin RA, Wawrousek EF, Piatigorsky J. Expression of the murine alpha B-crystallin gene is not restricted to the lens. *Mol Cell Biol.* 1989;9:1083–1091.
44. Atomi Y, Yamada S, Nishida T. Early changes of alpha B-crystallin mRNA in rat skeletal muscle to mechanical tension and denervation. *Biochem Biophys Res Commun.* 1991;181:1323–1330.
45. Bhat SP, Horwitz J, Srinivasan A, Ding L. AlphaB-crystallin exists as an independent protein in the heart and in the lens. *Eur J Biochem.* 1991;202:775–781.
46. Kato K, Shinohara H, Kurobe N, Goto S, Inaguma Y, Ohshima K. Immunoreactive alphaA-crystallin in rat non-lenticular tissues detected with a sensitive immunoassay method. *Biochim Biophys Acta.* 1991;1080:173–180.
47. Li DW, Liu JP, Mao YW, et al. Calcium-activated RAF/MEK/ERK signaling pathway mediates p53-dependent apoptosis and is abrogated by alpha B-crystallin through inhibition of RAS activation. *Mol Biol Cell.* 2005;16(9):4437–4453.
48. Mitra A, Basak T, Datta K, Naskar S, Sengupta S, Sarkar S. Role of alpha-crystallin B as a regulatory switch in modulating cardiomyocyte apoptosis by mitochondria or endoplasmic reticulum during cardiac hypertrophy and myocardial infarction. *Cell Death Dis.* 2013;4:e582.
49. Ying X, Zhang J, Wang Y, Wu N, Yew DT. Alpha-crystallin protected axons from optic nerve degeneration after crushing in rats. *J Mol Neurosci.* 2008;35(3):253–258.
50. Rose K, Schroer U, Volk GF, et al. Axonal regeneration in the organotypically cultured monkey retina: biological aspects, dependence on substrates and age-related proteomic profiling. *Restor Neurol Neurosci.* 2008;26(4–5):249–266.
51. Iwaki T, Kume-Iwaki A, Liem RK, Goldman JE. Alpha B-crystallin is expressed in non-lenticular tissues and accumulates in Alexander's disease brain. *Cell.* 1989; 57:71–78.
52. Duguid JR, Bohmont CW, Liu NG, Tourtellotte WW. Changes in brain gene expression shared by scrapie and Alzheimer disease. *Proc Natl Acad Sci USA.* 1989;86: 7260–7264.
53. Murano S, Thweatt R, Shmookler Reis RJ, Jones RA, Moerman EJ, Goldstein S. Diverse gene sequences are overexpressed in Werner syndrome fibroblasts undergoing premature replicative senescence. *Mol Cell Biol.* 1991;11:3905–3914.
54. Liu JP, Schlosser R, Ma WY, et al. Human alphaA- and alphaB-crystallins prevent UVA-induced apoptosis through regulation of PKCalpha, RAF/MEK/ERK and AKT signaling pathways. *Exp Eye Res.* 2004;79(6):393–403.
55. Kamradt MC, Chen F, Cryns VL. The small heat shock protein alpha B-crystallin negatively regulates cytochrome c- and caspase-8-dependent activation of caspase-3 by inhibiting its autoproteolytic maturation. *J Biol Chem.* 2001;276(19):16059–16063.
56. Mao YW, Liu JP, Xiang H, Li DW. Human alphaA- and alphaB-crystallins bind to Bax and Bcl-X(S) to sequester their translocation during staurosporine-induced apoptosis. *Cell Death Differ.* 2004;11(5):512–526.
57. Andley UP, Song Z, Wawrousek EF, Fleming TP, Bassnett S. Differential protective activity of alpha A- and alphaB-crystallin in lens epithelial cells. *J Biol Chem.* 2000; 275(47):36823–36831.
58. Mao YW, Xiang H, Wang J, Korsmeyer S, Reddan J, Li DW. Human bcl-2 gene attenuates the ability of rabbit lens epithelial cells against H2O2-induced apoptosis through down-regulation of the alpha B-crystallin gene. *J Biol Chem.* 2001;276(46): 43435–43445.
59. Li DW, Xiang H, Mao YW, et al. Caspase-3 is actively involved in okadaic acid-induced lens epithelial cell apoptosis. *Exp Cell Res.* 2001;266(2):279–291.

60. Kamradt MC, Lu M, Werner ME, et al. The small heat shock protein alpha B-crystallin is a novel inhibitor of TRAIL-induced apoptosis that suppresses the activation of caspase-3. *J Biol Chem*. 2005;280(12):11059–11066.
61. Andley UP, Patel HC, Xi JH. The R116C mutation in alpha A-crystallin diminishes its protective ability against stress-induced lens epithelial cell apoptosis. *J Biol Chem*. 2002;277(12):10178–10186.
62. Maloyan A, Sanbe A, Osinska H, et al. Mitochondrial dysfunction and apoptosis underlie the pathogenic process in alpha-B-crystallin desmin-related cardiomyopathy. *Circulation*. 2005;112(22):3451–3461.
63. Morozov V, Wawrousek EF. Caspase-dependent secondary lens fiber cell disintegration in alphaA-/alphaB-crystallin double-knockout mice. *Development*. 2006;133(5): 813–821.
64. Pasupuleti N, Matsuyama S, Voss O, et al. The anti-apoptotic function of human alphaA-crystallin is directly related to its chaperone activity. *Cell Death Dis*. 2010;1:e31.
65. Shin JH, Kim SW, Lim CM, Jeong JY, Piao CS, Lee JK. alphaB-crystallin suppresses oxidative stress-induced astrocyte apoptosis by inhibiting caspase-3 activation. *Neurosci Res*. 2009;64(4):355–361.
66. Morrison LE, Whittaker RJ, Klepper RE, Wawrousek EF, Glembotski CC. Roles for alphaB-crystallin and HSPB2 in protecting the myocardium from ischemia-reperfusion-induced damage in a KO mouse model. *Am J Physiol Heart Circ Physiol*. 2004;286(3):H847–H855.
67. Kantorow M, Hawse JR, Cowell TL, et al. Methionine sulfoxide reductase A is important for lens cell viability and resistance to oxidative stress. *Proc Natl Acad Sci USA*. 2004;101(26):9654–9659.
68. Marchetti MA, Pizarro GO, Sagher D, et al. Methionine sulfoxide reductases b1, b2, and b3 are present in the human lens and confer oxidative stress resistance to lens cells. *Invest Ophthalmol Vis Sci*. 2005;46(6):2107–2112.
69. Lee do Y, Lee MW, Lee HJ, et al. ERK1/2 activation attenuates TRAIL-induced apoptosis through the regulation of mitochondria-dependent pathway. *Toxicol In Vitro*. 2006;20(6):816–823.
70. McGreal RS, Brennan LA, Kantorow WL, et al. Chaperone-independent mitochondrial translocation and protection by alphaB-crystallin in RPE cells. *Exp Eye Res*. 2013;110:10–17.
71. Driessen HP, Herbrink P, Bloemendal H, de Jong WW. Primary structure of the bovine beta-crystallin Bp chain. Internal duplication and homology with gamma-crystallin. *Eur J Biochem*. 1981;121:83–91.
72. Wistow G, Slingsby C, Blundell T, Driessen H, de Jong W, Bloemendal H. Eye-lens proteins: the three-dimensional structure of beta-crystallin predicted from monomeric gamma-crystallin. *FEBS Lett*. 1981;133:9–16.
73. Lubsen NH, Aarts HJM, Schoenmakers JGG. The evolution of lenticular proteins: the beta- and gamma-crystallin supergene family. *Prog Biophys Mol Biol*. 1988;51:47–76.
74. Blundell T, Lindley P, Miller L, et al. The molecular structure and stability of the eye lens: X-ray analysis of gamma-crystallin II. *Nature*. 1981;289:771–777.
75. Gupta R, Srivastava K, Srivastava OP. Truncation of motifs III and IV in human lens betaA3-crystallin destabilizes the structure. *Biochemistry*. 2006;45(33):9964–9978.
76. Flaugh SL, Kosinski-Collins MS, King J. Interdomain side-chain interactions in human gammaD crystallin influencing folding and stability. *Protein Sci*. 2005;14(8):2030–2043.
77. FitzGerald PG, Casselman J. Discrimination between the lens fiber cell 114 kd cytoskeletal protein and alpha-actinin. *Curr Eye Res*. 1990;9:873–882.
78. Shimeld SM, Purkiss AG, Dirks RP, Bateman OA, Slingsby C, Lubsen NH. Urochordate betagamma-crystallin and the evolutionary origin of the vertebrate eye lens. *Curr Biol*. 2005;15(18):1684–1689.

79. Aravind P, Suman SK, Mishra A, Sharma Y, Sankaranarayanan R. Three-dimensional domain swapping in nitrollin, a single-domain betagamma-crystallin from nitrosospira multiformis, controls protein conformation and stability but not dimerization. *J Mol Biol*. 2009;385(1):163–177.
80. Jobby MK, Sharma Y. Caulollins from Caulobacter crescentus, a pair of partially unstructured proteins of betagamma-crystallin superfamily, gain structure upon binding calcium. *Biochemistry*. 2007;46(43):12298–12307.
81. Jobby MK, Sharma Y. Calcium-binding to lens betaB2- and betaA3-crystallins suggests that all beta-crystallins are calcium-binding proteins. *FEBS J*. 2007; 274(16):4135–4147.
82. Chen J, Toptygin D, Brand L, King J. Mechanism of the efficient tryptophan fluorescence quenching in human gammaD-crystallin studied by time-resolved fluorescence. *Biochemistry*. 2008;47(40):10705–10721.
83. Slingsby C, Bateman OA. Quarternary interactions in eye lens beta-crystallins: basic and acidic subunits of beta-crystallins favor heterologous association. *Biochemistry*. 1990;29:6592–6599.
84. Wistow G, Turnell B, Summers L, et al. X-ray analysis of the eye lens protein gamma-II crystallin at 1.9 A resolution. *J Mol Biol*. 1983;170(1):175–202.
85. Werten PJ, Lindner RA, Carver JA, de Jong WW. Formation of betaA3/betaB2-crystallin mixed complexes: involvement of N- and C-terminal extensions. *Biochim Biophys Acta*. 1999;1432(2):286–292.
86. Hejtmancik JF, Wingfield P, Chambers C, et al. Association properties of beta-B2- and betaA3-crystallin: ability to form dimers. *Protein Eng*. 1997;10(11):1347–1352.
87. Chan MP, Dolinska M, Sergeev YV, Wingfield PT, Hejtmancik JF. Association properties of betaB1- and betaA3-crystallins: ability to form heterotetramers. *Biochemistry*. 2008;47(42):11062–11069.
88. Ajaz MS, Ma Z, Smith DL, Smith JB. Size of human lens beta-crystallin aggregates are distinguished by N-terminal truncation of betaB1. *J Biol Chem*. 1997;272(17): 11250–11255.
89. Hejtmancik JF, Wingfield PT, Sergeev YV. beta-Crystallin association. *Exp Eye Res*. 2004;79(3):377–383.
90. Sergeev YV, Hejtmancik JF, Wingfield PT. Energetics of domain-domain interactions and entropy driven association of beta-crystallins. *Biochemistry*. 2004;43(2):415–424.
91. Mayr EM, Jaenicke R, Glockshuber R. Domain interactions and connecting peptides in lens crystallins. *J Mol Biol*. 1994;235:84–88.
92. Hope JN, Chen HC, Hejtmancik JF. Aggregation of betaA3-crystallin is independent of the specific sequence of the domain connecting peptide. *J Biol Chem*. 1994;269: 21141–21145.
93. Sergeev YV, Wingfield PT, Hejtmancik JF. Monomer-dimer equilibrium of normal and modified beta A3-crystallins: experimental determination and molecular modeling. *Biochemistry*. 2000;39(51):15799–15806.
94. Sergeev YV, Soustov LV, Chelnokov EV, et al. Increased sensitivity of amino-arm truncated {beta}A3-crystallin to UV-light-induced photoaggregation. *Invest Ophthalmol Vis Sci*. 2005;46(9):3263–3273.
95. Annunziata O, Pande A, Pande J, Ogun O, Lubsen NH, Benedek GB. Oligomerization and phase transitions in aqueous solutions of native and truncated human beta B1-crystallin. *Biochemistry*. 2005;44(4):1316–1328.
96. Kroone RC, Elliott GS, Ferszt A, Slingsby C, Lubsen NH, Schoenmakers JGG. The role of the sequence extensions in beta-crystallin assembly. *Protein Eng*. 1994;7: 1395–1399.
97. Werten PJL, Carver JA, Jaenicke R, de Jong WW. The elusive role of the N-terminal extension of betaA3- and betaA1-crystallin. *Protein Eng*. 1996;9(11):1021–1028.

98. Marin-Vinader L, Onnekink C, Van Genesen ST, Slingsby C, Lubsen NH. In vivo heteromer formation. Expression of soluble betaA4-crystallin requires coexpression of a heteromeric partner. *FEBS J.* 2006;273(14):3172–3182.
99. White HE, Driessen HPC, Slingsby C, Moss DS, Lindley PF. Packing interactions in the eye-lens. Structural analysis, internal symmetry and lattice interactions of bovine gamma IVa-crystallin. *J Mol Biol.* 1989;207:217–235.
100. Tanaka T, Benedek GB. Observation of protein diffusivity in intact human and bovine lenses with application to cataract. *Invest Ophthalmol.* 1975;14:449–456.
101. Quax-Jeuken Y, Driessen H, Leunissen J, Quax W, de Jong W, Bloemendal H. BetaS-crystallin: structure and evolution of a distinct member of the beta gamma-superfamily. *EMBO J.* 1985;4:2597–2602.
102. Zigler Jr JS, Russell P, Horwitz J, Reddy VN, Kinoshita JH. Further studies on low molecular weight crystallins: relationship between the bovine beta s, the human 24 kD protein and the gamma-crystallins. *Curr Eye Res.* 1986;5:395–401.
103. Slingsby C, Croft LR. Developmental changes in the low molecular weight proteins of the bovine lens. *Exp Eye Res.* 1973;17:369–376.
104. McDevitt DS, Croft LR. On the existence of gamma-crystallin in the bird lens. *Exp Eye Res.* 1977;25:473–481.
105. van Rens GLM, de Jong WW, Bloemendal H. One member of the gamma-crystallin gene family, gammaS, is expressed in birds. *Exp Eye Res.* 1991;53:135–138.
106. Hatfield JS, Skoff RP, Maisel H, Eng L, Bigner DD. The lens epithelium contains glial fibrillary acidic protein (GFAP). *J Neuroimmunol.* 1985;8:347–357.
107. Wistow GJ, Piatigorsky J. Lens crystallins: the evolution and expression of proteins for a highly specialized tissue. *Annu Rev Biochem.* 1988;57:479–504.
108. Piatigorsky J, Wistow GJ. Enzyme/crystallins: gene sharing as an evolutionary strategy. *Cell.* 1989;57:197–199.
109. Piatigorsky J. Crystallin genes: specialization by changes in gene regulation may precede gene duplication. *J Struct Funct Genomics.* 2003;3(1–4):131–137.
110. Piatigorsky J, Wistow GJ. The recruitment of crystallins: new functions precede gene duplication. *Science.* 1991;252:1078–1079.
111. Valapala M, Wilson C, Hose S, et al. Lysosomal-mediated waste clearance in retinal pigment epithelial cells is regulated by CRYBA1/betaA3/A1-crystallin via V-ATPase-MTORC1 signaling. *Autophagy.* 2014;10(3):480–496.
112. Zigler Jr JS, Zhang C, Grebe R, et al. Mutation in the betaA3/A1-crystallin gene impairs phagosome degradation in the retinal pigmented epithelium of the rat. *J Cell Sci.* 2011;124(pt 4):523–531.
113. Wignes JA, Goldman JW, Weihl CC, Bartley MG, Andley UP. p62 Expression and autophagy in alphaB-crystallin R120G mutant knock-in mouse model of hereditary cataract. *Exp Eye Res.* 2013;115:263–273.
114. Maddala R, Rao VP. alpha-Crystallin localizes to the leading edges of migrating lens epithelial cells. *Exp Cell Res.* 2005;306(1):203–215.
115. Zhang C, Gehlbach P, Gongora C, et al. A potential role for beta- and gamma-crystallins in the vascular remodeling of the eye. *Dev Dyn.* 2005;234(1):36–47.
116. Fan J, Dong L, Mishra S, Chen Y, FitzGerald P, Wistow G. A role for gammaS-crystallin in the organization of actin and fiber cell maturation in the mouse lens. *FEBS J.* 2012;279(16):2892–2904.
117. Zuo RJ, Zhao YC, Lei W, Wang TS, Wang BC, Yang ZM. Crystallin alphaB acts as a molecular guard in mouse decidualization: regulation and function during early pregnancy. *FEBS Lett.* 2014;588(17):2944–2951.
118. Brady JP, Garland DL, Green DE, Tamm ER, Giblin FJ, Wawrousek EF. AlphaB-crystallin in lens development and muscle integrity: a gene knockout approach. *Invest Ophthalmol Vis Sci.* 2001;42(12):2924–2934.

119. Brady JP, Garland D, Duglas-Tabor Y, Robison Jr WG, Groome A, Wawrousek EF. Targeted disruption of the mouse alpha A-crystallin gene induces cataract and cytoplasmic inclusion bodies containing the small heat shock protein alpha B-crystallin. *Proc Natl Acad Sci USA*. 1997;94:884–889.
120. Zhang J, Li J, Huang C, et al. Targeted knockout of the mouse betaB2-crystallin gene (Crybb2) induces age-related cataract. *Invest Ophthalmol Vis Sci*. 2008;49(12): 5476–5483.
121. Leonard M, Zhang L, Zhai N, et al. Modulation of N-cadherin junctions and their role as epicenters of differentiation-specific actin regulation in the developing lens. *Dev Biol*. 2011;349(2):363–377.
122. Alcala H, Maisel H. Biochemistry of lens plasma membranes and cytoskeleton. In: Maisel H, ed. *The Ocular Lens*. New York: Marcel Dekker Inc.; 1985:169–222.
123. Talian JC, Zelenka PS. Calpactin I in the differentiating embryonic chicken lens: mRNA levels and protein distribution. *Dev Biol*. 1991;143(1):68–77.
124. Watanabe M, Kobayashi H, Rutishauser U, Katar M, Alcala J, Maisel H. NCAM in the differentiation of embryonic lens tissue. *Dev Biol*. 1989;135:414–423.
125. Xu KY, Becker LC. Ultrastructural localization of glycolytic enzymes on sarcoplasmic reticulum vesticles. *J Histochem Cytochem*. 1998;46(4):419–427.
126. Gonen T, Sliz P, Kistler J, Cheng Y, Walz T. Aquaporin-0 membrane junctions reveal the structure of a closed water pore. *Nature*. 2004;429(6988):193–197.
127. Wistow GJ, Pisano MM, Chepelinsky AB. Tandem sequence repeats in transmembrane channel proteins. *Trends Biochem Sci*. 1991;16(5):170–171.
128. Zampighi GA, Simon SA, Hall JE. The specialized junctions of the lens. *Int Rev Cytol*. 1992;136:185–225.
129. Costello MJ, McIntosh TJ, Robertson JD. Distribution of gap junctions and square array junctions in the mammalian lens. *Invest Ophthalmol Vis Sci*. 1989;30(5):975–989.
130. Ball LE, Garland DL, Crouch RK, Schey KL. Post-translational modifications of aquaporin 0 (AQP0) in the normal human lens: spatial and temporal occurrence. *Biochemistry*. 2004;43(30):9856–9865.
131. Gonen T, Cheng Y, Sliz P, et al. Lipid-protein interactions in double-layered two-dimensional AQP0 crystals. *Nature*. 2005;438(7068):633–638.
132. Scheuring S, Buzhynskyy N, Jaroslawski S, Goncalves RP, Hite RK, Walz T. Structural models of the supramolecular organization of AQP0 and connexons in junctional microdomains. *J Struct Biol*. 2007;160(3):385–394.
133. Varadaraj K, Kushmerick C, Baldo GJ, Bassnett S, Shiels A, Mathias RT. The role of MIP in lens fiber cell membrane transport. *J Membr Biol*. 1999;170(3):191–203.
134. Shiels A, Bassnett S, Varadaraj K, et al. Optical dysfunction of the crystalline lens in aquaporin-0-deficient mice. *Physiol Genomics*. 2001;7(2):179–186.
135. Varadaraj K, Kumari S, Shiels A, Mathias RT. Regulation of aquaporin water permeability in the lens. *Invest Ophthalmol Vis Sci*. 2005;46(4):1393–1402.
136. Rose KM, Wang Z, Magrath GN, Hazard ES, Hildebrandt JD, Schey KL. Aquaporin 0-calmodulin interaction and the effect of aquaporin 0 phosphorylation. *Biochemistry*. 2008;47(1):339–347.
137. Jensen MO, Dror RO, Xu H, et al. Dynamic control of slow water transport by aquaporin 0: implications for hydration and junction stability in the eye lens. *Proc Natl Acad Sci USA*. 2008;105(38):14430–14435.
138. Harries WE, Akhavan D, Miercke LJ, Khademi S, Stroud RM. The channel architecture of aquaporin 0 at a 2.2-A resolution. *Proc Natl Acad Sci USA*. 2004; 101(39):14045–14050.
139. Yu XS, Yin X, Lafer EM, Jiang JX. Developmental regulation of the direct interaction between the intracellular loop of connexin 45.6 and the C terminus of major intrinsic protein (aquaporin-0). *J Biol Chem*. 2005;280(23):22081–22090.

140. Lindsey Rose KM, Gourdie RG, Prescott AR, Quinlan RA, Crouch RK, Schey KL. The C terminus of lens aquaporin 0 interacts with the cytoskeletal proteins filensin and CP49. *Invest Ophthalmol Vis Sci*. 2006;47(4):1562–1570.
141. Liu BF, Liang JJ. Confocal fluorescence microscopy study of interaction between lens MIP26/AQP0 and crystallins in living cells. *J Cell Biochem*. 2008;104(1): 51–58.
142. Goodenough DA. The crystalline lens. A system networked by gap junctional intercellular communication. *Semin Cell Biol*. 1992;3(1):49–58.
143. Mathias RT, White TW, Gong X. Lens gap junctions in growth, differentiation, and homeostasis. *Physiol Rev*. 2010;90(1):179–206.
144. Musil LS, Beyer EC, Goodenough DA. Expression of the gap junction protein connexin43 in embryonic chick lens: molecular cloning, ultrastructural localization, and post-translational phosphorylation. *J Membr Biol*. 1990;116(2):163–175.
145. Paul DL, Ebihara L, Takemoto LJ, Swenson KI, Goodenough DA. Connexin46, a novel lens gap junction protein, induces voltage-gated currents in nonjunctional plasma membrane of Xenopus oocytes. *J Cell Biol*. 1991;115(4):1077–1089.
146. White TW, Bruzzone R, Goodenough DA, Paul DL. Mouse Cx50, a functional member of the connexin family of gap junction proteins, is the lens fiber protein MP70. *Mol Biol Cell*. 1992;3(7):711–720.
147. Xia CH, Cheng C, Huang Q, et al. Absence of alpha3 (Cx46) and alpha8 (Cx50) connexins leads to cataracts by affecting lens inner fiber cells. *Exp Eye Res*. 2006;83(3): 688–696.
148. Slavi N, Rubinos C, Li L, et al. Connexin 46 (cx46) gap junctions provide a pathway for the delivery of glutathione to the lens nucleus. *J Biol Chem*. 2014;289(47): 32694–32702.
149. Cheng C, Xia CH, Li L, White TW, Niimi J, Gong X. Gap junction communication influences intercellular protein distribution in the lens. *Exp Eye Res*. 2008;86(6): 966–974.
150. Berthoud VM, Westphale EM, Grigoryeva A, Beyer EC. PKC isoenzymes in the chicken lens and TPA-induced effects on intercellular communication. *Invest Ophthalmol Vis Sci*. 2000;41(3):850–858.
151. Zampighi GA, Planells AM, Lin D, Takemoto D. Regulation of lens cell-to-cell communication by activation of PKCgamma and disassembly of Cx50 channels. *Invest Ophthalmol Vis Sci*. 2005;46(9):3247–3255.
152. Yin X, Liu J, Jiang JX. Lens fiber connexin turnover and caspase-3-mediated cleavage are regulated alternately by phosphorylation. *Cell Commun Adhes*. 2008; 15(1):1–11.
153. Allen DP, Low PS, Dola A, Maisel H. Band 3 and ankyrin homologues are present in eye lens: evidence for all major erythrocyte membrane components in same nonerythroid cell. *Biochem Biophys Res Commun*. 1987;149:266–275.
154. Lang RA, Herman K, Reynolds AB, Hildebrand JD, Plageman Jr TF. p120-catenin-dependent junctional recruitment of Shroom3 is required for apical constriction during lens pit morphogenesis. *Development*. 2014;141(16):3177–3187.
155. FitzGerald PG. Lens intermediate filaments. *Exp Eye Res*. 2009;88(2):165–172.
156. Nowak RB, Fischer RS, Zoltoski RK, Kuszak JR, Fowler VM. Tropomodulin1 is required for membrane skeleton organization and hexagonal geometry of fiber cells in the mouse lens. *J Cell Biol*. 2009;186(6):915–928.
157. Kuwabara T. Microtubules in the lens. *Arch Ophthalmol*. 1968;79:189–195.
158. Lee A, Fischer RS, Fowler VM. Stabilization and remodeling of the membrane skeleton during lens fiber cell differentiation and maturation. *Dev Dyn*. 2000;217(3): 257–270.
159. Rafferty NS, Scholz DL, Goldberg M, Lewyckyj M. Immunocytochemical evidence for an actin-myosin system in lens epithelial cells. *Exp Eye Res*. 1990;51:591–600.

160. Ireland M, Lieska N, Maisel H. Lens actin: purification and localization. *Exp Eye Res.* 1983;37:393–408.
161. Bagchi M, Katar M, Lo WK, Yost R, Hill C, Maisel H. ERM proteins of the lens. *J Cell Biochem.* 2004;92(3):626–630.
162. Kam Z, Volberg T, Geiger B. Mapping of adherens junction components using microscopic resonance energy transfer imaging. *J Cell Sci.* 1995;108:1051–1062.
163. Lo WK, Mills A, Kuck JF. Actin filament bundles are associated with fiber gap junctions in the primate lens. *Exp Eye Res.* 1994;58:189–196.
164. Ireland M, Maisel H. A family of lens fiber cell specific proteins. *Lens Eye Toxic Res.* 1989;6:623–638.
165. Kibbelaar MA, Ramaekers FC, Ringens PJ, et al. Is actin in eye lens a possible factor in visual accomodation. *Nature.* 1980;285:506–508.
166. Rafferty NS, Scholz DL. Comparative study of actin filament patterns in lens epithelial cells. Are these determined by the mechanisms of lens accommodation. *Curr Eye Res.* 1989;8:569–579.
167. Lu JY, Mohammed TA, Donohue ST, Al-Ghoul KJ. Distribution of basal membrane complex components in elongating lens fibers. *Mol Vis.* 2008;14: 1187–1203.
168. Al-Ghoul KJ, Lindquist TP, Kirk SS, Donohue ST. A novel terminal web-like structure in cortical lens fibers: architecture and functional assessment. *Anat Rec.* 2010;293(11):1805–1815.
169. Alcala J, Maisel H. Biochemistry of lens plasma membranes and cytoskeleton. In: Maisel H, ed. *The Ocular Lens: Structure, Function and Pathology.* New York: Marcel Dekker; 1985.
170. Sussman MA, McAvoy JW, Rudisill M, et al. Lens tropomodulin: developmental expression during differentiation. *Exp Eye Res.* 1996;63:223–232.
171. Lo WK, Shaw AP, Wen XJ. Actin filament bundles in cortical fiber cells of the rat lens. *Exp Eye Res.* 1997;65(5):691–701.
172. Ramaekers FCS, Bloemendal H. Cytoskeletal and contractile structures in lens cell differentiation. In: Bloemendal H, ed. *Molecular and Cellular Biology of the Eye Lens.* New York: John Wiley & Sons; 1981:85–136.
173. Menko AS, Bleaken BM, Libowitz AA, Zhang L, Stepp MA, Walker JL. A central role for vimentin in regulating repair function during healing of the lens epithelium. *Mol Biol Cell.* 2014;25(6):776–790.
174. Sredy J, Roy D, Spector A. Identification of two of the major phosphorylated polypeptides of the bovine lens utilizing a lens cAMP-dependent protein kinase system. *Curr Eye Res.* 1984;3:1423–1431.
175. Matsuyama M, Tanaka H, Inoko A, et al. Defect of mitotic vimentin phosphorylation causes microophthalmia and cataract via aneuploidy and senescence in lens epithelial cells. *J Biol Chem.* 2013;288(50):35626–35635.
176. Yang J, Luo L, Liu X, et al. Down regulation of the PEDF gene in human lens epithelium cells changed the expression of proteins vimentin and alphaB-crystallin. *Mol Vis.* 2010;16:105–112.
177. Sax CM, Farrell FX, Zehner ZE, Piatigorsky J. Regulation of vimentin gene expression in the ocular lens. *Dev Biol.* 1990;139:56–64.
178. Krimpenfort PJ, Schaart G, Peiper FR, et al. Tissue-specific expression of a vimentin-desmin hybrid gene in transgenic mice. *EMBO J.* 1988;7:941–947.
179. Boyer S, Maunoury R, Gomes D, de Nechaud B, Hill AM, Dupouey P. Expression of glial fibrillary acidic protein and vimentin in mouse lens epithelial cells during development in vivo and during proliferation and differentiation in vitro: comparison with the developmental appearance of GFAP in the mouse central nervous system. *J Neurosci Res.* 1990;27:55–64.

180. Sandilands A, Prescott AR, Carter JM, et al. Vimentin and CP49/filensin form distinct networks in the lens which are independently modulated during lens fibre cell differentiation. *J Cell Sci*. 1995;108:1397–1406.
181. FitzGerald PG, Gottlieb W. The Mr 115 kd fiber cell-specific protein is a component of the lens cytoskeleton. *Curr Eye Res*. 1989;8:801–811.
182. Hess JF, Casselman JT, FitzGerald PG. Gene structure and cDNA sequence identify the beaded filament protein CP49 as a highly divergent type I intermediate filament protein. *J Biol Chem*. 1996;271:6729–6735.
183. Fudge DS, McCuaig JV, Van Stralen S, et al. Intermediate filaments regulate tissue size and stiffness in the murine lens. *Invest Ophthalmol Vis Sci*. 2011;52(6):3860–3867.
184. Gokhin DS, Nowak RB, Kim NE, et al. Tmod1 and CP49 synergize to control the fiber cell geometry, transparency, and mechanical stiffness of the mouse lens. *PLoS One*. 2012;7(11):e48734.
185. Lu ZQ, Sun WH, Yan J, Jiang TX, Zhai SN, Li Y. Cigarette smoking, body mass index associated with the risks of age-related cataract in male patients in northeast China. *Int J Ophthalmol*. 2012;5(3):317–322.
186. Carter JM, Hutcheson AM, Quinlan RA. In vitro studies on the assembly properties of the lens proteins CP49, CP115: coassembly with alpha-crystallin but not with vimentin. *Exp Eye Res*. 1995;60:181–192.
187. Goulielmos G, Gounari F, Remington S, et al. Filensin and phakinin form a novel type of beaded intermediate filaments and coassemble de novo in cultured cells. *J Cell Biol*. 1996;132:643–655.
188. Goulielmos G, Remington S, Schwesinger F, Georgatos SD, Gounari F. Contributions of the structural domains of filensin in polymer formation and filament distribution. *J Cell Sci*. 1996;109:447–456.
189. Reddy VN. Dynamics of transport systems in the eye. Friedenwald Lecture. *Invest Ophthalmol Vis Sci*. 1979;18:1000–1018.
190. Trayhurn P, Van Heyningen R. The role of respiration in the energy metabolism of the bovine lens. *Biochem J*. 1972;129:507–509.
191. Hockwin O, Blum G, Korte I, Murata T, Radetzki W, Rast F. Studies on the citric acid cycle and its portion of glucose breakdown by calf and bovine lenses in vitro. *Ophthalmic Res*. 1971;2:143–148.
192. Kinoshita JH, Merola LO. The utilization of pyruvate and its conversion to glutamate in calf lens. *Exp Eye Res*. 1961;1:53–59.
193. Aviram A, Schalitt M, Kassem N, Groen JJ. Glucose utilization, glutathione, potassium and sodium content of isolated bovine lens. *Clin Chim Acta*. 1966;14:442–449.
194. Giblin FJ, McCready JP, Reddan JR, Dziedzic DC, Reddy VN. Detoxification of H_2O_2 by cultured rabbit lens epithelial cells: participation of the glutathione redox cycle. *Exp Eye Res*. 1985;40:827–840.
195. Giblin FJ, Schrimscher L, Chakrapani B, Reddy VN. Exposure of rabbit lens to hyperbaric oxygen in vitro: regional effects on GSH level. *Invest Ophthalmol Vis Sci*. 1988;29:1312–1319.
196. Seland JH, Chylack Jr LT. Acute glucose-derived osmotic stress in rabbit lenses. *Acta Ophthalmol (Copenh)*. 1986;64:533–539.
197. Burg MB, Kador PF. Sorbitol, osmoregulation, and the complications of diabetes. *J Clin Investig*. 1988;81:635–640.
198. Kinoshita JH. Cataracts in galactosemia: the Jonas Friedenwald Memorial Lecture. *Invest Ophthalmol*. 1965;4:786–799.
199. Kador PF, Kinoshita JH. Diabetic and galactosaemic cataracts. *Ciba Found Symp*. 1984;106:110–131.

200. Lee AY, Chung SK, Chung SS. Demonstration that polyol accumulation is responsible for diabetic cataract by the use of transgenic mice expressing the aldose reductase gene in the lens. *Proc Natl Acad Sci USA*. 1995;92(7):2780–2784.
201. Chylack Jr LT, Henriques HF, Cheng HM, Tung WH. Efficacy of Alrestatin, an aldose reductase inhibitor, in human diabetic and nondiabetic lenses. *Ophthalmology*. 1979;86:1579–1585.
202. Zhang P, Xing K, Randazzo J, Blessing K, Lou MF, Kador PF. Osmotic stress, not aldose reductase activity, directly induces growth factors and MAPK signaling changes during sugar cataract formation. *Exp Eye Res*. 2012;101:36–43.
203. Obrosova IG, Chung SS, Kador PF. Diabetic cataracts: mechanisms and management. *Diabetes Metab Res Rev*. 2010;26(3):172–180.
204. Linetsky M, Shipova E, Cheng R, Ortwerth BJ. Glycation by ascorbic acid oxidation products leads to the aggregation of lens proteins. *Biochim Biophys Acta*. 2008;1782(1): 22–34.
205. Simpanya MF, Ansari RR, Suh KI, Leverenz VR, Giblin FJ. Aggregation of lens crystallins in an in vivo hyperbaric oxygen guinea pig model of nuclear cataract: dynamic light-scattering and HPLC analysis. *Invest Ophthalmol Vis Sci*. 2005;46(12): 4641–4651.
206. Udupa PE, Sharma KK. Effect of oxidized betaB3-crystallin peptide (152-166) on thermal aggregation of bovine lens gamma-crystallins: identification of peptide interacting sites. *Exp Eye Res*. 2005;80(2):185–196.
207. Dillon J. UV-B as a pro-aging and pro-cataract factor. *Doc Ophthalmol*. 1994;88(3–4):339–344.
208. Obrosova IG, Fathallah L, Lang HJ. Interaction between osmotic and oxidative stress in diabetic precataractous lens: studies with a sorbitol dehydrogenase inhibitor. *Biochem Pharmacol*. 1999;58(12):1945–1954.
209. Spector A, Garner WH. Hydrogen peroxide and human cataract. *Exp Eye Res*. 1981;33:673–681.
210. Beebe DC, Holekamp NM, Siegfried C, Shui YB. Vitreoretinal influences on lens function and cataract. *Philos Trans R Soc Lond B Biol Sci*. 2011;366(1568): 1293–1300.
211. Brennan LA, McGreal RS, Kantorow M. Oxidative stress defense and repair systems of the ocular lens. *Front Biosci*. 2012;4:141–155.
212. Ganea E, Harding JJ. Glutathione-related enzymes and the eye. *Curr Eye Res*. 2006; 31(1):1–11.
213. Lou MF. Redox regulation in the lens. *Prog Retin Eye Res*. 2003;22(5):657–682.
214. Reddy VN. Glutathione and its function in the lens—an overview. *Exp Eye Res*. 1990;50:771–778.
215. Pau H, Graf P, Sies H. Glutathione levels in human lens: regional distribution in different forms of cataract. *Exp Eye Res*. 1990;50:17–20.
216. Craghill J, Cronshaw AD, Harding JJ. The identification of a reaction site of glutathione mixed-disulphide formation on gammaS-crystallin in human lens. *Biochem J*. 2004;379(pt 3):595–600.
217. Giblin FJ, Reddan JR, Schrimscher L, Dziedzic DC, Reddy VN. The relative roles of the glutathione redox cycle and catalase in the detoxification of H_2O_2 by cultured rabbit lens epithelial cells. *Exp Eye Res*. 1990;50:795–804.
218. Mancini MA, Unaker NJ, Giblin FJ, Reddan JR. Histochemical localization of catalase in cultured lens epithelial cells. *Ophthalmic Res*. 1989;21:369–373.
219. Raghavachari N, Krysan K, Xing K, Lou MF. Regulation of thioltransferase expression in human lens epithelial cells. *Invest Ophthalmol Vis Sci*. 2001;42(5):1002–1008.
220. Tomarev SI, Zinovieva RD. Squid major lens polypeptides are homologous to glutathione S-transferases subunits. *Nature*. 1988;336:86–88.

221. Spector A, Ma W, Wang RR, Yang Y, Ho YS. The contribution of GSH peroxidase-1, catalase and GSH to the degradation of H_2O_2 by the mouse lens. *Exp Eye Res.* 1997;64(3):477–485.
222. Dahm R, van Marle J, Quinlan RA, Prescott AR, Vrensen GF. Homeostasis in the vertebrate lens: mechanisms of solute exchange. *Philos Trans R Soc Lond B Biol Sci.* 2011;366(1568):1265–1277.
223. Unakar NJ, Tsui JY. Sodium-potassium-dependent ATPase I. cytochemical localization in normal and cataractous rat lenses. *Invest Ophthalmol Vis Sci.* 1980;19:630–641.
224. Palva M, Palkama A. Electron microscopical, histochemical and biochemical findings on the Na-ATPase activity in the epithelium of the rat lens. *Exp Eye Res.* 1976;22:229–236.
225. Neville MC, Paterson CA, Hamilton PM. Evidence for two sodium pumps in the crystalline lens of the rabbit eye. *Exp Eye Res.* 1978;27:637–648.
226. Garner MH, Kong Y. Lens epithelium and fiber Na, K-ATPases: distribution and localization by immunocytochemistry. *Invest Ophthalmol Vis Sci.* 1999;40(10): 2291–2298.
227. Delamere NA, Tamiya S. Lens ion transport: from basic concepts to regulation of Na,K-ATPase activity. *Exp Eye Res.* 2009;88(2):140–143.
228. Mathias RT, Kistler J, Donaldson P. The lens circulation. *J Membr Biol.* 2007;216(1):1–16.
229. Marian MJ, Li H, Borchman D, Paterson CA. Plasma membrane Ca^{2+}-ATPase expression in the human lens. *Exp Eye Res.* 2005;81(1):57–64.
230. Hightower KR, Leverenz V, Reddy VN. Calcium transport in the lens. *Invest Ophthalmol Vis Sci.* 1980;19:1059–1066.
231. Borchman D, Paterson CA, Delamere NA. Ca^{2+}-ATPase activity in the rabbit and bovine lens. *Curr Eye Res.* 1989;8:1049–1054.
232. Gao J, Sun X, Martinez-Wittinghan FJ, Gong X, White TW, Mathias RT. Connections between connexins, calcium, and cataracts in the lens. *J Gen Physiol.* 2004;124(4):289–300.
233. Sabah JR, Davidson H, McConkey EN, Takemoto L. In vivo passage of albumin from the aqueous humor into the lens. *Mol Vis.* 2004;10:254–259.
234. Gorthy WC, Snavely MR, Berrong ND. Some aspects of transport and digestion in the lens of the normal young adult rat. *Exp Eye Res.* 1971;12:112–119.
235. Lee CJ, Vroom JA, Fishman HA, Bent SF. Determination of human lens capsule permeability and its feasibility as a replacement for Bruch's membrane. *Biomaterials.* 2006;27(8):1670–1678.
236. Francois H, Rabaey M. Permeability of the lens capsule for the lens proteins. *Arch Ophthalmol.* 1958;36:837–844.
237. Goodenough DA, Dick JSB, Lyons JE. Lens metabolic cooperation: a study of mouse lens transport and permeability visualized with freeze-substitution autoradiography and electron microscopy. *J Cell Biol.* 1980;86:576–589.
238. Brown HG, Pappas GD, Ireland ME, Kuszak JR. Ultrastructural biochemical and immunologic evidence of receptor-mediated endocytosis in the crystalline lens. *Invest Ophthalmol Vis Sci.* 1990;31:2579–2592.
239. Sabah JR, Schultz BD, Brown ZW, Nguyen AT, Reddan J, Takemoto LJ. Transcytotic passage of albumin through lens epithelial cells. *Invest Ophthalmol Vis Sci.* 2007;48(3):1237–1244.
240. Kern HL, Ho CK. Localization and specificity of the transport system for sugars in the calf lens. *Exp Eye Res.* 1973;15:751–765.
241. Patterson JW. A review of glucose transport in the lens. *Invest Ophthalmol.* 1965;4:667–679.

242. Merriman-Smith BR, Krushinsky A, Kistler J, Donaldson PJ. Expression patterns for glucose transporters GLUT1 and GLUT3 in the normal rat lens and in models of diabetic cataract. *Invest Ophthalmol Vis Sci*. 2003;44(8):3458–3466.
243. Lachke SA, Maas RL. RNA granules and cataract. *Expert Rev Ophthalmol*. 2011; 6(5):497–500.
244. Brennan LA, Kantorow WL, Chauss D, et al. Spatial expression patterns of autophagy genes in the eye lens and induction of autophagy in lens cells. *Mol Vis*. 2012; 18:1773–1786.
245. Chauss D, Basu S, Rajakaruna S, et al. Differentiation state-specific mitochondrial dynamic regulatory networks are revealed by global transcriptional analysis of the developing chicken lens. *G3 (Bethesda)*. 2014;4(8):1515–1527.

CHAPTER TWELVE

Molecular Genetics of Cataract

Alan Shiels*^,1^, J. Fielding Hejtmancik†
*Department of Ophthalmology and Visual Sciences, Washington University School of Medicine, St. Louis, Missouri, USA
†Ophthalmic Genetics and Visual Function Branch, National Eye Institute, National Institutes of Health, Bethesda, Maryland, USA
[1]Corresponding author: e-mail address: Shiels@vision.wustl.edu

Contents

Abstract

Lens opacities or cataract(s) represent a universally important cause of visual impairment and blindness. Typically, cataract is acquired with aging as a complex disorder involving environmental and genetic risk factors. Cataract may also be inherited with an early onset either in association with other ocular and/or systemic abnormalities or as an isolated lens phenotype. Here we briefly review recent advances in gene discovery for inherited and age-related forms of cataract that are providing new insights into lens development and aging.

1. INTRODUCTION

The crystalline lens plays a central role in vertebrate eye development and refractive vision.[1] Loss of lens transparency, or cataract, is a frequently acquired cause of visual impairment in those over 40 years of age.[2] Despite advances in surgical treatment, age-related cataract remains a clinically important cause of visual impairment in economically developed countries

Progress in Molecular Biology and Translational Science, Volume 134
ISSN 1877-1173
http://dx.doi.org/10.1016/bs.pmbts.2015.05.004
© 2015 Elsevier Inc.
All rights reserved.

(www.preventblindness.org) and of blindness (51%) worldwide.[3] Clinical examination of age-related cataract by slit-lamp imaging reveals three classical types of opacities, namely, nuclear cataract, cortical cataract, and posterior subcapsular cataract. Each can occur in isolation or in combination (mixed cataract) and may progress to total opacification of the lens. Several grading systems have been devised that measure cataract severity and progression rate (e.g., Lens Opacities Classification System III).[4] Epidemiological studies have established that age-related cataract is a multifactorial disorder involving complex interactions between environmental risk factors (e.g., UV exposure, tobacco smoking) and genetic susceptibility loci.[5] Heritability estimates for age-related cataract range from 35% to 48% for nuclear opacities and 24% to 58% for cortical opacities.[6] However, the identities of the underlying genetic factors remain poorly characterized.

In addition to being acquired with aging, cataract may present in one or both eyes with an early onset starting at birth (congenital), during infancy, during childhood, or adolescence, and represents a significant cause of visual disability in the pediatric age-group worldwide.[7] Under slit-lamp examination, pediatric cataract exhibits considerable phenotypic variability with respect to location, size, shape, density, age at onset, rate of progression, and even color of opacities within the lens. There is no universally accepted classification system for pediatric cataract. However, several main types of pediatric opacities can be distinguished based on their lens location including nuclear, lamellar, sutural, polar or subcapsular, and total.[8] Congenital and infantile forms of cataract, especially when unilateral, represent a clinically important cause of impaired visual development as a result of deprivation amblyopia. Surgical treatment of pediatric cataract poses a long-term risk of postoperative complications including secondary glaucoma, nystagmus, and retinal detachment.[9–11] Etiological studies of pediatric cataract have estimated that genetic causes account for 10–39% of pediatric cataract/cases. However, this may be an underestimate since approximately 50–60% of cases are deemed idiopathic.[12–14]

Mendelian inheritance of cataract has been documented since the late 1800s and in the 1960s an inherited form of cataract (CAE1) that was closely linked with the Duffy blood-group locus (Fy) became the first monogenic disease assigned to an autosome (chromosome 1) in humans.[15] Since then, considerable progress has been made in mapping, identifying, and characterizing genes underlying inherited forms of cataract mostly as a result of linkage analysis in extended families.[16–19] Currently, if the keyword "cataract(s)" is used to search the Online Mendelian Inheritance in Man database (www.omim.org), over 600 entries are retrieved revealing a wide spectrum of cataract phenotypes along with many animal—particularly mouse—models. In addition to featuring as a secondary or variably associated symptom of many genetic syndromes and/or metabolic disorders,[20] inherited cataract can present as an isolated or primary lens phenotype—the latter accompanied by relatively mild ocular signs including microcornea, refractive error (e.g., myopia), and/or eye movement disorders (nystagmus, strabismus, amblyopia).[16–19]

Typically, inherited cataract presents bilaterally with an early onset (birth—40 years), and most cases are diagnosed as congenital, infantile, or juvenile and display considerable inter- and intrafamilial variation in clinical appearance. All three classical modes of Mendelian inheritance are represented; however, most families segregating isolated or primary cataract with few other ocular signs exhibit autosomal dominant transmission with high penetrance. In this chapter, we review recent advances in gene discovery for Mendelian forms of isolated or primary cataract and highlight the implications of these findings for identifying genetic determinants of the much more common forms of age-related cataract.

2. GENES UNDERLYING ISOLATED OR PRIMARY INHERITED CATARACT

According to OMIM, at least 42 loci have been identified for inherited forms of isolated or primary cataract (Table 1) with minimal other ocular signs (e.g., microcornea). Currently, 12 of these are "orphan" loci with no identified genes. For convenience, the 30 known genes may be arbitrarily divided into four groups based on subcellular localization and/or protein function, namely, cytoplasmic crystallins, membrane proteins, cytoskeletal proteins, and DNA/RNA-binding proteins.

2.1 Genes Encoding Crystallins

Crystallin (*CRY*) genes encode >90% of the lens cytoplasmic proteins and these highly abundant, long-lived, soluble proteins play a key role in establishing the gradient refractive index of the lens.[21] Around 100 different mutations in 12 crystallin genes segregating in over 100 families have been identified that account for about 40–50% of all autosomal dominant cataract reported so far. *CRYAA* and *CRYAB* encode the α-crystallins, two members of the small heat-shock protein (sHSP) family, that form large multimeric complexes (Mr ~500 kDa) with chaperone-like properties.[22]

Table 1 Loci and Genes for Cataract (CTRCT)

Cataract Phenotype	Locus	Inheritance	Associated Phenotypes	Gene	Phenotype MIM No.	Gene/Locus MIM No.
CTRCT1; multiple types	1q21.1	AD/AR	± Microcornea	*GJA8*	116200	600897
CTRCT2; multiple types	2q33.3	AD	± Microcornea	*CRYGC*	604307	123680
CTRCT3; multiple types	22q11.23	AD	± Microcornea	*CRYBB2*	601547	123620
CTRCT4; multiple types	2q33.3	AD	± Microcornea	*CRYGD*	115700	123690
CTRCT5; multiple types	16q21	AD/AR		*HSF4*	116800	602438
CTRCT6; multiple types	1p36.13	AD/AR	Age-related cortical	*EPHA2*	116600	176946
CTRCT7	17q24	AD		?	115660	?
CTRCT8; multiple types	1pter-p36.13	AD		?	115665	?
CTRCT9; multiple types	21q22.3	AD/AR	± Microcornea	*CRYAA*		123580
CTRCT10; multiple types	17q11.2	AD		*CRYBA1*	600881	123610

CTRCT11; multiple types	10q24.32	AD	Anterior segment mesenchymal dysgenesis, microphthalmia, neurodevelopmental abnormalities	*PITX3*	610623	602669
CTRCT12; multiple types	3q22.1	AD	Myopia?	*BFSP2*	611597	603212
CTRCT13	6p24	AR	Adult i (blood group) phenotype	*GCNT2*	110800	600429
CTRCT14; multiple types	13q12.1	AD		*GJA3*	601885	121015
CTRCT15; multiple types	12q13.3	AD		*MIP*	615274	154050
CTRCT16; multiple types	11q22.3	AD/AR	Myopathy, cardiomyopathy	*CRYAB*	613763	123590
CTRCT17; multiple types	22q12.1	AD/AR		*CRYBB1*	611544	6009291
CTRCT18	3p21.31	AR		*FYCO1*	610019	607182
CTRCT19	19q13.41	AR		*LIM2*	615277	154045
CTRCT20; multiple types	3q27.3	AD		*CRYGS*	116100	123730
CTRCT21; multiple types	16q22-q23	AD	± Microcornea	*MAF*	610202	177075
CTRCT22; multiple types	22q11.23	AD/AR		*CRYBB3*	609741	123630

Continued

Table 1 Loci and Genes for Cataract (CTRCT)—cont'd

Cataract Phenotype	Locus	Inheritance	Associated Phenotypes	Gene	Phenotype MIM No.	Gene/Locus MIM No.
CTRCT23	22q12.1	AD		*CRYBA4*	610425	123631
CTRCT24; anterior polar	17p13	AD		?	601202	?
CTRCT25	15q21-q22	AD		?	605728	?
CTRCT26; multiple types	9q13-q22	AR		?	605749	?
CTRCT27; nuclear progressive	2p12	AD		?	607304	?
CTRCT28	6p12-q12	Complex	Age-related cortical, susceptibility to	?	609026	?
CTRCT29; coralliform	2pter-p24	AD		?	115800	?
CTRCT30; pulverulent	10p13	AD		*VIM*	116300	193060
CTRCT31; multiple types	20q11.21	AD		*CHMP4B*	605387	610897
CTRCT32; multiple types	14q22-q23	AD		?	%115650	?
CTRCT33; cortical	20p12.1	AR		*BFSP1*	611391	603307

CTRCT34; multiple types	1p34.3-p32.2	AR	± Microcornea	?	612968	?
CTRCT35; congenital nuclear	19q13	AR		?	609376	?
CTRCT36	9q22.33	AR		*TDRD7*	613887	611258
CTRCT37; cerulean	12q24.2-q24.3	AD		?	614422	?
CTRCT38	7q34	AR	Sengers syndrome	*AGK*	614691	610345
CTRCT39; multiple types	2q34	AD		*CRYGB*	615188	123670
CTRCT40	Xp22.13	X-linked	Nance–Horan (cataract-dental) syndrome	*NHS*	302200	300457
CTRCT41	4p16.1	AD	Wolfram syndrome (DIDMOAD)	*WFS1*	116400	606201
CTRCT42	2q34	AD		*CRYBA2*	115900	600836

Mutations in *CRYAA* are variably associated with nuclear-type opacities and microcornea. By contrast, mutations in *CRYAB* are also associated with several myopathies consistent with its abundant expression in muscle where its binds and stabilizes desmin. Mutations in the genes for β-crystallins (*CRYBB1*, *BB2*, *BB3*, *BA1*, *BA2*, and *BA4*) and γ-crystallins (*CRYGB*, *GC*, *GD*, and *GS*) tend to disrupt folding of the structurally conserved Greek-key domains and/or alter the protein's surface properties resulting in reduced solubility and increased precipitation.[23]

2.2 Genes Encoding Membrane Proteins

An increasing number of transmembrane and membrane-associated proteins that mostly facilitate diverse transport, junctional, or kinase functions have been associated with inherited forms of cataract.[16–19] *GJA3* and *GJA8* encode the gap-junction proteins connexin-46 and connexin-50, respectively, that oligomerize to form hexameric gap-junction channels involved in lens intercellular communication (e.g., ions, electrolytes). Mutations in *GJA3* and *GJA8* are typically associated with nuclear and zonular-pulverulent opacities and account for about 20% of families with autosomal dominant cataract. Functional expression studies reveal that mutant connexins exhibit failed targeting to the cell surface and/or altered channel properties that compromise intercellular communication.[24] *MIP* encodes the aquaporin-0 water channel that, in addition to water transport, plays an important cell–cell adhesion role critical for lens integrity and transparency. Mutations in *MIP* underlie autosomal dominant cataract with variable morphologies and most are believed to cause abnormal retention of the mutant protein within the endoplasmic reticulum (ER).[25,26] *LIM2* encodes a member (MP20) of the peripheral myelin protein-22 (PMP22)_claudin family of transmembrane proteins that share functions in cell adhesion and junction formation, and is required for formation of the lens core syncytium. Despite its relative abundance and likely structural role in the lens, mutations in *LIM2* have so far been found to underlie autosomal recessive cataract suggesting loss-of-function effects.[27,28]

EPHA2 encodes a member of the ephrin receptor subfamily of protein-tyrosine kinases and through its interaction with src kinase has been implicated in lens cell migration. Mutations in *EPHA2* have been associated with autosomal dominant and recessive forms of cataract consistent with deleterious gain-of-function and loss-of-function mechanisms, respectively. Mutations clustered in the cytoplasmic sterile-alpha-motif domain underlying autosomal dominant cataract have been shown to destabilize the receptor and impair Akt-activated cell migration *in vitro*.[29–31] *AGK* encodes the mitochondrial membrane protein acylglycerol kinase a key enzyme in membrane-lipid metabolism, and a truncation mutation has been associated with autosomal recessive cataract in a Saudi family. Similar mutations in AGK have also been associated with "syndromic" cataract including Sengers syndrome and infantile mitochondrial disease.[32] *CHMP4B* encodes charged multivesicular body protein 4B (CHMP4B)—a core component of the endosome sorting complex required for transport-III. *CHMP4B* mutations are associated with posterior-polar cataract consistent with a role for endosome–lysosome pathway in lens homeostasis.[33] *WFS1* encodes the transmembrane protein, wolframin, that is located primarily in the ER, and is believed to play an important role in regulating ER stress and calcium homeostasis. Typically, homozygous mutations in *WFS1* underlie autosomal recessive Wolfram syndrome 1, which is characterized by diabetes insipidus, diabetes mellitus, optic atrophy, and deafness (DIDMOAD). However, a heterozygous missense mutation in *WFS1* has been associated with autosomal dominant congenital nuclear cataract without ocular or systemic abnormalities in an Irish family.[34]

2.3 Genes Encoding Cytoskeletal Proteins

Genes encoding key components of the lens cytoskeleton or proteins with functional ties to the cytoskeleton have been found to underlie inherited cataract. *BFSP1* (CP115 or filensin) and *BFSP2* (CP49 or phakanin) encode intermediate filament-like proteins that combine with α-crystallin to form beaded-filament structures found only in lens fiber cells.[35] Both autosomal dominant and recessive cataracts have been associated with mutations in *BFSP1* and *BFSP2*. The dominant opacities tend to appear as nuclear or lamellar with sutural involvement, whereas the recessive opacities tend to be cortical. A single mutation in the gene coding for the ubiquitous intermediate filament, vimentin (*VIM*) has also been linked with autosomal dominant pulverulent cataract.[36] *FYCO1* encodes a scaffolding protein that is active in microtubule transport of autophagic vesicles. Multiple *FYCO1* mutations underlie autosomal recessive cataract consistent with an important role for autophagy in lens transparency.[37] *NHS* encodes a regulator of actin remodeling and cell shape. Copy number variations (CNVs) in *NHS* cause X-linked cataract (CXN). However, nonsense and frameshift mutations in *NHS* cause Nance–Horan (cataract-dental) syndrome in males and mild sutural opacities in female carriers.[38,39]

2.4 Genes Encoding DNA- or RNA-Binding Proteins

Mutations in several transcription factor genes involved in eye development can present primarily as a cataract phenotype. *HSF4* regulates transcription of sHSPs including lens CRYAB.[40] Mutations in *HSF4* that underlie autosomal dominant cataract lie within the alpha-helical DNA-binding domain, whereas *HSF4* mutations underlying autosomal recessive cataract lie outside this functionally conserved domain. Three *HSF4* truncation mutations linked with recessive cataract have been shown to result in the loss of regulatory domains at the C-terminal end of the protein consistent with loss-of-function mechanisms.[41] *PITX3* encodes a paired-like homeodomain transcription factor that plays a key role in the regulation of genes involved in lens development including *MIP/AQP0*.[42] Mutations in *PITX3* underlie both dominant and recessive forms of cataract with or without anterior segment dysgenesis and microphthalmia. All known coding mutations are located outside the functionally conserved DNA-binding homeodomain. However, several frameshift mutations, including a recurrent 17 bp duplication, have been shown to disrupt the C-terminal *otp/aristaless/rax* (OAR) domain resulting in an altered DNA-binding profile and reduced transactivation activity.[43] *MAF* encodes a bZIP transcription factor that binds to MAF-responsive elements located in the promoter regions of target genes, including those for crystallins and MIP, with varying transactivation properties. *MAF* mutations, located within the conserved DNA-binding domain, underlie autosomal dominant cataract with or without microcornea and disease severity has been correlated with differential transactivation effects on several of the beta/gamma crystallin genes.[44]

Beyond DNA-binding transcription factors, mutations in the gene coding for an RNA-binding protein, tudor domain containing-7 (*TDRD7*), have been associated with autosomal recessive cataract.[45] TDRD7 localizes to a distinct subset of cytoplasmic RNA granules that interact with processing (P) bodies and Staufen-1 (STAU-1) ribonucleoproteins. These findings point to an important role for posttranscriptional regulation of mRNA processing and subcellular localization in lens homeostasis.

3. GENES ASSOCIATED WITH AGE-RELATED CATARACT

Case–control association studies have sought to find genetic susceptibility factors for age-related cataract. Several such studies using a candidate gene approach have found coding and noncoding variations in some of the same genes underlying inherited cataract that are also associated with age-related cataract. These genes include *EPHA2* (1p), *GJA3* (13q), *GJA8* (1q), *MIP* (12q), *HSF4* (16q), *LIM2* (19q), and *CRYAA* (21q).[46–50] However, only the *EPHA2* association has been replicated in different populations including with cortical cataract in Caucasians and Han Chinese, with cortical cataract and PSC in Indians, and with any cataract in Caucasians and Asians from China and India.[46,51–54] There was no association of *EPHA2* variants with nuclear cataract in any population.

Candidate genes underlying syndromic or systemic forms of cataract have also been associated with age-related cataract. *GALK1* (17q) encodes the first enzyme in galactose metabolism and causative mutations in this gene underlie autosomal recessive galactokinase 1-deficiency with cataract—the latter as a result of galactitol accumulation and subsequent osmotic stress. The "Osaka" coding variation in *GALK1* (p.A198V) results in enzyme instability and has been associated with increased risk of age-related cataract in the Japanese population.[55] *SLC16A12* (10q) encodes a creatine transporter and mutations in this gene underlie autosomal dominant juvenile cataract plus microcornea and renal glucosuria.[56] Variations in the 5′-untranslated region of *SLC16A12* have been associated with age-related cataract in Caucasians.[57] Mutations in *WRN* (8p), which encodes a genomic DNA helicase, cause autosomal recessive Werner syndrome—a premature aging (progeroid) disorder with early-onset cataract. Both CNVs and single nucleotide variants in *WRN* have been associated with age-related cortical cataract in Han Chinese.[58–60]

In addition to genes involved in hereditary cataract, variations in other candidate genes not directly associated with inherited forms of cataract have been tentatively implicated in age-related cataract. These include genes that function in antioxidant metabolism (*GSTM1*, *GSTT1*), lactose metabolism (*LCT*), drug metabolism (*NAT2*), folate metabolism (*MTHFR*), DNA repair (*XPD*), lipid/cholesterol transport (*APOE*), kinesin/microtubule motor transport (*KLC1*), actin-cytoskeleton regulation (*EZR*), chaperone activity (*HSP70*), and ephrin signaling (*EFNA5*).[61–70]

A limitation of candidate gene case–control studies is that they may be statistically underpowered to find genome-wide association and are often not replicated in different populations. A previous genome-wide linkage scan has reported several statistically significant susceptibility loci for age-related cortical cataract in Caucasians, including a locus on chromosome 6p12-q12 (CTRCT28); however, no specific genes were identified.[71] So far, only one of the above genes underlying inherited cataract has met genome-wide statistical significance for association with age-related cataract. A recent meta-analysis of genome-wide association studies (GWAS) in multiethnic Asians has identified variants near the 3′-end of *CRYAA* (21q) strongly associated with age-related nuclear cataract.[72] This study also identified intronic variations in *KCNAB1* (3q), which encodes a member of the potassium voltage-gated channel, shaker-related subfamily, that were significantly associated with age-related nuclear cataract along with 14 other loci showing suggestive association. Mutations in two of the latter genes, *SLC4A4* (4q) and *COL4A1* (13q), underlie syndromic forms of congenital cataract associated with renal disease. Finally, another GWAS based on electronic medical records linked to DNA biobanks from individuals of mostly European descent identified a novel range of potential susceptibility loci for age-related cataract including those harboring aldolase B (*ALDOB*), mitogen-activated protein kinase kinase kinase 1 (*MAP3K1*), and monocyte-specific enhancer factor 2C (*MEF2C*).[73] However, none of these susceptibility loci met genome-wide significance.

4. SUMMARY AND OUTLOOK

In addition to the plethora of genetic syndromes and metabolic disorders that are variably associated with cataract, at least 42 genes and loci have been found to underlie inherited forms of isolated or primary cataract. A review of the known causative genes suggests that they may be arbitrarily divided into two broad groups. The first group recapitulates important structure–function aspects of lens biology involving α- and β/γ-crystallins (e.g., *CRYAA*, *CRYBB2*, *CRYGD*), α-connexins (*GJA3*, *GJA8*), and other lens abundant membrane or cytoskeleton proteins (e.g., *MIP/AQP0*, *BFSP2*). The second group highlights developmental and/or regulatory processes or pathways in the lens involving several transcription factors (e.g., *HSF4*, *PITX3*) and an expanding cohort of functionally diverse genes (e.g., *EPHA2*, *TDRD7*, *FYCO1*).

Despite the increasing genetic heterogeneity of inherited cataract, gene discovery studies remain both scientifically and clinically important for several reasons. First, they provide a rational, "gene-centric" basis for clinical classification and molecular diagnosis of Mendelian forms of cataract. While morphological descriptions of lens opacities may assist in the diagnosis of certain genetic syndromes (e.g., sutural opacities in female NHS carriers), in most cases phenotype–genotype correlations are challenging as morphologically similar opacities may be caused by different genes and vice versa.

Genetic studies employing next-generation (massively parallel) exome sequencing now facilitate concurrent profiling of the known candidate genes for underlying mutations and may also enable the identification of novel genes for cataract. As exome sequencing becomes more cost effective, such studies of inherited cataract will provide personalized diagnosis and enhanced genetic counseling for affected families and sporadic cases. Second, they identify plausible candidate genes for genome-wide association (or linkage) studies seeking to discover genetic determinants of age-related cataract. Currently, there is increasing evidence that several genes underlying rare forms of inherited cataract can also influence susceptibility to the much more prevalent forms of age-related cataract (e.g., *EPHA2*, *CRYAA*). These observations raise the possibility of molecular genetic links between lens development and aging. Finally, they provide new insights into lens biology. Notably, the recent discovery of mutations in genes coding for EPHA2, TDRD7, and FYCO1 has provided the first evidence for the functional importance of ephrin signaling, posttranscriptional mRNA regulation, and the autophagy pathway, respectively, in human lens transparency. In the longer term, such discoveries may translate into novel therapeutic strategies that can delay or even prevent cataract onset.

ACKNOWLEDGMENTS

This work was supported by NIH/NEI grants EY012284 (to A.S.) and EY02687 (Core Grant for Vision Research), and by an unrestricted grant to the Department of Ophthalmology and Visual Sciences from Research to Prevent Blindness.

REFERENCES

1. Wormstone IM, Wride MA. The ocular lens: a classic model for development, physiology and disease. *Philos Trans R Soc Lond B Biol Sci*. 2011;366:1190–1192.
2. Asbell PA, Dualan I, Mindel J, Brocks D, Ahmad M, Epstein S. Age-related cataract. *Lancet*. 2005;365:599–609.
3. Pascolini D, Mariotti SP. Global estimates of visual impairment: 2010. *Br J Ophthalmol*. 2012;96:614–618.
4. Chylack Jr LT, Wolfe JK, Singer DM, et al. The lens opacities classification system III. The longitudinal study of cataract study group. *Arch Ophthalmol*. 1993;111:831–836.
5. McCarty CA, Taylor HR. The genetics of cataract. *Invest Ophthalmol Vis Sci*. 2001; 42(8):1677–1678.
6. Sanfilippo PG, Hewitt AW, Hammond CJ, Mackey DA. The heritability of ocular traits. *Surv Ophthalmol*. 2010;55:561–583.
7. Medsinge A, Nischal KK. Pediatric cataract: challenges and future directions. *Clin Ophthalmol*. 2015;9:77–90.
8. Amaya L, Taylor D, Russell-Eggitt I, Nischal KK, Lengyel D. The morphology and natural history of childhood cataracts. *Surv Ophthalmol*. 2003;48:125–144.
9. Lambert SR, Purohit A, Superak HM, Lynn MJ, Beck AD. Long-term risk of glaucoma after congenital cataract surgery. *Am J Ophthalmol*. 2013;156:355–361.

10. Young MP, Heidary G, VanderVeen DK. Relationship between the timing of cataract surgery and development of nystagmus in patients with bilateral infantile cataracts. *J AAPOS*. 2012;16:554–557.
11. Haargaard B, Andersen EW, Oudin A, et al. Risk of retinal detachment after pediatric cataract surgery. *Invest Ophthalmol Vis Sci*. 2014;55:2947–2951.
12. Rahi JS, Dezateux C. Congenital and infantile cataract in the United Kingdom: underlying or associated factors. British congenital cataract interest group. *Invest Ophthalmol Vis Sci*. 2000;41:2108–2114.
13. Haargaard B, Wohlfahrt J, Fledelius HC, Rosenberg T, Melbye M. A nationwide Danish study of 1027 cases of congenital/infantile cataracts: etiological and clinical classifications. *Ophthalmology*. 2004;111:2292–2298.
14. Lim Z, Rubab S, Chan YH, Levin AV. Pediatric cataract: the Toronto experience-etiology. *Am J Ophthalmol*. 2010;149:887–892.
15. Hejtmancik JF. The genetics of cataract: our vision becomes clearer. *Am J Hum Genet*. 1998;62:520–525.
16. Shiels A, Hejtmancik JF. Genetic origins of cataract. *Arch Ophthalmol*. 2007;125: 165–173.
17. Hejtmancik JF. Congenital cataracts and their molecular genetics. *Semin Cell Dev Biol*. 2008;19:134–149.
18. Shiels A, Bennett TM, Hejtmancik JF. Cat-Map: putting cataract on the map. *Mol Vis*. 2010;16:2007–2015.
19. Shiels A, Hejtmancik JF. Genetics of human cataract. *Clin Genet*. 2013;84:120–127.
20. Trumler AA. Evaluation of pediatric cataracts and systemic disorders. *Curr Opin Ophthalmol*. 2011;22:365–379.
21. Pierscionek BK, Regini JW. The gradient index lens of the eye: an opto-biological synchrony. *Prog Retin Eye Res*. 2012;31:332–349.
22. Clark AR, Lubsen NH, Slingsby C. sHSP in the eye lens: crystallin mutations, cataract and proteostasis. *Int J Biochem Cell Biol*. 2012;44:1687–1697.
23. Wistow G. The human crystallin gene families. *Hum Genomics*. 2012;6:26.
24. Beyer EC, Ebihara L, Berthoud VM. Connexin mutants and cataracts. *Front Pharmacol*. 2013;4:43.
25. Shiels A. Focus on molecules: major intrinsic protein. *Exp Eye Res*. 2012;101:107–108.
26. Lo WK, Biswas SK, Brako L, Shiels A, Gu S, Jiang JX. Aquaporin-0 targets interlocking domains to control the integrity and transparency of the eye lens. *Invest Ophthalmol Vis Sci*. 2014;55:1202–1212.
27. Shi Y, Barton K, De Maria A, Petrash JM, Shiels A, Bassnett S. The stratified syncytium of the vertebrate lens. *J Cell Sci*. 2009;122:1607–1615.
28. Maher GJ, Black GC, Manson FD. Focus on molecules: lens intrinsic membrane protein (LIM2/MP20). *Exp Eye Res*. 2012;103:115–116.
29. Park JE, Son AI, Hua R, Wang L, Zhang X, Zhou R. Human cataract mutations in EPHA2 SAM domain alter receptor stability and function. *PLoS One*. 2012;7: e36564.
30. Shi Y, De Maria A, Bennett T, Shiels A, Bassnett S. A role for epha2 in cell migration and refractive organization of the ocular lens. *Invest Ophthalmol Vis Sci*. 2012;53: 551–559.
31. Cheng C, Ansari MM, Cooper JA, Gong X. EphA2 and Src regulate equatorial cell morphogenesis during lens development. *Development*. 2013;140:4237–4245.
32. Aldahmesh MA, Khan AO, Mohamed JY, Alghamdi MH, Alkuraya FS. Identification of a truncation mutation of acylglycerol kinase (AGK) gene in a novel autosomal recessive cataract locus. *Hum Mutat*. 2012;33:960–962.
33. Shiels A, Bennett TM, Knopf HL, et al. CHMP4B, a novel gene for autosomal dominant cataracts linked to chromosome 20q. *Am J Hum Genet*. 2007;81:596–606.

34. Berry V, Gregory-Evans C, Emmett W, et al. Wolfram gene (WFS1) mutation causes autosomal dominant congenital nuclear cataract in humans. *Eur J Hum Genet.* 2013;21:1356–1360.
35. Song S, Landsbury A, Dahm R, Liu Y, Zhang Q, Quinlan RA. Functions of the intermediate filament cytoskeleton in the eye lens. *J Clin Invest.* 2009;119:1837–1848.
36. Muller M, Bhattacharya SS, Moore T, et al. Dominant cataract formation in association with a vimentin assembly disrupting mutation. *Hum Mol Genet.* 2009;18:1052–1057.
37. Chen J, Ma Z, Jiao X, et al. Mutations in FYCO1 cause autosomal-recessive congenital cataracts. *Am J Hum Genet.* 2011;88:827–838.
38. Brooks SP, Coccia M, Tang HR, et al. The Nance-Horan syndrome protein encodes a functional WAVE homology domain (WHD) and is important for co-ordinating actin remodelling and maintaining cell morphology. *Hum Mol Genet.* 2010;19:2421–2432.
39. Coccia M, Brooks SP, Webb TR, et al. X-linked cataract and Nance-Horan syndrome are allelic disorders. *Hum Mol Genet.* 2009;18:2643–2655.
40. Somasundaram T, Bhat SP. Developmentally dictated expression of heat shock factors: exclusive expression of HSF4 in the postnatal lens and its specific interaction with alphaB-crystallin heat shock promoter. *J Biol Chem.* 2004;279:44497–44503.
41. Merath K, Ronchetti A, Sidjanin DJ. Functional analysis of HSF4 mutations found in patients with autosomal recessive congenital cataracts. *Invest Ophthalmol Vis Sci.* 2013;54:6646–6654.
42. Sorokina EA, Muheisen S, Mlodik N, Semina EV. MIP/aquaporin 0 represents a direct transcriptional target of PITX3 in the developing lens. *PLoS One.* 2011;6:e21122.
43. Verdin H, Sorokina EA, Meire F, et al. Novel and recurrent PITX3 mutations in Belgian families with autosomal dominant congenital cataract and anterior segment dysgenesis have similar phenotypic and functional characteristics. *Orphanet J Rare Dis.* 2014;9:26.
44. Vanita V, Guo G, Singh D, Ott CE, Robinson PN. Differential effect of cataract-associated mutations in MAF on transactivation of MAF target genes. *Mol Cell Biochem.* 2014;396:137–145.
45. Lachke SA, Alkuraya FS, Kneeland SC, et al. Mutations in the RNA granule component TDRD7 cause cataract and glaucoma. *Science.* 2011;331:1571–1576.
46. Shiels A, Bennett TM, Knopf HL, et al. The EPHA2 gene is associated with cataracts linked to chromosome 1p. *Mol Vis.* 2008;14:2042–2055.
47. Shi Y, Shi X, Jin Y, et al. Mutation screening of HSF4 in 150 age-related cataract patients. *Mol Vis.* 2008;14:1850–1855.
48. Zhou Z, Wang B, Hu S, Zhang C, Ma X, Qi Y. Genetic variations in GJA3, GJA8, LIM2, and age-related cataract in the Chinese population: a mutation screening study. *Mol Vis.* 2011;17:621–626.
49. Zhou Z, Wang B, Luo Y, et al. Major intrinsic protein (MIP) polymorphism is associated with age-related cataract in Chinese. *Mol Vis.* 2011;17:2292–2296.
50. Validandi V, Reddy VS, Srinivas PN, et al. Temperature-dependent structural and functional properties of a mutant (F71L) alphaA-crystallin: molecular basis for early onset of age-related cataract. *FEBS Lett.* 2011;585:3884–3889.
51. Jun G, Guo H, Klein BE, et al. EPHA2 is associated with age-related cortical cataract in mice and humans. *PLoS Genet.* 2009;5:e1000584.
52. Tan W, Hou S, Jiang Z, Hu Z, Yang P, Ye J. Association of EPHA2 polymorphisms and age-related cortical cataract in a Han Chinese population. *Mol Vis.* 2011;17:1553–1558.
53. Sundaresan P, Ravindran RD, Vashist P, et al. EPHA2 polymorphisms and age-related cataract in India. *PLoS One.* 2012;7:e33001.
54. Yang J, Luo J, Zhou P, Fan Q, Luo Y, Lu Y. Association of the ephreceptor tyrosinekinase-type A2 (EPHA2) gene polymorphism rs3754334 with age-related cataract risk: a meta-analysis. *PLoS One.* 2013;8:e71003.

55. Okano Y, Asada M, Fujimoto A, et al. A genetic factor for age-related cataract: identification and characterization of a novel galactokinase variant, "Osaka," in Asians. *Am J Hum Genet.* 2001;68:1036–1042.
56. Abplanalp J, Laczko E, Philp NJ, et al. The cataract and glucosuria associated monocarboxylate transporter MCT12 is a new creatine transporter. *Hum Mol Genet.* 2013;22:3218–3226.
57. Zuercher J, Neidhardt J, Magyar I, et al. Alterations of the 5'untranslated region of SLC16A12 lead to age-related cataract. *Invest Ophthalmol Vis Sci.* 2010;51:3354–3361.
58. Su S, Yao Y, Zhu R, et al. The associations between single nucleotide polymorphisms of DNA repair genes, DNA damage, and age-related cataract: Jiangsu Eye Study. *Invest Ophthalmol Vis Sci.* 2013;54:1201–1207.
59. Jiang J, Zhou J, Yao Y, et al. Copy number variations of DNA repair genes and the age-related cataract: Jiangsu Eye Study. *Invest Ophthalmol Vis Sci.* 2013;54:932–938.
60. Jiang S, Hu N, Zhou J, et al. Polymorphisms of the WRN gene and DNA damage of peripheral lymphocytes in age-related cataract in a Han Chinese population. *Age.* 2013;35:2435–2444.
61. Sun L, Xi B, Yu L, et al. Association of glutathione S-transferases polymorphisms (GSTM1 and GSTT1) with senile cataract: a meta-analysis. *Invest Ophthalmol Vis Sci.* 2010;51:6381–6386.
62. Karas-Kuzelicki N, Pfeifer V, Lukac-Bajalo J. Synergistic effect of high lactase activity genotype and galactose-1-phosphate uridyl transferase (GALT) mutations on idiopathic presenile cataract formation. *Clin Biochem.* 2008;41:869–874.
63. Tamer L, Yilmaz A, Yildirim H, et al. N-acetyltransferase 2 phenotype may be associated with susceptibility to age-related cataract. *Curr Eye Res.* 2005;30:835–839.
64. Zetterberg M, Tasa G, Prince JA, et al. Methylenetetrahydrofolate reductase genetic polymorphisms in patients with cataract. *Am J Ophthalmol.* 2005;140:932–934.
65. Padma G, Mamata M, Reddy KR, Padma T. Polymorphisms in two DNA repair genes (XPD and XRCC1)—association with age related cataracts. *Mol Vis.* 2011;17:127–133.
66. Utheim OA, Ritland JS, Utheim TP, et al. Apolipoprotein E genotype and risk for development of cataract and age-related macular degeneration. *Acta Ophthalmol.* 2008;86:401–403.
67. Zhang L, Xu JW, Qu X, Liu DR, Liu P, Zhao XZ. Association of a rare haplotype in kinesin light chain 1 gene with age-related cataract in a Han Chinese population. *PLoS One.* 2013;8:e64052.
68. Lin Q, Zhou N, Zhang N, et al. Genetic variations and polymorphisms in the ezrin gene are associated with age-related cataract. *Mol Vis.* 2013;19:1572–1579.
69. Zhang Y, Gong J, Zhang L, Xue D, Liu H, Liu P. Genetic polymorphisms of HSP70 in age-related cataract. *Cell Stress Chaperones.* 2013;18:703–709.
70. Lin Q, Zhou N, Zhang N, Qi Y. Mutational screening of EFNA5 in Chinese age-related cataract patients. *Ophthalmic Res.* 2014;52:124–129.
71. Iyengar SK, Klein BE, Klein R, et al. Identification of a major locus for age-related cortical cataract on chromosome 6p12-q12 in the Beaver Dam Eye Study. *Proc Natl Acad Sci USA.* 2004;101:14485–14490.
72. Liao J, Su X, Chen P, et al. Meta-analysis of genome-wide association studies in multiethnic Asians identifies two loci for age-related nuclear cataract. *Hum Mol Genet.* 2014;23:6119–6128.
73. Ritchie MD, Verma SS, Hall MA, et al. Electronic medical records and genomics (eMERGE) network exploration in cataract: several new potential susceptibility loci. *Mol Vis.* 2014;20:1281–1295.

SECTION III

Refraction

CHAPTER THIRTEEN

RPE and Choroid Mechanisms Underlying Ocular Growth and Myopia

Yan Zhang[1], Christine F. Wildsoet
School of Optometry, University of California, Berkeley, California, USA
[1]Corresponding author: e-mail address: yanzhang@berkeley.edu

Contents

Abstract

Myopia is the most common type of refractive errors and one of the world's leading causes of blindness. Visual manipulations in animal models have provided convincing evidence for the role of environmental factors in myopia development. These models along with *in vitro* studies have provided important insights into underlying mechanisms. The key locations of the retinal pigment epithelium (RPE) and choroid make them plausible conduits for relaying growth regulatory signals originating in the retina to the sclera, which ultimately determines eye size and shape. Identifying the key signal molecules and their targets may lead to the development of new myopia control treatments. This section summarizes findings implicating the RPE and choroid in myopia development. For RPE and/or choroid, changes in morphology, activity of ion channels/transporters, as well as in gene and protein expression, have been linked to altered eye growth. Both tissues thus represent potential targets for novel therapies for myopia.

Progress in Molecular Biology and Translational Science, Volume 134
ISSN 1877-1173
http://dx.doi.org/10.1016/bs.pmbts.2015.06.014
© 2015 Elsevier Inc.
All rights reserved.

1. INTRODUCTION

Uncorrected refractive errors represent one of the world's leading causes of blindness and a significant contributor to the global burden of eye diseases.[1,2] For children and young adults, myopia, hyperopia, and astigmatism represent the categories of refractive errors encountered; these same conditions may be found in older adults, with presbyopia representing an additional potential cause of vision loss for this group.

Myopia (near-sightedness) describes the condition in which the image of a distant object is focused in front of the retina, resulting in blurred distance vision when left uncorrected. Myopia reflects the mismatch between the refracting power of the eye and its optical axial length. Most myopia is caused by excessive ocular elongation, with refracting power being near normal (Fig. 1).[3] Myopia carries an increased risk of a variety of sight-threatening pathologies, including myopic maculopathy, retinal detachment, choroidal neovascularization, cataract, and glaucoma, with high myopes (classically defined as spherical equivalent refractive errors equal to or greater than −6 D), being at greatest risk.[4,5]

Myopia is now the most common type of refractive error and one of the world's leading causes of functional blindness due to lack of access to optical corrections.[6] A figure of 41.6% for persons aged 12–54 years is given in the most recently published myopia prevalence data for the United States,[7] while even higher, epidemic levels of myopia have been reported for many Asia countries, e.g., 96.5% in young Korean males, along with increases in the average amount of myopia.[8,9] Thus, myopia now represents a significant public health problem worldwide, both socially and economically.[1,10] These climbing prevalence statistics are driving research aimed at effective therapeutic interventions to prevent the development of myopia and/or slow its progression.

It is now generally accepted that both genetic and environmental factors play roles in the development of human myopia.[5,11,12] Genetic studies of myopia, using linkage and genome-wide association approaches, have now identified multiple myopic loci and candidate genes for high myopia and the most common form of juvenile myopia.[12–14] Nonetheless, human epidemiological studies have also provided convincing evidence for environmental influences, with near work and outdoor activities being among the factors identified to affect myopia prevalence.[15]

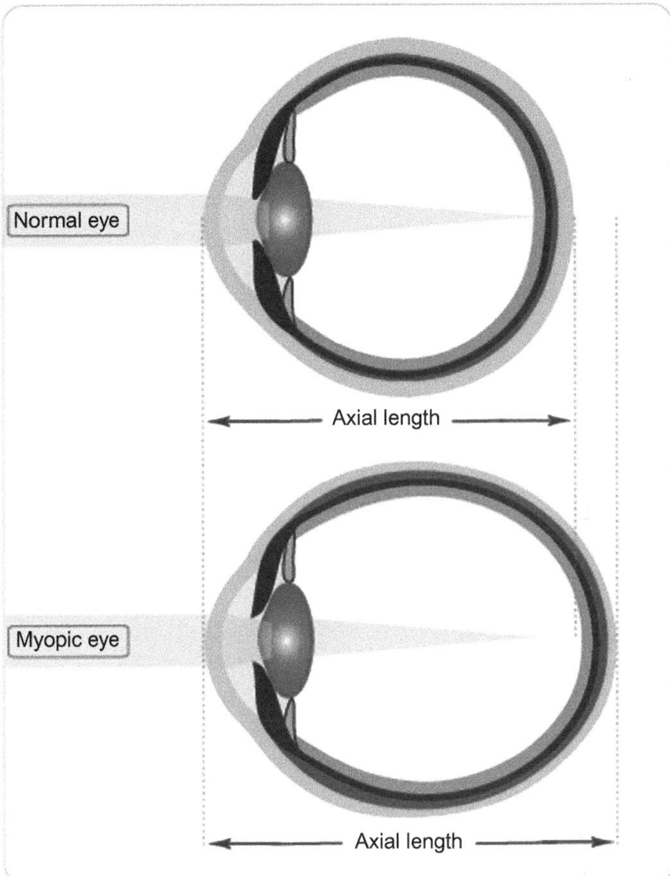

Figure 1 Schematic diagram illustrating the principal gross anatomical differences between human emmetropic and myopic eyes, the latter typically being longer and more prolate, with a longer vitreous chamber. (See the color plate.)

That environmental factors influence ocular growth regulation and thus refractive errors is further supported by animal studies in which the visual environment is manipulated to alter optical defocus and/or the quality of the retinal image. Specifically, both negative defocusing lenses, used to move the plane of focus behind the retina, and form-deprivation strategies, e.g., achieved using diffusers to cover the eyes, accelerate eye growth in young animals, thereby inducing myopia (Fig. 2). Chickens, guinea pigs, tree shrews, and monkeys represent the most widely studied models,[11] with the mouse and zebrafish also making appearances in select studies.[17,18]

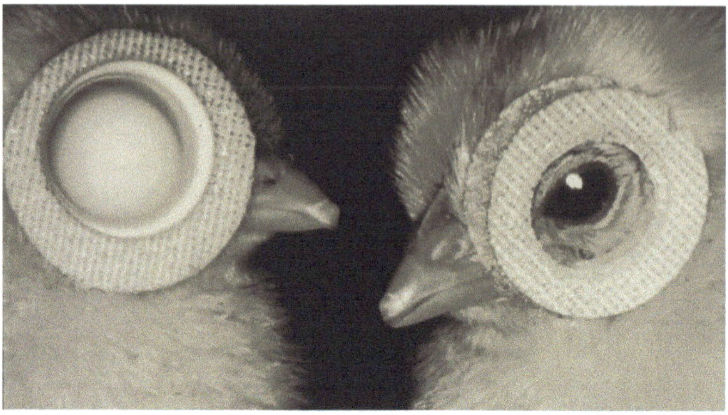

Figure 2 Chicks fitted with either a translucent, form-depriving goggle (left), or a defocusing spectacle lens (right). *From Ref. 16.*

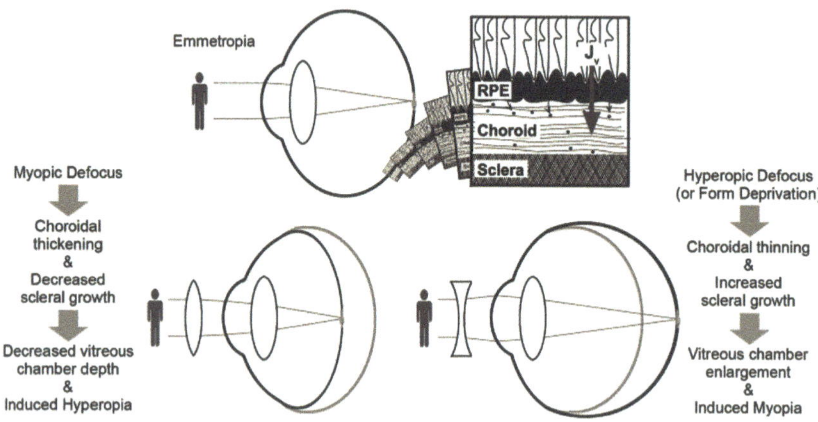

Figure 3 Schematic diagram summarizing models used to study eye growth regulation, including key ocular features of these models; plausible local ocular growth regulatory signal pathway included. *From Ref. 16.*

Such models represent important tools for investigating the molecular and cellular signaling pathways mediating ocular growth regulation, which may, in turn, lead to the development of new myopia therapies.[19,20] Fortuitously, ocular growth appears to be largely regulated by local ocular mechanisms.[11] Thus, related studies have focused on retino-scleral signaling cascades linking the retina, the presumed source of ocular growth signals, to the choroid and sclera, whose growth/remodeling ultimately determines the physical dimensions of the vitreous chamber and the location of the retina (Fig. 3).[21,22] Through investigations into the molecular and cellular components of these signaling pathways and the changes linked to altered ocular growth, and the cellular and biochemical events mediating the changes in the choroid and sclera, it is plausible that novel pharmacological treatments for controlling myopia be forthcoming.[11]

This chapter covers the roles of the retinal pigment epithelium (RPE) and choroid in ocular growth regulation and refractive error development, including myopia, and encompasses molecular, biochemical and cellular mechanisms. Please refer further to the following chapters, "Molecular and Biochemical Aspects of the Retina on Refraction" (written by Ranjay Chakraborty and Machelle Pardue), "Scleral Mechanisms Underlying Ocular Growth and Myopia" (written by Ravikath Metlapally and Christine F. Wildsoet), and "Genetics of Refraction and Myopia" (written by Qingjiong Zhang).

2. THE ROLE OF THE RPE IN EYE GROWTH REGULATION

The RPE is a monolayer of highly specialized pigmented cells, separating the neural retina from the vascular choroid. Being pigmented, these cells serve to absorb stray light within the eye. Being also interconnected by tight junctions, these cells represent a critical component of the blood–retina barrier (Fig. 4), with essential roles in the maintenance of retinal integrity.[23–25] These cells also show other specializations that reflect their role in maintaining retinal homeostasis. Thus, they are polarized, with asymmetric distributions of specialized transport proteins and channels over their apical and basolateral membranes, allowing for the tight regulation of exchange between the retina and the choroid of many molecules, including ions, water, nutrients, and waste products. In addition, the RPE plays a critical role in the maintenance of photoreceptor function; related functions include the phagocytosis of photoreceptor outer segments, which follows a diurnal cycle, and uptake and recycling of retinal as a critical step in photopigment regeneration. These various, well-recognized functions of the RPE have long been the subject of study. However, more recent research has uncovered additional functions for the RPE. For example, it is now known to be a major source of cytokines and growth factors, with important roles in maintaining retinal integrity, in establishing the immune privilege of the eye and potentially in early eye growth regulation.

In the context of myopia development, it has been demonstrated that postnatal eye growth is largely controlled locally. Evidence comes from lesioning studies involving optic nerve section and related pharmacological studies; thus, even when the retina–brain link is disrupted, myopia may be induced with appropriate experimental manipulations in both chickens and guinea pigs.[11,26,27] This model of local ocular growth regulation also

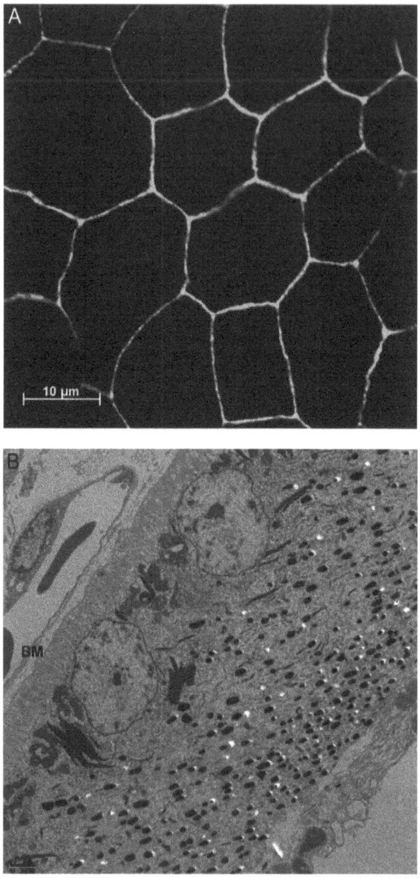

Figure 4 (A) Cultured human fetal RPE cells. Tight junctions, shown stained for ZO-1 (zonula occludens-1, in green), ensure that the RPE functions as an effective barrier between the retina and the choroid, with exchange tightly regulated through ion channels and transporters. (B) Transmission electron micrographs of chick RPE; cells show a distinct asymmetry, with nuclei located in the basal region adjacent Bruch's membrane (BM) and choriocapillaris (CC), and melanin granules more concentrated in the apical (retinal) region. (See the color plate.)

explains observations of localized ocular shape changes in response to localized manipulation of retinal images.[11] The RPE's key location between the retina and the choroid makes it a possible conduit for relaying growth regulatory signals originating in the retina to the choroid and sclera.[16] The RPE is known to have receptors for many of the signaling molecules that have been implicated in eye growth regulation, including dopamine (DA), acetylcholine, vasoactive intestinal peptide (VIP), and glucagon.[16,28–32]

In addition, the transepithelial transport of ions and/or fluid across the RPE may have implications for choroidal thickness, which appears to be one of the targets of ocular growth signals.[33] Finally, recent studies have demonstrated differential expression of a number of genes in the RPE of eyes undergoing experimental manipulations perturbing normal eye growth.[34,35]

2.1 Morphological Features of RPE in Myopic and "Recovering" Eyes and Potential Role in Myopia-Related Pathology

Morphological changes have been reported in the RPE of experimental animals with induced myopia.[36–38] For example, in form-deprived chicks, the increase in total area of the RPE layer in their myopic eyes was coupled to an increase in the surface area of individual RPE cells, by way of maintaining coverage of the expanded vitreous chamber.[36] Nonetheless, the majority of the RPE cells in the form-deprived eyes retained near-normal, hexagonal shapes despite their expanded surface areas.[36] Interestingly, while in normal eyes, RPE cell expansion was limited mainly in the peripheral regions, the expansion in surface area in myopic chick eyes encompassed all but a temporal region.[36] Enlarged RPE cells but not increases in cell numbers have also been reported in the lid-sutured eyes of a mammalian model, although the distribution of multinucleated RPE cells was reported to be significantly altered.[37,38] The preceding changes do not in themselves suggest pathological changes in the RPE. However, the observations of Liang *et al.* may be interpreted as indirect evidence of altered RPE function. Specifically, in chick eyes allowed to recovery from form-deprivation-induced myopia, they reported significant edema and altered basal in-foldings in the RPE along with thickening of Bruch's membrane.[39] Nonetheless, studies of the RPE of eyes with experimentally-induced myopia in the context of related pathologies have been very limited to-date, and the following sections will focus on its potential role in eye growth regulation.

2.2 Ion and Fluid Transport Across the RPE and Implications for Eye Growth Regulation

As noted earlier, RPE cells are interconnected by tight junctions, which prevent the free exchange of ions and water between the retina and the choroid. However, the RPE has specializations that allow for regulated fluid exchange between the retina and the choroid, and specifically, transport of ions and water from the subretinal space to the blood, as is critical for maintaining retinal homeostasis.[16,23] Among relevant channels present on the RPE are potassium (K^+) and chloride (Cl^-) channels, which are known to regulate transepithelial fluid movement. Many show asymmetric distribution, consistent with their functions. Thus, Cl^- channels and a cystic fibrosis transmembrane conductance regulator have been localized to the basolateral side of RPE cells, while Na^+, K^+-ATPase, the source of energy for transepithelial transport, is located on the apical membrane.[16,23,25,40] While the role of the RPE in maintaining retinal homeostasis is likely to be common to all species, species differences in the distribution of ion channels and function have been reported. For the chick, which is a widely used model in myopia research, a detailed summary for its ion and fluid transport in the RPE and its potential role in eye growth regulation is contained in reviews by Rymer *et al.* and Crewther.[16,41]

In the context of eye growth regulation, the roles of RPE ion and fluid channels are not well understood, although the observation of early, rapid changes in choroidal thickness during the development of myopia and hyperopia in chicks provides a plausible link between ion and fluid transport across the RPE and eye growth regulation.[21,42] Furthermore, potassium (K^+) and phosphate levels are reported to be decreased, and chloride (Cl^-) elevated, in the vitreous of form-deprivation myopic chicks.[43] The genes encoding the Cl^- transporter and channel were down-regulated in the RPE with lens-induced myopia.[33] These results open the possibility that the choroid thinning observed during the early phase of experimental myopia induction may be a product of decreased ion and fluid transport to the choroid. That the concentrations of K^+, Na^+, and Cl^- ions are reported to be elevated in RPE-photoreceptor outer segment regions of freeze-dried preparation from eyes allowed to recover from induced myopia, represents further indirect evidence implicating ion transport in eye growth regulation.[44,45] Energy dispersive X-ray microanalysis was used for the latter assays. Note also that these changes were detected only early in the recovery process, with levels more closely paralleling those of fellow eyes with longer recovery periods. More direct evidence implicating ion transport in eye growth regulation is provided by studies involving pharmacological manipulation in chicks.[46] Thus, intravitreal injection of barium chloride, a nonspecific potassium channel inhibitor, was found to inhibit the compensatory ocular growth responses to imposed optical defocus, be they elicited with positive lenses or negative lenses, and bumetanide, a selective sodium–potassium–chloride cotransporter inhibitor, selectively inhibited the response to negative lens.

The premise that altered ion transport across the RPE is a feature of eye growth regulation, as discussed above, leaves open the identity of signal molecules responsible. Among possibilities are one or more retinal neurotransmitters, including DA, which has been linked to eye growth regulation and for which there are receptors on the RPE.[47,48]

2.3 Neurotransmitters as Plausible Signal Molecules for RPE-Mediated Eye Growth Regulation

Retinal neurotransmitters including DA, acetylcholine, and glucagon have been the focus of many studies in relation to their roles in retinal functions in the context of eye growth regulation. Below, we consider the possibility that the RPE could be the site of action of such molecules, serving as signal molecules for eye growth regulation, on the basis that many of these molecules also have receptors on RPE and are known to affect RPE physiology.[16]

2.3.1 Dopamine

In the retina, DA, serving as a neurotransmitter and neuromodulator, plays important roles in retinal function.[49] This topic is explored in more detail in chapter "Molecular and Biochemical Aspects of the Retina on Refraction" (written by Ranjay Chakraborty and Machelle Pardue) of this volume. Here, the possible role of DA in eye growth regulation, acting via the RPE is discussed, along with related physiology.

There are five subtypes of DA receptors (D1–D5), which, based on their biochemical and pharmacological properties, have been further categorized into D1-like (D1, D5) and D2-like (D2–D4) subfamilies.[50,51] DA receptors are known to be widely distributed in the eye, including on neural cells within the retina and RPE.[32,49,52–54] DA receptors have been identified on human RPE cells as well as teleost, chick, cat, and bovine RPE.[31,32,52,53] Both D1 and D2 DA receptors have been identified in cultured human RPE.[55,56] In chicks, D2/3 receptors have been identified on the basal side of the RPE. However, due to the use of *in situ* hybridization and immunocytochemistry in the study in chick, the presence of DA receptors on the apical surface of RPE cells was left unresolved due to heavy pigmentation in this region.[32]

In vivo animal studies, mostly involving the chick model, have consistently reported inhibitory effects on experimental myopia of DA receptor agonists, typically delivered intravitreally.[57–59] Indirect support for the possibility that RPE is the site of action of these drugs is provided by two studies in chicks. First, [^{3}H]-spiperone, a D2-receptor antagonist, was demonstrated to reach the RPE when administered by either intravitreal or subconjunctival injection.[58] Second, *in vitro* electrophysiology studies using retina–RPE–choroid preparations showed that RPE function was altered differently with retinal versus choroidal perfusion of DA. Note also that DA has been named as a likely modulation of basolateral Cl^- channels in the RPE, acting through different populations of DA receptors on the apical and basolateral membranes of RPE.[47]

2.3.2 Acetylcholine

Acetylcholine (ACh) receptors fall into two broad categories, of membrane receptors, metabotropic muscarinic acetylcholine receptors (mAChRs) and ionotropic nicotinic acetylcholine receptors (nAChR).[60] Of these two types of receptors, muscarinic receptors are a family of G-protein-coupled receptors, which comprise five receptor subtypes (M1–M5) in mammals. In chicks, only M2–M5 receptor subtypes have been identified.[61,62] In addition to serving as a retinal neurotransmitter, ACh appears to play an important role in the developing retina.[63,64] There is now substantial evidence that ACh, acting via both muscarinic and nicotinic receptors (mAChR and nAChR), is involved in eye growth regulation.[16,48,65,66] Discussion here is limited to the more widely studied mAChR-mediated effects.

Studies linking ACh and muscarinic receptors with eye growth regulation and specifically myopia are many. In humans, atropine, a nonselective mAChR antagonist, has a long history of use for myopia control, dating back to the middle of the last century. Today, it remains the mostly widely used pharmacological agent clinically for this purpose, despite evidence of rebound effects after treatment is terminated, at least for higher doses.[65,67,68] Animal models studied in this context include the mouse, chicks, guinea pigs, tree shrews, and monkeys, with consistent antimyopia effects of muscarinic receptor antagonists being reported.[16,66,69,70] For example, in chicks, both intravitreal and subconjunctival injection of atropine were found to inhibit the development of form-deprivation and lens-induced myopia.[59,71,72] Other, more selective antimuscarinic drugs including pirenzepine and himbacine have also been reported to inhibit the development of myopia, with M4 receptors being favored as the receptor subtype mediating these effects based on the selectivity profiles of the latter drugs, i.e., M1/M4 (mammal) or M2/M4 (chick) for pirenzepine and M2/M4 for himbacine.[66,72–75]

As with DA receptors, the wide distribution of muscarinic receptors throughout ocular tissues offers multiple candidate tissues, and the sites of action for the antimyopia effects of muscarinic receptor antagonists include but are not limited to retina, choroid, and sclera.[16,71,72] The RPE also expresses muscarinic receptors and thus is a candidate tissue.[76,77] In chick RPE, the relevant receptor subtypes, M2, M3, and M4 receptors, have all been identified.[61] Physiological actions mediated by mAChRs activation include increased phosphoinositide turnover and intracellular Ca^{2+}, as seen in cultured human RPE.[28,77,78] However, at this time, other supporting evidence for the RPE being the site of action of the described antimyopia actions is lacking.

In addition to the two neurotransmitter receptors discussed above, the RPE is also reported to possess receptors for a number of other neurotransmitters, including glucagon and VIP, which have been implicated in eye growth regulation.[16,29,30,79] In the case of glucagon, the RPE has been considered a plausible site of action in two separate studies in chicks,[80,81] and in the case of VIP, the observation of polarized secretion of macromolecules with application of VIP to cultured RPE cells provides a plausible mechanism for growth regulation of the nearby choroid and/or sclera.[82]

2.4 RPE-Derived Growth Factors and Cytokines as Plausible Eye Growth Signal Molecules or Regulators

The RPE represents a major source of growth factors and cytokines, including insulin-like growth factor-1, transforming growth factor-beta (TGF-β), fibroblast growth factor (FGF), and vascular endothelial growth factor (VEGF).[16,23] Synthesized locally and subsequently secreted, they have been attributed roles in the maintenance of the structure and homeostasis of retina and choroid. Note that depending on the direction of their secretion, i.e., toward the retina and/or choroid side, these growth factors have potential to effect changes that are limited to the retina or choroid/sclera.

So far, there has been only limited study of the roles of RPE-derived growth factors and cytokines in postnatal eye growth regulation,[34,35] and results were not always conclusive due to the approaches used. For example, the combined tissue collection of the RPE and retina in early gene expression studies involving the chick precludes conclusion about the site of observed changes in expression.[83,84] Nonetheless, the report of downregulation of *Bone Morphogenetic Protein (BMP2)* gene expression in retina/RPE after form deprivation of chicks for either 6 h or 3 days is consistent with findings in isolated chick RPE with prolonged negative lens treatment.[85] These studies made use of high-throughput gene expression profiling with DNA microarray screening. In follow-up studies of isolated chick RPE, the gene expression of *BMP*-2, -4, and -7 was found to be bidirectionally regulated, with opposite signs of imposed defocus eliciting opposite responses (decreased with negative lenses, increased with positive lenses), even with exposures as short as 2 h and 2 days.[34,35] The rapidity of these responses suggests roles for these BMPs in the initiation and early phases of altered eye growth. Additional RPE gene expression studies have implicated in defocus-induced "myopic growth," TGF-β2, which belongs to the same superfamily of growth factors as BMPs, and has been previously linked to eye growth regulation in studies involving chicks and tree shrews.[86–89] Further characterization of the signal pathways involving these RPE-derived growth factors/cytokines, to identify both up- and downstream components, may well uncover novel treatments for myopia control.

3. THE ROLE OF CHOROID IN EYE GROWTH REGULATION

The choroid lies between the RPE and the sclera. In most species, it can be divided into five layers histologically, starting from the inner (retinal) side: Bruch's membrane, choroicapillaris, Haller's layer, Sattler's layer, and the suprachoroidea, with all but the first layer being largely vascular,[21] although in birds, the suprachoroid contains large, endothelium-lined spaces (lacunae), which resemble lymphatic vessels.[90,91] The choroid also contains a variety of nonvascular resident cells, including melanocytes, fibroblasts, nonvascular smooth muscle cells, and immunocompetent cells, supported by collagenous and elastic elements.[21] Traditionally, the choroid has been assigned as its major functions, supply of oxygen and nutrients to the outer retina, light absorption (pigmented choroid), thermoregulation, and modulation of intraocular pressure. However, recent studies also point to a role for the choroid in ocular focus adjustment, including emmetropization, and thus eye growth regulation, opening up the possibility of novel therapeutic approaches for myopia control.[21] Elucidating underlying signal pathways and mechanisms are essential first steps.

Referred to sometimes as choroidal accommodation, changes in choroidal thickness in response to imposed defocus were first described in young chicks, which also show the most dramatic changes of all animals studied. These changes serve to move the retina toward the altered plane of focus. Thus, while the choroid of young chicks is about 250 μm thick centrally and 100 μm thick peripherally, similar to mammals and primates, in response to substantial imposed myopic defocus, e.g., with +15 D lenses, the choroid of the chick eye increases its thickness significantly, effecting a correspondingly large, compensatory change in refraction.[21,42,92–94] With imposed hyperopic defocus, the choroid thins instead of thickening, pulling the retina backward toward the altered image plane. In refractive terms, the net effects are induced hyperopia and myopia, respectively. Form deprivation, which also induces myopia, also causes choroidal thinning, although here, the adjustment to the position of the retina serves no compensatory role. These changes in choroidal thickness occur very rapidly, being detectible with high-frequency ultrasonography in a matter of minutes in young chicks.[11,34,42,94,95] Similar choroidal responses have been documented in other animals, including guinea pigs, marmosets, macaques, and humans most recently, although the scale of the changes are much smaller than those observed in chicks in all cases.[21,96]

The mechanisms underlying the above choroidal thickness changes remain to be fully elucidated and it is possible that different mechanisms underlie the thickening and thinning responses. To-date, related changes in blood flow and structure have been described in chicks, along with bidirectional changes in the permeability of the choroidal vasculature, i.e., decreased during form deprivation and increased during recovery from form-deprivation myopia.[97–99] The protein content of suprachoroidal fluids has also been reported to be decreased in form-deprived eyes, and increased after normal vision is restored, consistent with anatomical localization of the thickness changes to the outer choroidal lacunae in chicks.[21,97] Presumably, these proteins serve as osmotic agents to regulate the water content and thus thickness of the outer choroid, with proteoglycans being among identified molecules reported to be elevated in eyes wearing positive lenses or in recovery (after diffuser removal).[94] It has also been speculated that non-vascular smooth muscle cells contribute to choroidal thickness changes, by contracting or relaxing as appropriate.[94] The extent to which observed changes in choroidal blood flow contribute to thickness changes via related changes in vessel diameters remains to be clarified, but may be significant, at least in mammals and primates whose choroids appear to lack lacunae. The role of the RPE as a regulator of one or more of these events also remains to be established. At the most basic level, it is possible that the RPE, by regulating ion and fluid exchange between the retina and the choroid, contributes to the regulation of choroidal thickness.[16,21] Alternatively, more complex signal cascades may be involved. For example, two DA receptor agonists, apomorphine and quinpirole, administered by intravitreal injection, have been linked to transient choroidal thickening; both also inhibit lens-induced myopia and both have potential access to receptors on the RPE, although retinal sites of action are plausible alternatives.[100] Likewise, retinal glucagon has been linked to altered eye growth, and intravitreal injection of exogenous glucagon is reported to modulate the choroidal thickness changes induced by visual manipulations.[101]

Of available animal models for myopia, the chick has been mostly widely studied in terms of regulatory mechanisms, with glucagon and retinoic acid (RA) being the subject of a number of studies. The pictures for both are complex. In the case of glucagon, the chick choroid as well as retina expresses glucagon and its receptors,[30] and choroidal glucagon protein levels are reported to increase with short term (up to 1 day) positive lens wear, and be unaltered by negative lenses.[102] Furthermore, insulin, which generally has opposing actions to glucagon, also appears to modulate choroidal thickness, apparently through an RPE-dependent mechanism, as demonstrated *in vitro* with chick eyecup preparations, in which added insulin thinned the choroid, in the presence of either RPE or RPE-conditioned medium.[101,103] There is also strong evidence implicating retinal and choroidal RA in eye growth regulation.[21] In relation to the choroid, RA shows bidirectional changes in response to visual manipulations that slow (positive lens and removal of diffusers) or accelerate (negative lens or diffuser) eye growth.[104] Choroidal expression of the RA-synthesizing enzyme, retinaldehyde dehydrogenase 2, also exhibits differential regulation with negative and positive lens treatment, as well as recovery from form deprivation.[105,106]

In addition to serving as a focusing mechanism, the choroid may also play an important role in regulating scleral growth and remodeling. Modulation of scleral proteoglycan synthesis appears to be one of the targets of choroidal RA.[106,107] In addition, the choroid expresses and synthesizes a variety of growth factors and enzymes, including bFGF, TGF-β, tissue plasminogen activator (t-PA), and matrix metalloproteinases, all of which have been linked to scleral remodeling and/or eye growth regulation.[22,105,108–111] For example, during the development of myopia, *TGF-β* gene has been shown to be differentially expressed in the choroid in chicks, albeit not in tree shrews.[105,111] Despite the difference between chicks and tree shrews noted in relation to choroidal *TGF-β* gene expression, other studies of gene expression in both tree shrews and marmosets point to involvement of the choroid in eye growth regulation. Microarray gene profiling applied to RPE/choroid preparations from marmosets undergoing lens treatment of opposite signs, revealed altered expression of a number of the 204 screened genes, including protein tyrosine phosphatase receptor type B, *TGF-β-induced*, and *FGF-2*.[112] Interestingly, in the tree shrew, similar differential gene expression patterns were observed in the choroid with three different visual manipulations (negative lens, form deprivation, and continuous darkness), implying a common myopia-inducing molecular signaling cascade, at least within the choroid.[113,114]

Recent studies in chick also provide an interesting perspective on the potential role of members of the VEGFs family in eye growth regulation. They are best known for their roles in angiogenesis, and VEGF antagonists such as bevacizumab, an antibody against human VEGF, are now widely used clinically in the treatment of wet age-related maculopathy. However, recent years have seen an expansion of their clinical use to include other macular pathologies, including myopic maculopathy.[115] Thus, the findings that members of the VEGF family and their receptors are expressed in chick choroid, and intravitreal injection of bevacizumab inhibits both the development of form-deprivation myopia and the choroidal thickening during the recovery from form-deprivation myopia in chicks implies a fundamental role for this family in regulating choroidal function.[116,117]

Further studies into the role of choroid in eye growth regulation and underlying signal pathways and mechanisms may lead to the development of new therapeutic approaches for myopia treatment through the modulation of choroidal functions.

Please refer further to chapter, "Scleral Mechanisms Underlying Ocular Growth and Myopia" (written by Ravi Metlapally and Christine F. Wildsoet).

REFERENCES

1. Ono K, Hiratsuka Y, Murakami A. Global inequality in eye health: country-level analysis from the Global Burden of Disease Study. *Am J Public Health*. 2010;100:1784–1788.
2. Resnikoff S, Pascolini D, Mariotti SP, Pokharel GP. Global magnitude of visual impairment caused by uncorrected refractive errors in 2004. *Bull World Health Organ*. 2008;86:63–70.
3. McBrien NA, Millodot M. A biometric investigation of late onset myopic eyes. *Acta Ophthalmol (Copenh)*. 1987;65:461–468.
4. Saw SM, Gazzard G, Shih-Yen EC, Chua WH. Myopia and associated pathological complications. *Ophthalmic Physiol Opt*. 2005;25:381–391.
5. Flitcroft DI. The complex interactions of retinal, optical and environmental factors in myopia aetiology. *Prog Retin Eye Res*. 2012;31:622–660.
6. Pascolini D, Mariotti SP. Global estimates of visual impairment: 2010. *Br J Ophthalmol*. 2012;96:614–618.
7. Vitale S, Sperduto RD, Ferris 3rd FL. Increased prevalence of myopia in the United States between 1971–1972 and 1999–2004. *Arch Ophthalmol*. 2009;127:1632–1639.
8. Pan CW, Ramamurthy D, Saw SM. Worldwide prevalence and risk factors for myopia. *Ophthalmic Physiol Opt*. 2012;32:3–16.
9. Jung SK, Lee JH, Kakizaki H, Jee D. Prevalence of myopia and its association with body stature and educational level in 19-year-old male conscripts in Seoul, South Korea. *Invest Ophthalmol Vis Sci*. 2012;53:5579–5583.

10. Vitale S, Cotch MF, Sperduto R, Ellwein L. Costs of refractive correction of distance vision impairment in the United States, 1999–2002. *Ophthalmology*. 2006;113: 2163–2170.
11. Wallman J, Winawer J. Homeostasis of eye growth and the question of myopia. *Neuron*. 2004;43:447–468.
12. Wojciechowski R. Nature and nurture: the complex genetics of myopia and refractive error. *Clin Genet*. 2011;79:301–320.
13. Solouki AM, Verhoeven VJ, van Duijn CM, et al. A genome-wide association study identifies a susceptibility locus for refractive errors and myopia at 15q14. *Nat Genet*. 2012;42:897–901.
14. Hysi PG, Young TL, Mackey DA, et al. A genome-wide association study for myopia and refractive error identifies a susceptibility locus at 15q25. *Nat Genet*. 2012;42: 902–905.
15. Morgan IG, Ohno-Matsui K, Saw SM. Myopia. *Lancet*. 2012;379:1739–1748.
16. Rymer J, Wildsoet CF. The role of the retinal pigment epithelium in eye growth regulation and myopia: a review. *Vis Neurosci*. 2005;22:251–261.
17. Pardue MT, Faulkner AE, Fernandes A, et al. High susceptibility to experimental myopia in a mouse model with a retinal on pathway defect. *Invest Ophthalmol Vis Sci*. 2008;49:706–712.
18. Veth KN, Willer JR, Collery RF, et al. Mutations in zebrafish lrp2 result in adult-onset ocular pathogenesis that models myopia and other risk factors for glaucoma. *PLoS Genet*. 2011;7:e1001310.
19. Sivak J. The cause(s) of myopia and the efforts that have been made to prevent it. *Clin Exp Optom*. 2012;95:572–582.
20. Aller TA. Clinical management of progressive myopia. *Eye (Lond)*. 2014;28:147–153.
21. Nickla DL, Wallman J. The multifunctional choroid. *Prog Retin Eye Res*. 2010;29: 144–168.
22. Rada JA, Shelton S, Norton TT. The sclera and myopia. *Exp Eye Res*. 2006;82: 185–200.
23. Strauss O. The retinal pigment epithelium in visual function. *Physiol Rev*. 2005;85: 845–881.
24. Boulton M, Dayhaw-Barker P. The role of the retinal pigment epithelium: topographical variation and ageing changes. *Eye (Lond)*. 2001;15:384–389.
25. Marmor M, Wolfensberger T. *The Retinal Pigment Epithelium: Function and Disease*. 1st ed. New York, NY: Oxford University Press; 1998:1–745.
26. Wildsoet CF. Active emmetropization—evidence for its existence and ramifications for clinical practice. *Ophthalmic Physiol Opt*. 1997;17:279–290.
27. McFadden SA, Wildsoet CF. Mammalian eyes need an intact optic nerve to detect the sign of defocus during emmetropisation. *ARVO Abstract*. 2009;50. E-Abstract 1620.
28. Friedman Z, Hackett SF, Campochiaro PA. Human retinal pigment epithelial cells possess muscarinic receptors coupled to calcium mobilization. *Brain Res*. 1988;446:11–16.
29. Koh SW, Chader GJ. Elevation of intracellular cyclic AMP and stimulation of adenylate cyclase activity by vasoactive intestinal peptide and glucagon in the retinal pigment epithelium. *J Neurochem*. 1984;43:1522–1526.
30. Buck C, Schaeffel F, Simon P, Feldkaemper M. Effects of positive and negative lens treatment on retinal and choroidal glucagon and glucagon receptor mRNA levels in the chicken. *Invest Ophthalmol Vis Sci*. 2004;45:402–409.
31. Versaux-Botteri C, Gibert JM, Nguyen-Legros J, Vernier P. Molecular identification of a dopamine D1b receptor in bovine retinal pigment epithelium. *Neurosci Lett*. 1997;237:9–12.
32. Rohrer B, Stell WK. Localization of putative dopamine D2-like receptors in the chick retina, using in situ hybridization and immunocytochemistry. *Brain Res*. 1995;695:110–116.

33. Zhang H, Wong CL, Shan SW, et al. Characterisation of Cl(-) transporter and channels in experimentally induced myopic chick eyes. *Clin Exp Optom*. 2011;94: 528–535.
34. Zhang Y, Liu Y, Wildsoet CF. Bidirectional, optical sign-dependent regulation of BMP2 gene expression in chick retinal pigment epithelium. *Invest Ophthalmol Vis Sci*. 2012;53:6072–6080.
35. Zhang Y, Liu Y, Ho C, Wildsoet CF. Effects of imposed defocus of opposite sign on temporal gene expression patterns of BMP4 and BMP7 in chick RPE. *Exp Eye Res*. 2013;109:98–106.
36. Lin T, Grimes PA, Stone RA. Expansion of the retinal pigment epithelium in experimental myopia. *Vis Res*. 1993;33:1881–1885.
37. Harman AM, Hoskins R, Beazley LD. Experimental eye enlargement in mature animals changes the retinal pigment epithelium. *Vis Neurosci*. 1999;16:619–628.
38. Fleming PA, Harman AM, Beazley LD. Changing topography of the RPE resulting from experimentally induced rapid eye growth. *Vis Neurosci*. 1997;14:449–461.
39. Liang H, Crewther SG, Crewther DP, Pirie B. Morphology of the recovery from form deprivation myopia in the chick. *Aust NZ J Ophthalmol*. 1996;24:41–44.
40. Maminishkis A, Chen S, Jalickee S, et al. Confluent monolayers of cultured human fetal retinal pigment epithelium exhibit morphology and physiology of native tissue. *Invest Ophthalmol Vis Sci*. 2006;47:3612–3624.
41. Crewther DP. The role of photoreceptors in the control of refractive state. *Prog Retin Eye Res*. 2000;19:421–457.
42. Wildsoet C, Wallman J. Choroidal and scleral mechanisms of compensation for spectacle lenses in chicks. *Vis Res*. 1995;35:1175–1194.
43. Seko Y, Shimokawa H, Pang J, Tokoro T. Disturbance of electrolyte balance in vitreous of chicks with form-deprivation myopia. *Jpn J Ophthalmol*. 2000;44:15–19.
44. Crewther SG, Liang H, Junghans BM, Crewther DP. Ionic control of ocular growth and refractive change. *Proc Natl Acad Sci USA*. 2006;103:15663–15668.
45. Liang H, Crewther SG, Crewther DP, Junghans BM. Structural and elemental evidence for edema in the retina, retinal pigment epithelium, and choroid during recovery from experimentally induced myopia. *Invest Ophthalmol Vis Sci*. 2004;45: 2463–2474.
46. Crewther SG, Murphy MJ, Crewther DP. Potassium channel and NKCC cotransporter involvement in ocular refractive control mechanisms. *PLoS One*. 2008;3:e2839.
47. Gallemore RP, Steinberg RH. Effects of dopamine on the chick retinal pigment epithelium. Membrane potentials and light-evoked responses. *Invest Ophthalmol Vis Sci*. 1990;31:67–80.
48. Stone RA, Sugimoto R, Gill AS, Liu J, Capehart C, Lindstrom JM. Effects of nicotinic antagonists on ocular growth and experimental myopia. *Invest Ophthalmol Vis Sci*. 2001;42:557–565.
49. Reis RA, Ventura AL, Kubrusly RC, de Mello MC, de Mello FG. Dopaminergic signaling in the developing retina. *Brain Res Rev*. 2007;54:181–188.
50. Kebabian JW, Calne DB. Multiple receptors for dopamine. *Nature*. 1979;277:93–96.
51. Vallone D, Picetti R, Borrelli E. Structure and function of dopamine receptors. *Neurosci Biobehav Rev*. 2000;24:125–132.
52. Dearry A, Burnside B. Stimulation of distinct D2 dopaminergic and alpha 2-adrenergic receptors induces light-adaptive pigment dispersion in teleost retinal pigment epithelium. *J Neurochem*. 1988;51:1516–1523.
53. Bruinink A, Dawis S, Niemeyer G, Lichtensteiger W. Catecholaminergic binding sites in cat retina, pigment epithelium and choroid. *Exp Eye Res*. 1986;43:147–151.
54. Nguyen-Legros J, Versaux-Botteri C, Vernier P. Dopamine receptor localization in the mammalian retina. *Mol Neurobiol*. 1999;19:181–204.

55. Dearry A, Falardeau P, Shores C, Caron MG. D2 dopamine receptors in the human retina: cloning of cDNA and localization of mRNA. *Cell Mol Neurobiol*. 1991;11: 437–453.
56. Zhang Y, Maminishkis A, Zhi C, et al. Apomorphine regulates TGF-β1 and TGF-β2 expression in human fetal retinal pigment epithelial cells. *Invest Ophthalmol Vis Sci*. 2009;50. E-Abstract 3845.
57. Stone RA, Lin T, Laties AM, Iuvone PM. Retinal dopamine and form-deprivation myopia. *Proc Natl Acad Sci USA*. 1989;86:704–706.
58. Rohrer B, Spira AW, Stell WK. Apomorphine blocks form-deprivation myopia in chickens by a dopamine D2-receptor mechanism acting in retina or pigmented epithelium. *Vis Neurosci*. 1993;10:447–453.
59. Schmid KL, Wildsoet CF. Inhibitory effects of apomorphine and atropine and their combination on myopia in chicks. *Optom Vis Sci*. 2004;81:137–147.
60. Fukuda K, Kubo T, Akiba I, Maeda A, Mishina M, Numa S. Molecular distinction between muscarinic acetylcholine receptor subtypes. *Nature*. 1987;327:623–625.
61. Fischer AJ, McKinnon LA, Nathanson NM, Stell WK. Identification and localization of muscarinic acetylcholine receptors in the ocular tissues of the chick. *J Comp Neurol*. 1998;392:273–284.
62. Yin GC, Gentle A, McBrien NA. Muscarinic antagonist control of myopia: a molecular search for the M1 receptor in chick. *Mol Vis*. 2004;10:787–793.
63. Hutchins JB. Acetylcholine as a neurotransmitter in the vertebrate retina. *Exp Eye Res*. 1987;45:1–38.
64. Ford KJ, Feller MB. Assembly and disassembly of a retinal cholinergic network. *Vis Neurosci*. 2012;29:61–71.
65. Chua WH, Balakrishnan V, Chan YH, et al. Atropine for the treatment of childhood myopia. *Ophthalmology*. 2006;113:2285–2291.
66. Stone RA, Pardue MT, Iuvone PM, Khurana TS. Pharmacology of myopia and potential role for intrinsic retinal circadian rhythms. *Exp Eye Res*. 2013;114:35–47.
67. Song YY, Wang H, Wang BS, Qi H, Rong ZX, Chen HZ. Atropine in ameliorating the progression of myopia in children with mild to moderate myopia: a meta-analysis of controlled clinical trials. *J Ocul Pharmacol Ther*. 2011;27:361–368.
68. Tong L, Huang XL, Koh AL, Zhang X, Tan DT, Chua WH. Atropine for the treatment of childhood myopia: effect on myopia progression after cessation of atropine. *Ophthalmology*. 2009;116:572–579.
69. Barathi VA, Beuerman RW. Molecular mechanisms of muscarinic receptors in mouse scleral fibroblasts: prior to and after induction of experimental myopia with atropine treatment. *Mol Vis*. 2011;17:680–692.
70. Zou L, Liu R, Zhang X, et al. Upregulation of regulator of G-protein signaling 2 in the sclera of a form deprivation myopic animal model. *Mol Vis*. 2014;20:977–987.
71. McBrien NA, Moghaddam HO, Reeder AP. Atropine reduces experimental myopia and eye enlargement via a nonaccommodative mechanism. *Invest Ophthalmol Vis Sci*. 1993;34:205–215.
72. Stone RA, Lin T, Laties AM. Muscarinic antagonist effects on experimental chick myopia. *Exp Eye Res*. 1991;52:755–758.
73. Luft WA, Ming Y, Stell WK. Variable effects of previously untested muscarinic receptor antagonists on experimental myopia. *Invest Ophthalmol Vis Sci*. 2003;44:1330–1338.
74. Cottriall CL, Truong HT, McBrien NA. Inhibition of myopia development in chicks using himbacine: a role for M(4) receptors? *Neuroreport*. 2001;12:2453–2456.
75. Leech EM, Cottriall CL, McBrien NA. Pirenzepine prevents form deprivation myopia in a dose dependent manner. *Ophthalmic Physiol Opt*. 1995;15:351–356.
76. Salceda R. Muscarinic receptors binding in retinal pigment epithelium during rat development. *Neurochem Res*. 1994;19:1207–1210.

77. Osborne NN, FitzGibbon F, Schwartz G. Muscarinic acetylcholine receptor-mediated phosphoinositide turnover in cultured human retinal pigment epithelium cells. *Vis Res.* 1991;31:1119–1127.
78. Sekiguchi-Tonosaki M, Obata M, Haruki A, Himi T, Kosaka J. Acetylcholine induces Ca^{2+} signaling in chicken retinal pigmented epithelial cells during dedifferentiation. *Am J Physiol.* 2009;296:C1195–C1206.
79. Ekman R, Tornqvist K. Glucagon and VIP in the retina. *Invest Ophthalmol Vis Sci.* 1985;26:1405–1409.
80. Beloukhina N, Vessey K, Stell WK. Glucagon prevents myopia via distal retina or RPE. *IOVS.* 2005;46. ARVO E-Abstract 3337.
81. Zhu X, Liu Y, Wallman J. Glucagon increases choroidal thickness of chick eyes by acting on the retinal pigment epithelium. *IOVS.* 2005;46. ARVO E-Abstract 3338.
82. Koh SW. VIP stimulation of polarized macromolecule secretion in cultured chick embryonic retinal pigment epithelium. *Exp Cell Res.* 1991;197:1–7.
83. McGlinn AM, Baldwin DA, Tobias JW, Budak MT, Khurana TS, Stone RA. Form-deprivation myopia in chick induces limited changes in retinal gene expression. *Invest Ophthalmol Vis Sci.* 2007;48:3430–3436.
84. Stone RA, Khurana TS. Gene profiling in experimental models of eye growth: clues to myopia pathogenesis. *Vis Res.* 2010;50:2322–2333.
85. Zhang Y, Liu Y, Xu J, Nimri N, Wildsoet CF. Microarray analysis of RPE gene expression in chicks during long-term imposed hyperopic defocus. *Invest Ophthalmol Vis Sci.* 2010;51. E-Abstract 3680.
86. Zhang Y, Liu Y, Ho C, Hammond D, Wildsoet CF. Differential expression of BMP7, TGF-β2, and noggin in chick RPE after imposed optical defocus. *Invest Ophthalmol Vis Sci.* 2012;53. E-Abstract 3458.
87. Wagner DO, Sieber C, Bhushan R, Borgermann JH, Graf D, Knaus P. BMPs: from bone to body morphogenetic proteins. *Sci Signal.* 2010;3. mr1.
88. Rohrer B, Stell WK. Basic fibroblast growth factor (bFGF) and transforming growth factor beta (TGF-beta) act as stop and go signals to modulate postnatal ocular growth in the chick. *Exp Eye Res.* 1994;58:553–561.
89. McBrien NA. Regulation of scleral metabolism in myopia and the role of transforming growth factor-beta. *Exp Eye Res.* 2013;114:128–140.
90. De Stefano ME, Mugnaini E. Fine structure of the choroidal coat of the avian eye. Vascularization, supporting tissue and innervation. *Anat Embryol.* 1997;195:393–418.
91. De Stefano ME, Mugnaini E. Fine structure of the choroidal coat of the avian eye. Lymphatic vessels. *Invest Ophthalmol Vis Sci.* 1997;38:1241–1260.
92. Ramrattan RS, van der Schaft TL, Mooy CM, de Bruijn WC, Mulder PG, de Jong PT. Morphometric analysis of Bruch's membrane, the choriocapillaris, and the choroid in aging. *Invest Ophthalmol Vis Sci.* 1994;35:2857–2864.
93. Krebs W, Krebs I. *Primate Retina and Choroid: Atlas of Fine Structure in Man and Monkey.* New York: Springer-Verlag; 1991.
94. Wallman J, Wildsoet C, Xu A, et al. Moving the retina: choroidal modulation of refractive state. *Vis Res.* 1995;35:37–50.
95. Zhu X, Park TW, Winawer J, Wallman J. In a matter of minutes, the eye can know which way to grow. *Invest Ophthalmol Vis Sci.* 2005;46:2238–2241.
96. Chakraborty R, Read SA, Collins MJ. Monocular myopic defocus and daily changes in axial length and choroidal thickness of human eyes. *Exp Eye Res.* 2012;103:47–54.
97. Pendrak K, Papastergiou GI, Lin T, Laties AM, Stone RA. Choroidal vascular permeability in visually regulated eye growth. *Exp Eye Res.* 2000;70:629–637.
98. Junghans BM, Crewther SG, Liang H, Crewther DP. A role for choroidal lymphatics during recovery from form deprivation myopia? *Optom Vis Sci.* 1999;76:796–803.

99. Hirata A, Negi A. Morphological changes of choriocapillaris in experimentally induced chick myopia. *Graefes Arch Clin Exp Ophthalmol.* 1998;236:132–137.
100. Nickla DL, Totonelly K, Dhillon B. Dopaminergic agonists that result in ocular growth inhibition also elicit transient increases in choroidal thickness in chicks. *Exp Eye Res.* 2010;91:715–720.
101. Zhu X, Wallman J. Opposite effects of glucagon and insulin on compensation for spectacle lenses in chicks. *Invest Ophthalmol Vis Sci.* 2009;50:24–36.
102. Feldkaemper MP, Schaeffel F. Evidence for a potential role of glucagon during eye growth regulation in chicks. *Vis Neurosci.* 2002;19:755–766.
103. Sheng C, Zhu X, Wallman J. In vitro effects of insulin and RPE on choroidal and scleral components of eye growth in chicks. *Exp Eye Res.* 2013;116:439–448.
104. Mertz JR, Wallman J. Choroidal retinoic acid synthesis: a possible mediator between refractive error and compensatory eye growth. *Exp Eye Res.* 2000;70:519–527.
105. Simon P, Feldkaemper M, Bitzer M, Ohngemach S, Schaeffel F. Early transcriptional changes of retinal and choroidal TGFbeta-2, RALDH-2, and ZENK following imposed positive and negative defocus in chickens. *Mol Vis.* 2004;10:588–597.
106. Rada JA, Hollaway LR, Lam W, Li N, Napoli JL. Identification of RALDH2 as a visually regulated retinoic acid synthesizing enzyme in the chick choroid. *Invest Ophthalmol Vis Sci.* 2012;53:1649–1662.
107. Troilo D, Nickla DL, Mertz JR, Summers Rada JA. Change in the synthesis rates of ocular retinoic acid and scleral glycosaminoglycan during experimentally altered eye growth in marmosets. *Invest Ophthalmol Vis Sci.* 2006;47:1768–1777.
108. Hu W, Criswell MH, Fong SL, et al. Differences in the temporal expression of regulatory growth factors during choroidal neovascular development. *Exp Eye Res.* 2009;88:79–91.
109. Lambert V, Munaut C, Jost M, et al. Matrix metalloproteinase-9 contributes to choroidal neovascularization. *Am J Pathol.* 2002;161:1247–1253.
110. Wang Y, Gillies C, Cone RE, O'Rourke J. Extravascular secretion of t-PA by the intact superfused choroid. *Invest Ophthalmol Vis Sci.* 1995;36:1625–1632.
111. Jobling AI, Wan R, Gentle A, Bui BV, McBrien NA. Retinal and choroidal TGF-beta in the tree shrew model of myopia: isoform expression, activation and effects on function. *Exp Eye Res.* 2009;88:458–466.
112. Shelton L, Troilo D, Lerner MR, Gusev Y, Brackett DJ, Rada JS. Microarray analysis of choroid/RPE gene expression in marmoset eyes undergoing changes in ocular growth and refraction. *Mol Vis.* 2008;14:1465–1479.
113. He L, Frost MR, Siegwart Jr JT, Norton TT. Gene expression signatures in tree shrew choroid during lens-induced myopia and recovery. *Exp Eye Res.* 2014;123:56–71.
114. He L, Frost MR, Siegwart Jr JT, Norton TT. Gene expression signatures in tree shrew choroid in response to three myopiagenic conditions. *Vis Res.* 2014;102:52–63.
115. Ruiz-Moreno JM, Montero JA. Intravitreal bevacizumab to treat myopic choroidal neovascularization: 2-year outcome. *Graefes Arch Clin Exp Ophthalmol.* 2010;248:937–941.
116. Feldkaemper M, Schaeffel F, Fuchs L. Vascular endothelial growth factor A, C, D and vascular endothelial growth factor receptor 1, 2, 3 mRNA expression in the chicken retina, RPE and choroid. *Invest Ophthalmol Vis Sci.* 2013;54. E-Abstract 5171.
117. Mathis U, Ziemssen F, Schaeffel F. Effects of a human VEGF antibody (Bevacizumab) on deprivation myopia and choroidal thickness in the chicken. *Exp Eye Res.* 2014;127:161–169.

CHAPTER FOURTEEN

Scleral Mechanisms Underlying Ocular Growth and Myopia

Ravi Metlapally[1], **Christine F. Wildsoet**
School of Optometry, University of California, Berkeley, California, USA
[1]Corresponding author: e-mail address: metlapally@berkeley.edu

Contents

Abstract

In the regulation of ocular growth, scleral events critically determine eye size and thus the refractive status of the eye. Increased scleral matrix remodeling can lead to exaggerated eye growth causing myopia and additionally increased risk of ocular pathological complications. Thus, therapies targeting these changes in sclera hold potential to limit such complications since sclera represents a relatively safe and accessible drug target. Understanding the scleral molecular mechanisms underlying ocular growth is essential to identifying plausible therapeutic targets in the sclera. This section provides a brief update on molecular studies that pertain to the sclera in the context of ocular growth regulation and myopia.

1. INTRODUCTION

The sclera, the outermost layer of the eye, performs many key functions. Apart from offering protection to the interior ocular structures, it determines the final shape and size of the eye, serves as the anchor for extraocular muscles, and provides support for the ciliary muscle, which subserves lenticular accommodation. It also provides channels for blood vessels and nerves serving intraocular structures and allows for the exchange of fluids, including aqueous humor entering the choroid via the uveoscleral pathway. The sclera is predominantly made up of collagen, with interspersed

Progress in Molecular Biology and Translational Science, Volume 134
ISSN 1877-1173
http://dx.doi.org/10.1016/bs.pmbts.2015.05.005
© 2015 Elsevier Inc.
All rights reserved.

fibroblasts that produce and maintain its extracellular matrix (ECM). It undergoes several changes during the development and progression of myopia, which are at first subtle at the gross anatomical level, but may give way to more substantial changes accompanied by scleral thinning and weakening, leading to pathological complications involving the normally protected retina and choroid—such as maculopathies, retinal schisis, and detachment, with highly myopic eyes being most at risk. A thinner and biomechanically weaker peripapillary sclera in myopia can affect the biomechanics of the lamina cribrosa explaining the increased susceptibility for glaucomatous optic nerve damage.[1,2]

2. STRUCTURAL AND BIOMECHANICAL CHANGES IN MYOPIA

Structural and biomechanical changes in the myopic sclera of human eyes are well documented; apart from being thinner than normal, its glycosaminoglycan and collagen contents are reduced and its fibril assembly disorganized, rendering it biomechanically weaker.[3–6] Humans with moderate as well as high or pathological myopia have been shown to have thinner than normal scleras, with reductions in the thickness of the posterior sclera up to 31% of the normal mature human sclera being reported in pathological myopia.[4,7] The most often studied biomechanical property of the sclera is creep rate, which represents the extension of the sclera over time when a constant load is applied. Increases in creep rate are consistent with decreases in biomechanical stability, and for sclera tissue from myopic eyes, increases of over 200% compared to values for normal eyes have been documented.[8,9] Such biomechanical changes in the sclera facilitate elongation of the eye in myopia, and in their more exaggerated form, may lead to posterior staphylomas, corresponding to localized mechanical failure of the sclera. Scleral reinforcement surgery, which currently represents the only treatment option for such complications, is not without risk of complications.[10–12] A seemingly and increasingly plausible option would be to inhibit the scleral changes underlying the excessive elongation in myopia. Indeed, the main rationale behind the research directed at understanding scleral molecular mechanisms underlying myopia development has been to devise less invasive therapies to promote natural matrix deposition, and to improve scleral strength by way of resisting ocular elongation and slowing myopia progression. Research directed at the visual and retinal events that trigger such anomalous growth and the nature of the signal cascades generated is discussed elsewhere in this chapter. Relevant scleral studies are summarized here.

3. MOLECULAR CHANGES IN MYOPIA

The structural and biomechanical changes in the myopic sclera alluded to above are products of biochemical and molecular changes in the sclera. Detailed characterization of molecular changes in the sclera has been possible with the establishment of animal models for myopia (reviewed in detail in Ref. 13). Note that much of this research has focused on mammalian models, mostly the tree shrew, because their sclera is most like the human (primate) sclera, comprising a single fibrous layer as opposed to the bilayered sclera of the chick. In addition to confirming findings from studies in humans of decreased glycosaminoglycan and collagen content in myopia,[4] such studies have linked scleral changes in myopic eyes to altered expression of a number of genes, including collagen (predominantly type-I), matrix metalloproteases (MMPs), tissue inhibitors of MMPs (TIMPs), FGF receptor-1, TGFβ, and integrins (reviewed in detail in Ref. 14).[15–17] Studies involving the tree shrew model have shown reduced scleral hydroxyproline and levels of sulfated glycosaminoglycans at the posterior pole of myopic eyes, suggesting reduced collagen accumulation and proteoglycans, respectively.[18] This reduction in scleral collagen content is accompanied by reduction in collagen fibril diameter.[19,20] Collagen constitutes 85–90% of the total scleral protein, with collagen type-I showing by far the highest expression (>99%) of the numerous collagen subtypes.[21] Nonetheless, other collagen subtypes, such as type III, V, and VI, have also been identified in human sclera,[22] and as many as 11 collagen subtypes (I, III, V, VI, VII, VIII, IX, XIII, XV, XVI, and XVII) have been reported in tree shrew sclera.[23] The various collagen subtypes do not appear to be uniformly affected in myopia. However, a reduction in the collagen subtype ratio (V/I) has been linked to myopia, with speculation that it may be important in determining alterations in fibril diameter in myopia.[23] In tree shrews, higher levels of the active form of MMP-2, an enzyme associated with breakdown of collagen and proteoglycans, and decreased levels of TIMP-1 have also been reported in myopic scleras.[24,25] In addition, downregulation of collagen binding integrin subunits α1, α2, and β1, as well as TGFβ isoforms, particularly TGFβ1, and upregulation of FGF receptor-1 have been linked to myopia development.[16,26–28] Apart from these studies tying growth factors to scleral remodeling, recent studies in the guinea pig model have implicated two second messengers, cyclic AMP and cyclic GMP; specifically, both scleral cyclic AMP and cyclic GMP levels were found to be increased in form deprivation myopia and subconjunctival injections of activators of these second messengers induced myopic shifts in normal animals, while their inhibitors inhibited form deprivation myopia.[29,30]

A point of relevance is that most of the molecular changes in the sclera discussed above are reversible. For example, it has been shown that when myopia-inducing experimental treatments are terminated, the expression levels of several of these molecules generally return to baseline in eyes undergoing recovery from induced myopia.[14,25,31] Furthermore, biomechanical testing showed that scleral strips from tree shrew eyes recovering from induced myopia were less extensible than those from myopic eyes, implying some recovery of biomechanical properties as well.[8] Together these observations provide strong rationale for intervening at a molecular level to "rescue" the myopic sclera.

4. RECENT ADVANCES IN MOLECULAR STUDIES

The studies summarized above heavily relied on real-time PCR as well as standard biochemical assays. However, with rapid advances in technologies, it is now possible to study genome-wide gene expression profiles. Thus, related studies in human donor tissue have further opened up the scleral molecular landscape, identifying many new genes in the sclera that have never been studied thus far, and signifying that although collagen occupies the most "real estate" (90%), there may be several other important players participating in maintaining normal scleral health.[32–34] Likewise, recent studies involving the tree shrew model have identified many new candidate genes belonging to families such as cell signaling/matrix-degrading/structural proteins in myopia development.[26,35] Genes reported to be upregulated in the sclera during myopia development include TGFβ receptor-3, TGFβ-induced protein ig-h3, and MMP14, while TGFβ1, TGFβ2, thrombospondin-1, tenascin, osteonectin, osteopontin, TIMP-3, and a disintegrin and metalloproteinase with thrombospondin motifs 5 (ADAMTS5) were downregulated.

Across many fields, the roles of micro (mi)-RNAs, which represent small noncoding RNA molecules, are attracting increasing interests as potential therapeutic targets. Thus far, they have been profiled in the human sclera and implicated in regulation of early (fetal) ocular growth.[36] Specifically, several collagen-specific miRNAs show increased expression in fetal eyes exhibiting active growth (enlargement). Drawing parallels between fetal and myopic eye growth—they both involve active growth, it seems reasonable to posit roles for these small molecules in the regulation of myopic eye growth, by modulating the expression levels of various scleral mRNAs (genes). Understanding the roles of miRNAs in myopia awaits characterization of scleral genome-wide expression profiles in a suitable animal model of myopia. Such studies will also offer a rich resource for identifying key gene signaling pathways and exploring specific molecular mechanisms, with the potential for identifying scleral gene targets for novel myopia therapies. Preliminary findings from scleral miRNA as well as mRNA profiling studies in the mouse myopia model indicate that miRNAs of the let-7 family, previously implicated in matrix remodeling in other tissues, are upregulated in eyes exposed to form deprivation. Notable signaling pathways showing overrepresentation of genes include intermediate filament organization, scaffold protein binding, detection of stimuli, calcium ion, and G-protein pathways.[37]

5. POTENTIAL THERAPEUTIC APPROACHES

Thus far, research into drug therapies aimed at the sclera for myopia prevention or retardation is still nascent, and to-date no single molecule has been proven conclusively to act via a known/established scleral mechanism. Nonetheless, the sclera is listed among possible sites of action of the only two pharmacological treatments in current clinical use for myopia, topical atropine, which is widely used in East Asia, and oral 7-methylxanthine, which is approved for use in Denmark.[38–40] Other exploratory treatments targeting the sclera involve collagen cross-linking procedures, of which some are already being applied clinically to stabilize corneas, and biopolymers designed as tissue scaffolds to support new ECM production and/or alter scleral remodeling.[41,42]

The sclera is an attractive target tissue for myopia therapy for a number of reasons: as noted above, it defines the final size and shape of the eye, the changes in myopia have been relatively well characterized, and it is a relatively safe and accessible drug target. Understanding molecular mechanisms and identifying regulators of scleral remodeling are critical for devising strategies to maintain or improve scleral resistance to ocular elongation. At this time, there is still much to be explored and understood in this field, due to the complex nature of genome-wide gene regulatory networks and the as yet poorly understood role of genetic susceptibility in myopia (discussed

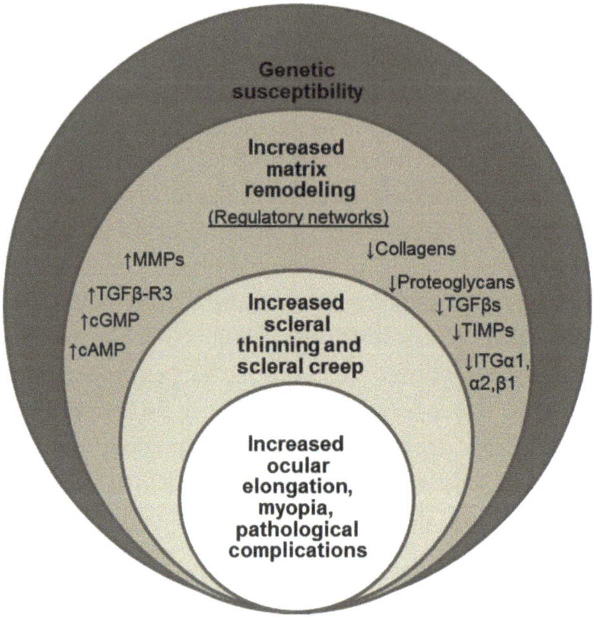

Figure 1 Schematic of scleral remodeling mechanisms and implications in myopia.

elsewhere in this chapter) (Fig. 1). Moving forward with targeted scleral therapies will also benefit from complementary research into drug delivery systems compatible with local and sustained action.

REFERENCES

1. Jonas JB, Jonas SB, Jonas RA, Holbach L, Panda-Jonas S. Histology of the parapapillary region in high myopia. *Am J Ophthalmol.* 2011;152(6):1021–1029.
2. Norman RE, Flanagan JG, Sigal IA, Rausch SM, Tertinegg I, Ethier CR. Finite element modeling of the human sclera: influence on optic nerve head biomechanics and connections with glaucoma. *Exp Eye Res.* 2011;93(1):4–12.
3. Curtin BJ. Physiopathologic aspects of scleral stress–strain. *Trans Am Ophthalmol Soc.* 1969;67:417–461.
4. Avetisov ES, Savitskaya NF, Vinetskaya MI, Iomdina EN. A study of biochemical and biomechanical qualities of normal and myopic eye sclera in humans of different age groups. *Metab Pediatr Syst Ophthalmol.* 1983;7(4):183–188.
5. Curtin BJ, Iwamoto T, Renaldo DP. Normal and staphylomatous sclera of high myopia. An electron microscopic study. *Arch Ophthalmol.* 1979;97(5):912–915.
6. Curtin BJ, Teng CC. Scleral changes in pathological myopia. *Trans Am Acad Opthalmol Otolaryngol.* 1958;62(6):777–788, discussion 788–790.
7. Cheng HM, Singh OS, Kwong KK, Xiong J, Woods BT, Brady TJ. Shape of the myopic eye as seen with high-resolution magnetic resonance imaging. *Optom Vis Sci.* 1992;69(9):698–701.
8. Siegwart Jr JT, Norton TT. Regulation of the mechanical properties of tree shrew sclera by the visual environment. *Vision Res.* 1999;39(2):387–407.

9. Phillips JR, Khalaj M, McBrien NA. Induced myopia associated with increased scleral creep in chick and tree shrew eyes. *Invest Ophthalmol Vis Sci.* 2000;41(8):2028–2034.
10. Avetisov ES, Tarutta EP, Iomdina EN, Vinetskaya MI, Andreyeva LD. Nonsurgical and surgical methods of sclera reinforcement in progressive myopia. *Acta Ophthalmol Scand.* 1997;75(6):618–623.
11. Ward B, Tarutta EP, Mayer MJ. The efficacy and safety of posterior pole buckles in the control of progressive high myopia. *Eye (Lond).* 2009;23(12):2169–2174.
12. Chen M, Dai J, Chu R, Qian Y. The efficacy and safety of modified Snyder-Thompson posterior scleral reinforcement in extensive high myopia of Chinese children. *Graefes Arch Clin Exp Ophthalmol.* 2013;251(11):2633–2638.
13. Wallman J, Winawer J. Homeostasis of eye growth and the question of myopia. *Neuron.* 2004;43(4):447–468.
14. McBrien NA, Gentle A. Role of the sclera in the development and pathological complications of myopia. *Prog Retin Eye Res.* 2003;22(3):307–338.
15. Barathi VA, Beuerman RW. Molecular mechanisms of muscarinic receptors in mouse scleral fibroblasts: prior to and after induction of experimental myopia with atropine treatment. *Mol Vis.* 2011;17:680–692.
16. McBrien NA, Metlapally R, Jobling AI, Gentle A. Expression of collagen-binding integrin receptors in the mammalian sclera and their regulation during the development of myopia. *Invest Ophthalmol Vis Sci.* 2006;47(11):4674–4682.
17. Rada JA, Perry CA, Slover ML, Achen VR. Gelatinase A and TIMP-2 expression in the fibrous sclera of myopic and recovering chick eyes. *Invest Ophthalmol Vis Sci.* 1999;40(13):3091–3099.
18. Norton TT, Rada JA. Reduced extracellular-matrix in mammalian sclera with induced myopia. *Vision Res.* 1995;35(9):1271–1281.
19. Curtin BJ, Iwamoto T, Renaldo DP. Normal and staphylomatous sclera of high myopia: an electron microscopy study. *Arch Ophthalmol.* 1979;97:912–915.
20. McBrien NA, Cornell LM, Gentle A. Structural and ultrastructural changes to the sclera in a mammalian model of high myopia. *Invest Ophthalmol Vis Sci.* 2001;42(10):2179–2187.
21. Norton TT, Miller EJ. Collagen and protein levels in sclera during normal development, induced myopia, and recovery in tree shrews. *Invest Ophthalmol Vis Sci.* 1995;36(4):S760.
22. Marshall GE, Lee WR. Distribution of collagen types I-VI in aged human cornea and sclera compared. *Invest Ophthalmol Vis Sci.* 1993;34:1202.
23. Gentle A, Liu Y, Martin JE, Conti GL, McBrien NA. Collagen gene expression and the altered accumulation of scleral collagen during the development of high myopia. *J Biol Chem.* 2003;278(19):16587–16594.
24. Guggenheim JA, McBrien NA. Form-deprivation myopia induces activation of scleral matrix metalloproteinase-2 in tree shrew. *Invest Ophthalmol Vis Sci.* 1996;37(7): 1380–1395.
25. Siegwart Jr JT, Norton TT. The time course of changes in mRNA levels in tree shrew sclera during induced myopia and recovery. *Invest Ophthalmol Vis Sci.* 2002;43(7):2067–2075.
26. Gao H, Frost MR, Siegwart Jr JT, Norton TT. Patterns of mRNA and protein expression during minus-lens compensation and recovery in tree shrew sclera. *Mol Vis.* 2011;17:903–919.
27. Gentle A, McBrien NA. Retinoscleral control of scleral remodelling in refractive development: a role for endogenous FGF-2? *Cytokine.* 2002;18(6):344–348.
28. Jobling A, Nguyen M, Gentle A, McBrien NA. Isoform-specific changes in scleral TGF-beta expression and the regulation of collagen synthesis during myopia progression. *J Biol Chem.* 2004;279:18121–18126.
29. Fang F, Pan M, Yan T, et al. The role of cGMP in ocular growth and the development of form-deprivation myopia in guinea pigs. *Invest Ophthalmol Vis Sci.* 2013;54(13): 7887–7902.

30. Tao Y, Pan M, Liu S, et al. cAMP level modulates scleral collagen remodeling, a critical step in the development of myopia. *PLoS One*. 2013;8(8):e71441.
31. McBrien NA, Lawlor P, Gentle A. Scleral remodeling during the development of and recovery from axial myopia in the tree shrew. *Invest Ophthalmol Vis Sci*. 2000;41(12): 3713–3719.
32. Cui W, Bryant MR, Sweet PM, McDonnell PJ. Changes in gene expression in response to mechanical strain in human scleral fibroblasts. *Exp Eye Res*. 2004;78(2): 275–284.
33. Young TL, Hawthorne F, Feng S, et al. Whole genome expression profiling of normal human fetal and adult ocular tissues. *Exp Eye Res*. 2013;116:265–278.
34. Young TL, Scavello GS, Paluru PC, Choi JD, Rappaport EF, Rada JA. Microarray analysis of gene expression in human donor sclera. *Mol Vis*. 2004;10:163–176.
35. Guo L, Frost MR, He L, Siegwart Jr JT, Norton TT. Gene expression signatures in tree shrew sclera in response to three myopiagenic conditions. *Invest Ophthalmol Vis Sci*. 2013;54(10):6806–6819.
36. Metlapally R, Gonzalez P, Hawthorne FA, Tran-Viet KN, Wildsoet CF, Young TL. Scleral micro-RNA signatures in adult and fetal eyes. *PLoS One*. 2013;8(10):e78984.
37. Metlapally R, Park H, Wang K, et al. Genome-wide scleral micro- and messenger-RNA profiling in the mouse myopia model. *ARVO, Meeting Abstracts*. 2014;55(5):3588.
38. Trier K, Munk Ribel-Madsen S, Cui D, Brogger CS. Systemic 7-methylxanthine in retarding axial eye growth and myopia progression: a 36-month pilot study. *J Ocul Biol Dis Infor*. 2008;1(2–4):85–93.
39. Chia A, Chua WH, Cheung YB, et al. Atropine for the treatment of childhood myopia: safety and efficacy of 0.5%, 0.1%, and 0.01% doses (Atropine for the Treatment of Myopia 2). *Ophthalmology*. 2012;119(2):347–354.
40. Cui D, Trier K, Zeng J, et al. Effects of 7-methylxanthine on the sclera in form deprivation myopia in guinea pigs. *Acta Ophthalmol*. 2011;89(4):328–334.
41. Wollensak G, Iomdina E, Dittert DD, Salamatina O, Stoltenburg G. Cross-linking of scleral collagen in the rabbit using riboflavin and UVA. *Acta Ophthalmol Scand*. 2005;83(4):477–482.
42. Su J, Wall ST, Healy KE, Wildsoet CF. Scleral reinforcement through host tissue integration with biomimetic enzymatically degradable semi-interpenetrating polymer network. *Tissue Eng Part A*. 2010;16(3):905–916.

CHAPTER FIFTEEN

Molecular and Biochemical Aspects of the Retina on Refraction

Ranjay Chakraborty, Machelle T. Pardue[1]
Department of Ophthalmology, Emory University School of Medicine, Atlanta, Georgia, USA
[1]Corresponding author: e-mail address: mpardue@emory.edu

Contents

Abstract

Mutant mouse models with specific visual pathway defects offer an advantage to comprehensively investigate the role of specific pathways/neurons involved in refractive development. In this review, we will focus on recent studies using mouse models that have provided insight into retinal pathways and neurotransmitters controlling refractive development. Specifically, we will examine the contributions of rod and cone photoreceptors and the ON and OFF retinal pathways to visually driven eye growth with emphasis on dopaminergic mechanisms.

Progress in Molecular Biology and Translational Science, Volume 134
ISSN 1877-1173
http://dx.doi.org/10.1016/bs.pmbts.2015.06.013
© 2015 Elsevier Inc.
All rights reserved.

1. RETINA

The retina is a complex ocular structure that converts wavelengths of light into neuronal signals that become perceived visual images. The mammalian retina is composed of approximately 55 morphologically distinct cell types, each with a different function.[1] From outer to inner retina, photoreceptors and horizontal cells, bipolar cells, amacrine cells, and ganglion cells constitute the major neuronal populations in the mammalian eye[2,3] (Fig. 1). While describing how the retinal neurons are interconnected and the integrated eloquence by which a visual signal is created is beyond the scope of this review, several excellent reviews are available (Refs. 1,2,5).

The retina is required for regulating visually driven eye growth, yet our knowledge of what visual signals and retinal pathways control eye growth is lacking. In this review, we will focus on a few key retinal cells and neurotransmitters that have been implicated in myopia, and show how

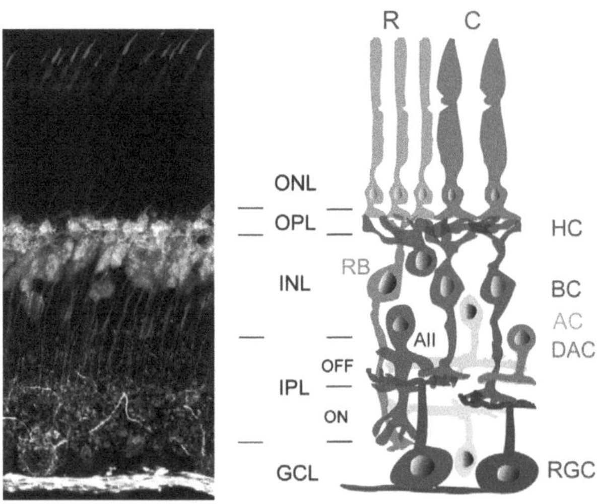

Figure 1 Retinal anatomy and circuitry. During visual processing, the output from the rod and cone photoreceptors in the outer nuclear layer (ONL) is decomposed into a number of different parallel information channels by synapsing in the outer plexiform layer (OPL) to different cells in the inner nuclear layer (INL; bipolar, amacrine, and horizontal cells). The output from these inner retinal cells is sampled by different retinal ganglion cells in the inner plexiform layer (IPL). Finally, depending on the type of information, specific ganglion cells in the ganglion cell layer (GCL) transmit the signal to higher visual structures in the brain. R, rod; C, cone; RB, rod bipolar cell; BC, cone bipolar cell; HC, horizontal cell; AC, amacrine cell; AII, AII amacrine cell; DAC, dopaminergic amacrine cell. *Image modified from Ref. 4.* (See the color plate.)

investigations to explore complex signaling pathways and retinal circuits using the mouse model have provided some remarkable insights into the retinal mechanisms of refractive development.

2. RETINAL INPUT ESSENTIAL FOR OCULAR GROWTH

As previously discussed (Refs. 6–18) a wide range of animal studies have shown that the visual environment influences refractive development of the eye. It has been established that eye growth regulation occurs at the retinal level in response to both diffusers and defocus lenses. Previous studies have shown that severing the optic nerve in young chicks does not prevent the development of myopia in response to both negative lenses[19,20] and diffusers.[21] Furthermore, in both chicks[22,8] and primates,[23] if partial diffusers are imposed on only half of the visual field, only that corresponding half of the eye elongates and becomes myopic. Similarly, chick eyes compensate for both negative[22,24] and positive[24] lenses imposed on local retinal areas using hemifield spectacle lenses with ocular growth restricted to the defocused parts of the visual field. These studies demonstrate retinal, and not cortical processing, is sufficient to regulate refractive eye growth.

However, there is some evidence that cortical processing may also influence eye growth. While higher order visual processing was not required for compensation to hyperopic defocus in chickens, either optic nerve[21] or both optic nerve and ciliary nerve sections[20] induced hyperopia in chickens with normal visual input.[21,20] Together these findings suggest that higher order processing within the visual system may influence ocular growth, but is not a requirement for visually driven ocular growth. Given that the visual mechanisms regulating refractive development localize principally to the retina, any defect in visual transmission through the retina could potentially influence ocular growth, and may lead to development of refractive errors.

Several studies have suggested the role of various retinal cell types/pathways and neurotransmitters in normal refractive development of the eye. In chickens, physiological and morphological changes in photoreceptors are associated with experimentally induced myopia.[25] In addition, differential eye growth under both normal and FD conditions in response to neurotoxins blocking responses from the photoreceptors,[26] ON and OFF pathways,[27,28] and the inner retina[29] have been shown. Several retinal neurotransmitters such as dopamine (DA),[30,31] glucagon,[32] acetylcholine,[33] nitric oxide,[34,35] and retinoic acid[36,37] have also been implicated in defocus-induced ocular growth in animals. Abnormalities in visual transmission through the retina may result from mutations in retinal neurons/pathways, changes in various retinal neurotransmitters associated with the mutation, or a combination of both factors. While these experiments demonstrate the influence of the retina in normal ocular development, these experimental approaches do not ensure complete and selective blockage of a single pathway or neuronal type. Mutant mouse models with specific visual pathway defects offer an advantage to comprehensively investigate the role of specific pathways/neurons involved in refractive development. In this review, we will focus on recent studies using mouse models that have provided insight into retinal pathways and neurotransmitters controlling refractive development.

3. MOUSE— A NOVEL ANIMAL MODEL TO EXPLORE RETINAL MECHANISM OF REFRACTIVE DEVELOPMENT

In the recent years, there has been a growing interest in using mouse models for investigating complex signaling pathways and retinal circuitries, and their influence on ocular refractive development.[38–49] Mouse models offer the advantage of altering both genes and environment in the same animal by using various knockout models that are generated by manipulating the mouse genome, combined with altered visual input with lenses or form deprivation. Additionally, close resemblance of the mouse retinal structure to humans, short gestational period, and large litter sizes make them an excellent experimental model for refractive development research. However, small eye size, absence of fovea, nocturnal behavior, poor visual acuity,[50] and large depth of field[51] are some limitations of using murine models for refractive development studies (see review, Ref. 52). Despite these limitations, the mouse eye responds to visual form deprivation with temporal properties and magnitude comparable with other mammalian models (see review, Ref. 52).

Studying the effects of visual manipulations in various mouse mutants provides a unique opportunity to examine the role of gene/environment interactions in refractive development. A number of previous studies have examined the effects of a specific gene defect using various mutant mice under normal and form-deprived visual conditions.[41,42,47,48,53–58] The mouse model provides the opportunity to investigate how the gene defect and/or the associated changes in the levels of retinal neurotransmitters alter refractive development with normal visual input, as well as the influence of the gene defect on myopia susceptibility. In view of these points, probing genetic and environmental interactions is the most promising aspect of using the mouse to provide important insights into the mechanisms regulating eye growth.

4. RETINAL NEUROTRANSMITTERS AND REFRACTIVE DEVELOPMENT

Before describing studies of mouse models with retinal neuron defects, it is important to acknowledge the rich diversity of neurotransmitters present in the retina that have been associated with experimentally induced refractive errors in animal models. In order for the rate of ocular growth to change, visual stimulation from the retina must activate signaling pathways that modulate scleral growth. Evidence from pharmacological and genetic studies suggests that several signaling pathways control refractive eye growth.[59–61] For instance, the expression level of ZENK in chickens or the mammalian homolog, Egr-1, has been shown to increase or decrease with hyperopic or myopic eye growth, respectively.[62,63]

A large body of previous studies has examined muscarinic acetylcholine receptor mechanisms in refractive development of the eye. Both non-selective (such as atropine) and partially selective (such as pirenzepine) muscarinic antagonists have been shown to have inhibitory effects on experimental myopia in chickens and mammals (see review, Refs. 60,64). Clinically, atropine[65–67] and pirenzepine[68–70] have also been used to slow myopia progression in children. In both laboratory animals and children, antimyopia effect of atropine has been found to be independent of the drug's action on accommodation.[64] Recent results using a mutant mouse suggest that loss of the muscarinic cholinergic receptor gene, M_2, or pharmacological blocking of M_2 will provide resistance to myopia by altering the scleral collagen composition.[58]

Another possible signaling pathway candidate is adenosine which is known to be regulated by light and alters collagen synthesis.[56] All adenosine receptor subtypes are expressed in the retina, choroid, and sclera of the mammalian eye,[71] and have been suggested to play an important role in the regulation of eye growth in both mammalian[71,72] and human eyes.[73] The adenosine A_{2A} receptor KO mice have relative myopia compared to wild-type (WT) littermates and altered scleral ultrastructure.[56]

Nitric oxide, a gaseous neurotransmitter synthesized by both the retina and the choroid,[74] is thought to be part of the signal cascade mediating ocular growth inhibition in response to myopic defocus.[75] Studies on chickens have shown pharmacological inhibition of nitric oxide to prevent the increase in choroidal thickness normally induced by myopic defocus, resulting in myopic eye growth.[34,35,76] However, some earlier chicken studies have reported inconsistent results compared to these newer studies on the effects of nitric oxide in ocular growth regulation of chickens.[77,78] The differences in results could be due to differences in drug concentrations used by these studies. Furthermore, a transient suppression of retinal nitric oxide synthase (NOS, enzyme catalyzing the production of nitric oxide) activity was observed in guinea pigs with acute form deprivation, whereas chronic form deprivation (about 14–21 days) was associated with significant upregulation of NOS levels in the posterior eye.[79,80] Other potential candidates that either inhibit (such as glucagon[32]) or promote (such as retinoic acid[36,37]) ocular growth in response to defocus have been suggested to be part of this signaling pathway.

4.1 Role of Dopamine in Refractive Development and Myopia Susceptibility

In the retina, DA is synthesized and released by a subset of amacrine–interplexiform cells, known as the dopaminergic amacrine cells.[81,82] Dopaminergic neurons convert tyrosine to L-3,4-dihydroxyphenylalanine (L-DOPA) via tyrosine hydroxylase and L-DOPA to DA via aromatic L-amino acid decarboxylase.[82] DA is released by the neuron and metabolized to 3,4-dihydroxyphenylacetic acid (DOPAC), which is the main DA metabolite in the rodent retina.[83]

DA has been implicated as a stop signal in refractive eye growth.[61] In both chicken and mammalian models, form deprivation decreases DA biosynthesis,[30,84,85] whereas DA mimetic treatment (L-DOPA or receptor agonists) prevents form-deprivation myopia (see Ref. 61 for review; Refs. 30,31,86). Furthermore, particular DA receptor activation may be important for refractive development signaling, as D2R has been shown to be important for the inhibitory effects of form deprivation,[87,88] lens defocus,[89] and bright lighting[90,91] in chickens. In the mammalian guinea pig model of spontaneous myopia, D1-like receptors inhibit and D2-like receptors promote myopic growth.[92] However, the loss of D2R in mice prevents the form-deprivation response.[55] It should be noted that some recent reports indicate unaltered retinal DA activity in mice with form deprivation.[93] Together, these results indicate that more studies are needed to elucidate how DA modulates refractive development and myopia. Section 6 reviews dopaminergic modulation of myopia susceptibility in retinal mouse mutants (discussed below).

5. RETINAL NEURONS/PATHWAYS AND REFRACTIVE DEVELOPMENT IN MUTANT MICE

5.1 Photoreceptor Input to Myopia

Since the photoreceptors form the first layer of photosensory neurons in the retina, it is plausible that photoreceptors are involved in mechanisms sensing defocus and/or communicating that error signal across the retina to the RPE and the choroid. In fact, studies have suggested that, in emmetropia, the focal plane is located at the photoreceptor inner segments, and both the alignment and directionality of photoreceptors are important components for retinal blur detection.[25] Over the years, morphological changes of photoreceptors (elongation of rod outer segments),[94] reduction in photoreceptor cell density,[95] changes in the outer segment shedding under various lighting conditions,[96,97] and changes in electroretinogram[98] associated with experimentally induced myopia have all pointed toward the possible role of photoreceptors in refractive development of the eye. While a number of studies have suggested that cone pathways are likely to dictate the signaling needed for proper eye development, there is also some evidence for the involvement of rods in regulating ocular growth. The requirement of a high-acuity retinal image (largely attributed to cone mediated signaling) for emmetropization,[99] development of myopia under dim lighting conditions (when cones are less sensitive) in chickens,[100] increased and decreased susceptibility to form-deprivation myopia, respectively, in cone- and rod-dominated animal models,[101,102] and reduced response to experimental myopia in chickens treated with formoguanamine, a photoreceptor neurotoxin that causes significant damage to cone outer segments, all suggest the importance of cone activity in emmetropization.[26] However, normal response to form-deprivation myopia in monkeys treated with laser ablation at the cone-rich fovea,[103] and similar myopic responses in monkeys with form deprivation imposed on the rod-dominated peripheral regions or the entire visual field[104] suggest that cone pathways may not completely dominate the signaling for mammalian eye growth, a finding further supported by the following mouse studies of experimental myopia.

5.1.1 rd1 $^{-/-}$ *and* rd10 $^{-/-}$ *Mice—Photoreceptor Degeneration Models*

In a recent study, Park *et al.*[47] found that under normal visual conditions, retinal degeneration causes significantly hyperopic refractive errors and shorter axial lengths in *rd1*$^{-/-}$ and *rd10*$^{-/-}$ mice compared to WT mice.

Pde6b$^{rd1/rd1}$ (*rd1*$^{-/-}$, Refs. 105–107) and *Pde6b*$^{rd10/rd10}$ (*rd10*$^{-/-}$, Refs. 108,109) mice with a mutation in the *Pde6b* gene are two frequently used mouse models of photoreceptor degeneration. *Pde6b* mutation disrupts encoding of the β-subunit of cyclic nucleotide phosophodiesterase-6,[110,111] a mutation also seen in patients with RP.[112,113]

Interestingly, with form deprivation, both degeneration strains show faster and greater susceptibility to form-deprivation myopia than WT mice (~6–7 D in 2 weeks in *rd1*$^{-/-}$ and *rd10*$^{-/-}$ mice vs. ~3 D in 4–5 weeks in *rd1*$^{+/+}$ and *rd10*$^{+/+}$; Fig. 2).[47] In agreement with previous studies,[114,115] retinal DA levels were altered. Levels of the DA metabolite, DOPAC, and the DOPAC/DA ratios (a measure of DA turnover) were significantly lower in *rd1*$^{-/-}$ and *rd10*$^{-/-}$ mice compared to the WT mice during normal visual experience, indicating a lower DA metabolism in degenerated mouse retinas.[47] Furthermore, both degeneration models exhibited a high correlation between lower basal levels of retinal DOPAC and greater susceptibility to form-deprivation myopia (Fig. 3). These results indicate that retinal degeneration by itself may not cause myopia, but may reduce DA

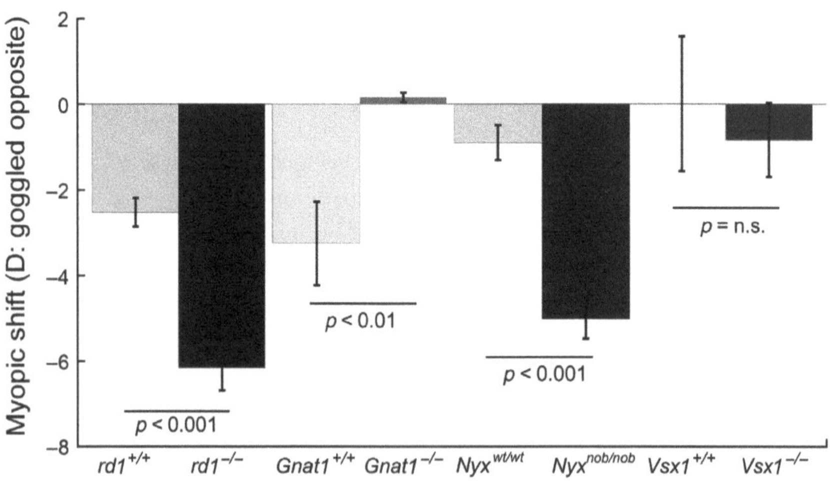

Figure 2 Myopic shift (goggled opposite eye) across several mutant mouse strains with photoreceptor or ON/OFF retinal pathway defects after 2 weeks of form deprivation. Note that the mutant mice are on different background strains and thus, the WT controls are shown for each strain. *Rd1*$^{-/-}$ mice and *Nyx*$^{nob/nob}$ mice show significantly greater myopic shift than their WT controls. In contrast, *Gnat1*$^{-/-}$ mice with nonfunctional rods did not show a shift with form-deprivation myopia. *Vsx1*$^{+/+}$ mice are on a 129SV background and do not respond to form deprivation. The loss of *Vsx1* did not change the susceptibility to form-deprivation myopia. (See the color plate.)

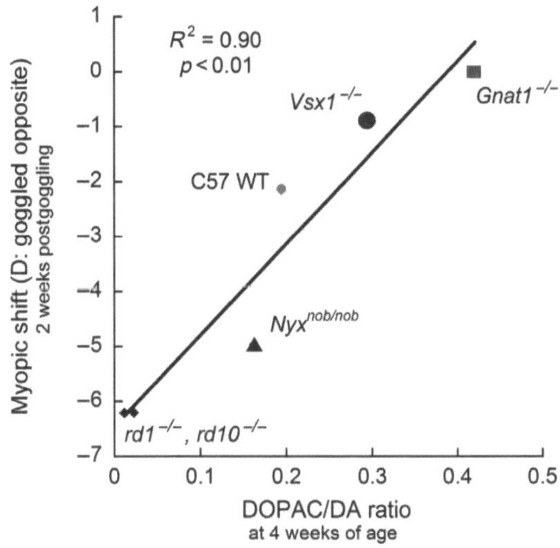

Figure 3 Myopic shift after 2 weeks of form deprivation in several different mutant strains with retinal defects plotted against the DOPAC/DA ratio at 4 weeks of age when the form deprivation was initiated. These results suggest that DA turnover at the time of goggling may influence the susceptibility to form-deprivation myopia.

metabolism in the retina, which may lead to an increased susceptibility to myopia under myopigenic visual environments.

5.1.2 Gnat1 $^{-/-}$ *Mice—Nonfunctional Rod Model*

The importance of rod pathways in visual processing under different light conditions and their potential role in refractive error development is demonstrated by the presence of myopia in human patients with congenital stationary night blindness due to abnormal visual transmission between rods and ON bipolar cells.[116]

Recently, Park *et al.*[53] reported abnormal refractive development in *Gnat1*$^{-/-}$ mice with nonfunctional rod photoreceptors (genetic mutation in rhodopsin-associated G protein, transducin α-1, Ref. 117). Unlike normal refractive development in WT mice, *Gnat1*$^{-/-}$ mice do not show a relative increase in hyperopia with age, instead the refractive curve in *Gnat1*$^{-/-}$ mice remains stable throughout the developmental period measured from 4 to 12 weeks. Additionally, *Gnat1*$^{-/-}$ mice are unresponsive to form deprivation and do not develop a myopic shift (Fig. 2). In this study, the authors examined retinal DA and DOPAC levels across age, and found significantly lower (and stable) retinal DOPAC levels in *Gnat1*$^{-/-}$ mice compared to WT mice throughout the period of development.

Furthermore, *Gnat1*$^{-/-}$ retinas exhibit significantly greater DA turnover (measured from DOPAC/DA ratio) at early ages of refractive development, which rapidly decline from the second week of development.[53] These results suggest that functional rods are critical to normal refractive development and form-deprivation response in mice, and that DA metabolism and tonic levels of DA during ocular development are important predictors of susceptibility to form-deprivation myopia in murine eyes. Future studies using *Gnat2*$^{-/-}$ mutants with nonfunctional cones[118] may increase our understanding of how different photoreceptors might regulate normal refractive development under different ambient lighting conditions.

5.2 ON and OFF Pathway Contributions to Myopia

Effects of ON and OFF pathways on eye development have been examined using various neurotoxins that specifically block the ON and OFF responses to light.[27,28] In chickens, elimination of the OFF pathway using intravitreal injections of the D-isomer of a Müller cell gliotoxin α-amino adipic acid (DαAAA) resulted in an enhanced rate of axial elongation under normal visual conditions, but a slower ocular growth rate with form deprivation.[27] Conversely, inhibition of the ON channel with the L-isomer (LαAAA) caused a reduction in axial eye growth of normal eyes, but increased eye growth in form-deprived animals. Similar to chickens, blocking of the ON pathway with D,L-2-amino-4-phosphonobutyric (a selective ON pathway inhibitor) has been shown to cause a significant reduction in axial eye growth of cats.[119] Genetic mutations of neurons or receptor in the ON and OFF pathways in mutant mouse models represent a novel approach to investigating the role of the ON and OFF signaling in refractive development.

5.2.1 Nyx$^{nob/nob}$ *Mice—ON Pathway Defect Model*

Pardue *et al.*[48] examined the refractive development and DA levels of the *Nyx*$^{nob/nob}$ mouse,[120] which carries a null mutation in *Nyx*,[121] resulting in a loss of function of the ON pathway.[120] *Nyx* encodes the protein nyctalopin, which is located on the postsynaptic side of the photoreceptor to ON bipolar cell synapse.[122] ON and OFF channels of the visual system are imperative for processing contrast sensitivity information,[123,124] an important prerequisite for a high-resolution retinal image.

Under normal unmanipulated visual conditions, the loss of *Nyx* causes only slightly more hyperopic refractions in *Nyx*$^{nob/nob}$ mice compared to WT mice.[48] However, imposing form deprivation results in a significantly rapid myopic shift in *Nyx*$^{nob/nob}$ mice compared to *Nyx*$^{wt/wt}$ mice (Fig. 2).

Additionally, during normal visual development, DA and DOPAC levels were significantly lower in the $Nyx^{nob/nob}$ mice in comparison with the $Nyx^{wt/wt}$ mice. These results indicate that low endogenous DA levels or blurred visual input secondary to the ON pathway defect may increase the susceptibility to myopia development in the mouse eye.

5.2.2 Vsx1 $^{-/-}$ *Mice—OFF Pathway Defect Model*

Chakraborty *et al.*[54] examined the role of OFF pathway signaling in refractive development of the eye using the $Vsx1^{-/-}$ mice on a 129S1/Sv background,[125] which carry a null mutation in the visual system homeobox 1 gene, *Vsx1*.[126] The detection of *Vsx1* in the mouse retina at postnatal day 5 in the developing bipolar cell region,[126] and a reduction in immunolabeling at the axonal termini of various OFF cone bipolar cells (and a few ON bipolar cells) in adult $Vsx1^{-/-}$ retinas[125,127,128] suggest that *Vsx1* is essential for late terminal differentiation and functioning of OFF cone bipolar cells. However, Chakraborty *et al.* found that a selective impairment of the retinal OFF visual pathway caused by the *Vsx1* mutation does not significantly alter the normal refractive development in $Vsx1^{-/-}$ mice compared to the $Vsx1^{+/+}$ mice, potentially due to normal visual transmission through other *Vsx1*-independent ON and OFF bipolar cells in the retina.[54] Interestingly, both $Vsx1^{+/+}$ and $Vsx1^{-/-}$ mice do not respond to imposed form deprivation (Fig. 2). Furthermore, at 4 weeks of age, 129S1/Sv mice ($Vsx1^{+/+}$) exhibit a significantly elevated retinal DA turnover compared to the commonly used C57BL/6J mice, which may prevent against form-deprivation myopia in both 129S1/Sv $Vsx1^{-/-}$ and $Vsx1^{+/+}$ mice.[54] Although, these results indicate that OFF pathway signaling may not be critically important for normal refractive development in mice, future studies with mouse mutants that express complete loss of the OFF pathway are required to investigate this further.

5.3 Amacrine and Ganglion Cell Contributions to Refractive Development

There is also some evidence that the inner retina (especially the amacrine cells and the retinal ganglion cells, RGCs) may play some role in refractive development of the eye. Blocking amacrine cell function using cell-specific neurotoxins, such as 6-hydroxy-dopamine (dopaminergic amacrine cells, Ref. 129) and ethylcholine mustard aziridinium ion (cholinergic amacrine cells, Ref. 130), does not alter refractive compensation to imposed defocus in chickens. However, other neurotoxins, like kainic acid[29,131] causing a nonspecific damage to the inner retina (including amacrine cells) at higher doses, lead to increased ocular growth (mostly in the posterior chamber) under normal conditions, and a reduction in myopic eye growth under form-deprived conditions. These findings warrant further investigation using amacrine cell knockout mutants.[132] Finally, blockade of RGC function does not prevent form-deprivation myopia in both chickens[133] and tree shrews,[134] suggesting that the outer retina perhaps plays a major role in defocus detection and signaling for ocular development.[25]

5.4 Dopamine Modulation of Myopia Susceptibility in Retinal Mouse Mutants

In addition to acute changes in DA with abnormal visual input, results from mutant mice also suggest that basal levels of DA turnover (as indicated by DOPAC/DA ratio) may influence susceptibility to form-deprivation myopia. For example, mouse models of retinal degeneration ($rd1^{-/-}$ and $rd10^{-/-}$) have decreased DOPAC and DOPAC/DA levels throughout life and increased susceptibility to form-deprivation myopia.[47] Since DA is released via ON pathway stimulation,[135] mutations in the ON pathway would decrease retinal DA levels. Such is the case in $Nyx^{wt/wt}$ mice which have reduced DA and DOPAC levels and increased susceptibility to myopia.[48] Alternatively, mice with high levels of retinal DA and/or DOPAC have reduced susceptibility to myopia, such as $Gnat1^{-/-}$[53] and 129SV $Vsx1^{+/+}$ mice.[54] Figure 3 shows the relationship between DOPAC/DA levels at 4 weeks of age and the susceptibility to subsequent form-deprivation myopia. These results suggest that DA turnover during early development may "preset" the susceptibility to myopigenic stimulation. Further research is needed to determine how DA alters refractive eye growth so that therapeutic approaches can be developed.

6. CONCLUSIONS

In conclusion, the retina plays an important role in regulating visually driven ocular growth in mammals. The mouse is an extremely useful animal model to examine retinal mechanisms controlling eye growth. Using genetic mouse mutants, genes controlling specific retinal receptors, neurotransmitters, and cell types can be selectively probed to examine their role in normal refractive development, as well as under altered visual conditions.

Mutations in different retinal neurons/signaling pathways have differential effects on normal and visually deprived refractive development of the eye. The refractive phenotypes observed in different retinal mutations may result from the mutation itself, changes in various retinal neurotransmitters associated with the mutation (such as changes in DA levels), or a combination of both factors. In mice, although photoreceptors are important for normal refractive development, rod pathways in particular (both functional rods and ON pathway) appear to be extremely critical for both normal refractive development as well as response to visual form deprivation. While cone pathways have previously been implicated in normal ocular refractive development, it requires further investigation using mouse mutants (with mutations in functional cones or OFF cone pathways) to determine if their role is essential.

In rodents, alterations in endogenous retinal DA (or DOPAC levels) associated with various mutations are important determinants of susceptibility to form-deprivation myopia. It should be noted that this chapter specifically reviewed the changes in retinal DA levels and their potential implications in refractive error development in mice. However, the mammalian retina is a hub for many other neurotransmitters (such as nitric oxide, glucagon, retinoic acid, vasoactive intestinal peptide, etc.). Changes in other neurotransmitters with experimental myopia and their interaction with retinal DA during refractive error development are yet to be explored, and beyond the scope of this chapter. Further understanding of how abnormal visual signals from the retina are transmitted downstream through the RPE and the choroid, causing long-term changes in the sclera are important for designing therapeutic interventions for myopia control.

REFERENCES

1. Masland RH. Neuronal diversity in the retina. *Curr Opin Neurobiol*. 2001;11(4):431–436.
2. Masland RH. The neuronal organization of the retina. *Neuron*. 2012;76(2):266–280.
3. Jeon CJ, Strettoi E, Masland RH. The major cell populations of the mouse retina. *J Neurosci*. 1998;18(21):8936–8946.
4. Hoon M, Okawa H, Santina LD, Wong RO. Functional architecture of the retina: development and disease. *Prog Retin Eye Res*. 2014;42:44–84.
5. Masland RH. The fundamental plan of the retina. *Nat Neurosci*. 2001;4(9):877–886.
6. Wallman J, Winawer J. Homeostasis of eye growth and the question of myopia. *Neuron*. 2004;43(4):447–468.
7. Wallman J, Turkel J, Trachtman J. Extreme myopia produced by modest change in early visual experience. *Science*. 1978;201(4362):1249–1251.
8. Wallman J, Wildsoet C, Xu A, et al. Moving the retina: choroidal modulation of refractive state. *Vis Res*. 1995;35(1):37–50.
9. Schaeffel F, Glasser A, Howland HC. Accommodation, refractive error and eye growth in chickens. *Vis Res*. 1988;28(5):639–657.

10. Howlett MH, McFadden SA. Form-deprivation myopia in the guinea pig (*Cavia porcellus*). *Vis Res*. 2006;46(1–2):267–283.
11. Howlett MH, McFadden SA. Spectacle lens compensation in the pigmented guinea pig. *Vis Res*. 2009;49(2):219–227.
12. Smith 3rd EL, Hung LF. Form-deprivation myopia in monkeys is a graded phenomenon. *Vis Res*. 2000;40(4):371–381.
13. Smith 3rd EL, Hung LF. The role of optical defocus in regulating refractive development in infant monkeys. *Vis Res*. 1999;39(8):1415–1435.
14. Troilo D, Judge SJ. Ocular development and visual deprivation myopia in the common marmoset (Callithrix jacchus). *Vis Res*. 1993;33(10):1311–1324.
15. Metlapally S, McBrien NA. The effect of positive lens defocus on ocular growth and emmetropization in the tree shrew. *J Vis*. 2008;8(3):1.1–1.12.
16. Van Sluyters RC. Recovery from monocular stimulus deprivation amblyopia in the kitten. *Ophthalmology*. 1978;85(5):478–488.
17. Gao Q, Liu Q, Ma P, Zhong X, Wu J, Ge J. Effects of direct intravitreal dopamine injections on the development of lid-suture induced myopia in rabbits. *Graefes Arch Clin Exp Ophthalmol*. 2006;244(10):1329–1335.
18. Shen W, Vijayan M, Sivak JG. Inducing form-deprivation myopia in fish. *Invest Ophthalmol Vis Sci*. 2005;46(5):1797–1803.
19. Wildsoet C, Wallman J. Choroidal and scleral mechanisms of compensation for spectacle lenses in chicks. *Vis Res*. 1995;35(9):1175–1194.
20. Wildsoet CF. Neural pathways subserving negative lens-induced emmetropization in chicks—insights from selective lesions of the optic nerve and ciliary nerve. *Curr Eye Res*. 2003;27(6):371–385.
21. Troilo D, Gottlieb MD, Wallman J. Visual deprivation causes myopia in chicks with optic nerve section. *Curr Eye Res*. 1987;6(8):993–999.
22. Wallman J, Gottlieb MD, Rajaram V, Fugate-Wentzek LA. Local retinal regions control local eye growth and myopia. *Science*. 1987;237(4810):73–77.
23. Smith 3rd EL, Huang J, Hung LF, Blasdel TL, Humbird TL, Bockhorst KH. Hemiretinal form deprivation: evidence for local control of eye growth and refractive development in infant monkeys. *Invest Ophthalmol Vis Sci*. 2009;50(11):5057–5069.
24. Diether S, Schaeffel F. Local changes in eye growth induced by imposed local refractive error despite active accommodation. *Vis Res*. 1997;37(6):659–668.
25. Crewther DP. The role of photoreceptors in the control of refractive state. *Prog Retin Eye Res*. 2000;19(4):421–457.
26. Westbrook AM, Crewther SG, Liang H, et al. Formoguanamine-induced inhibition of deprivation myopia in chick is accompanied by choroidal thinning while retinal function is retained. *Vis Res*. 1995;35(14):2075–2088.
27. Crewther DP, Crewther SG. Pharmacological modification of eye growth in normally reared and visually deprived chicks. *Curr Eye Res*. 1990;9(8):733–740.
28. Crewther DP, Crewther SG, Xie RZ. Changes in eye growth produced by drugs which affect retinal ON or OFF responses to light. *J Ocul Pharmacol Ther*. 1996;12(2):193–208.
29. Ehrlich D, Sattayasai J, Zappia J, Barrington M. Effects of selective neurotoxins on eye growth in the young chick. *Ciba Found Symp*. 1990;155:63–84, discussion 84–88.
30. Stone RA, Lin T, Laties AM, Iuvone PM. Retinal dopamine and form-deprivation myopia. *Proc Natl Acad Sci USA*. 1989;86(2):704–706.
31. Iuvone PM, Tigges M, Stone RA, Lambert S, Laties AM. Effects of apomorphine, a dopamine receptor agonist, on ocular refraction and axial elongation in a primate model of myopia. *Invest Ophthalmol Vis Sci*. 1991;32(5):1674–1677.
32. Feldkaemper MP, Schaeffel F. Evidence for a potential role of glucagon during eye growth regulation in chicks. *Vis Neurosci*. 2002;19(6):755–766.

33. Stone RA, Lin T, Laties AM. Muscarinic antagonist effects on experimental chick myopia. *Exp Eye Res*. 1991;52(6):755–758.
34. Nickla DL, Wilken E, Lytle G, Yom S, Mertz J. Inhibiting the transient choroidal thickening response using the nitric oxide synthase inhibitor l-NAME prevents the ameliorative effects of visual experience on ocular growth in two different visual paradigms. *Exp Eye Res*. 2006;83(2):456–464.
35. Nickla DL, Wildsoet CF. The effect of the nonspecific nitric oxide synthase inhibitor NG-nitro-L-arginine methyl ester on the choroidal compensatory response to myopic defocus in chickens. *Optom Vis Sci*. 2004;81(2):111–118.
36. McFadden SA, Howlett MH, Mertz JR, Wallman J. Acute effects of dietary retinoic acid on ocular components in the growing chick. *Exp Eye Res*. 2006;83(4):949–961.
37. McFadden SA, Howlett MH, Mertz JR. Retinoic acid signals the direction of ocular elongation in the guinea pig eye. *Vis Res*. 2004;44(7):643–653.
38. Tejedor J, de la Villa P. Refractive changes induced by form deprivation in the mouse eye. *Invest Ophthalmol Vis Sci*. 2003;44(1):32–36.
39. Qian YS, Chu RY, Hu M, Hoffman MR. Sonic hedgehog expression and its role in form-deprivation myopia in mice. *Curr Eye Res*. 2009;34(8):623–635.
40. Schmucker C, Schaeffel F. In vivo biometry in the mouse eye with low coherence interferometry. *Vis Res*. 2004;44(21):2445–2456.
41. Schippert R, Burkhardt E, Feldkaemper M, Schaeffel F. Relative axial myopia in Egr-1 (ZENK) knockout mice. *Invest Ophthalmol Vis Sci*. 2007;48(1):11–17.
42. Wisard J, Faulkner A, Chrenek MA, et al. Exaggerated eye growth in IRBP-deficient mice in early development. *Invest Ophthalmol Vis Sci*. 2011;52(8):5804–5811.
43. Barathi VA, Boopathi VG, Yap EP, Beuerman RW. Two models of experimental myopia in the mouse. *Vis Res*. 2008;48(7):904–916.
44. Zhou X, Shen M, Xie J, et al. The development of the refractive status and ocular growth in C57BL/6 mice. *Invest Ophthalmol Vis Sci*. 2008;49(12):5208–5214.
45. Jiang M, Wu PC, Fini ME, et al. Single-shot dimension measurements of the mouse eye using SD-OCT. *Ophthalmic Surg Lasers Imaging*. 2012;43(3):252–256.
46. Tkatchenko TV, Shen Y, Tkatchenko AV. Analysis of postnatal eye development in the mouse with high-resolution small animal magnetic resonance imaging. *Invest Ophthalmol Vis Sci*. 2010;51(1):21–27.
47. Park H, Tan CC, Faulkner A, et al. Retinal degeneration increases susceptibility to myopia in mice. *Mol Vis*. 2013;19:2068–2079.
48. Pardue MT, Faulkner AE, Fernandes A, et al. High susceptibility to experimental myopia in a mouse model with a retinal on pathway defect. *Invest Ophthalmol Vis Sci*. 2008;49(2):706–712.
49. Chou TH, Kocaoglu OP, Borja D, et al. Postnatal elongation of eye size in DBA/2J mice compared with C57BL/6J mice: in vivo analysis with whole-eye OCT. *Invest Ophthalmol Vis Sci*. 2011;52(6):3604–3612.
50. Prusky GT, Alam NM, Beekman S, Douglas RM. Rapid quantification of adult and developing mouse spatial vision using a virtual optomotor system. *Invest Ophthalmol Vis Sci*. 2004;45(12):4611–4616.
51. Remtulla S, Hallett PE. A schematic eye for the mouse, and comparisons with the rat. *Vis Res*. 1985;25(1):21–31.
52. Pardue MT, Stone RA, Iuvone PM. Investigating mechanisms of myopia in mice. *Exp Eye Res*. 2013;114:96–105.
53. Park Hn, Jabbar SB, Tan CC, et al. Visually-driven ocular growth in mice requires functional rod photoreceptors. *Invest Ophthalmol Vis Sci*. 2014;55(10):6272–6279.
54. Chakraborty R, Park Hn, Aung MH, et al. Comparison of refractive development and retinal dopamine in OFF pathway mutant and C57BL/6J wild-type mice. *Mol Vis*. 2014;20:1318–1327.

55. Huang F, Yan T, Shi F, et al. Activation of dopamine D2 receptor is critical for the development of form-deprivation myopia in the C57BL/6 mouse. *Invest Ophthalmol Vis Sci*. 2014;55(9):5537–5544.
56. Zhou X, Huang Q, An J, et al. Genetic deletion of the adenosine A2A receptor confers postnatal development of relative myopia in mice. *Invest Ophthalmol Vis Sci*. 2010;51(9):4362–4370.
57. Zhou G, Strom RC, Giguere V, Williams RW. Modulation of retinal cell populations and eye size in retinoic acid receptor knockout mice. *Mol Vis*. 2001;7:253–260.
58. Barathi VA, Kwan JL, Tan QS, et al. Muscarinic cholinergic receptor (M2) plays a crucial role in the development of myopia in mice. *Dis Model Mech*. 2013;6(5):1146–1158.
59. Stone RA, Khurana TS. Gene profiling in experimental models of eye growth: clues to myopia pathogenesis. *Vis Res*. 2010;50(23):2322–2333.
60. Ganesan P, Wildsoet CF. Pharmaceutical intervention for myopia control. *Expert Rev Ophthalmol*. 2010;5(6):759–787.
61. Feldkaemper M, Schaeffel F. An updated view on the role of dopamine in myopia. *Exp Eye Res*. 2013;114:106–119.
62. Ashby R, Kozulin P, Megaw PL, Morgan IG. Alterations in ZENK and glucagon RNA transcript expression during increased ocular growth in chickens. *Mol Vis*. 2010;16:639–649.
63. Ashby RS, Zeng G, Leotta AJ, Tse DY, McFadden SA. Egr-1 mRNA expression is a marker for the direction of mammalian ocular growth. *Invest Ophthalmol Vis Sci*. 2014;55(9):5911–5921.
64. Stone RA, Pardue MT, Iuvone PM, Khurana TS. Pharmacology of myopia and potential role for intrinsic retinal circadian rhythms. *Exp Eye Res*. 2013;114:35–47.
65. Walline JJ, Lindsley K, Vedula SS, Cotter SA, Mutti DO, Twelker JD. Interventions to slow progression of myopia in children. *Cochrane Database Syst Rev*. 2011;12: CD004916.
66. Chua WH, Balakrishnan V, Chan YH, et al. Atropine for the treatment of childhood myopia. *Ophthalmology*. 2006;113(12):2285–2291.
67. Li SM, Wu SS, Kang MT, et al. Atropine slows myopia progression more in Asian than white children by meta-analysis. *Optom Vis Sci*. 2014;91(3):342–350.
68. Siatkowski RM, Cotter S, Miller JM, et al. Safety and efficacy of 2% pirenzepine ophthalmic gel in children with myopia: a 1-year, multicenter, double-masked, placebo-controlled parallel study. *Arch Ophthalmol*. 2004;122(11):1667–1674.
69. Siatkowski RM, Cotter SA, Crockett RS, et al. Two-year multicenter, randomized, double-masked, placebo-controlled, parallel safety and efficacy study of 2% pirenzepine ophthalmic gel in children with myopia. *J AAPOS*. 2008;12(4):332–339.
70. Tan DT, Lam DS, Chua WH, Shu-Ping DF, Crockett RS, Asian Pirenzepine Study Group. One-year multicenter, double-masked, placebo-controlled, parallel safety and efficacy study of 2% pirenzepine ophthalmic gel in children with myopia. *Ophthalmology*. 2005;112(1):84–91.
71. Cui D, Trier K, Zeng J, Wu K, Yu M, Ge J. Adenosine receptor protein changes in guinea pigs with form deprivation myopia. *Acta Ophthalmol*. 2010;88(7):759–765.
72. Cui D, Trier K, Zeng J, et al. Effects of 7-methylxanthine on the sclera in form deprivation myopia in guinea pigs. *Acta Ophthalmol*. 2011;89(4):328–334.
73. Chen X, Xue A, Chen W, et al. Assessment of exonic single nucleotide polymorphisms in the adenosine A2A receptor gene to high myopia susceptibility in Chinese subjects. *Mol Vis*. 2011;17:486–491.
74. Fischer AJ, Stell WK. Nitric oxide synthase-containing cells in the retina, pigmented epithelium, choroid, and sclera of the chick eye. *J Comp Neurol*. 1999;405(1):1–14.
75. Nickla DL, Wallman J. The multifunctional choroid. *Prog Retin Eye Res*. 2010;29(2): 144–168.

76. Nickla DL, Damyanova P, Lytle G. Inhibiting the neuronal isoform of nitric oxide synthase has similar effects on the compensatory choroidal and axial responses to myopic defocus in chicks as does the non-specific inhibitor L-NAME. *Exp Eye Res.* 2009;88(6):1092–1099.
77. Fujikado T, Kawasaki Y, Fujii J, et al. The effect of nitric oxide synthase inhibitor on form-deprivation myopia. *Curr Eye Res.* 1997;16(10):992–996.
78. Fujikado T, Tsujikawa K, Tamura M, Hosohata J, Kawasaki Y, Tano Y. Effect of a nitric oxide synthase inhibitor on lens-induced myopia. *Ophthalmic Res.* 2001;33(2):75–79.
79. Wu J, Liu Q, Yang X, Yang H, Wang XM, Zeng JW. Time-course of changes to nitric oxide signaling pathways in form-deprivation myopia in guinea pigs. *Brain Res.* 2007;1186:155–163.
80. Wu J, Liu Q, Yang X, Yang H, Wang XM, Zeng JW. Changes of nitric oxide synthase and cyclic guanosine mono-phosphate in form deprivation myopia in guinea pigs. *Chin Med J (Engl).* 2007;120(24):2238–2244.
81. Dowling JE, Ehinger B. The interplexiform cell system. I. Synapses of the dopaminergic neurons of the goldfish retina. *Proc R Soc Lond B Biol Sci.* 1978;201(1142):7–26.
82. Witkovsky P. Dopamine and retinal function. *Doc Ophthalmol.* 2004;108(1):17–40.
83. Nir I, Haque R, Iuvone PM. Diurnal metabolism of dopamine in the mouse retina. *Brain Res.* 2000;870(1–2):118–125.
84. Pendrak K, Nguyen T, Lin T, Capehart C, Zhu X, Stone RA. Retinal dopamine in the recovery from experimental myopia. *Curr Eye Res.* 1997;16(2):152–157.
85. Rohrer B, Iuvone PM, Stell WK. Stimulation of dopaminergic amacrine cells by stroboscopic illumination or fibroblast growth factor (bFGF, FGF-2) injections: possible roles in prevention of form-deprivation myopia in the chick. *Brain Res.* 1995;686(2):169–181.
86. Iuvone PM, Tigges M, Fernandes A, Tigges J. Dopamine synthesis and metabolism in rhesus monkey retina: development, aging, and the effects of monocular visual deprivation. *Vis Neurosci.* 1989;2(5):465–471.
87. Rohrer B, Spira AW, Stell WK. Apomorphine blocks form-deprivation myopia in chickens by a dopamine D2-receptor mechanism acting in retina or pigmented epithelium. *Vis Neurosci.* 1993;10(3):447–453.
88. McCarthy CS, Megaw P, Devadas M, Morgan IG. Dopaminergic agents affect the ability of brief periods of normal vision to prevent form-deprivation myopia. *Exp Eye Res.* 2007;84(1):100–107.
89. Nickla DL, Totonelly K. Dopamine antagonists and brief vision distinguish lens-induced- and form-deprivation-induced myopia. *Exp Eye Res.* 2011;93(5):782–785.
90. Cohen Y, Peleg E, Belkin M, Polat U, Solomon AS. Ambient illuminance, retinal dopamine release and refractive development in chicks. *Exp Eye Res.* 2012;103:33–40.
91. Ashby RS, Schaeffel F. The effect of bright light on lens compensation in chicks. *Invest Ophthalmol Vis Sci.* 2010;51(10):5247–5253.
92. Jiang L, Long K, Schaeffel F, et al. Effects of dopaminergic agents on progression of naturally occurring myopia in albino guinea pigs (Cavia porcellus). *Invest Ophthalmol Vis Sci.* 2014;55(11):7508–7519.
93. Wu XH, Li YY, Zhang PP, et al. Unaltered retinal dopamine levels in a C57BL/6 mouse model of form-deprivation myopia. *Invest Ophthalmol Vis Sci.* 2015;56(2):967–977.
94. Liang H, Crewther DP, Crewther SG, Barila AM. A role for photoreceptor outer segments in the induction of deprivation myopia. *Vis Res.* 1995;35(9):1217–1225.
95. Beresford JA, Crewther SG, Crewther DP. Anatomical correlates of experimentally induced myopia. *Aust N Z J Ophthalmol.* 1998;26(suppl 1):S84–S87.

96. Basinger S, Hoffman R, Matthes M. Photoreceptor shedding is initiated by light in the frog retina. *Science*. 1976;194(4269):1074–1076.
97. Currie JR, Hollyfield JG, Rayborn ME. Rod outer segments elongate in constant light: darkness is required for normal shedding. *Vis Res*. 1978;18(8):995–1003.
98. Fujikado T, Kawasaki Y, Suzuki A, Ohmi G, Tano Y. Retinal function with lens-induced myopia compared with form-deprivation myopia in chicks. *Graefes Arch Clin Exp Ophthalmol*. 1997;235(5):320–324.
99. Nevin ST, Schmid KL, Wildsoet CF. Sharp vision: a prerequisite for compensation to myopic defocus in the chick? *Curr Eye Res*. 1998;17(3):322–331.
100. Lauber JK, Kinnear A. Eye enlargement in birds induced by dim light. *Can J Ophthalmol*. 1979;14(4):265–269.
101. McBrien NA, Moghaddam HO, New R, Williams LR. Experimental myopia in a diurnal mammal (Sciurus carolinensis) with no accommodative ability. *J Physiol*. 1993;469:427–441.
102. Nathan J, Crewther SG, Crewther DP, Kiely PM. Effects of retinal image degradation on ocular growth in cats. *Invest Ophthalmol Vis Sci*. 1984;25(11):1300–1306.
103. Smith 3rd EL, Ramamirtham R, Qiao-Grider Y, et al. Effects of foveal ablation on emmetropization and form-deprivation myopia. *Invest Ophthalmol Vis Sci*. 2007;48(9):3914–3922.
104. Smith 3rd EL, Hung LF, Huang J. Relative peripheral hyperopic defocus alters central refractive development in infant monkeys. *Vis Res*. 2009;49(19):2386–2392.
105. Pittler SJ, Keeler CE, Sidman RL, Baehr W. PCR analysis of DNA from 70-year-old sections of rodless retina demonstrates identity with the mouse rd defect. *Proc Natl Acad Sci USA*. 1993;90(20):9616–9619.
106. Carter-Dawson LD, LaVail MM, Sidman RL. Differential effect of the rd mutation on rods and cones in the mouse retina. *Invest Ophthalmol Vis Sci*. 1978;17(6):489–498.
107. LaVail MM, Matthes MT, Yasumura D, Steinberg RH. Variability in rate of cone degeneration in the retinal degeneration (rd/rd) mouse. *Exp Eye Res*. 1997;65(1):45–50.
108. Chang B, Hawes NL, Hurd RE, Davisson MT, Nusinowitz S, Heckenlively JR. Retinal degeneration mutants in the mouse. *Vis Res*. 2002;42(4):517–525.
109. Gargini C, Terzibasi E, Mazzoni F, Strettoi E. Retinal organization in the retinal degeneration 10 (rd10) mutant mouse: a morphological and ERG study. *J Comp Neurol*. 2007;500(2):222–238.
110. Pittler SJ, Baehr W. Identification of a nonsense mutation in the rod photoreceptor cGMP phosphodiesterase beta-subunit gene of the rd mouse. *Proc Natl Acad Sci USA*. 1991;88(19):8322–8326.
111. Chang B, Hawes NL, Pardue MT, et al. Two mouse retinal degenerations caused by missense mutations in the beta-subunit of rod cGMP phosphodiesterase gene. *Vis Res*. 2007;47(5):624–633.
112. McLaughlin ME, Ehrhart TL, Berson EL, Dryja TP. Mutation spectrum of the gene encoding the beta subunit of rod phosphodiesterase among patients with autosomal recessive retinitis pigmentosa. *Proc Natl Acad Sci USA*. 1995;92(8):3249–3253.
113. McLaughlin ME, Sandberg MA, Berson EL, Dryja TP. Recessive mutations in the gene encoding the beta-subunit of rod phosphodiesterase in patients with retinitis pigmentosa. *Nat Genet*. 1993;4(2):130–134.
114. Nir I, Iuvone PM. Alterations in light-evoked dopamine metabolism in dystrophic retinas of mutant rds mice. *Brain Res*. 1994;649(1–2):85–94.
115. Hankins M, Ikeda H. Early abnormalities of retinal dopamine pathways in rats with hereditary retinal dystrophy. *Doc Ophthalmol*. 1994;86(3):325–334.
116. Miyake Y, Yagasaki K, Horiguchi M, Kawase Y, Kanda T. Congenital stationary night blindness with negative electroretinogram. A new classification. *Arch Ophthalmol*. 1986;104(7):1013–1020.

117. Calvert PD, Krasnoperova NV, Lyubarsky AL, et al. Phototransduction in transgenic mice after targeted deletion of the rod transducin alpha-subunit. *Proc Natl Acad Sci USA*. 2000;97(25):13913–13918.
118. Chang B, Dacey MS, Hawes NL, et al. Cone photoreceptor function loss-3, a novel mouse model of achromatopsia due to a mutation in Gnat2. *Invest Ophthalmol Vis Sci*. 2006;47(11):5017–5021.
119. Smith 3rd EL, Fox DA, Duncan GC. Refractive-error changes in kitten eyes produced by chronic on-channel blockade. *Vis Res*. 1991;31(5):833–844.
120. Pardue MT, McCall MA, LaVail MM, Gregg RG, Peachey NS. A naturally occurring mouse model of X-linked congenital stationary night blindness. *Invest Ophthalmol Vis Sci*. 1998;39(12):2443–2449.
121. Gregg RG, Mukhopadhyay S, Candille S, et al. Identification of the gene and the mutation responsible for the mouse nob phenotype. *Invest Ophthalmol Vis Sci*. 2003;44(1):378–384.
122. Morgans CW, Ren G, Akileswaran L. Localization of nyctalopin in the mammalian retina. *Eur J Neurosci*. 2006;23(5):1163–1171.
123. Schiller PH. The ON and OFF channels of the visual system. *Trends Neurosci*. 1992;15(3):86–92.
124. Schiller PH, Sandell JH, Maunsell JH. Functions of the ON and OFF channels of the visual system. *Nature*. 1986;322(6082):824–825.
125. Chow RL, Volgyi B, Szilard RK, et al. Control of late off-center cone bipolar cell differentiation and visual signaling by the homeobox gene Vsx1. *Proc Natl Acad Sci USA*. 2004;101(6):1754–1759.
126. Chow RL, Snow B, Novak J, et al. Vsx1, a rapidly evolving paired-like homeobox gene expressed in cone bipolar cells. *Mech Dev*. 2001;109(2):315–322.
127. Ohtoshi A, Wang SW, Maeda H, et al. Regulation of retinal cone bipolar cell differentiation and photopic vision by the CVC homeobox gene Vsx1. *Curr Biol*. 2004;14(6):530–536.
128. Cheng CW, Chow RL, Lebel M, et al. The Iroquois homeobox gene, Irx5, is required for retinal cone bipolar cell development. *Dev Biol*. 2005;287(1):48–60.
129. Schwahn HN, Schaeffel F. Chick eyes under cycloplegia compensate for spectacle lenses despite six-hydroxy dopamine treatment. *Invest Ophthalmol Vis Sci*. 1994;35(9):3516–3524.
130. Fischer AJ, Miethke P, Morgan IG, Stell WK. Cholinergic amacrine cells are not required for the progression and atropine-mediated suppression of form-deprivation myopia. *Brain Res*. 1998;794(1):48–60.
131. Wildsoet CF, Pettigrew JD. Kainic acid-induced eye enlargement in chickens: differential effects on anterior and posterior segments. *Invest Ophthalmol Vis Sci*. 1988;29(2):311–319.
132. Smith ML, Souza FG, Bruce KS, Strang CE, Morley BJ, Keyser KT. Acetylcholine receptors in the retinas of the alpha7 nicotinic acetylcholine receptor knockout mouse. *Mol Vis*. 2014;20:1328–1356.
133. McBrien NA, Moghaddam HO, Cottriall CL, Leech EM, Cornell LM. The effects of blockade of retinal cell action potentials on ocular growth, emmetropization and form deprivation myopia in young chicks. *Vis Res*. 1995;35(9):1141–1152.
134. Norton TT, Essinger JA, McBrien NA. Lid-suture myopia in tree shrews with retinal ganglion cell blockade. *Vis Neurosci*. 1994;11(1):143–153.
135. Boatright JH, Gordon JR, Iuvone PM. Inhibition of endogenous dopamine release in amphibian retina by L-2-amino-4-phosphonobutyric acid (L-AP4) and trans-2-aminocyclopentane-1,3-dicarboxylate (ACPD). *Brain Res*. 1994;649(1–2):339–342.

CHAPTER SIXTEEN

Genetics of Refraction and Myopia

Qingjiong Zhang[1]
State Key Lab of Ophthalmology, Zhongshan Ophthalmic Center, Sun Yat-sen University, Guangzhou, PR China
[1]Corresponding author: e-mail address: zhangqji@mail.sysu.edu.cn

Contents

Abstract

Both genetic and environmental factors play roles in the development of refractive errors. Identification of genes involved in refractive errors may help in elucidating the underlying molecular mechanism related to both genetic defects and environmental pressure. Recent development of techniques for genome wide analysis provides unique opportunity in dissecting the genetic basis related to refractive errors. This chapter tries to give a brief overview on the recent progress of genetic study of refractive errors, especially myopia.

1. GENETIC CONTRIBUTION TO REFRACTION

Refractive error includes myopia, hyperopia, astigmatism, and presbyopia. It is the most common cause of reduced visual acuity. Both genetic and environmental factors are involved in the development of refractive error.[1–3] Family aggregation, pedigree analysis, and twin studies all support the importance of genetic factors. Investigation of large families with high myopia or high hyperopia demonstrated inheritance as autosomal-dominant, autosomal-recessive, or X-linked traits. Most common types of refractive error are more likely to be inherited as complex traits, while extreme forms

Progress in Molecular Biology and Translational Science, Volume 134
ISSN 1877-1173
http://dx.doi.org/10.1016/bs.pmbts.2015.05.007
© 2015 Elsevier Inc.
All rights reserved.

are more likely to result from monogenic defects, especially those of early onset or those accompanied by other ocular and/or systemic anomalies.

Identifying the molecular basis of refractive error helps us to understand the molecular mechanisms and subsequently helps to design rational therapeutic intervention for this condition. Genetic study of refractive error provides a unique opportunity to detect key molecules that may play important roles in the development of refractive error. Variants in these molecules may lead directly to refractive error. On the other hand, these molecules may serve as targets for refractive changes induced by environmental factors. Progress in genome research and in analyzing tools for genome variants greatly enhances the identification of genetic factors that may participate in refractive error.

Traditional genetic studies such as linkage analysis were mostly based on analysis of families with extreme forms of refractive error, such as high myopia and high hyperopia. Recently, genome-wide association studies (GWAS) have been used to analyze different form of refractive error. Exome sequencing is becoming a frequent tool in disclosing genetic factors responsible for extreme form of refractive error. So far, a number of genes have been identified to be candidates responsible for refractive error.

2. MOLECULAR GENETICS OF HYPEROPIA

Infants are usually born with physiologic hyperopia that decreases by emmetropization within the first few years of life. A small percentage of people may have high hyperopia with refraction above +5D.

GWAS have detected a number of loci associated with hyperopia, including two loci of genome-wide significant association around 15q14 and 8q12 that overlap with known loci for myopia.[4]

High hyperopia may present alone or associated with other diseases. The vast majority of cases of high hyperopia represent physiologic high hyperopia that is not associated with other ocular or systemic anomalies. However, pathologic high hyperopia may present as a part of a complex of eye or systemic abnormalities, such as nanophthalmos, microphthalmia, anterior segment malformations, Leber congenital amaurosis, Down's syndrome, and fragile X syndrome.[5–9] Pedigree analyses have demonstrated inheritance of hyperopia as autosomal-dominant and autosomal-recessive traits. Several loci for high hyperopia have been mapped, including NNO1 (OMIM 600165) at chromosome 11p for autosomal-dominant nanophthalmos,[6] *MFRP* of NNO2 (OMIM 606227) at chromosome 11q23.3 for autosomal-recessive nanophthalmos,[10] NNO3 (OMIM 611897) at chromosome 2q11-q14 for autosomal-dominant congenital simple microphthalmia,[11] *TMEM98* of NNO4 (OMIM 615949) at chromosome 17p12-q12 for autosomal-dominant nanophthalmos,[12,13] and *PRSS56* (OMIM 613858) at chromosome 2q37.1 for autosomal-recessive posterior microphthalmos.[14,15] Families with physiologic high hyperopia are not uncommon, but the loci or genes responsible for them are yet to be identified.

3. MOLECULAR GENETICS OF MYOPIA

Genetic factors for both common myopia and high myopia have been extensively studied in recent years. A great number of loci have been reported for either common myopia or high myopia. However, the exact genetic variants contributing to myopia development are still unknown for the vast majority of myopia, both common and high myopia.

3.1 Genetic Loci for Common Myopia

Association studies on candidate genes alone without previous GWAS evidence, whether on common myopia or high myopia, will not be discussed here because: (1) most such studies are based on either a limited number of cases with low power or with a less significant level of association and (2) as a result, some of them have resulted in controversial conclusions.

A number of GWAS have been reported to disclose the genetic factors predisposing to refractive error or common myopia.[4,16–20] These studies identified more than 20 novel loci for refractive error and common myopia. Of these, the most striking findings were from two large-scale GWASs of myopia loci either based on the analysis of refractive error[18] or based on the age of myopia onset.[19] One study that identified 24 new loci is a GWAS meta-analysis involving 37,382 individuals from 27 studies of European ancestry and 8376 individuals from 5 Asian cohorts.[18] The other study, which detected 22 loci, is a GWAS of 45,771 participants of European ancestry in the U.S. population.[19] Significant association ($p < 5 \times 10^{-8}$) was identified between refractive error or age of myopia onset and single nucleotide polymorphisms (SNPs) near the following 12 genes in both of these two studies: *PRSS56* (OMIM 613858), *BMP3* (OMIM 112263), *KCNQ5* (OMIM 607357), *LAMA2* (OMIM 156225), *TOX* (OMIM 606863), *TJP2* (OMIM 607709), *RDH*5 (OMIM 601617), *ZIC2* (OMIM 603073), *RASGRF1* (OMIM 606600), *GJD2* (OMIM 607058), *RBFOX1* (OMIM 605104), and *SHISA6* (HGNC 34491). These studies provide firm evidence of the polygenic nature of refractive error and common myopia.[21] However, variants in these loci may only explain a small fraction of refractive error, perhaps as low as 3.4%.[21] Most alleles with significant association are of high frequency, suggesting that individual alleles influencing refractive error have small effects.[19] It might be interesting to know if rare variants with larger effect sizes might be present in or near these genes. Recently, whole-exome sequencing on 298 probands with early onset high myopia did not find evidence to support the association of myopia with rare variants in these genes.[22] So far, the exact molecular basis of genetic factors involving in common myopia waits to be identified, although a number of significant loci have been confirmed.

3.2 Genetic Loci and Genes for High Myopia

3.2.1 GWAS Study of High Myopia

SNPs in several loci or genes have been detected to associate with high myopia by GWAS, including 11q24.1,[23] *CTNND2* (OMIM 604275),[24] 4q25,[25] 13q12.12,[26] 1q41,[27] *ZFHX1B* (OMIM 605802),[28] and *VIPR2* (OMIM 601970) and *SNTB1* (OMIM 600026).[29] Two of these loci, 4q25 and *VIPR2*, were located in two myopia loci [MYP11 (OMIM 609994) and MYP4, respectively] that have been mapped based on linkage analysis of families with high myopia.[30,31] However, MYP4 has been retracted and replaced by MYP17 (OMIM 608367).[32] Limited replication studies of some of these loci provide additional supportive evidence. Unexpectedly, neither of these loci has been replicated in other GWAS of high myopia or in that of common myopia.[21] These studies may need further replication and the molecular basis underlying such association needs to be further clarified.

3.2.2 Loci or Genes for Mendelian High Myopia

A proportion of high myopia is transmitted as a monogenic disease with Mendelian inheritance, especially in those individuals in whom high myopia presented in early childhood without significant environmental effects of close work. Mendelian high myopia may be transmitted as an autosomal-dominant, autosomal-recessive, or X-linked trait. These types of myopia may occur alone or may consistently be accompanied with other ocular or systemic diseases. Linkage studies, candidate gene analysis and exome sequencing have provided a series of firm evidence of monogenic involvement. However, genetic association studies on these types of high myopia will not be discussed here.

3.2.2.1 Nonsyndromic High Myopia

Families with nonsyndromic high myopia are seen fairly frequently in the ophthalmological clinic. A number of genetic loci and causative genes have been identified for this type of high myopia on the basis of linkage analysis, exome sequencing, or candidate gene analysis.

There are 11 novel loci for which the genes have not been determined, including: (1) nine loci for autosomal-dominant high myopia [MYP2 (OMIM 160700) on chromosome 18p11.31,[33] MYP3 (OMIM 603221) on chromosome 12q,[34] MYP5 (OMIM 608474) on chromosome 17q21-q22,[35] MYP11 (OMIM 609994) on chromosome 4q22-q27,[30] MYP12 (OMIM 609995) on chromosome 2q37.1,[36] MYP15 (OMIM 612717) on chromosome 10q21.1,[37] MYP16 (OMIM 612554) on chromosome 5p15.33-p15.2,[38] MYP17 (OMIM 608367) on chromosome 7p15,[32] MYP19 (OMIM 613969) on chromosome 5p15.1-p13.3,[39] and MYP20 (OMIM 614166) on chromosome 13q12.12[26]]; (2) one locus for autosomal-recessive high myopia [MYP18 (OMIM 255500) on chromosome 14q21-q24][40]; and (3) one locus for X-linked high myopia [MYP13 (OMIM 300613) on Xq23-q25].[41,42]

Mutations in six genes have been reported to cause nonsyndromic high myopia, including *SCO2* at MYP6 (OMIM 602474, chromosome 22q12),[43] *ZNF644* at MYP21 (OMIM 614159, chromosome 1p22.2),[44] *LRPAP1* at MYP23 (OMIM 104225, chromosome 4p16),[45] *SLC39A5* at MYP24 (OMIM 608730, chromosome 12q13),[46] *LEPREL1* (OMIM 610341, chromosome 3q28),[47,48] and *CTSH* (OMIM 116820, chromosome 15q25.1).[45] Of the six genes, mutations in *SCO2*, *ZNF644*, and *SLC39A5* are associated with autosomal-dominant high myopia, while mutations in *LRPAP1*, *LEPREL1*, and *CTSH* are associated with autosomal-recessive high myopia. Analysis of all six genes based on whole-exome sequencing data only identified 9 candidate variants in 9 of 298 probands with early onset high myopia,[49] suggesting that the genetic defects responsible for most cases of high myopia remain to be determined.

In addition, variants in several other genes have been analyzed in patients with high myopia. Although mutations in these genes have been identified in a few cases with high myopia, variants in these genes are not consistently associated with high myopia. Thus, further comprehensive study is required before claiming the association of such mutations with high myopia, and these genes will not be described here in detail.

3.2.2.2 Syndromic High Myopia

High myopia has been consistently observed as an accompany signs in certain ocular or systemic diseases for many of which the causative genes are well known. Ocular or systemic diseases with which high myopia is always observed include: (1) retinitis pigmentosa caused by mutations in *RP2* (OMIM 300757)[50–52] and *RPGR* (OMIM 312610)[53–56]; (2) congenital night blindness caused by mutations in *NYX* (OMIM 300278),[57] *CACNA1F* (OMIM 300110),[58,59] *GRM6* (OMIM 604096),[60,61] and *LRIT3* (OMIM 615004)[62]; (3) Stickler syndrome associated with mutations in *COL2A1* (OMIM 120140)[63–65] or *COL11A1* (OMIM 120280)[66–68]; (4) Marfan syndrome with mutations in *FBN1* (OMIM 134797)[69,70]; (5) Weill–Marchesani syndrome with mutations in *ADAMTS10* (OMIM 608990)[71,72] and *ADAMTS17* (OMIM 607511)[71]; (6) Knobloch syndrome with mutations in *COL18A1* (OMIM 120328)[73,74]; (7) the syndrome of microcornea, myopic chorioretinal atrophy, and telecanthus with mutations in *ADAMTS18* (OMIM 607512)[75]; and (8) Cohen syndrome with mutations in *VPS13B* (OMIM 607817).[76,77] Interestingly, mutations in *NYX* have also been reported to cause high myopia without night blindness.[78,79] This opens the possibility that additional mutations in other genes associated with syndromic high myopia may potentially be candidates for nonsyndromic high myopia. This idea has been further extended in MYP1 (OMIM 310460).

MYP1, located in Xq28, was the first locus identified for high myopia. The original family mapped to this locus demonstrated syndromic high myopia, characterized by high myopia, cone dysfunction, and color vision deficiency.[80,81] Several additional families with similar phenotypes were identified subsequently.[82–84] A large family with nonsyndromic high myopia has been mapped to MYP1.[85] Recently, unique haplotypes (LIAVA and LVAVA) in *OPN1LW* (OMIM 300822) have been reported to be responsible for syndromic high myopia mapped to MYP1.[84] Our recent study demonstrates that unique variants in *OPN1LW* are also associated with nonsyndromic high myopia.

4. HUMAN VARIANTS IN GENES FOR EXPERIMENTAL MYOPIA

Reduced or increased expression of many genes has been implicated in experimental myopia in the chicken, rat, and monkey.[86,87] Of them, alteration of ZENK expression (human *EGR1* OMIM 128990) consistently has been shown to be involved in ocular growth and refraction. Upregulation of ZENK in retinal glucagon amacrine cells is assumed to create a STOP signal to inhibit axial eye growth, whereas downregulation of ZENK is associated with axial eye growth in animals.[88–90] Egr1 knockout mice had longer eyes and a relative myopic shift in refraction.[91] However, there is no evidence showing association of human myopia with *ERG1* variants. Similarly, whether most genes implicated in experimental myopia in animal studies have a role in human myopia are still uncertain.

5. SUMMARY

The study of experimental myopia study in the past decades has enriched our understanding of the environmental effects on development of myopia as well as of the interaction of gene expression in response to environmental changes. It has demonstrated that genetic changes actively or passively underlie the molecular basis of myopia development. However, how genes responsible for experimental myopia play roles in human myopia remains largely unknown. The development of techniques for whole-genome analysis provides an opportunity to investigate the interaction of genetic variants with human myopia systematically, and great progress in identification of genes responsible for human myopia has been made in recent years. These successes open windows for dissecting the molecular mechanisms of myopia in human beings. However, the exact genetic factors responsible for most myopia are largely unknown at present. It is expected that great number of new myopia genes and their functional networks will be identified in the near future, especially with the increasing application of new techniques such as whole-exome sequencing, whole-genome sequencing, whole-transcriptome sequencing, and proteome sequencing.

REFERENCES

1. Stambolian D. Genetic susceptibility and mechanisms for refractive error. *Clin Genet.* 2013;84(2):102–108.
2. Wojciechowski R. Nature and nurture: the complex genetics of myopia and refractive error. *Clin Genet.* 2011;79(4):301–320.
3. Young TL, Metlapally R, Shay AE. Complex trait genetics of refractive error. *Arch Ophthalmol.* 2007;125(1):38–48.
4. Simpson CL, Wojciechowski R, Oexle K, et al. Genome-wide meta-analysis of myopia and hyperopia provides evidence for replication of 11 loci. *PLoS One.* 2014;9(9): e107110.
5. Waring 3rd GO, Rodrigues MM, Laibson PR. Anterior chamber cleavage syndrome. A stepladder classification. *Surv Ophthalmol.* 1975;20(1):3–27.
6. Othman MI, Sullivan SA, Skuta GL, et al. Autosomal dominant nanophthalmos (NNO1) with high hyperopia and angle-closure glaucoma maps to chromosome 11. *Am J Hum Genet.* 1998;63(5):1411–1418.
7. Foxman SG, Heckenlively JR, Bateman JB, Wirtschafter JD. Classification of congenital and early onset retinitis pigmentosa. *Arch Ophthalmol.* 1985;103(10):1502–1506.
8. Maino DM, Schlange D, Maino JH, Caden B. Ocular anomalies in fragile X syndrome. *J Am Optom Assoc.* 1990;61(4):316–323.
9. Catalano RA. Down syndrome. *Surv Ophthalmol.* 1990;34(5):385–398.

10. Sundin OH, Leppert GS, Silva ED, et al. Extreme hyperopia is the result of null mutations in MFRP, which encodes a Frizzled-related protein. *Proc Natl Acad Sci USA*. 2005;102(27):9553–9558.
11. Li H, Wang JX, Wang CY, et al. Localization of a novel gene for congenital nonsyndromic simple microphthalmia to chromosome 2q11-14. *Hum Genet*. 2008;122(6):589–593.
12. Awadalla MS, Burdon KP, Souzeau E, et al. Mutation in TMEM98 in a large white kindred with autosomal dominant nanophthalmos linked to 17p12-q12. *JAMA Ophthalmol*. 2014;132(8):970–977.
13. Sun W, Zhang Q. Does the association between TMEM98 and nanophthalmos require further confirmation? *JAMA Ophthalmol*. 2014;133(3):358–359.
14. Nair KS, Hmani-Aifa M, Ali Z, et al. Alteration of the serine protease PRSS56 causes angle-closure glaucoma in mice and posterior microphthalmia in humans and mice. *Nat Genet*. 2011;43(6):579–584.
15. Gal A, Rau I, El Matri L, et al. Autosomal-recessive posterior microphthalmos is caused by mutations in PRSS56, a gene encoding a trypsin-like serine protease. *Am J Hum Genet*. 2011;88(3):382–390.
16. Hysi PG, Young TL, Mackey DA, et al. A genome-wide association study for myopia and refractive error identifies a susceptibility locus at 15q25. *Nat Genet*. 2010;42(10): 902–905.
17. Solouki AM, Verhoeven VJ, van Duijn CM, et al. A genome-wide association study identifies a susceptibility locus for refractive errors and myopia at 15q14. *Nat Genet*. 2010;42(10):897–901.
18. Verhoeven VJ, Hysi PG, Wojciechowski R, et al. Genome-wide meta-analyses of multiancestry cohorts identify multiple new susceptibility loci for refractive error and myopia. *Nat Genet*. 2013;45(3):314–318.
19. Kiefer AK, Tung JY, Do CB, et al. Genome-wide analysis points to roles for extracellular matrix remodeling, the visual cycle, and neuronal development in myopia. *PLoS Genet*. 2013;9(2):e1003299.
20. Cheng CY, Schache M, Ikram MK, et al. Nine loci for ocular axial length identified through genome-wide association studies, including shared loci with refractive error. *Am J Hum Genet*. 2013;93(2):264–277.
21. Hysi PG, Wojciechowski R, Rahi JS, Hammond CJ. Genome-wide association studies of refractive error and myopia, lessons learned, and implications for the future. *Invest Ophthalmol Vis Sci*. 2014;55(5):3344–3351.
22. Li J, Jiang D, Xiao X, et al. Evaluation of 12 myopia-associated genes in Chinese patients with high myopia. *Invest Ophthalmol Vis Sci*. 2015;56(2):722–729.
23. Nakanishi H, Yamada R, Gotoh N, et al. A genome-wide association analysis identified a novel susceptible locus for pathological myopia at 11q24.1. *PLoS Genet*. 2009;5(9): e1000660.
24. Li YJ, Goh L, Khor CC, et al. Genome-wide association studies reveal genetic variants in CTNND2 for high myopia in Singapore Chinese. *Ophthalmology*. 2011;118(2):368–375.
25. Li Z, Qu J, Xu X, et al. A genome-wide association study reveals association between common variants in an intergenic region of 4q25 and high-grade myopia in the Chinese Han population. *Hum Mol Genet*. 2011;20(14):2861–2868.
26. Shi Y, Qu J, Zhang D, et al. Genetic variants at 13q12.12 are associated with high myopia in the Han Chinese population. *Am J Hum Genet*. 2011;88(6):805–813.
27. Fan Q, Barathi VA, Cheng CY, et al. Genetic variants on chromosome 1q41 influence ocular axial length and high myopia. *PLoS Genet*. 2012;8(6):e1002753.
28. Khor CC, Miyake M, Chen LJ, et al. Genome-wide association study identifies ZFHX1B as a susceptibility locus for severe myopia. *Hum Mol Genet*. 2013;22(25): 5288–5294.

29. Shi Y, Gong B, Chen L, et al. A genome-wide meta-analysis identifies two novel loci associated with high myopia in the Han Chinese population. *Hum Mol Genet.* 2013;22(11):2325–2333.
30. Zhang Q, Guo X, Xiao X, Jia X, Li S, Hejtmancik JF. A new locus for autosomal dominant high myopia maps to 4q22-q27 between D4S1578 and D4S1612. *Mol Vis.* 2005;11:554–560.
31. Naiglin L, Gazagne C, Dallongeville F, et al. A genome wide scan for familial high myopia suggests a novel locus on chromosome 7q36. *J Med Genet.* 2002;39(2):118–124.
32. Paget S, Julia S, Vitezica ZG, Soler V, Malecaze F, Calvas P. Linkage analysis of high myopia susceptibility locus in 26 families. *Mol Vis.* 2008;14:2566–2574.
33. Young TL, Ronan SM, Drahozal LA, et al. Evidence that a locus for familial high myopia maps to chromosome 18p. *Am J Hum Genet.* 1998;63(1):109–119.
34. Young TL, Ronan SM, Alvear AB, et al. A second locus for familial high myopia maps to chromosome 12q. *Am J Hum Genet.* 1998;63(5):1419–1424.
35. Paluru P, Ronan SM, Heon E, et al. New locus for autosomal dominant high myopia maps to the long arm of chromosome 17. *Invest Ophthalmol Vis Sci.* 2003;44(5): 1830–1836.
36. Paluru PC, Nallasamy S, Devoto M, Rappaport EF, Young TL. Identification of a novel locus on 2q for autosomal dominant high-grade myopia. *Invest Ophthalmol Vis Sci.* 2005;46(7):2300–2307.
37. Nallasamy S, Paluru PC, Devoto M, Wasserman NF, Zhou J, Young TL. Genetic linkage study of high-grade myopia in a Hutterite population from South Dakota. *Mol Vis.* 2007;13:229–236.
38. Lam CY, Tam PO, Fan DS, et al. A genome-wide scan maps a novel high myopia locus to 5p15. *Invest Ophthalmol Vis Sci.* 2008;49(9):3768–3778.
39. Ma JH, Shen SH, Zhang GW, et al. Identification of a locus for autosomal dominant high myopia on chromosome 5p13.3-p15.1 in a Chinese family. *Mol Vis.* 2010;16: 2043–2054.
40. Yang Z, Xiao X, Li S, Zhang Q. Clinical and linkage study on a consanguineous Chinese family with autosomal recessive high myopia. *Mol Vis.* 2009;15:312–318.
41. Zhang Q, Guo X, Xiao X, Jia X, Li S, Hejtmancik JF. Novel locus for X linked recessive high myopia maps to Xq23-q25 but outside MYP1. *J Med Genet.* 2006;43(5):e20.
42. Zhang Q, Li S, Xiao X, Jia X, Guo X. Confirmation of a genetic locus for X-linked recessive high myopia outside MYP1. *J Hum Genet.* 2007;52(5):469–472.
43. Tran-Viet KN, Powell C, Barathi VA, et al. Mutations in SCO2 are associated with autosomal-dominant high-grade myopia. *Am J Hum Genet.* 2013;92(5):820–826.
44. Shi Y, Li Y, Zhang D, et al. Exome sequencing identifies ZNF644 mutations in high myopia. *PLoS Genet.* 2011;7(6):e1002084.
45. Aldahmesh MA, Khan AO, Alkuraya H, et al. Mutations in LRPAP1 are associated with severe myopia in humans. *Am J Hum Genet.* 2013;93(2):313–320.
46. Guo H, Jin X, Zhu T, et al. SLC39A5 mutations interfering with the BMP/TGF-beta pathway in non-syndromic high myopia. *J Med Genet.* 2014;51(8):518–525.
47. Mordechai S, Gradstein L, Pasanen A, et al. High myopia caused by a mutation in LEPREL1, encoding prolyl 3-hydroxylase 2. *Am J Hum Genet.* 2011;89(3):438–445.
48. Guo H, Tong P, Peng Y, et al. Homozygous loss-of-function mutation of the LEPREL1 gene causes severe non-syndromic high myopia with early-onset cataract. *Clin Genet.* 2014;86(6):575–579.
49. Jiang D, Li J, Xiao X, et al. Detection of mutations in LRPAP1, CTSH, LEPREL1, ZNF644, SLC39A5, and SCO2 in 298 families with early-onset high myopia by exome sequencing. *Invest Ophthalmol Vis Sci.* 2015;56(1):339–345.
50. Kaplan J, Pelet A, Martin C, et al. Phenotype-genotype correlations in X linked retinitis pigmentosa. *J Med Genet.* 1992;29(9):615–623.

51. Flaxel CJ, Jay M, Thiselton DL, et al. Difference between RP2 and RP3 phenotypes in X linked retinitis pigmentosa. *Br J Ophthalmol.* 1999;83(10):1144–1148.
52. Jayasundera T, Branham KE, Othman M, et al. RP2 phenotype and pathogenetic correlations in X-linked retinitis pigmentosa. *Arch Ophthalmol.* 2010;128(7):915–923.
53. Yokoyama A, Maruiwa F, Hayakawa M, et al. Three novel mutations of the RPGR gene exon ORF15 in three Japanese families with X-linked retinitis pigmentosa. *Am J Med Genet.* 2001;104(3):232–238.
54. Banin E, Mizrahi-Meissonnier L, Neis R, et al. A non-ancestral RPGR missense mutation in families with either recessive or semi-dominant X-linked retinitis pigmentosa. *Am J Med Genet A.* 2007;143A(11):1150–1158.
55. Prokisch H, Hartig M, Hellinger R, Meitinger T, Rosenberg T. A population-based epidemiological and genetic study of X-linked retinitis pigmentosa. *Invest Ophthalmol Vis Sci.* 2007;48(9):4012–4018.
56. Al-Maskari A, O'Grady A, Pal B, McKibbin M. Phenotypic progression in X-linked retinitis pigmentosa secondary to a novel mutation in the RPGR gene. *Eye (Lond).* 2009;23(3):519–521.
57. Xiao X, Jia X, Guo X, Li S, Yang Z, Zhang Q. CSNB1 in Chinese families associated with novel mutations in NYX. *J Hum Genet.* 2006;51(7):634–640.
58. Hemara-Wahanui A, Berjukow S, Hope CI, et al. A CACNA1F mutation identified in an X-linked retinal disorder shifts the voltage dependence of Cav1.4 channel activation. *Proc Natl Acad Sci USA.* 2005;102(21):7553–7558.
59. Vincent A, Wright T, Day MA, Westall CA, Heon E. A novel p.Gly603Arg mutation in CACNA1F causes Aland island eye disease and incomplete congenital stationary night blindness phenotypes in a family. *Mol Vis.* 2011;17:3262–3270.
60. Xu X, Li S, Xiao X, Wang P, Guo X, Zhang Q. Sequence variations of GRM6 in patients with high myopia. *Mol Vis.* 2009;15:2094–2100.
61. Sergouniotis PI, Robson AG, Li Z, et al. A phenotypic study of congenital stationary night blindness (CSNB) associated with mutations in the GRM6 gene. *Acta Ophthalmol.* 2012;90(3):e192–e197.
62. Zeitz C, Jacobson SG, Hamel CP, et al. Whole-exome sequencing identifies LRIT3 mutations as a cause of autosomal-recessive complete congenital stationary night blindness. *Am J Hum Genet.* 2013;92(1):67–75.
63. Ahmad NN, McDonald-McGinn DM, Zackai EH, et al. A second mutation in the type II procollagen gene (COL2AI) causing stickler syndrome (arthro-ophthalmopathy) is also a premature termination codon. *Am J Hum Genet.* 1993;52(1):39–45.
64. Wilkin DJ, Liberfarb R, Davis J, et al. Rapid determination of COL2A1 mutations in individuals with Stickler syndrome: analysis of potential premature termination codons. *Am J Med Genet.* 2000;94(2):141–148.
65. Liberfarb RM, Levy HP, Rose PS, et al. The Stickler syndrome: genotype/phenotype correlation in 10 families with Stickler syndrome resulting from seven mutations in the type II collagen gene locus COL2A1. *Genet Med.* 2003;5(1):21–27.
66. Zlotogora J, Sagi M, Schuper A, Leiba H, Merin S. Variability of Stickler syndrome. *Am J Med Genet.* 1992;42(3):337–339.
67. Annunen S, Korkko J, Czarny M, et al. Splicing mutations of 54-bp exons in the COL11A1 gene cause Marshall syndrome, but other mutations cause overlapping Marshall/Stickler phenotypes. *Am J Hum Genet.* 1999;65(4):974–983.
68. Richards AJ, McNinch A, Martin H, et al. Stickler syndrome and the vitreous phenotype: mutations in COL2A1 and COL11A1. *Hum Mutat.* 2010;31(6): E1461–E1471.
69. Montgomery RA, Geraghty MT, Bull E, et al. Multiple molecular mechanisms underlying subdiagnostic variants of Marfan syndrome. *Am J Hum Genet.* 1998;63(6): 1703–1711.

70. Dietz HC, McIntosh I, Sakai LY, et al. Four novel FBN1 mutations: significance for mutant transcript level and EGF-like domain calcium binding in the pathogenesis of Marfan syndrome. *Genomics*. 1993;17(2):468–475.
71. Morales J, Al-Sharif L, Khalil DS, et al. Homozygous mutations in ADAMTS10 and ADAMTS17 cause lenticular myopia, ectopia lentis, glaucoma, spherophakia, and short stature. *Am J Hum Genet.* 2009;85(5):558–568.
72. Li J, Jia X, Li S, Fang S, Guo X. Mutation survey of candidate genes in 40 Chinese patients with congenital ectopia lentis. *Mol Vis*. 2014;20:1017–1024.
73. Sertie AL, Sossi V, Camargo AA, Zatz M, Brahe C, Passos-Bueno MR. Collagen XVIII, containing an endogenous inhibitor of angiogenesis and tumor growth, plays a critical role in the maintenance of retinal structure and in neural tube closure (Knobloch syndrome). *Hum Mol Genet.* 2000;9(13):2051–2058.
74. Suzuki OT, Sertie AL, Der Kaloustian VM, et al. Molecular analysis of collagen XVIII reveals novel mutations, presence of a third isoform, and possible genetic heterogeneity in Knobloch syndrome. *Am J Hum Genet.* 2002;71(6):1320–1329.
75. Aldahmesh MA, Alshammari MJ, Khan AO, Mohamed JY, Alhabib FA, Alkuraya FS. The syndrome of microcornea, myopic chorioretinal atrophy, and telecanthus (MMCAT) is caused by mutations in ADAMTS18. *Hum Mutat.* 2013;34(9):1195–1199.
76. Kolehmainen J, Wilkinson R, Lehesjoki AE, et al. Delineation of Cohen syndrome following a large-scale genotype-phenotype screen. *Am J Hum Genet.* 2004;75(1):122–127.
77. Hennies HC, Rauch A, Seifert W, et al. Allelic heterogeneity in the COH1 gene explains clinical variability in Cohen syndrome. *Am J Hum Genet.* 2004;75(1):138–145.
78. Zhang Q, Xiao X, Li S, et al. Mutations in NYX of individuals with high myopia, but without night blindness. *Mol Vis*. 2007;13:330–336.
79. Yip SP, Li CC, Yiu WC, et al. A novel missense mutation in the NYX gene associated with high myopia. *Ophthalmic Physiol Opt.* 2013;33(3):346–353.
80. Haim M, Fledelius HC, Skarsholm. X-linked myopia in Danish family. *Acta Ophthalmol (Copenh)*. 1988;66(4):450–456.
81. Schwartz M, Haim M, Skarsholm D. X-linked myopia: Bornholm eye disease. Linkage to DNA markers on the distal part of Xq. *Clin Genet.* 1990;38(4):281–286.
82. Young TL, Deeb SS, Ronan SM, et al. X-linked high myopia associated with cone dysfunction. *Arch Ophthalmol.* 2004;122(6):897–908.
83. Michaelides M, Johnson S, Bradshaw K, et al. X-linked cone dysfunction syndrome with myopia and protanopia. *Ophthalmology*. 2005;112(8):1448–1454.
84. McClements M, Davies WI, Michaelides M, et al. Variations in opsin coding sequences cause x-linked cone dysfunction syndrome with myopia and dichromacy. *Invest Ophthalmol Vis Sci*. 2013;54(2):1361–1369.
85. Guo X, Xiao X, Li S, Wang P, Jia X, Zhang Q. Nonsyndromic high myopia in a Chinese family mapped to MYP1: linkage confirmation and phenotypic characterization. *Arch Ophthalmol.* 2010;128(11):1473–1479.
86. Stone RA, Khurana TS. Gene profiling in experimental models of eye growth: clues to myopia pathogenesis. *Vision Res*. 2010;50(23):2322–2333.
87. Smith 3rd EL, Hung LF, Arumugam B. Visual regulation of refractive development: insights from animal studies. *Eye (Lond)*. 2014;28(2):180–188.
88. Fischer AJ, McGuire JJ, Schaeffel F, Stell WK. Light- and focus-dependent expression of the transcription factor ZENK in the chick retina. *Nat Neurosci*. 1999;2(8):706–712.
89. Wildsoet CF, Wong RO. A far-sighted view of myopia. *Nat Med*. 1999;5(8):879–880.
90. Ashby R, McCarthy CS, Maleszka R, Megaw P, Morgan IG. A muscarinic cholinergic antagonist and a dopamine agonist rapidly increase ZENK mRNA expression in the form-deprived chicken retina. *Exp Eye Res*. 2007;85(1):15–22.
91. Schippert R, Burkhardt E, Feldkaemper M, Schaeffel F. Relative axial myopia in Egr-1 (ZENK) knockout mice. *Invest Ophthalmol Vis Sci*. 2007;48(1):11–17.

SECTION IV

Ocular Immunity

CHAPTER SEVENTEEN

Molecular Genetic Advances in Uveitis

Shengping Hou*, Aize Kijlstra†, Peizeng Yang*,[1]
*The First Affiliated Hospital of Chongqing Medical University, Chongqing Key Lab of Ophthalmology, Chongqing Eye Institute, Chongqing, PR China
†University Eye Clinic Maastricht, Maastricht, The Netherlands
[1]Corresponding author: e-mail address: peizengycmu@126.com

Contents

Abstract

Uveitis is usually considered as an intraocular inflammation characterized by variety of clinical features. Behcet's disease (BD), Vogt–Koyanagi–Harada (VKH) syndrome, acute anterior uveitis (AAU), and birdshot chorioretinopathy (BCR) are examples of non-infectious forms of uveitis. Although the precise pathogenesis remains unclear, accumulating evidence shows that complex genetic backgrounds coupled with an aberrant immune response may be implicated in the development of uveitis. The complement and pattern recognition systems are both important factors of the innate immune system and are involved in the pathogenesis of uveitis. Copy number variants (CNVs) of complement component 4 have been found to be associated with BD and VKH syndrome, but not with AAU. Several CNVs and gene polymorphisms of toll-like receptors were found to be associated with BD. Leukocytes are an important part of the adaptive immune system and various molecules on these cells play an important role in the development of uveitis. Genes encoding for human leukocyte antigens (HLAs) have been shown to be associated with certain uveitis entities, including BD (*HLA-B51*), VKH syndrome (*HLA-DR4, DRB1/DQA1*), AAU (*HLA-B27*), and BCR (*HLA-A29*). Genome wide association studies showed that the IL-23R locus was a shared risk factor for

Progress in Molecular Biology and Translational Science, Volume 134
ISSN 1877-1173
http://dx.doi.org/10.1016/bs.pmbts.2015.04.009
© 2015 Elsevier Inc.
All rights reserved.

multiple uveitis entities including BD, AAU, and VKH syndrome. In addition, various other non-HLA genes are also associated with BD or VKH syndrome, such as IL-10, STAT4, STAT3, and UBAC2. These studies support the hypothesis that genetic factors play a key role in the pathogenesis of uveitis.

ABBREVIATIONS

AAU acute anterior uveitis
BCR birdshot chorioretinopathy
BD Behcet's disease
C4 complement component 4
CCR1 C–C chemokine receptor type 1
CNVs copy number variants
GWASs genome-wide association studies
HLAs human leukocyte antigens
IL-6 interleukin-6
OPN osteopontin
PBMCs peripheral blood mononuclear cells
SNP single nucleotide polymorphism
STAT signal transducer and activator of transcription
TLRs toll-like receptors
VKH Vogt–Koyanagi–Harada

1. INTRODUCTION

Uveitis is defined as inflammation of the uveal tract, which is classically composed of the iris, ciliary body, and choroid, and is an important cause of blindness of young and middle-aged individuals in the world. It is important to note that uveitis usually causes permanent and irreversible blindness, since the inflammatory response that damages the retina and optic nerve cannot be repaired by the currently available medical technology. Epidemiological studies show that uveitis accounts for 10–15% of the blindness in the Western world.[1]

Uveitis can be caused by infectious and noninfectious mechanisms, and the latter entities are often associated with extraocular involvement. Behcet's disease (BD) is a chronic multisystemic inflammatory disorder characterized by nongranulomatous uveitis, genital ulcers, recurrent oral ulcers, and skin lesions. This disease is commonly seen along the ancient Silk Road,[2] from Asia to the Mediterranean basin and includes countries such as China, Turkey,

Korea, and Japan. Vogt–Koyanagi–Harada (VKH) syndrome on the other hand is an autoimmune disease characterized by a bilateral granulomatous panuveitis and systemic disorders including vitiligo, poliosis, alopecia, central nervous system, and auditory signs. VKH syndrome mainly affects certain pigmented races, such as Asians and Native Americans.[3,4] Acute anterior uveitis (AAU), the most common form of uveitis, is characterized by sudden-onset, self-limiting, and recurrent inflammation involving the iris and ciliary body.[5] It is considered as the most frequent extra-articular feature in seronegative arthritis including ankylosing spondylitis (AS). AAU can affect up to one-third of patients with spondyloarthritis.[6–8] Birdshot chorioretinopathy (BCR) is a noninfectious uveitis entity that is confined to the eye. It is a chronic, bilateral, and posterior uveitis with a distinctive clinical presentation and has an almost 100% genetic association with *HLA-A29*.[9]

Although the etiology and pathogenesis of BD, VKH syndrome, AAU, and BCR is not completely clear, the strong association with HLA and regional distribution of these diseases shows that immunogenetic predisposition is involved in the development of these diseases.

The immune response is generally divided into the innate and the adaptive immune system. Innate immunity is immediately mobilized upon infection and acts as the first line of defense against pathogen invasion. The adaptive immune system on the other hand serves as a second line of defense against reexposure to the same pathogen. The complement system and pattern recognition system are considered to be components of the innate immune response. Leukocytes such as T and B cells are important players of the adaptive immune system and participate in the development of uveitis.

HLA genes have been shown to be associated with BD (*HLA-B51*), VKH syndrome (*HLA-DR4, DRB1/DQA1*), AAU (*HLA-B27*), and BCR (*HLA-A29*). Recently, genome-wide association studies (GWASs) from Japan, Turkey, and China have shown that not only genes of the HLA system but that variants of many non-HLA genes such as interleukin-10 (*IL-10*), *IL-23R/IL-12RB2*, *STAT4* genes,[10–12] are also involved in uveitis, thereby supporting the hypothesis that genetic factors contribute to the pathogenesis of this disease. Interestingly, Hou *et al.*[13] also found that the IL-23R locus was associated with VKH syndrome in a Chinese Han population. Additionally, this locus also showed an association with AAU.[14] Other genetic variants and copy number variants (CNVs) in IL-17 pathway-related genes also showed an association with BD and VKH syndrome. The aforementioned evidence supports a strong genetic basis for uveitis entities such as BD, VKH syndrome, AAU, and BCR. This is confirmed by a strong association with immunogenetic loci, especially HLA antigens, repeatability in different ethnic populations, and the sharing of risk factors among various forms of uveitis. The aim of this chapter is to provide further details concerning the pathogenesis of uveitis entities such as BD, VKH syndrome, AAU, and BCR with an emphasis on new insights in the area of immunogenetics (genetic associations are summarized in Tables 1 and 2).

2. MOLECULAR GENETIC STUDIES ON THE GENES INVOLVED IN THE INNATE IMMUNE SYSTEM AND THEIR RELATION TO UVEITIS

Toll-like receptors (TLRs) are an important family of cellular receptors involved in the innate immune response. They are triggered by a variety of evolutionary conserved molecules from microbial pathogens. It has been proposed that the activity of these receptors may vary among genetic variants of TLRs and that certain immune disorders may show an association with these genetic polymorphisms.[68] Eight SNPs in *TLR2*, *TLR4*, *TLR8*, and *TLR9* were examined to test the association with BD, VKH syndrome, and AAU.[53] The results only showed association of genetic variants in TLR2 with BD. No association was observed with VKH syndrome and AAU, despite the fact that large patient samples were included in these studies. TLR2 mRNA expression in peripheral blood mononuclear cells (PBMCs) was increased in healthy individuals carrying the risk genotype of *TLR2* following stimulation with peptidoglycan.[53] *TLR4* SNPs were also shown to be associated with BD in another study.[69] The association of CNVs in several TLRs such as *TLR1*, *TLR2*, *TLR3*, *TLR5*, *TLR6*, *TLR7*, *TLR9*, and *TLR10* with BD, VKH syndrome, and AAU was also examined in a Chinese Han population,[70] and the results showed a significantly increased frequency of more than one copy of *TLR7* in male BD patients and more than two copies in female BD patients, although no association was seen with VKH syndrome or AAU. Functional studies showed that expression of *TLR7* mRNA was increased in PBMCs from individuals carrying a high copy number of *TLR7*.[70]

Receptors on NK cells such as KIR have been shown to be a susceptibility factor for VKH syndrome.[71] Additionally, the inhibitory receptor CD94/NKG2A and its haplotypes were found to be associated with BD.[72]

Table 1 Summary of the Genes Involved in Behcet's Disease

Genes	Odds Ratio	95% Confidence Interval	Ethnic Group	References
C4	2.84	1.90–4.24	Chinese	15
CCR1/CCR3	0.28	0.2–0.4	Chinese	16
CCR5	2.37	1.1–5.1	Italian	17
CD40	1.98	1.38–2.83	Chinese	18
CPVL	2.26	1.47–3.45	Turkish	19
eNOS	1.88	1.27–2.49	Turkish	20
	3.2	1.4–7.3	Korean	21
	1.26	2.13–3.62	Tunisian	22
ERAP1	4.56	2.88–7.22	Turkish	23
FCRL3	0.7	0.5–0.9	Chinese	24
ICAM1	1.26	2.13–3.62	Tunisian	25
	4.2	1.9–9.3	Italian	26
IL-1β	3.63	1.23–12.97	Turkish	27
IL-4	3.40	1.72–7.12	Turkish	28
IL-6	3.5	1.2–10.0	Korean	29
IL-10	1.20	1.02–1.40	Iran	30
	1.45	1.34–1.58	Turkish, Arab, Greek, UK, Korean, Japanese	11
	1.45	1.32–1.60	Japanese, Turkish, Korean	10
IL-12B	1.8	1.0–3.3	Japanese	31
IL-18	1.48	1.10–1.97	Turkish	32
IL-23R-IL-12RB2	1.51	1.27–1.78	Iran	30
	1.28	1.18–1.39	Turkish	11
	1.35	0.95–1.91	Japanese, Turkish, Korean	10
	1.86	1.39–2.49	Chinese	33
IRF-1	3.71	1.778–7.770	Korean	34
KIAA1529	2.04	1.45–2.88	Turkish	35
LOC100129342	1.84	1.32–2.58	Turkish	19

Continued

Table 1 Summary of the Genes Involved in Behcet's Disease—cont'd

Genes	Odds Ratio	95% Confidence Interval	Ethnic Group	References
MCP1	1.51	1.05–2.17	Chinese	36
MDR1	3.03	1.41–6.54	Turkish	37
MIF	1.46	1.19–1.79	Chinese	38
miR-146a	1.33	1.17–1.52	Chinese	39
MMP2	0.6	0.44–0.87	Korean	40
MMP9	0.371	0.152–0.905	Korean	41
MTHFR	1.70	1.23–2.35	Turkish	42
NRAMP1	1.88	1.21–2.93	Turkish	43
PDGFRL	0.59	0.49–0.72	Chinese	44
PTPN22	2.4	1.2–4.7	UK, Middle East	45
SLC11A1	0.60	0.37–0.95	Korean	46
STAT3	1.712	1.238–2.369	Chinese	47
STAT4	1.45	1.3–1.6	Chinese	12,48
	1.27	1.17–1.37	Turkish	23
SUMO4	23.40	2.33–235.54	Korean	49
	1.41	1.01–1.97	Tunisian	50
	1.7	1.3–2.2	Chinese	51
TGFBR3	0.617	0.441–0.863	Chinese	52
TLR2	1.47	1.26–1.71	Chinese	53
TLR4	1.96	1.26–3.26	Korean	54
	1.67	1.08–2.60	Japanese	55
TNF-α	1.68	1.10–2.56	Moroccan	56
	3.08	1.73–5.47	Iranian Azeri Turk	57
TNFAIP3	2.03	1.65–2.49	Chinese	58
UBAC2	1.5	1.2–1.7	Chinese	59
UBASH3B	1.71	1.23–2.38	Turkish	19
VDR	1.89	1.32–2.71	Tunisians	60
VEGF	0.10	0.011–0.875	Korean	61

Table 2 Summary of the Associated Genes with VKH Syndrome

Genes	Odd Ratio	95% Confidence Interval	Ethnic	References
CTLA4	0.762	0.631–0.922	Chinese	62
JAK1	0.71	0.61–0.83	Chinese	63
OPN	1.830	1.200–2.789	Chinese	64
IL-17F	1.52	1.13–2.06	Chinese	65
IL-23R-C1ORF141	1.82	1.60–2.05	Chinese	13
STAT4	1.777	1.126–2.807	Chinese	66
TNFAIP3	1.6	1.3–1.9	Chinese	67

3. MOLECULAR GENETICS STUDIES ON THE GENES INVOLVED IN THE ADAPTIVE IMMUNE SYSTEM AND UVEITIS

3.1 HLA Genes

HLA molecules play a crucial role in antigen presentation and genetic diversity is thought to increase species survival following microbial infection. Many diseases have been shown to be associated with certain HLA variants. Accumulating evidence has confirmed that *HLA-B51* was strongly associated with BD in different ethnic groups.[10,11,51,73,74] In addition, other genes within the HLA region such as MICA also showed a strong association with BD.[75,76] Several HLA genes including *HLA-DR4*, *HLA-DR53*, and *DRB1/DQA1* were shown to be strongly associated with VKH syndrome in a variety of ethnic groups.[13,77–79] HLA class I molecules *HLA-B27* and *HLA-A29* were found to be associated with AAU[80] and BCR,[81] respectively. The first GWAS in BCR revealed a strong genetic predisposition for the *ERAP2* gene. ERAP2 affects antigen processing for HLA class I, and this finding provides a novel mechanism explaining the genetic predisposition for this disease.[82]

3.2 Genetic Advances on Genes Involved in Th1 Cell Pathways

Previous studies showed that Th1 cells play an important role in the development of uveitis as evidenced by an increased expression of T-bet and IFN-γ in uveitis patients.[83] Several candidate association studies were performed based on the important role of Th1 cells in uveitis. Signal transducer and activator of transcription protein 4 (STAT4) is a critical factor for Th1 cell development. Multiple SNPs in the *STAT4* region were found to be associated with BD in a Chinese Han population.[12] Further study indicated that the rs897200 risk genotype in *STAT4* played a pathogenic role through upregulating the *STAT4* transcription and IL-17 production.[12] The association of *STAT4* with BD was recently confirmed in a Turkish population,[23] suggesting that *STAT4* is a shared risk factor for BD in different ethnic cohorts.

C–C chemokine receptor type 1 (CCR1) and CCR3 were found to be involved in the accumulation and activation of Th1 cells. Recent studies showed that the *CCR1/CCR3* gene was associated with BD in Chinese Han and Turkish populations.[16,23] Functional studies showed that carriers of the protective allele of *CCR1* have a higher expression of *CCR1* and an increased migration of monocytes,[23] suggesting that impaired clearance of pathogens may contribute to the pathogenesis of BD.

Cytotoxic T-lymphocyte antigen 4 (CTLA-4) mediates an inhibitory signal to T cells and results in decreased IL-2 production. In a recent study, Du *et al.*[62,84] investigated the association of *CTLA-4* gene polymorphisms with BD and VKH syndrome in a Chinese Han population. The results showed that the frequency of the G allele at the +49 site and the frequency of haplotype −1661A:−318C:+49G:CT60G contribute to the risk for VKH syndrome, but not for BD.[62,84]

Interleukins are important cytokines that can play a role in the development of uveitis. Li *et al.*[85] found that rs3212227 in the *IL-12B* gene was associated with two uveitis entities including VKH syndrome and BD. IL-18 is an important proinflammatory cytokine stimulating the production of IFN-γ in collaboration with IL-12 by Th1 cells and was found to be associated with BD in Turkish and Korean populations.[86]

TNF-α has been implicated in the pathogenesis of BD. Additionally, anti-TNF-α represents an important treatment modality for BD patients.[87] Meta-analyses confirmed an association of *TNF-α* polymorphisms with BD in various ethnic populations.[56,57,88]

3.3 Genetic Advances on the Genes Involved in Th17 Cell Pathways

Recent evidence supports the critical role of Th17 cells in mediating chronic inflammatory disorders such as BD and VKH syndrome[89,90] and suggests the involvement of genes related to the Th17 cell. A recent study identified the association of *IL-23R* genes with BD in a Chinese Han population.[33]

The results showed that two SNPs in *IL-23R* were associated with susceptibility to BD. GWASs also confirmed this association with BD in a variety of ethnic populations including Japanese, Turkish, and Iranian patients.[10,11,30]

CD40 is expressed on antigen-presenting cells and produces the second signal of cell activation through interaction with CD40L.[91] This interaction has been shown to play a role in various immune diseases.[92] Chen *et al.*[92] found that the TT genotype of rs4810485 or rs1883832 of *CD40* was associated with BD, but not with VKH syndrome.[93] The interaction of CD40 with CD40L is regulated by tumor necrosis factor receptor-associated factor (TRAF), and recent studies showed that three SNPs in *TRAF5* including rs6540679, rs12569232, and rs10863888, and rs13210247 in *TRAF3IP2* were associated with BD and VKH syndrome.[94] Additionally, increased expression of TRAF5 and increased production of TNF-α and IL-6 were found in individuals carrying the risk genotype of *TRAF5* rs6540679.[94]

Most recently, Hou *et al.*[13] performed a GWAS in a group of 1538 VKH patients and 5603 unaffected individuals and found an association between non-HLA genes including *IL-23R-C1ORF141* and *ADO/ZNF365/EGR2* with VKH syndrome. The five non-HLA genes were all found to be expressed in human iris tissue. *IL-23R* was also expressed in the ciliary body, and *EGR2* was expressed in the ciliary body and choroid. The risk G allele of rs117633859 in *IL-23R* exhibited low transcriptional activation and was associated with diminished *IL-23R* mRNA expression.[13]

Osteopontin (OPN, secreted phosphoprotein 1) may enhance T cell survival and proliferation, and is involved in the development of uveitis by promoting the responses of Th17 cells. Chu *et al.*[64] found that the serum level of OPN was increased in VKH patients and showed an association of *OPN* polymorphisms and its receptors with this disease. Gene variants of *STAT3*, which is also involved in Th17 cell function, have been shown as a risk factor for BD.[47] Additionally, the *JAK1* and *MCP-1* genes were also shown to be associated with BD in a Chinese Han population.[36,47,95] Other genes involved in Th17 pathways such as *IL-6* and *IL-1β* also showed an association with BD.[27,29] These studies suggest that genetic variants of cytokines associated with Th17 pathways may play an important role in the development of BD and VKH syndrome.

Polymorphisms of *IL-23R* and *ERAP1* reached genome-wide significance in a recent GWAS on 1711 patients with AAU and 10,000 control individuals.[14]

3.4 Genetic Advances on the Genes Involved in Treg Cell Pathways

MiR-146a is prevalently expressed in Treg cells and known as a negative regulator of innate immunity in a variety of immune diseases.[96] Zhou *et al.*[39] examined the association of polymorphisms in miR-146a with BD and VKH syndrome in a Chinese Han population and found that a polymorphism in this gene was associated with BD, but not with VKH syndrome.

Yu *et al.*[97] examined the association of 10 SNPs in genes involved in Treg cell pathways in patients with AAU and found significantly increased frequencies of the *FoxO1* SNP rs2297626 AA genotype and A allele in this disease.

4. THE ROLE OF CNVs IN UVEITIS

The complement system is an important element of the innate immune system and is involved in the pathogenesis of uveitis. Recent studies showed that total serum complement component 4 (C4) level was increased in patients with active BD as compared with controls, but not in AAU patients.[15] Copy number variation analysis identified a significant association between the presence of more than two copies of C4A with BD. While HLA-B51, which is located on the same chromosome as C4, also showed a strong association with BD in the Chinese Han population, logistic regression showed that the C4A copy number variation was an independent risk factor for BD from HLA-B51. A high copy number of C4 was positively associated with the expression level of C4A and the production of IL-6.[15] However, another study showed that the expression of total C4 in serum was significantly decreased in patients with VKH when compared with controls. CNV analysis showed a significantly decreased frequency of more than two copies of C4A or more than four copies of total C4 in VKH patients relative to controls. The association of C4 with VKH syndrome was also independent of HLA-DR4.[98] The different role of C4 in VKH versus BD may be due to the fact that BD is considered an autoinflammatory disease mediated by an aberrant response against microbial antigens, whereas VKH is an autoimmune disease directed against melanocyte antigens.[99,100]

Hou *et al.*[101] found that increased frequencies of having more than two copies of *IL-17F* and *IL-23A* were associated with BD or VKH syndrome. In contrast, CNVs in *IL-17A* and *IL-23R* were not found to be associated with BD or VKH syndrome. IL-17F protein levels were positively related with gene copy numbers of *IL-17F*, and individuals with high gene copy numbers of *IL-17F* show an enhanced *in vitro* proliferation of PBMCs.

5. SUMMARY

Uveitis is a complex disease and a collective term for a variety of different entities characterized by different clinical findings sometimes isolated to the eye but often also associated with inflammation of extraocular sites. Its pathogenesis is dependent on combination of a certain genetic background with an imbalance in the regulation of the immune response. The field of uveitis continues to advance, increasing numbers of genetic risk factors associated with uveitis will be identified, and these results will help to unravel the pathogenesis of this disease. However, most of the uveitis pathogenic genes have yet to be discovered, and genetic heterogeneity needs to be further validated in different populations. Novel methods and multiple-ethnic population studies are needed to elucidate the exact pathogenesis of uveitis. The understanding of novel pathogenic mechanisms of uveitis may provide a foundation to develop new strategies in the treatment of uveitis.

REFERENCES

1. Wakefield D, Chang JH. Epidemiology of uveitis. *Int Ophthalmol Clin*. 2005; 45(2):1–13.
2. Evereklioglu C. Current concepts in the etiology and treatment of Behcet disease. *Surv Ophthalmol*. 2005;50(4):297–350.
3. Murakami S, Inaba Y, Mochizuki M, Nakajima A, Urayama A. A nation-wide survey on the occurrence of Vogt-Koyanagi-Harada disease in Japan. *Nihon Ganka Gakkai Zasshi*. 1994;98(4):389–392.
4. Moorthy RS, Inomata H, Rao NA. Vogt-Koyanagi-Harada syndrome. *Surv Ophthalmol*. 1995;39(4):265–292.
5. Chang JH, Wakefield D. Uveitis: a global perspective. *Ocul Immunol Inflamm*. 2002; 10(4):263–279.
6. Gran JT, Husby G. Clinical, epidemiologic, and therapeutic aspects of ankylosing spondylitis. *Curr Opin Rheumatol*. 1998;10(4):292–298.
7. Wendling D. Uveitis in seronegative arthritis. *Curr Rheumatol Rep*. 2012; 14(5):402–408.
8. Ji SX, Yin XL, Yuan RD, Zheng Z, Huo Y, Zou H. Clinical features of ankylosing spondylitis associated with acute anterior uveitis in Chinese patients. *Int J Ophthalmol*. 2012;5(2):164–166.
9. Priem HA, Kijlstra A, Noens L, Baarsma GS, De Laey JJ, Oosterhuis JA. HLA typing in birdshot chorioretinopathy. *Am J Ophthalmol*. 1988;105(2):182–185.
10. Mizuki N, Meguro A, Ota M, et al. Genome-wide association studies identify IL23R-IL12RB2 and IL10 as Behcet's disease susceptibility loci. *Nat Genet*. 2010; 42(8):703–706.

11. Remmers EF, Cosan F, Kirino Y, et al. Genome-wide association study identifies variants in the MHC class I, IL10, and IL23R-IL12RB2 regions associated with Behcet's disease. *Nat Genet.* 2010;42(8):698–702.
12. Hou S, Yang Z, Du L, et al. Identification of a susceptibility locus in STAT4 for Behcet's disease in Han Chinese in a genome-wide association study. *Arthritis Rheum.* 2012;64(12):4104–4113.
13. Hou S, Du L, Lei B, et al. Genome-wide association analysis of Vogt-Koyanagi-Harada syndrome identifies two new susceptibility loci at 1p31.2 and 10q21.3. *Nat Genet.* 2014;46(9):1007–1011.
14. Robinson PC, Claushuis TA, Cortes A, et al. Genetic dissection of acute anterior uveitis reveals similarities and differences in associations observed with ankylosing spondylitis. *Arthritis Rheumatol.* 2015;67(1):140–151.
15. Hou S, Qi J, Liao D, et al. Copy number variations of complement component C4 are associated with Behcet's disease but not with ankylosing spondylitis associated with acute anterior uveitis. *Arthritis Rheum.* 2013;65(11):2963–2970.
16. Hou S, Xiao X, Li F, Jiang Z, Kijlstra A, Yang P. Two-stage association study in Chinese Han identifies two independent associations in CCR1/CCR3 locus as candidate for Behcet's disease susceptibility. *Hum Genet.* 2012;131(12):1841–1850.
17. Atzeni F, Boiardi L, Casali B, et al. CC chemokine receptor 5 polymorphism in Italian patients with Behcet's disease. *Rheumatology (Oxford, England).* 2012;51(12): 2141–2145.
18. Chen F, Hou S, Jiang Z, et al. CD40 gene polymorphisms confer risk of Behcet's disease but not of Vogt-Koyanagi-Harada syndrome in a Han Chinese population. *Rheumatology.* 2012;51(1):47–51.
19. Fei Y, Webb R, Cobb BL, Direskeneli H, Saruhan-Direskeneli G, Sawalha AH. Identification of novel genetic susceptibility loci for Behcet's disease using a genome-wide association study. *Arthritis Res Ther.* 2009;11(3):R66.
20. Oksel F, Keser G, Ozmen M, et al. Endothelial nitric oxide synthase gene Glu298Asp polymorphism is associated with Behcet's disease. *Clin Exp Rheumatol.* 2006;24(5 suppl 42):S79–S82.
21. Kim JU, Chang HK, Lee SS, et al. Endothelial nitric oxide synthase gene polymorphisms in Behcet's disease and rheumatic diseases with vasculitis. *Ann Rheum Dis.* 2003;62(11):1083–1087.
22. Ben Dhifallah I, Houman H, Khanfir M, Hamzaoui K. Endothelial nitric oxide synthase gene polymorphism is associated with Behcet's disease in Tunisian population. *Hum Immunol.* 2008;69(10):661–665.
23. Kirino Y, Bertsias G, Ishigatsubo Y, et al. Genome-wide association analysis identifies new susceptibility loci for Behcet's disease and epistasis between HLA-B*51 and ERAP1. *Nat Genet.* 2013;45(2):202–207.
24. Li K, Zhao M, Hou S, Du L, Kijlstra A, Yang P. Association between polymorphisms of FCRL3, a non-HLA gene, and Behcet's disease in a Chinese population with ophthalmic manifestations. *Mol Vis.* 2008;14:2136–2142.
25. Ben Dhifallah I, Karray EF, Sassi F, Hamzaoui K. Intercellular adhesion molecule 1 K469E gene polymorphism is associated with presence of skin lesions in Tunisian Behcet's disease patients. *Tissue Antigens.* 2010;75(1):74–78.
26. Boiardi L, Salvarani C, Casali B, et al. Intercellular adhesion molecule-1 gene polymorphisms in Behcet's disease. *J Rheumatol.* 2001;28(6):1283–1287.
27. Ozcimen AA, Dilek K, Bingol U, et al. IL-1 cluster gene polymorphisms in Turkish patients with Behcet's disease. *Int J Immunogenet.* 2011;38(4):295–301.
28. Oral HB, Dilek K, Ozcimen AA, et al. Interleukin-4 gene polymorphisms confer Behcet's disease in Turkish population. *Scand J Immunol.* 2011;73(6):594–601.
29. Chang HK, Jang WC, Park SB, et al. Association between interleukin 6 gene polymorphisms and Behcet's disease in Korean people. *Ann Rheum Dis.* 2005;64(2):339–340.

30. Xavier JM, Shahram F, Davatchi F, et al. Association study of IL10 and IL23R-IL12RB2 in Iranian patients with Behcet's disease. *Arthritis Rheum.* 2012;64(8): 2761–2772.
31. Yanagihori H, Oyama N, Nakamura K, Mizuki N, Oguma K, Kaneko F. Role of IL-12B promoter polymorphism in Adamantiades-Behcet's disease susceptibility: an involvement of Th1 immunoreactivity against Streptococcus Sanguinis antigen. *J Invest Dermatol.* 2006;126(7):1534–1540.
32. Htoon J, Nadig A, Hughes T, et al. IL18 polymorphism is associated with Behcet's disease but not lupus in patients from Turkey. *J Rheumatol.* 2011;38(5):962–963.
33. Jiang Z, Yang P, Hou S, et al. IL-23R gene confers susceptibility to Behcet's disease in a Chinese Han population. *Ann Rheum Dis.* 2010;69(7):1325–1328.
34. Lee YJ, Kang SW, Song JK, et al. Associations between interferon regulatory factor-1 polymorphisms and Behcet's disease. *Hum Immunol.* 2007;68(9):770–778.
35. Zerbib J, Seddon JM, Richard F, et al. rs5888 variant of SCARB1 gene is a possible susceptibility factor for age-related macular degeneration. *PLoS One.* 2009;4(10):6.
36. Hou S, Yang P, Du L, et al. Monocyte chemoattractant protein-1 -2518 A/G single nucleotide polymorphism in Chinese Han patients with ocular Behcet's disease. *Hum Immunol.* 2010;71(1):79–82.
37. Rustemoglu A, Gul U, Gumus-Akay G, et al. MDR1 gene polymorphisms may be associated with Behcet's disease and its colchicum treatment response. *Gene.* 2012; 505(2):333–339.
38. Zheng X, Wang D, Hou S, et al. Association of macrophage migration inhibitory factor gene polymorphisms with Behcet's disease in a Han Chinese population. *Ophthalmology.* 2012;119(12):2514–2518.
39. Zhou Q, Hou S, Liang L, et al. MicroRNA-146a and Ets-1 gene polymorphisms in ocular Behcet's disease and Vogt-Koyanagi-Harada syndrome. *Ann Rheum Dis.* 2014;73:170–176.
40. Park KS, Min Y, Park SR, et al. Matrix metalloproteinase-2, -9, -12, and tissue inhibitor of metalloproteinase 2 gene polymorphisms and cutaneous expressions in patients with Behcet's disease. *Tissue Antigens.* 2012;79(5):333–339.
41. Lee YJ, Kang SW, Baek HJ, et al. Association between matrix metalloproteinase 9 promoter polymorphisms and Behcet's disease. *Hum Immunol.* 2010;71(7):717–722.
42. Karakus N, Yigit S, Kalkan G, et al. Association between the methylene tetrahydrofolate reductase gene C677T mutation and colchicine unresponsiveness in Behcet's disease. *Mol Vis.* 2012;18:1696–1700.
43. Ates O, Dalyan L, Hatemi G, Hamuryudan V, Topal-Sarikaya A. Genetic susceptibility to Behcet's syndrome is associated with NRAMP1 (SLC11A1) polymorphism in Turkish patients. *Rheumatol Int.* 2009;29(7):787–791.
44. Hou S, Xiao X, Zhou Y, et al. Genetic variant on PDGFRL associated with Behcet disease in Chinese Han populations. *Hum Mutat.* 2013;34(1):74–78.
45. Baranathan V, Stanford MR, Vaughan RW, et al. The association of the PTPN22 620W polymorphism with Behcet's disease. *Ann Rheum Dis.* 2007;66(11):1531–1533.
46. Kim SK, Jang WC, Park SB, et al. SLC11A1 gene polymorphisms in Korean patients with Behcet's disease. *Scand J Rheumatol.* 2006;35(5):398–401.
47. Hu K, Hou S, Jiang Z, Kijlstra A, Yang P. JAK2 and STAT3 polymorphisms in a Han Chinese population with Behcet's disease. *Invest Ophthalmol Vis Sci.* 2012; 53(1):538–541.
48. Hou S, Kijlstra A, Yang P. The genetics of Behcet's disease in a Chinese population. *Front Med.* 2012;6(4):354–359.
49. Park G, Kim HS, Choe JY, Kim SK. SUMO4 C438T polymorphism is associated with papulopustular skin lesion in Korean patients with Behcet's disease. *Rheumatol Int.* 2012;32(10):3031–3037.

50. Kamoun M, Ben Dhifallah I, Karray E, Zakraoui L, Hamzaoui K. Association of small ubiquitin-like modifier 4 (SUMO4) polymorphisms in a Tunisian population with Behcet's disease. *Clin Exp Rheumatol*. 2010;28(4 suppl 60):S45–S49.
51. Hou S, Yang P, Du L, et al. SUMO4 gene polymorphisms in Chinese Han patients with Behcet's disease. *Clin Immunol*. 2008;129(1):170–175.
52. Chen Y, Yang P, Li F, et al. Association analysis of TGFBR3 gene with Vogt-Koyanagi-Harada disease and Behcet's disease in the Chinese Han population. *Curr Eye Res*. 2012;37(4):312–317.
53. Fang J, Hu R, Hou S, et al. Association of TLR2 gene polymorphisms with ocular Behcet's disease in a Chinese Han population. *Invest Ophthalmol Vis Sci*. 2013;54(13):8384–8392.
54. Horie Y, Meguro A, Ota M, et al. Association of TLR4 polymorphisms with Behcet's disease in a Korean population. *Rheumatology (Oxford, England)*. 2009;48(6):638–642.
55. Meguro A, Ota M, Katsuyama Y, et al. Association of the toll-like receptor 4 gene polymorphisms with Behcet's disease. *Ann Rheum Dis*. 2008;67(5):725–727.
56. Radouane A, Oudghiri M, Chakib A, Bennani S, Touitou I, Barat-Houari M. SNPs in the TNF-alpha gene promoter associated with Behcet's disease in Moroccan patients. *Rheumatology (Oxford, England)*. 2012;51(9):1595–1599.
57. Bonyadi M, Jahanafrooz Z, Esmaeili M, et al. TNF-alpha gene polymorphisms in Iranian Azeri Turkish patients with Behcet's disease. *Rheumatol Int*. 2009;30(2):285–289.
58. Li H, Liu Q, Hou S, et al. TNFAIP3 gene polymorphisms confer risk for Behcet's disease in a Chinese Han population. *Hum Genet*. 2013;132(3):293–300.
59. Hou S, Shu Q, Jiang Z, et al. Replication study confirms the association between UBAC2 and Behcet's disease in two independent Chinese sets of patients and controls. *Arthritis Res Ther*. 2012;14(2):R70.
60. Karray EF, Ben Dhifallah I, Ben Abdelghani K, et al. Associations of vitamin D receptor gene polymorphisms FokI and BsmI with susceptibility to rheumatoid arthritis and Behcet's disease in Tunisians. *Joint Bone Spine*. 2012;79(2):144–148.
61. Nam EJ, Han SW, Kim SU, et al. Association of vascular endothelial growth factor gene polymorphisms with Behcet disease in a Korean population. *Hum Immunol*. 2005;66(10):1068–1073.
62. Du L, Yang P, Hou S, et al. Association of the CTLA-4 gene with Vogt-Koyanagi-Harada syndrome. *Clin Immunol*. 2008;127(1):43–48.
63. Hu K, Hou S, Li F, Xiang Q, Kijlstra A, Yang P. JAK1, but not JAK2 and STAT3, confers susceptibility to Vogt-Koyanagi-Harada (VKH) syndrome in a Han Chinese population. *Invest Ophthalmol Vis Sci*. 2013;54(5):3360–3365.
64. Chu M, Yang P, Hu R, et al. Elevated serum osteopontin levels and genetic polymorphisms of osteopontin are associated with Vogt-Koyanagi-Harada disease. *Invest Ophthalmol Vis Sci*. 2011;52(10):7084–7089.
65. Shu Q, Yang P, Hou S, et al. Interleukin-17 gene polymorphism is associated with Vogt-Koyanagi-Harada syndrome but not with Behcet's disease in a Chinese Han population. *Hum Immunol*. 2010;71(10):988–991.
66. Hu K, Yang P, Jiang Z, Hou S, Du L, Li F. STAT4 polymorphism in a Chinese Han population with Vogt-Koyanagi-Harada syndrome and Behcet's disease. *Hum Immunol*. 2010;71(7):723–726.
67. Li H, Liu Q, Hou S, et al. TNFAIP3 gene polymorphisms in a Chinese Han population with Vogt-Koyanagi-Harada syndrome. *PLoS One*. 2013;8(3):e59515.
68. Cook DN, Pisetsky DS, Schwartz DA. Toll-like receptors in the pathogenesis of human disease. *Nat Immunol*. 2004;5(10):975–979.
69. Kirino Y, Zhou Q, Ishigatsubo Y, et al. Targeted resequencing implicates the familial Mediterranean fever gene MEFV and the toll-like receptor 4 gene TLR4 in Behcet disease. *Proc Natl Acad Sci USA*. 2013;110(20):8134–8139.

70. Fang J, Chen L, Tang J, et al. Association between copy number variation of TLR7 and ocular Behcet's disease in a Chinese Han population. *Invest Ophthalmol Vis Sci.* 2015;56(3):1517–1523.
71. Levinson RD. Killer immunoglobulin-like receptor genes in uveitis. *Ocul Immunol Inflamm.* 2011;19(3):192–201.
72. Seo J, Park JS, Nam JH, et al. Association of CD94/NKG2A, CD94/NKG2C, and its ligand HLA-E polymorphisms with Behcet's disease. *Tissue Antigens.* 2007;70(4): 307–313.
73. de Menthon M, Lavalley MP, Maldini C, Guillevin L, Mahr A. HLA-B51/B5 and the risk of Behcet's disease: a systematic review and meta-analysis of case–control genetic association studies. *Arthritis Rheum.* 2009;61(10):1287–1296.
74. Gul A, Ohno S. HLA-B*51 and Behcet disease. *Ocul Immunol Inflamm.* 2012;20(1): 37–43.
75. Yabuki K, Mizuki N, Ota M, et al. Association of MICA gene and HLA-B*5101 with Behcet's disease in Greece. *Invest Ophthalmol Vis Sci.* 1999;40(9):1921–1926.
76. Ota M, Mizuki N, Katsuyama Y, et al. The critical region for Behcet disease in the human major histocompatibility complex is reduced to a 46-kb segment centromeric of HLA-B, by association analysis using refined microsatellite mapping. *Am J Hum Genet.* 1999;64(5):1406–1410.
77. Zhao M, Jiang Y, Abrahams IW. Association of HLA antigens with Vogt-Koyanagi-Harada syndrome in a Han Chinese population. *Arch Ophthalmol.* 1991;109(3): 368–370.
78. Zhang XY, Wang XM, Hu TS. Profiling human leukocyte antigens in Vogt-Koyanagi-Harada syndrome. *Am J Ophthalmol.* 1992;113(5):567–572.
79. Islam SM, Numaga J, Fujino Y, et al. HLA class II genes in Vogt-Koyanagi-Harada disease. *Invest Ophthalmol Vis Sci.* 1994;35(11):3890–3896.
80. Feltkamp TE. HLA-B27 and acute anterior uveitis. *Curr Eye Res.* 1990;9(suppl): 213–218.
81. Gasch AT, Smith JA, Whitcup SM. Birdshot retinochoroidopathy. *Br J Ophthalmol.* 1999;83(2):241–249.
82. Kuiper J, Rothova A, de Boer J, Radstake T. The immunopathogenesis of birdshot chorioretinopathy; a bird of many feathers. *Prog Retin Eye Res.* 2015;44:99–110.
83. Li B, Yang P, Chu L, et al. T-bet expression in the iris and spleen parallels disease expression during endotoxin-induced uveitis. *Graefe's Arch Clin Exp Ophthalmol.* 2007;245(3):407–413.
84. Du L, Kijlstra A, Yang P. Immune response genes in uveitis. *Ocul Immunol Inflamm.* 2009;17(4):249–256.
85. Li X, Bai L, Fang J, et al. Genetic variations of IL-12B, IL-12Rbeta1, IL-12Rbeta2 in Behcet's disease and VKH syndrome. *PLoS One.* 2014;9(5):e98373.
86. Jang WC, Park SB, Nam YH, et al. Interleukin-18 gene polymorphisms in Korean patients with Behcet's disease. *Clin Exp Rheumatol.* 2005;23(4 suppl 38):S59–S63.
87. Arida A, Fragiadaki K, Giavri E, Sfikakis PP. Anti-TNF agents for Behcet's disease: analysis of published data on 369 patients. *Semin Arthritis Rheum.* 2011;41(1):61–70.
88. Touma Z, Farra C, Hamdan A, et al. TNF polymorphisms in patients with Behcet disease: a meta-analysis. *Arch Med Res.* 2010;41(2):142–146.
89. Chi W, Yang P, Li B, et al. IL-23 promotes CD4+ T cells to produce IL-17 in Vogt-Koyanagi-Harada disease. *J Allergy Clin Immunol.* 2007;119(5):1218–1224.
90. Chen L, Yang P, Zhou H, et al. Diminished frequency and function of CD4+ CD25high regulatory T cells associated with active uveitis in Vogt-Koyanagi-Harada syndrome. *Invest Ophthalmol Vis Sci.* 2008;49(8):3475–3482.
91. Schoenberger SP, Toes RE, van der Voort EI, Offringa R, Melief CJ. T-cell help for cytotoxic T lymphocytes is mediated by CD40-CD40L interactions. *Nature.* 1998;393(6684):480–483.

92. Toubi E, Shoenfeld Y. The role of CD40-CD154 interactions in autoimmunity and the benefit of disrupting this pathway. *Autoimmunity*. 2004;37(6–7):457–464.
93. Chen F, Hou S, Jiang Z, et al. CD40 gene polymorphisms confer risk of Behcet's disease but not of Vogt-Koyanagi-Harada syndrome in a Han Chinese population. *Rheumatology (Oxford, England)*. 2011;51(1):47–51.
94. Xiang Q, Chen L, Hou S, et al. TRAF5 and TRAF3IP2 gene polymorphisms are associated with Behcet's disease and Vogt-Koyanagi-Harada syndrome: a case–control study. *PLoS One*. 2014;9(1):e84214.
95. Hou S, Qi J, Zhang Q, et al. Genetic variants in the JAK1 gene confer higher risk of Behcet's disease with ocular involvement in Han Chinese. *Hum Genet*. 2013; 132(9):1049–1058.
96. Tang Y, Luo X, Cui H, et al. MicroRNA-146A contributes to abnormal activation of the type I interferon pathway in human lupus by targeting the key signaling proteins. *Arthritis Rheum*. 2009;60(4):1065–1075.
97. Yu H, Liu Y, Zhang L, et al. FoxO1 gene confers genetic predisposition to acute anterior uveitis with ankylosing spondylitis. *Invest Ophthalmol Vis Sci*. 2014;55(12): 7970–7974.
98. Hou S, Qi J, Liao D, et al. High C4 gene copy numbers protects against Vogt-Koyanagi-Harada syndrome in Chinese Han. *Br J Ophthalmol*. 2014;98(12):1733–1737.
99. Gul A. Behcet's disease as an autoinflammatory disorder. *Curr Drug Targets Inflamm Allergy*. 2005;4(1):81–83.
100. Read RW, Rao NA, Cunningham ET. Vogt-Koyanagi-Harada disease. *Curr Opin Ophthalmol*. 2000;11(6):437–442.
101. Hou S, Liao D, Zhang J, et al. Genetic variations of IL17F and IL23A show associations with Behcet's disease and Vogt-Koyanagi-Harada syndrome. *Ophthalmology*. 2015; 122(3):518–523.

SECTION V

Aqueous Flow and Intraocular Pressure

CHAPTER EIGHTEEN

Intraocular Pressure and the Mechanisms Involved in Resistance of the Aqueous Humor Flow in the Trabecular Meshwork Outflow Pathways

Ernst R. Tamm[1], Barbara M. Braunger, Rudolf Fuchshofer
Institute of Human Anatomy and Embryology, University of Regensburg, Regensburg, Germany
[1]Corresponding author: e-mail address: ernst.tamm@vkl.uni-regensburg.de

Contents

Abstract

Intraocular pressure (IOP), the critical risk factor for glaucoma, is generated and maintained by the aqueous humor circulation system. Aqueous humor is secreted from the epithelial layers of the ciliary body and exits the eye through the trabecular meshwork or the uveoscleral outflow pathways. IOP builds up in response to a resistance to aqueous humor flow in the trabecular outflow pathways. The trabecular outflow resistance is localized in the inner wall region, which comprises the juxtacanalicular connective tissue (JCT) and the inner wall endothelium of Schlemm's canal (SC). Outflow resistance in this region is lowered through the relaxation of contractile myofibroblast-like cells in trabecular meshwork and the adjacent scleral spur, or the contraction of the ciliary muscle. In primary open-angle glaucoma, the most frequent form of glaucoma, outflow resistance of the inner wall region is typically higher than normal. There is evidence that the increase in resistance is related to characteristic biological changes in the resident cells of the JCT, which more and more acquire the structural and functional characteristics of contractile myofibroblasts. The changes involve an augmentation of their actin cytoskeleton and of their surrounding fibrillary extracellular matrix, which connects to JCT cells via integrins. This scenario leads to an overall stiffening of the inner wall region, and is modulated by transforming growth factor-β/

Progress in Molecular Biology and Translational Science, Volume 134
ISSN 1877-1173
http://dx.doi.org/10.1016/bs.pmbts.2015.06.007
© 2015 Elsevier Inc.
All rights reserved.

connective tissue growth factor signaling. Essentially comparable changes appear to occur in SC endothelial cells. Stiffening of JCT and SC cells is very likely a critical causative factor for the increase in trabecular outflow resistance in POAG.

1. INTRAOCULAR PRESSURE AND AQUEOUS HUMOR OUTFLOW

Glaucoma, the leading cause of irreversible blindness throughout the world,[1] is a chronic, progressive optic nerve neuropathy, in which optic nerve axons are damaged at the optic disc, the site of their exit from the eye. In several prospective randomized multi-center studies, intraocular pressure (IOP) has been identified as the critical causative risk factor for glaucoma and, consequently, the reduction of IOP delays or prevents the damage of optic nerve axons.[2–8] This chapter will review the mechanisms that are thought to generate IOP in the normal eye and in eyes affected with the most common form of glaucoma, primary open-angle glaucoma (POAG). In addition, we will outline a unifying concept to explain the molecular changes that change the biology of the aqueous humor outflow pathways to affect IOP in POAG.[9]

IOP depends on the aqueous humor circulation system, in which aqueous humor is secreted from the epithelial layers of the ciliary body into the posterior chamber[10] and exits the eye in the chamber angle via the conventional or trabecular outflow pathway, or the unconventional or uveoscleral outflow pathway (Fig. 1).[11] The trabecular outflow pathway comprises the trabecular meshwork (made up by the uveal and corneoscleral meshworks), the juxtacanalicular connective tissue, the endothelial lining of Schlemm's canal, the collecting channels, and the aqueous veins (Fig. 1).[11,12] After having passed through the trabecular outflow pathways, aqueous humor drains into the episcleral venous system.[13] In the unconventional or uveoscleral outflow pathways, aqueous humor diffuses through the interstitial spaces between the ciliary muscle bundles, and through the supraciliary and suprachoroidal spaces to reach the capillaries of the ciliary body or the lymphatic vessels of the orbit.[14] Accurate measurements of the rate of uveoscleral outflow are difficult to perform in the living eye.[14] In human cadaver eyes, the rate of uveoscleral outflow was measured to constitute 4–14% of total outflow.[15] Under physiological conditions, only the trabecular outflow pathways are relevant for the generation and maintenance of IOP.[16] The trabecular outflow pathways provide resistance against the flow of the aqueous humor and consequently lead to the generation of

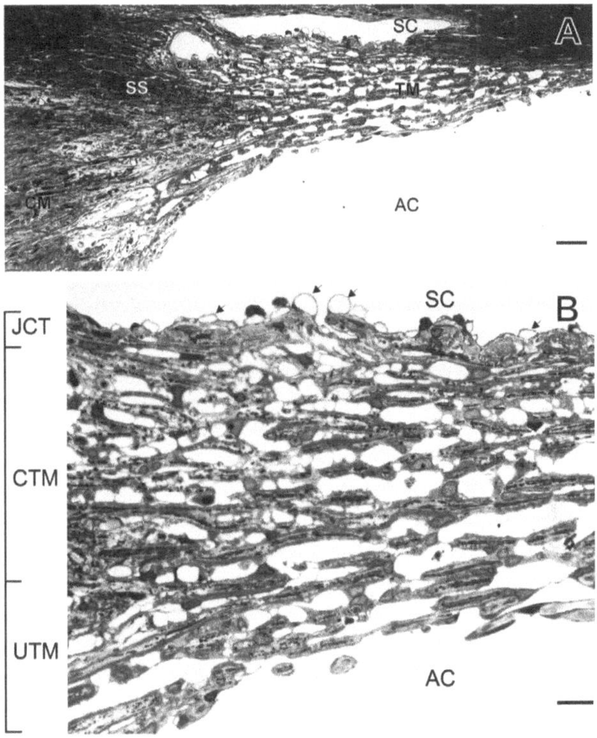

Figure 1 (A) Light micrograph of a meridional section through the trabecular meshwork. (B) is a magnification of (A). In the plane of this section, Schlemm's canal (SC) has two lumens that are separated by a septum. TM, trabecular meshwork; SS, scleral spur; CM, ciliary muscle; AC, anterior chamber; JCT, juxtacanalicular tissue; CTM, corneoscleral trabecular meshwork; UTM, uveal trabecular meshwork. Arrows in (B) point to giant vacuoles in the inner wall endothelium of SC. Magnification bars: 20 μm (A), 5 μm (B). *From Ref. 11.* (See the color plate.)

IOP. In eyes with POAG, outflow resistance in the trabecular outflow pathways is usually higher than in age-matched normal control eyes.[16] Passage of aqueous humor in the trabecular outflow pathways occurs as bulk flow driven by pressure gradient only and if IOP is in a steady state, the outflow of the aqueous humor equals its secretion from the epithelium of the ciliary body.[16]

2. TRABECULAR MESHWORK

The trabecular meshwork comprises connective tissue lamellae that are covered by flat, epithelial-like trabecular meshwork cells, which rest on a basal lamina and are of neural crest origin.[11,17] Anteriorly, the trabecular lamellae attach to the peripheral cornea in a region termed Schwalbe's line. Posteriorly, the trabecular lamellae are connected to the stroma of the ciliary body and iris at their junction, and to the scleral spur (Fig. 1). There is considerable evidence that the region close to Schwalbe's line serves as a niche for adult stem/progenitor cells that are capable of dividing and repopulating the trabecular meshwork after injury.[18] The outermost part of the trabecular meshwork, the juxtacanalicular tissue, lies immediately adjacent to the endothelium of Schlemm's canal. The juxtacanalicular tissue is not built up by lamellae, but largely consists of loose connective tissue with two to five layers of cells showing no epithelial, but rather a mesenchymal morphology, quite comparable to that of fibroblasts. The cells of the juxtacanalicular tissue are surrounded by an extracellular matrix that consists of fibrillar components and of an amorphous ground substance of hyaluronan and proteoglycans.[19] Throughout the juxtacanalicular tissue, a network of elastic fibers spreads, the cribriform plexus, which spans tangentially to the endothelium of Schlemm's canal. [20,21] The elastic fibers of the cribriform plexus consist of an elastin-containing core and of banded sheath material with a periodicity of about 50 nm. The sheath material contains collagen type VI and fibronectin.[20,22] The fibers of the cribriform plexus are connected to the endothelial cells of Schlemm's canal by fine fibrils that emerge from the sheath material (Fig. 2).

3. SCHLEMM'S CANAL

Schlemm's canal is a modified capillary blood vessel that transiently forms intra- and intercellular pores (Fig. 2), which can acquire unusually large diameters of up to 1 μm.[16] As result, the hydraulic conductivity of Schlemm's canal is the highest in the body and even higher than that in the fenestrated capillaries of the glomeruli in the kidney or the sinusoids in the liver.[16] Aqueous humor passes Schlemm's canal endothelial cells in a basal-luminal direction, which results in the formation of cellular outpouchings of Schlemm's canal cells (so-called giant vacuoles) in response to the pressure gradient associated with aqueous humor flow. The contacts of Schlemm's canal endothelial cells with the connecting fibrils of the cribriform plexus prevent the detachment of the endothelium in response to aqueous humor flow. Schlemm's canal pores likely develop from 60 nm mini-pores that are covered by a diaphragm containing plasmalemma-vesicle associated protein (PLVAP).[23,24] The molecular processes of pore formation are ill understood, but PLVAP is likely involved, as the formation of vascular pores is largely impaired in PLVAP-deficient mice.[23,25]

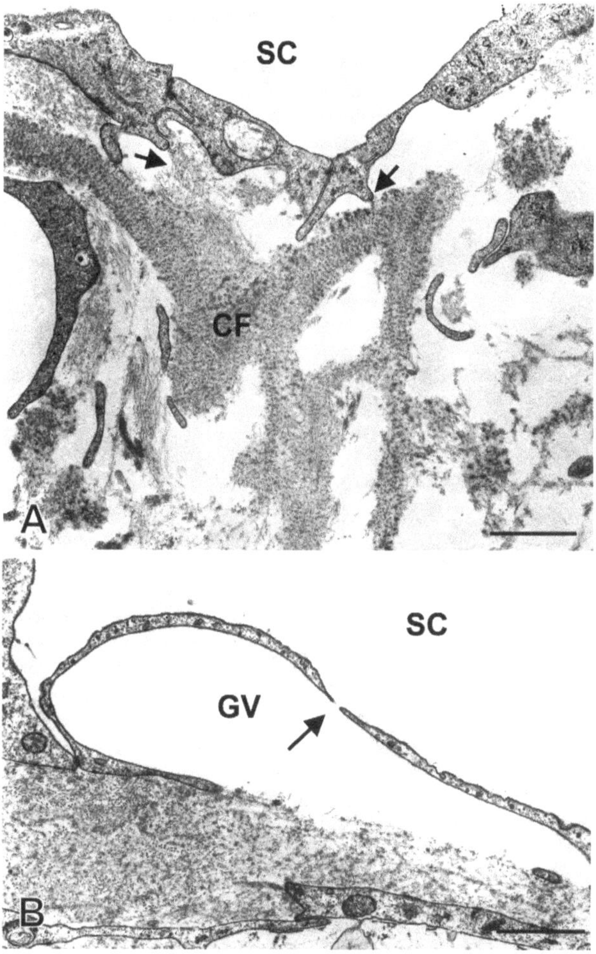

Figure 2 Transmission electron microscopy of the SC inner wall region. (A) Connecting fibers (CF) in the juxtacanalicular tissue, which emerge from the cribriform elastic plexus, connect the plexus with the inner wall endothelium of Schlemm's canal (SC). The connection with the inner wall endothelium is made via the banded sheath material of the fibers or via fine fibrils that emerge from it (arrows). (B) A giant vacuole (GV) in the inner wall of SC forms an intracellular pore (arrow). Magnification bars: 1 μm. *From Ref. 11.*

4. OUTFLOW RESISTANCE

It is generally accepted that the outflow resistance of the trabecular outflow pathways is localized in the inner wall region comprising the juxtacanalicular tissue and the inner wall endothelium of Schlemm's canal.[16] There are three potential mechanisms, which may be involved in the generation of trabecular outflow resistance and which will be discussed here. (1) Outflow resistance largely depends on the quality and quantity of the extracellular matrix in the extracellular spaces of the juxtacanalicular tissue. (2) Outflow resistance depends on the hydraulic conductivity of the inner endothelial layer of Schlemm's canal. (3) Outflow resistance is generated through a synergistic interaction between the extracellular outflow pathways of the juxtacanalicular tissue and the pores of Schlemm's canal endothelium.

Support for a critical role of the juxtacanalicular tissue extracellular matrix in generating outflow resistance and for mechanism (1) comes from experiments, which indicate that the perfusion of anterior segment organ cultures with enzymes that degrade extracellular matrix compounds, lowers outflow resistance.[26] Arguments against this mechanism are based on studies using morphometry in conjunction with conventional transmission electron microscopy.[27,28] The studies show that the area in the juxtacanalicular outflow pathways occupied by electron-dense extracellular matrix is much too small to generate the physiological outflow resistance. Still, the procedures that are applied to process tissue for transmission electron microscopy cause a substantial loss of nonfibrillar extracellular matrix components such as hyaluronan and proteoglycans. In fact, by quick-freeze deep-etch electron microscopy considerable higher amounts of extracellular matrix in the juxtacanalicular tissue were observed than with conventional electron microscopy.[29]

Results from studies in which the endothelial pores were visualized by scanning electron microscopy (SEM) and their sizes quantitatively evaluated argue against a critical role of Schlemm's canal endothelium and against mechanism (2) in generating outflow resistance.[30] According to the data of those studies, only 10% of the trabecular outflow resistance would be generated in Schlemm's canal endothelium. In contrast, more recent observations show that the number of pores in Schlemm's canal endothelium is influenced by the duration of chemical fixation,[31,32] indicating that the number of pores *in vivo* might actually be significantly smaller than seen with SEM.

The idea behind mechanism (3) and the concept of a synergistic functional interplay between juxtacanalicular tissue and Schlemm's canal endothelium is based on the hypothesis that the extracellular matrix of the juxtacanalicular tissue funnels the aqueous humor right to the pores in Schlemm's canal endothelium (funneling hypothesis).[33] Funneling results in a restriction of aqueous humor flow and thereby generates outflow resistance. Outflow resistance is reduced if either the number of pores in

Schlemm's canal endothelium or the flow pathways in the juxtacanalicular tissue increase and attenuate the funneling effect. Currently, the funneling hypothesis appears to be in agreement with the available experimental data.

5. CONTRACTILE MECHANISMS IN THE TRABECULAR OUTFLOW PATHWAYS

There are two contractile systems that influence the architecture of the trabecular outflow pathways to modify outflow resistance in the inner wall region. The ciliary muscle is attached to the scleral spur, the posterior attachment of the trabecular meshwork, via tendon-like structures.[34,35] Consequently, contraction of the ciliary muscle results in a changed geometry of the trabecular outflow pathways and reduces outflow resistance in the inner wall region. Experimental disinsertion of ciliary muscle's anterior insertion to the scleral spur completely abolishes its effects on outflow resistance and IOP in primates.[36]

The second contractile system comprises cells in the trabecular meshwork and the juxtacanalicular tissue, which show properties of contractile myofibroblasts.[37] The cells are particularly numerous in the posterior part of both regions and close to their attachment to the scleral spur (scleral spur cells) (Fig. 3).[34] The cells in this region show all the structural characteristics of myofibroblasts, like a cytoplasm that contains numerous myofilaments or cellular processes that form tendon-like contacts with the neighboring extracellular matrix fibers. Moreover, the cells contain high amounts of α-smooth muscle actin, the characteristic actin isoform of smooth muscle cells. A major difference between the myofibroblast-like cells in the trabecular outflow pathways and the cells of the neighboring ciliary muscle is the fact that the myofibroblasts are orientated in a circumferential direction and thus right perpendicularly to the longitudinally orientated bundles of the ciliary muscle (Fig. 3B). Mechanosensitive neuronal endings are frequently observed in this region and are most likely part of a proprioceptive system controlling the tone of both contractile systems.[39] Experimental data obtained in the trabecular outflow pathways of bovine eyes, where myofibroblast-like cells are especially numerous,[40] indicate that an elevated tone of contractile cells in the trabecular outflow pathways results in an elevated outflow resistance in the trabecular meshwork while their relaxation leads to a reduction of outflow resistance.[41] The same appears to be true for the trabecular outflow resistance in the primate eye.[42] Apparently, ciliary muscle and myofibroblast-like cells of the trabecular outflow system act as

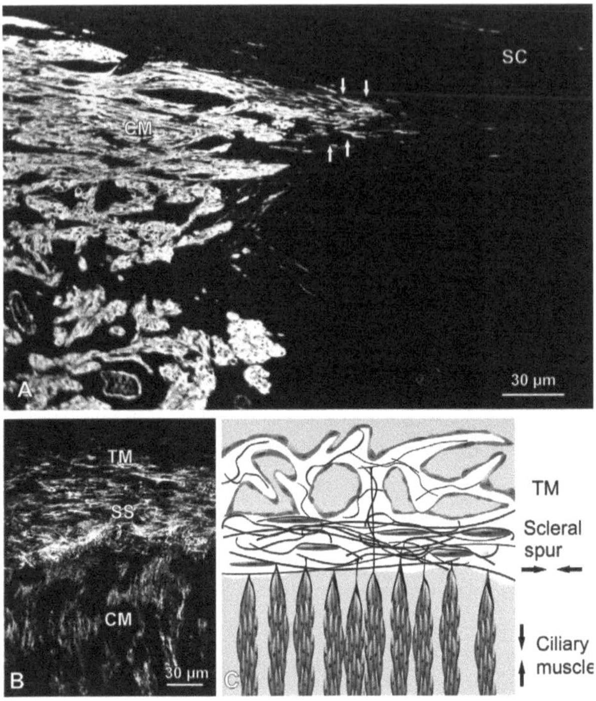

Figure 3 Meridional (A) and tangential section (B) through ciliary muscle (CM), scleral spur (SS), and trabecular meshwork (TM) stained with antibodies against α-smooth muscle actin. (A) Ciliary muscle cells and vascular smooth muscle cells stain positively with antibodies against α-smooth muscle actin. Arrows indicate the scleral spur, where all cells show intense immunoreactivity for α-smooth muscle actin. (B) Tangential section of scleral spur, trabecular meshwork, and ciliary muscle. Positively stained cells oriented in a circular direction are seen throughout the entire spur tissue. While ciliary muscle cells also stain positive, no staining is seen in the trabecular meshwork. (C) Schematic drawing (orientation analogue to B) depicting the different orientation of the contractile systems that influence the geometry of the trabecular outflow pathways. The anterior longitudinal or meridional ciliary muscle bundles, which attach to scleral spur and trabecular meshwork are oriented perpendicularly to the circumferential or equatorially oriented contractile myofibroblasts of posterior trabecular meshwork (TM) and scleral spur. Arrows indicate movements following contraction or relaxation. *Panels (A) and (B): From Ref. 34; Panel (C): From Ref. 38.*

functional antagonists while modulating outflow resistance. The likely explanation for their different function is their different orientation at an angle of 90° to each, which, upon contraction, leads to a modification of trabecular outflow pathway geometry in an antagonistic manner (Fig. 3C).

Studies in genetically modified mice provided evidence for a critical autoregulative role of nitric oxide that is released from the resident cells of the trabecular meshwork outflow pathways as part of an intrinsic mechano-regulatory system. Accordingly, an increasing mechanical load of the trabecular meshwork, e.g., caused by increased IOP increases the trabecular release of nitric oxide and, subsequently, decreases outflow resistance and IOP due to the nitric oxide-mediated relaxation of the cells of the trabecular outflow system.[43]

6. RESISTANCE OF THE TRABECULAR OUTFLOW PATHWAYS IN PRIMARY OPEN-ANGLE GLAUCOMA

The juxtacanalicular tissue of patients with POAG develops very characteristic structural changes in the course of the disease.[44] The quality and quantity of the elastic fibers sheath material in the cribriform plexus changes, and there is an increase of material originally described as "sheath-derived plaques."[45,46] The increase in the amounts of sheath-derived plaque material in the juxtacanalicular tissue correlates with the severity of optic nerve damage in POAG.[47] However, the amount of sheath-derived plaque material does not correlate with the level of IOP in individual eyes of patients with POAG,[47] a finding that strongly indicates that the increase in sheath-derived material is rather a symptom, but not the cause for the increase in trabecular outflow resistance in POAG. The increase in extracellular matrix in the juxtacanalicular tissue occurs in parallel to an increase in the levels of transforming growth factor (TGF)-β2 in the aqueous humor of most patients with POAG.[48] Cultured trabecular meshwork increase their extracellular matrix synthesis upon treatment with TGF-βs,[48] and may well do so *in vivo* in response to higher levels of TGF-β2. In addition, treatment with TGF-βs increases the contractile properties of the trabecular meshwork actin cytoskeleton.[48] Overall, the effects of TGF-β are modulated by a complex homeostatic signaling system of molecules that stimulate or inhibit its activity in the trabecular outflow pathways.[49]

Connective tissue growth factor (CTGF) appears to be the downstream molecule that mediates the effects of TGF-βs on trabecular meshwork extracellular matrix synthesis and the actin cytoskeleton of trabecular meshwork cells.[50,51] CTGF is critically required for the effects of TGF-β on extracellular matrix synthesis and for increasing the contractile properties of mesenchymal cells including that of the trabecular meshwork.[51,52] In general, the force generated by the actin cytoskeleton of mesenchymal cells is transmitted to the surrounding fibrillar extracellular matrix components via integrin-based cell-matrix contacts. A switch to a myofibroblast-like phenotype, a scenario that includes the augmentation of the actin cytoskeleton and the directly associated fibrillar extracellular matrix, is apparently caused by high activity of TGF-β signaling and mediated via the downstream mediator CTGF.

This concept of a TGF-β/CTGF-mediated switch as the critical causative factor for the increase in outflow resistance and IOP in eyes with POAG is directly supported by results from recent studies in genetically engineered rodents. Transgenic mice with ectopic overexpression of CTGF in their eyes develop POAG characterized by an increase in IOP that correlates with the loss of optic nerve axons.[51] The trabecular meshwork outflow pathways of the mice accumulate fibrillar extracellular matrix components and α-smooth muscle actin positive cells as sign of an increase in actin-based contractility. If the mice are treated with a Rho-kinase inhibitor that interferes with actin contractility, IOP returns within hours to levels seen in normal animals. The findings clearly indicate that the effects of CTGF on IOP are mediated through the modification of the trabecular meshwork actin cytoskeleton. We hypothesize that high activity of TGF-β signaling in POAG causes, via the downstream mediator CTGF, a similar change in the phenotype of trabecular outflow cells in human patients as seen in CTGF-overexpressing mice with glaucoma. In response to high activity of TGF-β signaling, the cells switch to a myofibroblast-like phenotype, a scenario, which strengthens simultaneously both their actin cytoskeleton and their directly associated extracellular matrix fibrils (Fig. 4). Because of the increased tone and the accumulation of extracellular matrix material, the trabecular meshwork becomes stiffer and loses its capability to respond to endogenous signals that cause its relaxation and consequently decrease outflow resistance under physiologic conditions. Experiments using atomic force microscopy support this concept as the trabecular meshwork of patients with POAG is stiffer than that of healthy age-matched controls.[53] Additional support for the hypothesis comes from a recent study in rats with a vector-based expression of a constitutively active RhoAGTPase (RhoAV14) in the trabecular outflow pathways.[54] The rats develop high IOP that is associated with an increase in extracellular matrix and F-actin in the trabecular meshwork outflow pathways. The changes are ameliorated following topical application of a Rho-kinase inhibitor.[54] Quite intriguingly, the switch to a stiffer cellular phenotype in POAG is apparently not restricted to the cells of the juxtacanalicular tissue, but happens simultaneously in Schlemm's canal endothelial cells, the other cell type in the trabecular meshwork outflow pathways that is very likely involved in the generation of outflow resistance. Schlemm's canal cells of glaucomatous eyes

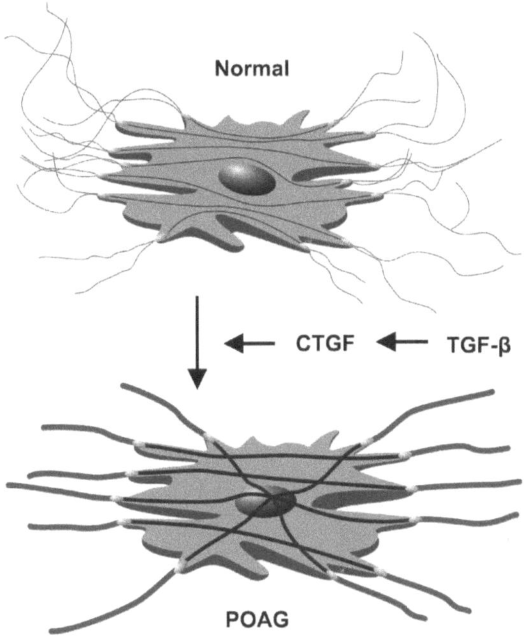

Figure 4 Schematic drawing of the actin cytoskeleton and the fibrillar extracellular matrix of cells in the juxtacanalicular tissue. The force generated by the actin cytoskeleton (red) is transmitted to the fibrillar extracellular matrix (green) by integrin-based cell-matrix contacts (yellow). The high activity of TGF-β signaling in POAG causes, via the downstream mediator CTGF, a switch to a myofibroblast-like phenotype including an augmentation of the actin cytoskeleton and its directly associated extracellular fibrillar matrix. Overall, the changes cause an increase in TM rigidity and aqueous humor outflow resistance. *From Ref. 38.* (See the color plate.)

were recently shown to have an increased cytoskeletal stiffness that leads to reduced pore formation in the cells, changes that likely contribute to an increase in outflow resistance.[55] The stiffness positively correlates with elevated altered expression of several key genes, particularly of CTGF.[55] Increase in cellular stiffness may be the common causative theme for the increase in outflow resistance in the trabecular outflow pathways.

REFERENCES

1. Resnikoff S, Pascolini D, Etya'ale D, et al. Global data on visual impairment in the year 2002. *Bull World Health Organ.* 2004;82(11):844–851.
2. Leske MC, Heijl A, Hussein M, Bengtsson B, Hyman L, Komaroff E. Factors for glaucoma progression and the effect of treatment: the early manifest glaucoma trial. *Arch Ophthalmol.* 2003;121(1):48–56.

3. Anderson DR. Collaborative normal tension glaucoma study. *Curr Opin Ophthalmol.* 2003;14(2):86–90.
4. Kass MA, Heuer DK, Higginbotham EJ, et al. The ocular hypertension treatment study: a randomized trial determines that topical ocular hypotensive medication delays or prevents the onset of primary open-angle glaucoma. *Arch Ophthalmol.* 2002;120(6): 701–713, discussion 829–730.
5. Gordon MO, Beiser JA, Brandt JD, et al. The ocular hypertension treatment study: baseline factors that predict the onset of primary open-angle glaucoma. *Arch Ophthalmol.* 2002;120(6):714–720, discussion 829–730.
6. The AGIS Investigators. The advanced glaucoma intervention study (AGIS): 7. The relationship between control of intraocular pressure and visual field deterioration. *Am J Ophthalmol.* 2000;130(4):429–440.
7. Collaborative Normal-Tension Glaucoma Study Group. Comparison of glaucomatous progression between untreated patients with normal-tension glaucoma and patients with therapeutically reduced intraocular pressures. *Am J Ophthalmol.* 1998;126(4):487–497.
8. Collaborative Normal-Tension Glaucoma Study Group. The effectiveness of intraocular pressure reduction in the treatment of normal-tension glaucoma. Collaborative normal-tension glaucoma study group. *Am J Ophthalmol.* 1998;126(4):498–505.
9. Tamm ER. Glaucoma opinion: a unifying concept for trabecular meshwork pathology in primary open angle glaucoma. *Int Glaucoma Rev.* 2013;15(2):45–47.
10. Tamm ER. The role of the ciliary body in aqueous humor dynamics: structural aspects. In: Besharse J, Dana R, Dartt D, eds. *Encyclopedia of the Eye.* Oxford: Academic Press; 2009:179–186.
11. Tamm ER. The trabecular meshwork outflow pathways: structural and functional aspects. *Exp Eye Res.* 2009;88(4):648–655.
12. Tamm ER. The trabecular meshwork outflow pathways. Functional morphology and surgical aspects. In: Shaarawy TM, Sherwood MB, Hitchings RA, Crowston JG, eds. 1st ed. Philadelphia: Saunders, Elsevier; 2009:31–44. Glaucoma; vol II.
13. Tamm ER, Toris CB, Crowston JG, et al. Basic science of intraocular pressure. In: Weinreb RN, Brandt JD, Garway-Heath D, Medeiros F, eds. *Intraocular Pressure. Reports and Consensus Statements of the 4th Global AIGS Consensus Meeting on Intraocular Pressure*, Amsterdam: Kugler Publications; 2007:1–14.
14. Alm A, Nilsson SF. Uveoscleral outflow—a review. *Exp Eye Res.* 2009;88(4):760–768.
15. Bill A, Phillips C. Uveoscleral drainage of aqueous humor in human eyes. *Exp Eye Res.* 1971;12:275.
16. Johnson M. What controls aqueous humour outflow resistance? *Exp Eye Res.* 2006;82(4):545–557.
17. Cvekl A, Tamm ER. Anterior eye development and ocular mesenchyme: new insights from mouse models and human diseases. *Bioessays.* 2004;26(4):374–386.
18. Braunger BM, Ademoglu B, Koschade SE, et al. Identification of adult stem cells in Schwalbe's line region of the primate eye. *Invest Ophthalmol Vis Sci.* 2014;55(11): 7499–7507.
19. Acott TS, Kelley MJ. Extracellular matrix in the trabecular meshwork. *Exp Eye Res.* 2008;86(4):543–561.
20. Lütjen-Drecoll E, Rohen JW. Functional morphology of the trabecular meshwork. In: Tasman W, Jaeger EA, eds. *Duane's Foundations of Clinical Ophthalmology.* Philadelphia: J.B. Lippincott Company; 2001:1–30.
21. Lütjen-Drecoll E. Functional morphology of the trabecular meshwork in primate eyes. *Prog Retin Eye Res.* 1999;18(1):91–119.
22. Hann CR, Springett MJ, Wang X, Johnson DH. Ultrastructural localization of collagen IV, fibronectin, and laminin in the trabecular meshwork of normal and glaucomatous eyes. *Ophthalmic Res.* 2001;33(6):314–324.

23. Herrnberger L, Ebner K, Junglas B, Tamm ER. The role of plasmalemma vesicle-associated protein (PLVAP) in endothelial cells of Schlemm's canal and ocular capillaries. *Exp Eye Res.* 2012;105:27–33.
24. Herrnberger L, Seitz R, Kuespert S, Bösl MR, Fuchshofer R, Tamm ER. Lack of endothelial diaphragms in fenestrae and caveolae of mutant Plvap-deficient mice. *Histochem Cell Biol.* 2012;138:709–724.
25. Herrnberger L, Hennig R, Kremer W, et al. Formation of fenestrae in murine liver sinusoids depends on plasmalemma vesicle-associated protein and is required for lipoprotein passage. *PLoS One.* 2014;9(12):e115005.
26. Bradley JM, Vranka J, Colvis CM, et al. Effect of matrix metalloproteinases activity on outflow in perfused human organ culture. *Invest Ophthalmol Vis Sci.* 1998;39(13): 2649–2658.
27. Ethier CR, Kamm RD, Palaszewski BA, Johnson MC, Richardson TM. Calculations of flow resistance in the juxtacanalicular meshwork. *Invest Ophthalmol Vis Sci.* 1986;27(12): 1741–1750.
28. Seiler T, Wollensak J. The resistance of the trabecular meshwork to aqueous humor outflow. *Graefes Arch Clin Exp Ophthalmol.* 1985;223:88–91.
29. Gong H, Ruberti J, Overby D, Johnson M, Freddo TF. A new view of the human trabecular meshwork using quick-freeze, deep-etch electron microscopy. *Exp Eye Res.* 2002;75(3):347–358.
30. Bill A, Svedbergh B. Scanning electron microscopic studies of the trabecular meshwork and the canal of Schlemm—an attempt to localize the main resistance to outflow of aqueous humor in man. *Acta Ophthalmol (Copenh).* 1972;50(3):295–320.
31. Ethier CR, Coloma FM, Sit AJ, Johnson M. Two pore types in the inner-wall endothelium of Schlemm's canal. *Invest Ophthalmol Vis Sci.* 1998;39(11):2041–2048.
32. Sit AJ, Coloma FM, Ethier CR, Johnson M. Factors affecting the pores of the inner wall endothelium of Schlemm's canal. *Invest Ophthalmol Vis Sci.* 1997;38(8): 1517–1525.
33. Overby DR, Stamer WD, Johnson M. The changing paradigm of outflow resistance generation: towards synergistic models of the JCT and inner wall endothelium. *Exp Eye Res.* 2009;88(4):656–670.
34. Tamm E, Flügel C, Stefani FH, Rohen JW. Contractile cells in the human scleral spur. *Exp Eye Res.* 1992;54:531–543.
35. Rohen JW, Lütjen E, Bárány EH. The relation between the ciliary muscle and the trabecular meshwork and its importance for the effect of miotics on aqueous outflow resistance. A study in two contrasting monkey species, *Macaca irus* and *Cercopithecus aethiops*. *Albrecht Von Graefes Arch Klin Exp Ophthalmol.* 1967;172:23–47.
36. Kaufman PL, Bárány EH. Loss of acute pilocarpine effect on outflow facility following surgical disinsertion and retrodisplacement of the ciliary muscle from the scleral spur in the cynomolgus monkey. *Invest Ophthalmol Vis Sci.* 1976;15:793–807.
37. Hinz B, Phan SH, Thannickal VJ, Galli A, Bochaton-Piallat ML, Gabbiani G. The myofibroblast: one function, multiple origins. *Am J Pathol.* 2007;170(6):1807–1816.
38. Tamm ER. Functional morphology of the outflow pathways of aqueous humor and their changes in open angle glaucoma. *Ophthalmologe.* 2013;110(11):1026–1035.
39. Tamm ER, Flügel C, Stefani FH, Lütjen-Drecoll E. Nerve endings with structural characteristics of mechanoreceptors in the human scleral spur. *Invest Ophthalmol Vis Sci.* 1994;35:1157–1166.
40. Flügel C, Tamm E, Lütjen-Drecoll E. Different cell populations in bovine trabecular meshwork: an ultrastructural and immunohistochemical study. *Exp Eye Res.* 1991;52: 681–690.
41. Wiederholt M, Thieme H, Stumpff F. The regulation of trabecular meshwork and ciliary muscle contractility. *Prog Retin Eye Res.* 2000;19(3):271–295.

42. Tian B, Gabelt BT, Geiger B, Kaufman PL. The role of the actomyosin system in regulating trabecular fluid outflow. *Exp Eye Res*. 2009;88(4):713–717.
43. Stamer WD, Lei Y, Boussommier-Calleja A, Overby DR, Ethier CR. eNOS, a pressure-dependent regulator of intraocular pressure. *Invest Ophthalmol Vis Sci*. 2011;52(13):9438–9444.
44. Tektas OY, Lütjen-Drecoll E. Structural changes of the trabecular meshwork in different kinds of glaucoma. *Exp Eye Res*. 2009;88(4):769–775.
45. Rohen JW, Futa R, Lütjen-Drecoll E. The fine structure of the cribriform meshwork in normal and glaucomatous eyes as seen in tangential sections. *Invest Ophthalmol Vis Sci*. 1981;21:574–585.
46. Rohen JW, Witmer R. Electron microscopic studies on the trabecular meshwork in glaucoma simplex. *Albrecht Von Graefes Arch Klin Exp Ophthalmol*. 1972;183:251–266.
47. Gottanka J, Johnson DH, Martus P, Lutjen-Drecoll E. Severity of optic nerve damage in eyes with POAG is correlated with changes in the trabecular meshwork. *J Glaucoma*. 1997;6(2):123–132.
48. Fuchshofer R, Tamm ER. The role of TGF-beta in the pathogenesis of primary open-angle glaucoma. *Cell Tissue Res*. 2012;347(1):279–290.
49. Fuchshofer R, Tamm ER. Modulation of extracellular matrix turnover in the trabecular meshwork. *Exp Eye Res*. 2009;88(4):683–688.
50. Junglas B, Yu AH, Welge-Lussen U, Tamm ER, Fuchshofer R. Connective tissue growth factor induces extracellular matrix deposition in human trabecular meshwork cells. *Exp Eye Res*. 2009;88(6):1065–1075.
51. Junglas B, Kuespert S, Seleem AA, et al. Connective tissue growth factor causes glaucoma by modifying the actin cytoskeleton of the trabecular meshwork. *Am J Pathol*. 2012;180(6):2386–2403.
52. Fuchshofer R, Ullmann S, Zeilbeck LF, Baumann M, Junglas B, Tamm ER. Connective tissue growth factor modulates podocyte actin cytoskeleton and extracellular matrix synthesis and is induced in podocytes upon injury. *Histochem Cell Biol*. 2011;136(3):301–319.
53. Last JA, Pan T, Ding Y, et al. Elastic modulus determination of normal and glaucomatous human trabecular meshwork. *Invest Ophthalmol Vis Sci*. 2011;52(5):2147–2152.
54. Pattabiraman PP, Rinkoski T, Poeschla E, Proia A, Challa P, Rao PV. RhoA GTPase-induced ocular hypertension in a rodent model is associated with increased fibrogenic activity in the trabecular meshwork. *Am J Pathol*. 2015;185(2):496–512.
55. Overby DR, Zhou EH, Vargas-Pinto R, et al. Altered mechanobiology of Schlemm's canal endothelial cells in glaucoma. *Proc Natl Acad Sci USA*. 2014;111(38):13876–13881.

CHAPTER NINETEEN

Glaucoma Genes and Mechanisms

Janey L. Wiggs[1]
Harvard Medical School, and Massachusetts Eye and Ear Infirmary, Boston, Massachusetts, USA
[1]Corresponding author e-mail address: Janey_Wiggs@meei.harvard.edu

Contents

Progress in Molecular Biology and Translational Science, Volume 134
ISSN 1877-1173
http://dx.doi.org/10.1016/bs.pmbts.2015.04.008
© 2015 Elsevier Inc.
All rights reserved.

Abstract

Genetic studies have yielded important genes contributing to both early-onset and adult-onset forms of glaucoma. The proteins encoded by the current collection of glaucoma genes participate in a broad range of cellular processes and biological systems. Approximately half the glaucoma-related genes function in the extracellular matrix, however proteins involved in cytokine signaling, lipid metabolism, membrane biology, regulation of cell division, autophagy, and ocular development also contribute to the disease pathogenesis. While the function of these proteins in health and disease are not completely understood, recent studies are providing insight into underlying disease mechanisms, a critical step toward the development of gene-based therapies. In this review, genes known to cause early-onset glaucoma or contribute to adult-onset glaucoma are organized according to the cell processes or biological systems that are impacted by the function of the disease-related protein product.

1. INTRODUCTION

Glaucoma is a collection of disorders that result in degeneration of the optic nerve. Common forms of glaucoma (primary open-angle glaucoma (POAG), normal-tension glaucoma (NTG), exfoliation glaucoma (XFG), and angle-closure glaucoma) are leading causes of irreversible blindness worldwide. Angle-closure glaucoma, caused by anatomical narrowing of the iridocorneal angle with subsequent blockage of the trabecular outflow pathways, is particularly common in Asia.[1] XFG develops in patients with exfoliation syndrome (XFS) characterized by the deposition of a heterogeneous mix of aggregated macromolecules throughout the ocular anterior segment.[2] POAG is the most common subtype of glaucoma and is defined as glaucoma occurring in the absence of any secondary features such as exfoliation. Approximately one-third of patients with open-angle glaucoma have NTG with progressive optic nerve degeneration despite intraocular pressure (IOP) in the normal range. Early-onset glaucoma (developing before the age of 40) includes juvenile open-angle glaucoma, developmental glaucoma (related to abnormal development of the ocular anterior segment), and congenital glaucoma (developing at birth or with the first three years of life).

Glaucoma has significant heritability with early-onset forms inherited as Mendelian autosomal dominant or recessive traits and adult-onset (after age 40) diseases inherited as complex traits.[3] Mutations in genes causing early-onset glaucoma are rare and have a large biological impact, while variants contributing to various forms of adult-onset glaucoma are common and individually have incremental effects on disease pathogenesis (Fig. 1). Genes responsible for early-onset glaucoma have been discovered using linkage analysis of large families while genes contributing to adult-onset disease have been identified using genome-wide association studies (GWAS) that typically require very large numbers of glaucoma cases and controls.

Current molecular techniques and approaches, especially GWAS and next-generation sequencing, are successfully identifying glaucoma-associated genes. While the function of these genes in health and disease are not completely understood, a number of cellular processes and systems with relevance to disease development are emerging. In this review, genes known to cause early-onset glaucoma or contribute to adult-onset glaucoma

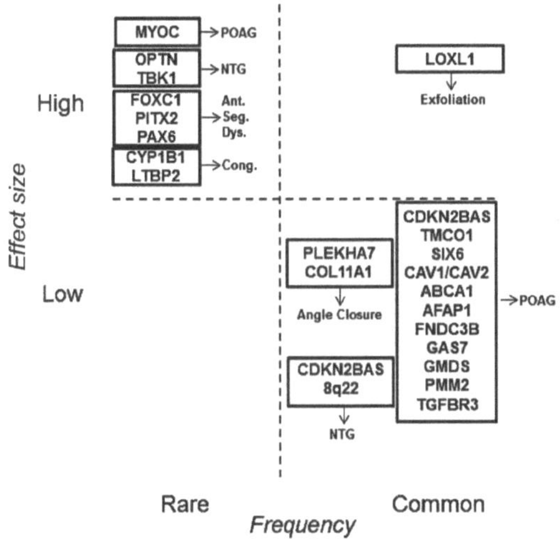

Figure 1 Frequency and effect size of gene variants in glaucoma. Mutations in genes causing early-onset Mendelian forms of glaucoma are rare but have large biological effects (*MYOC, OPTN, TBK1, FOXC1, PITX2, PAX6, CYP1B1, LTBP1*). Variants in genes influencing the susceptibility to adult-onset forms of glaucoma with complex inheritance are generally relatively common and individually have small biological effects (*CDKN2BAS, TMCO1, SIX6, CAV1/CAV2, ABCA1, AFAP1, FNDC3B, GAS7, PLEKHA7, GMDS, PMM2, TGFBR3, COL11A1, 8q22*). *COL15A1* and *COL18A1*, modifiers of early-onset glaucoma, have intermediate frequency and effect size (not shown in this figure). Variants in *LOXL1* contributing to exfoliation syndrome are common but also have large biological effects.

are organized according to the cell processes or biological systems that are impacted by the function of the disease-related protein product (Table 1). The types of glaucoma caused by (early-onset disease) or associated with (adult-onset disease) are listed in Table 2. The biological processes and systems implicated in glaucoma by genetic studies may be future targets of novel therapies that could prevent disease development.

2. ENDOPLASMIC RETICULUM STRESS RESPONSE

The endoplasmic reticulum (ER) stress response can be caused by overexpression of genes or gene mutations that lead to protein aggregation or other processes that prevent the nascent polypeptide from progressing through the ER.[92] Genetic abnormalities that disrupt the normal function of the ER can lead to the unfolded protein response which can trigger cell death.[93] ER-related cell death contributes to several diseases, including neurodegenerative disorders, renal disease, and ocular disease.[94–96] Missense mutations in MYOC, coding for myocilin, are likely to cause ER stress through the misfolded protein response[4,5,70].

2.1 *MYOC* (Myocilin)

MYOC mutations are a cause of early-onset POAG (juvenile-onset open-angle glaucoma). The majority of disease-causing mutations are dominantly inherited missense alleles. There is a range in disease onset with some mutations causing disease during the first decade of life (PRO370LEU, TYR357HIS) while others cause later-onset.[71] A nonsense mutation (GLN368X) is a common mutation in individuals with later-onset disease.[72] Interestingly, whole-gene deletions or a nonsense mutation near the N-terminal do not cause disease suggesting that the underlying disease mechanism is not loss of function but dominant negative or gain of function.[97–99] Using an *in vitro* system, *MYOC* mutations were shown to decrease protein solubility and mutant protein solubility was correlated with disease severity.[100] In a Tyr357His transgenic mouse (one of the most insoluble mutations), mutant myocilin accumulates in the ER causing the misfolded protein response[5] and elevation of IOP. Sodium 4-phenylbutyrate, a molecular chaperone known to relieve the misfolded protein response in urea cycle disorders,[101] also relieved ER stress and lowered IOP in this animal model.[4] Reagents that relieve the presumed misfolded protein response in humans could be developed as novel therapies for patients with glaucoma caused by *MYOC* mutations.

Table 1 Biological Systems and Processes Involved in Inherited Glaucoma

Biological Process or System	**Gene**	**Protein**	**Key References**
Endoplasmic reticulum stress, unfolded protein response	*MYOC*	Myocilin	4,5
Extracellular matrix, cell junctions, and cell adhesion	*LTBP2*	Latent TGF-binding protein 2	6–9
	LOXL1	Lysyl oxidase like 1	10
	FNDC3B	Fibronectin type III domain containing 3B	11
	AFAP1	Actin filament-associated protein 1	12
	PLEKHA7	Pleckstrin homology domain-containing protein 7	13
	COL11A1	Collagen type XI, alpha1	14,15
	COL15A1	Collagen type XV, alpha1	16–18
	COL18A1	Collagen XVIII, alpha1	16–18
TGF beta signaling	*CDKN2BAS*	Cyclin-dependent kinase inhibitor 2B antisense	19–21
	TGFBR3	TGFbeta receptor 3	
TNF-alpha signaling	*OPTN*	Optineurin	22,23
	TBK1	Tank-binding kinase 1	24,25
Regulation of autophagy	*OPTN*	Optineurin	26
	TBK1	Tank-binding kinase 1	27
Lipid metabolism	*ABCA1*	ATP-binding cassette, subfamily A (ABC1) member 1	28,29
eNOS signaling and Caveolae	*CAV1/ CAV2*	Caveolins 1 and 2	30,31
Fructose and Mannose Metabolism	*GMDS*	GDP-mannose 4.6 dehydratase	32,33
	PMM2	Phosphomannomutase	

Continued

Table 1 Biological Systems and Processes Involved in Inherited Glaucoma—cont'd

Biological Process or System	Gene	Protein	Key References
Regulation of cell division	*GAS7*	Growth arrest-specific 7	34–38
	TMCO1	Transmembrane and coiled-coil domains-1	39
	CDKN2BAS	Cyclin-dependent kinase inhibitor 2B antisense	19–21
Regulation of ocular development	*FOXC1*	Forkhead box C1	40,41
	PITX2	Paired-like homeodomain 2	42,43
	PAX6	Paired box 6	44
	CYP1B1	Cytochrome P450, family 1, subfamily B, polypeptide 1	45,46
	LTBP2	Latent TGF-binding protein 2	47
	SIX6	SIX homeobox 6	48,49
Cerebrospinal fluid pressure	*8q22*	NA	50

Table 2 Glaucoma Genes and Diseases

Disease	Gene	Protein	Key References
Congenital glaucoma	*CYP1B1*	Cytochrome P450, family 1, subfamily B, polypeptide 1	51–53
	LTBP2	Latent TGF-binding protein 2	54,55
Anterior segment dysgenesis, Rieger syndrome	*FOXC1*	Forkhead box C1	56–59
	PITX2	Paired-like homeodomain 2	60–62
Aniridia	*PAX6*	Paired box 6	63–65
Microsperophakia, Weill Marchasani	*LTBP2*	Latent TGF-binding protein 2	66–68
Microphthalmia and cataract	*SIX6*	SIX homeobox 6	69

Table 2 Glaucoma Genes and Diseases—cont'd

Disease	Gene	Protein	Key References
Juvenile-onset primary open-angle glaucoma	*MYOC*	Myocilin	5,70–72
	COL15A1	Collagen type XV, alpha1	73
	COL18A1	Collagen XVIII, alpha1	73
Adult-onset primary open-angle glaucoma	*ABCA1*	ATP-binding cassette, subfamily A (ABC1) member 1	12,74
	AFAP1	Actin filament-associated protein 1	12
	CAV1/ CAV2	Caveolins 1 and 2	75–79
	CDKN2BAS	Cyclin-dependent kinase inhibitor 2B antisense	80
	FNDC3B	Fibronectin type III domain containing 3B	81,82
	GAS7	Growth arrest-specific 7	78,83
	GMDS	GDP-mannose 4,6 dehydratase	12
	PMM2	Phosphomannomutase	74
	SIX6	SIX homeobox 6	50,84
	TGFBR3	TGFbeta receptor 3	85
	TMCO1	Transmembrane and coiled-coil domains-1	80
Familial normal-tension glaucoma	*OPTN*	Optineurin	86–88
	TBK1	Tank-binding kinase 1	25
Adult-onset normal-tension glaucoma	*8q22*	NA	50
	CDKN2BAS	Cyclin-dependent kinase inhibitor 2B antisense	50,89,90
Angle-closure glaucoma	*COL11A1*	Collagen type XI, alpha1	14
	PLEKHA7	Pleckstrin homology domain-containing protein 7	14
Exfoliation syndrome	*LOXL1*	Lysyl oxidase like 1	91

3. EXTRACELLULAR MATRIX, CELL JUNCTIONS, AND CELL ADHESION

Processes involving extracellular matrix, especially in the trabecular meshwork, can influence aqueous outflow and elevation of IOP.[102] A number of glaucoma-related genes code for proteins that function in the extracellular matrix. Additionally, two glaucoma-associated genes influence cell junctions and several extracellular collagens also contribute to glaucoma pathogenesis.

3.1 *LTBP2* (Latent TGF-Binding Protein 2)

LTBP2 loss of function mutations can cause a range of ocular phenotypes including autosomal recessive congenital glaucoma,[54,55] microsperophakia,[66,67] and Weill Marchesani syndrome.[68] Recent work suggests that impaired *LTBP2* function primarily causes abnormal development of the ciliary zonules resulting in lens dislocations and other abnormalities.[47,103] *LTBP2* codes for latent TGF-binding protein 2, an extracellular matrix protein that is associated with microfibrils.[6] LTBP2 also has cell-adhesive properties[7,8] and a functional role in elastic fiber assembly.[9] A possible role in TGF beta signaling is not well established despite the protein homology to LTBP1 which does bind TGF beta.[104]

3.2 *LOXL1* (Lysyl Oxidase Like 1)

Common variants in *LOXL1* are significantly associated with XFS, an ocular condition characterized by the distribution of aggregated macromolecules throughout the eye including the trabecular outflow pathways.[91] XFS is the leading cause of secondary open-angle glaucoma (XFG) worldwide.[105] Lysyl oxidase like 1 is necessary for proper elastin formation and maintenance, a critical component of the extracellular matrix.[10] Initially, *LOXL1* missense alleles were associated with disease in populations worldwide,[106] however subsequent studies showed that the missense alleles are not likely to impact the amine oxidase activity of the enzyme.[107] It is likely that dysregulation of *LOXL1* is associated with disease development[108] and *LOXL1* variants influencing gene expression could influence disease risk. A *LOXL1* null mouse has some, but not all, features of XFS.[109]

3.3 *FNDC3B* (Fibronectin Type III Domain Containing 3B)

Recent GWAS have identified common variants within and near *FNDC3B* associated with elevated IOP[81] and thin central corneal thickness,[82] both risk factors for POAG. *FNDC3B* codes for an extracellular matrix protein involved in several signaling pathways, including PI3-kinase/Akt, Rb1 and TGFβ signaling.[11] TGF beta is known to induce extracellular matrix remodeling and can alter the cytoskeleton though both the canonical Smad and noncanonical signaling pathways.[110] Experiments have shown that variation in TGF beta2 can cause elevation of IOP in an *ex vivo* perfusion organ culture model, and also elevation of IOP in rodent eyes.[111]

3.4 *AFAP1* (Actin Filament-Associated Protein 1)

The protein encoded by *AFAP1* binds to actin filaments and promotes crosslinking. Modulation of the actin cytoskeleton is known to contribute to the regulation of aqueous outflow and IOP.[112] While not technically in the extracellular matrix, actin formation and stability can have a significant effect on cell shape, cell adhesion, and responsiveness to external stimuli presented by the extracellular matrix. A recent GWAS involving 4702 POGA cases and 9695 controls identified significant association of common *AFAP1* variants with POAG.[12] Immunohistochemistry showed that AFAP1 was present in both anterior segment and poster segment tissues, particularly trabecular meshwork and retinal astrocytes.

3.5 *PLEKHA7* (Pleckstrin Homology Domain-Containing Protein 7)

PLEKHA7 codes for a protein necessary for formation of adherens junctions that can control paracellular permeability.[13] This protein is expressed in ocular ciliary body and choroid and common variants in the gene have been associated with angle-closure glaucoma.[14] Hypothetically *PLEKHA7* may be related to angle closure though a mechanism involving compromise of the normal barrier to fluid leakage and/or aberrant fluid dynamics.

3.6 *COL11A1* (Collagen Type XI, Alpha1)

COL11A1 encodes one of the two α chains of type XI collagen. Common variants in *COL11A1* are associated with angle-closure glaucoma in Asian populations.[14] Rare *COL11A1* high-effect mutations cause Marshall syndrome, Stickler syndrome, type 2 or Stickler-like syndrome.[15] These syndromes include axial myopia, probably caused by abnormal fibrillar collagen matrix in the sclera. As angle-closure glaucoma patients generally are hyperopic, the associated *COL11A1* variants may influence (or are in linkage disequilibrium with variants that influence) *COL11A1* expression resulting in smaller hyperopic eyes predisposed to angle closure. Alternatively, as *COL11A1* is also expressed in the trabecular meshwork a direct effect on aqueous outflow is also possible.

3.7 *COL15A1* (Collagen Type XV, Alpha1) and *COL18A1* (Collagen XVIII, Alpha1)

COL15A1 and *COL18A1* code for multiplexin collagens type XV and XVIII. These two proteins are highly homologous and are localized to the extracellular matrix and basement membranes in multiple ocular tissues including the trabecular meshwork and Schlemm's canal.[16–18] Variants in both *COL15A1* and *COL18A1* appear to act as disease modifiers influencing the age of disease onset in families with early-onset glaucoma.[73] Interestingly, the *COL18A1* variant was only found in families who also carried the disease-causing *MYOC* mutation Gln368X, one of the *MYOC* mutations related to milder disease. This result suggests that there could be a specific interaction between *COL18A1* and *MYOC*, however further study is necessary to confirm this. The effect of the *COL15A1* and *COL18A1* variants could impair the stability of the trabecular outflow pathways including Schlemm's canal thereby reducing aqueous outflow and raising IOP which results in more severe disease at an earlier age compared to a family member with the primary mutation who does not also carry one of the collagen variants. More work will be necessary to confirm this hypothesis.

4. TGF BETA SIGNALING

4.1 *CDKN2BAS* (Cyclin-Dependent Kinase Inhibitor 2B Antisense)

TGF beta signaling is well known to contribute to processes involving the ocular anterior segment and IOP[113] as well as the glaucoma-related neurodegenerative processes involving the optic nerve.[114] Generally, TGF beta inhibits cell cycle progression resulting in terminal differentiation or in some situations, apoptosis. In astrocytes, TGF beta signaling is SMAD dependent and increased TGF beta can lead to an increase in *CDKN2B* which is an inhibitor of *CDK4/6* (cyclin dependent kinase 4 and 6), which are necessary for cell cycle progression (Fig. 2). Excess *CDKN2B* inhibits cell cycle

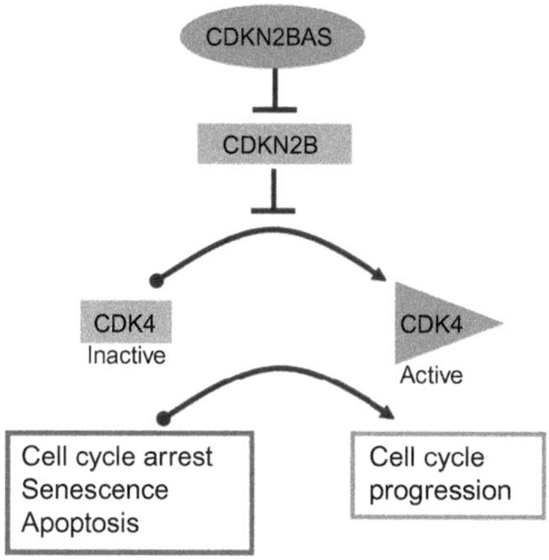

Figure 2 *CDKN2BAS* and cell cycle progression. *CDKN2BAS* (*Cyclin dependent kinase inhibitor 2B antisense*) is a long noncoding antisense RNA that regulates expression of *CDKN2B (cyclin dependent kinase inhibitor 2B)*, coding for an inhibitor of CDK4 (cyclin-dependent kinase 4) necessary for cell cycle progression.[19–21]

progression leading to apoptosis.[115] *CDKN2BAS* is an antisense RNA (also known as ANRIL) that regulates expression of CDKN2B among other molecules.[19–21] Common variants in the CDKN2BAS region are associated with POAG overall and in particular the NTG subgroup.[50,80,89,90] The minor allele of the *CDKN2BAS* variant with most robust association is protective. The role of CDKN2BAS in glaucoma is not yet defined. One hypothesis is that the associated CDKN2BAS allele (minor allele) could result in decreased expression of *CDKN2B* which would increase activity of *CDK4/6* promoting cell cycle progression (Fig. 2). Interestingly, the opposite allele of the variant associated with glaucoma is associated with glioma (lack of appropriate reduction in cell cycle progression) which provides some support for this hypothesis.[116] It is also possible that the associated SNPs are in linkage disequilibrium with other genomic variants within CDKN2BAS or other nearby genes that have a direct role in disease development.[117] Further experimentation is needed to clarify the role of this important molecule in glaucoma.

4.2 *TGFBR3* (TGFbeta Receptor 3)

Recently an association between a SNP near the gene coding for TGFbeta receptor 3 (TGFBR3) was identified for POAG in a mult-ethnic cohort, further supporting a role for TGFbeta signaling in this disease.[85]

5. TUMOR NECROSIS FACTOR-ALPHA SIGNALING

5.1 *OPTN* (Optineurin), *TBK1* (Tank-Binding Kinase 1)

Tumor necrosis factor alpha (TNF-alpha) is a proinflammatory cytokine that may contribute to retinal ganglion cell death in glaucoma.[118,119] TNF-alpha binding to its receptor initiates a cascade of events that can activate NFkBeta. Under normal conditions *OPTN* (coding for optineurin) is a negative regulator of NFkBeta activation.[22,23] A rare *OPTN* missense mutation (E50K) causes familial NTG,[86–88] and some *OPTN* mutations are also known to cause familial amyotrophic lateral sclerosis.[120] Mutant forms of *OPTN* do not efficiently inhibit TNF-alpha stimulated NFkBeta transcription which may lead to increased transcription of proapoptotic genes and cell death.[121] Interestingly, a second protein responsible for familial NTG, TBK1 (Tank-binding kinase 1) interacts with OPTN and this interaction is enhanced in an optineurin mutant (E50K) suggesting that binding of TBK1 to mutant OPTN could prevent the protein from inhibiting NFkBeta activation.[24,25] CLYD (cylindromatosis turban tumor syndrome protein), another negative inhibitor of TNF-alpha induced NFkB activation has also been shown to interact with Optineurin and the interaction is also increased by OPTN mutations.[23] Together, these results suggest that TNF-alpha induced NFkB transcription is detrimental to ganglion cells in glaucoma and that loss of inhibitors normally present to modulate this process (OPTN, CLYD) can result in severe familial optic nerve disease.

6. REGULATION OF AUTOPHAGY

6.1 *OPTN* (Optineurin), *TBK1* (Tank-Binding Kinase 1)

In addition to roles in TNF-alpha signaling there is also evidence that *OPTN* and *TBK1* regulate autophagy, a process that eliminates accumulation or proteins, organelles and other cellular debris.[122] Autophagy has been observed in glaucoma animal models.[123,124] Both *TBK1* and *OPTN* are capable of regulating autophagy. When *OPTN* is upregulated or mutated, autophagy is activated in neuronal cells.[26] Phosphorylation of OPTN by TBK1 promotes the recruitment of microtubule-associated protein 1 light chain 3 beta (*MAP1LC3B, LC3B*) a critical step in the formation of autophagosomes and initiation of autophagy.[27] The OPTN E50K–TBK1 enhanced interaction also promotes protein instability that can lead to autophagy.[125] Using iPSCs derived from patients with

TBK1 duplication increased expression of LC3-II a key marker of activation of autophagy was observed.[126] Data from *OPTN* and *TBK1* experiments suggests that autophagy may be an important pathway in the development of NTG.

7. LIPID METABOLISM

7.1 *ABCA1* (ATP-Binding Cassette, Subfamily A (*ABC1*) Member 1)

Recent GWAS have implicated *ABCA1* (coding for ATP-binding cassette, subfamily A (ABC1) member 1) in POAG.[12,74] Variants in this gene are also associated with IOP in normal populations.[81] The protein encoded by this gene is a major regulator of cellular cholesterol and phospholipid homeostasis.[28,29] ABCA1 is expressed in ocular tissues relevant to glaucoma including iris, ciliary body, retina, optic nerve head, optic nerve, and trabecular meshwork.[12] Previous studies using the DBA/2J mouse glaucoma model identified Abca1 in a cluster of transcripts with varied expression in response to ganglion cell death.[127] ABCA1 expression has also been reported to be higher in leukocytes from glaucoma patients.[128] A role for lipid metabolism in glaucoma is also supported by the protective effect of statins in patients with hyperlipidemia.[129]

8. ENDOTHELIAL NITRIC OXIDE SYNTHETASE SIGNALING AND CAVEOLAE

8.1 *CAV1/CAV2* (Caveolins 1 and 2)

Caveolae are invaginations of the plasma membrane formed primarily by the caveolin proteins.[30] These are especially common in vascular endothelial cells but can be present in many vertebrate cell types. Common variants near *CAV1*, coding for caveolin 1 have been associated with POAG in populations worldwide.[75–79] While dysregulation of *CAV1* has many downstream effects, one consequence of *CAV1* deficiency is activation of endothelial nitric oxide synthetase (eNOS) with subsequent increase in nitric oxide (NO).[31] NO can modulate the tone of luminal structures with adjacent smooth muscle including blood vessels and ocular structures such as Schlemm's canal and juxtacanalicular trabecular outflow pathways.[130,131] eNOS may play a role in the etiology of glaucoma; it is found in the human outflow pathway[132] and the vasculature supplying retinal ganglion cells,[133] which may affect the regulation of IOP and blood flow to the optic nerve, respectively. NOS3 (nitric oxide synthetase 3, coding for eNOS) may interact with estrogen to contribute to POAG in women.[134] Recently, *CAV1* variants were shown to be preferentially associated with the POAG subgroup with initial paracentral visual field loss, a phenotypic feature more common in POAG patients with evidence of vascular dysregulation.[135] Another potential role for caveolae and *CAV1* in POAG is the formation of "giant vaculoles" noted in trabecular meshwork cells in the setting of elevated IOP.[136]

9. FRUCTOSE AND MANNOSE METABOLISM

Two genes (GMDS and PMM2) coding for enzymes in the fructose and mannose metabolism pathway have been associated with POAG. Interestingly, highly penetrant alleles in both of these genes cause congenital glycosylation disorders[32] and both genes code for enzymes involved in different steps of the overall fructose-mannose metabolism pathway (KEGG pathway hsa00051). One of the products of this pathway is N-glycans that have a role in a number of cellular processes including targeting of proteins to lysosomes for degradation.[33]

9.1 *GMDS* (GDP-mannose 4,6 dehydratase)

GMDS has been associated with POAG in a GWAS of Caucasians with European ancestry.[12]

9.2 *PMM2* (Phosphomannomutase)

PMM2 has been associated with POAG in a GWAS of Asians.[74]

10. REGULATION OF CELL DIVISION

Several genes coding for proteins that can regulate cell division contribute to adult-onset forms of glaucoma including POAG and NTG.

10.1 *GAS7* (Growth Arrest-Specific 7)

GAS7 is a member of the growth arrest-specific family of genes expressed in terminally differentiated tissues.[34] This member of the growth arrest-specific gene family is expressed primarily in terminally differentiated brain cells and mature cerebellar Purkinje neurons but also in terminally differentiated fibroblasts.[35] GAS7 can induce neurite outgrowth in terminal neurons[36] and is also involved in some developmental processes such as osteoblast cell differentiation from mesenchymal cells.[37,38] Common variants near GAS7 were initially associated with elevated IOP in normal populations[78,83] and subsequently also associated with POAG.[12,50] RT-PCR shows that GAS7 is expressed in trabecular meshwork cells as well as retina and optic nerve.[38] Considering the expression of GAS7 primarily in neurons and its function in neurite outgrowth a role for the protein in IOP regulation is not clear. As part of the neurite outgrowth function GAS7 interacts with actin and microfilaments, and it is possible that this interaction also occurs in trabecular meshwork cells or other cells involved in aqueous humor dynamics.[137]

10.2 *TMCO1* (Transmembrane and Coiled-Coil Domains-1)

Common variants near *TMCO1* on chromosome 1q24 were initially associated with IOP in normal populations[38] and subsequently with POAG in Caucasians of European ancestry.[80] Loss of function mutations cause a recessive condition involving craniofacial dysmorphism, skeletal anomalies, and mental retardation that has been termed, "TMCO1 syndrome."[138,139] Coding sequence mutations do not appear to contribute to glaucoma, including the pedigrees affected by TMCO1 syndrome. The protein sequence is very highly conserved across mammalian species suggesting a critical biological function, and it is expressed in many human tissues including ocular tissues.[39] Using a GFP-TMCO1 fusion protein, the protein was localized to the ER[140] and mitochondria.[141] More recently, using immunohistochemistry *TMCO1* localized to nucleoli suggesting that the protein could have a role in aging through cell-cycle regulation.[39]

10.3 *CDKN2BAS* (Cyclin-Dependent Kinase Inhibitor 2B Antisense)

CDKN2BAS codes for a long noncoding antisense RNA that negatively regulates the expression of *CDKN2B*, coding for an inhibitor of cyclin-dependent kinases 4 and 6 necessary for cell cycle progression.[19–21] Common variants in the *CDKN2BAS* genomic region are strongly associated with POAG and NTG suggesting that cell-cycle regulation is an important feature of disease development.[50,80,89,90] Unlike *TMCO1* and *GAS7* which are associated with IOP as well as POAG, *CDKN2BAS* was initially associated with the cup-to-disc ratio in normal populations[84] and is more robustly associated with NTG compared with POAG,[50,80,89,90] suggesting that the primary influence is on the optic nerve. The cells involved are not yet known.

11. REGULATION OF OCULAR DEVELOPMENT

A number of genes responsible for early-onset forms of glaucoma regulate ocular development. Mutations in these genes cause ocular dysgenesis, primarily of the anterior segment structures, resulting in elevated IOP and subsequent damage to the optic nerve. Currently one adult-onset gene, *SIX6* associated with POAG, codes for a protein involved in ocular development.

11.1 *FOXC1* (Forkhead Box C1)

FOXC1 codes for a member of the family of forkhead domain proteins involved in developmental processes in many human tissues.[142] In the human eye, *FOXC1* mutations cause a spectrum of phenotypic abnormalities that includes iris hypoplasia and other features of anterior segment dysgenesis.[56,57] In addition, some *FOXC1* mutations appear to cause hearing loss, and *FOXC1* may contribute to De Hauwere syndrome characterized by anterior segment dysgenesis, hypertelorism, retardation, hypotonia, hearing loss, femoral head anomalies, and hydrocephalus.[58,59] Disease-causing mutations in *FOXC1* include missense changes in the forkhead domain, nonsense and frameshift mutations, and whole-gene deletions and duplications.[57] Mutations resulting in disease cause a loss of protein function particularly of the transactivation domain[40,41] and are inherited as an autosomal dominant trait with variable penetrance. *FOXC1* interacts with *PITX2* (see later) and *PITX2* can negatively regulate *FOXC1* transactivity.[143,144] Moreover, patients who have mutations in both genes have more severe disease.[145] In additional to hearing defects, patients with anterior segment dysgenesis caused by *FOXC1* mutations may also have heart defects.[146]

11.2 *PITX2* (Paired-Like Homeodomain 2)

PITX2 is a member of the bicoid class of homeodomain transcription factors that are necessary for embryonic development.[42,43] *PITX2* mutations were initially identified as a cause of classic Rieger syndrome defined by characteristic teeth and umbilical abnormalities as well as ocular anterior segment dysgenesis.[60–62] *PITX2* is also necessary for pituitary development[147] and *PITX2* variants have been associated with cardiac abnormalities, in particular atrial fibrillation.[148] *PITX2* mutations causing Rieger syndrome are loss of function and are inherited as an autosomal dominant trait with variable penetrance.[57] Deletion of an upstream regulatory region can also cause disease.[149] As noted above, *PITX2* interacts with *FOXC1* defining an important pathway for ocular development.[143,144] Induction of *PITX2* expression requires the Wnt/Dvl/beta-catenin pathway that leads to cell-type-specific proliferation.[150] Approximately 50% of patients with ocular dysgenesis caused by either *PITX2* or *FOXC1* mutations have glaucoma[57] characterized by high IOP. Glaucoma is likely caused by abnormal development of the trabecular outflow pathways, and in particular Schlemm's canal.[151]

11.3 *PAX6* (Paired Box 6)

PAX6 plays a critical role in ocular development.[44] *PAX6* loss of function mutations cause Aniridia, characterized by abnormal development of the iris,[63] as well as Peter's anomaly[64] and dominant forms of corneal keratitis.[65] Approximately 50% of patients with ocular developmental abnormalities due to *PAX6* mutations also are affected by early-onset glaucoma.[63] Interestingly, deletions of the downstream *PAX6* regulatory region are relatively common disease-causing mutations.[152] Large deletions that include *PAX6* can also involve the gene responsible for Wilm's tumor[153] and patients with ocular phenotypes suggestive of Aniridia or other conditions related to *PAX6* defects should have renal ultrasound screening.

11.4 *CYP1B1* (Cytochrome P450, Family 1, Subfamily B, Polypeptide 1)

CYP1B1 loss of function mutations are the most common cause of autosomal recessive congenital glaucoma worldwide.[51] Reported *CYP1B1* mutations include missense, frameshift, premature stop codons, small insertion/deletions, and large deletions.[51–53] *CYP1B1* codes for cytochrome P-450 1B1, a member of the large cytochrome P450 family. P450 1B1 is known to metabolize complex molecules such as polycyclic aromatic hydrocarbons and 17-β-estradiol.[154–156] The role of the protein in congenital glaucoma is not clear; however, it has been hypothesized that the P-450 1B1 activity is responsible for metabolism of a compound involved in ocular development.[45,46] Recently, CYP1B1 mutations have also been shown to contribute to glaucoma in older children (juvenile-onset).[157–160] Mutations in juvenile-onset children (onset between the ages of 3 and 20) are primarily missense alleles which may confer some residual enzyme activity (hypomorph alleles).[161] A recent study has found that the carrier frequency of CYP1B1 mutations in two populations in the United States is higher than expected based on the disease incidence of congenital glaucoma only. In particular, the frequency of missense alleles was higher than expected suggesting that missense mutations may contribute to disease other than congenital glaucoma and this could be juvenile-onset glaucoma and glaucoma related to ocular dysgenesis.[162]

11.5 *LTBP2* (Latent TGF-Binding Protein 2)

As noted earlier, *LTBP2* loss of function mutations can cause a variety of ocular conditions which result from abnormal development of the ciliary zonules.[47,54,55,66–68,103] *LTBP2* null mice develop lens abnormalities including lens dislocation, but not primary congenital glaucoma, suggesting that lens dislocation could underlie the development of glaucoma in humans with LTBP2 mutations.[47]

11.6 *SIX6* (SIX Homeobox 6)

SIX6 is one member of a human gene family originally identified by homology to the Drosophila sine oculis (so) gene required for eye development.[48] All six members of the human *SIX* family have a DNA-binding homeobox domain as well as a SIX domain which binds effector molecules. The human *SIX* genes also appear to regulate eye development through transcriptional activation of downstream genes.[49] *SIX6* expression is restricted to the eye and pituitary[163] and loss of function mutations in this gene are a cause of isolated microphthalmia with cataract type 2 (MCOPCT2).[69] Common *SIX6* variants, including a common missense mutation, are associated with POAG.[50,84] The missense change His141Asn, has also been associated with retinal nerve fiber layer thickness suggesting that the associated gene variants increase susceptibility to POAG by limiting the development of the retinal ganglion cells.[164] Using zebrafish and a morpholino knock-down complementation assay His141Asn and several other missense alleles were found to reduce the size of the optic nerve providing further support for the hypothesis that *SIX6* risk variants disrupt the development of the neural retina, leading to a reduced number of retinal ganglion cells which increases the risk of glaucoma-associated vision loss.[165,166]

12. CEREBROSPINAL FLUID PRESSURE

12.1 *8q22* Regulatory Region

A NTG GWAS identified significant association with common SNPs located in an evolutionarily conserved genomic region on chromosome

8q22.[50] This region contains regulatory sites annotated by ENCODE[167] as enhancers with highest activity in the choroid plexus (produces cerebrospinal fluid) and the ocular ciliary body (produces aqueous humor). Of interest, recent studies have suggested that low cerebral spinal fluid pressure may create a deleterious gradient across the lamina cribrosa in NTG mimicking a similar gradient induced by higher IOP in typical high-pressure POAG.[168] The genes influenced by the enhancers are not yet known although SNPs in this region may impact TGF beta signaling.[50]

13. SUMMARY

Genetic studies have yielded important genes contributing to both early-onset and adult-onset forms of glaucoma. The proteins encoded by the current collection of glaucoma genes participate in a broad range of cellular processes and biological systems. Extracellular matrix proteins are especially prevalent among glaucoma genes; however, proteins involved in cytokine signaling, lipid metabolism, membrane biology, fructose and mannose metabolism, regulation of cell division, autophagy, and ocular development also contribute to disease pathogenesis. The genes currently known to contribute to glaucoma account for only a fraction of the overall disease heritability,[169] and GWAS with larger and better characterized patient cohorts and current next-generation sequencing approaches are needed for novel gene discovery. Delineating the complete genetic architecture of glaucoma will make it possible to develop sensitive and specific gene-based tests that could identify individuals at risk for disease before irreversible damage to the optic nerve occurs. The discovery of disease-related genes will also provide new insights into the underlying molecular mechanisms responsible for glaucoma, an important step toward achieving novel gene-based preventative and protective therapies.

REFERENCES

1. Tham YC, Li X, Wong TY, Quigley HA, Aung T, Cheng CY. Global prevalence of glaucoma and projections of glaucoma burden through 2040: a systematic review and meta-analysis. *Ophthalmology*. 2014;121(11):2081–2090.
2. Elhawy E, Kamthan G, Dong CQ, Danias J. Pseudoexfoliation syndrome, a systemic disorder with ocular manifestations. *Hum Genomics*. 2012;6:22.
3. Wang R, Wiggs JL. Common and rare genetic risk factors for glaucoma. *Cold Spring Harb Perspect Med*. 2014;4(12):a017244.
4. Zode GS, Bugge KE, Mohan K, et al. Topical ocular sodium 4-phenylbutyrate rescues glaucoma in a myocilin mouse model of primary open-angle glaucoma. *Invest Ophthalmol Vis Sci*. 2012;53(3):1557–1565.

5. Zode GS, Kuehn MH, Nishimura DY, et al. Reduction of ER stress via a chemical chaperone prevents disease phenotypes in a mouse model of primary open angle glaucoma. *J Clin Invest.* 2011;121(9):3542–3553.
6. Kuchtey J, Kuchtey RW. The microfibril hypothesis of glaucoma: implications for treatment of elevated intraocular pressure. *J Ocul Pharmacol Ther.* 2014;30(2-3): 170–180.
7. Hyytiäinen M, Keski-Oja J. Latent TGF-beta binding protein LTBP-2 decreases fibroblast adhesion to fibronectin. *J Cell Biol.* 2003;163(6):1363–1374.
8. Vehviläinen P, Hyytiäinen M, Keski-Oja J. Latent transforming growth factor-beta-binding protein 2 is an adhesion protein for melanoma cells. *J Biol Chem.* 2003;278(27):24705–24713.
9. Sideek MA, Menz C, Parsi MK, Gibson MA. LTBP-2 competes with tropoelastin for binding to fibulin-5 and heparin, and is a negative modulator of elastinogenesis. *Matrix Biol.* 2014;34:114–123.
10. Liu X, Zhao Y, Gao J, et al. Elastic fiber homeostasis requires lysyl oxidase-like 1 protein. *Nat Genet.* 2004;36(2):178–182.
11. Cai C, Rajaram M, Zhou X, et al. Activation of multiple cancer pathways and tumor maintenance function of the 3q amplified oncogene FNDC3B. *Cell Cycle.* 2012;11(9):1773–1781.
12. Gharahkhani P, Burdon KP, Fogarty R, et al. Common variants near ABCA1, AFAP1 and GMDS confer risk of primary open-angle glaucoma. *Nat Genet.* 2014;46(10): 1120–1125.
13. Paschoud S, Jond L, Guerrera D, Citi S. PLEKHA7 modulates epithelial tight junction barrier function. *Tissue Barriers.* 2014;2(1):e28755.
14. Vithana EN, Khor CC, Qiao C, et al. Genome-wide association analyses identify three new susceptibility loci for primary angle closure glaucoma. *Nat Genet.* 2012;44(10):1142–1146.
15. Richards AJ, McNinch A, Martin H, et al. Stickler syndrome and the vitreous phenotype: mutations in COL2A1 and COL11A1. *Hum Mutat.* 2010;31(6): E1461–E1471.
16. Ylikärppä R, Eklund L, Sormunen R, et al. Double knockout mice reveal a lack of major functional compensation between collagens XV and XVIII. *Matrix Biol.* 2003;22(5):443–448.
17. Liton PB, Liu X, Stamer WD, Challa P, Epstein DL, Gonzalez P. Specific targeting of gene expression to a subset of human trabecular meshwork cells using the chitinase 3-like 1 promoter. *Invest Ophthalmol Vis Sci.* 2005;46(1):183–190.
18. Hurskainen M, Eklund L, Hägg PO, et al. Abnormal maturation of the retinal vasculature in type XVIII collagen/endostatin deficient mice and changes in retinal glial cells due to lack of collagen types XV and XVIII. *FASEB J.* 2005;19(11):1564–1566.
19. Pasmant E, Sabbagh A, Vidaud M, Bièche I. ANRIL, a long, noncoding RNA, is an unexpected major hotspot in GWAS. *FASEB J.* 2011;25(2):444–448.
20. Jarinova O, Stewart AF, Roberts R, et al. Functional analysis of the chromosome 9p21.3 coronary artery disease risk locus. *Arterioscler Thromb Vasc Biol.* 2009;29(10): 1671–1677.
21. Harismendy O, Notani D, Song X, et al. 9p21 DNA variants associated with coronary artery disease impair interferon-γ signalling response. *Nature.* 2011;470(7333): 264–268.
22. Sudhakar C, Nagabhushana A, Jain N, Swarup G. NF-kappaB mediates tumor necrosis factor alpha-induced expression of optineurin, a negative regulator of NF-kappaB. *PLoS One.* 2009;4(4):e5114.
23. Nagabhushana A, Bansal M, Swarup G. Optineurin is required for CYLD-dependent inhibition of TNFα-induced NF-κB activation. *PLoS One.* 2011;6(3):e17477.

24. Morton S, Hesson L, Peggie M, Cohen P. Enhanced binding of TBK1 by an optineurin mutant that causes a familial form of primary open angle glaucoma. *FEBS Lett.* 2008;582(6):997–1002.
25. Fingert JH, Robin AL, Stone JL, et al. Copy number variations on chromosome 12q14 in patients with normal tension glaucoma. *Hum Mol Genet.* 2011;20(12):2482–2494.
26. Shen X, Ying H, Qiu Y, et al. Processing of optineurin in neuronal cells. *J Biol Chem.* 2011;286(5):3618–3629.
27. Galluzzi L, Kepp O, Kroemer G. Autophagy and innate immunity ally against bacterial invasion. *EMBO J.* 2011;30(16):3213–3214.
28. Van Eck M. ATP-binding cassette transporter A1: key player in cardiovascular and metabolic disease at local and systemic level. *Curr Opin Lipidol.* 2014;25(4):297–303.
29. Oram JF, Lawn RM. ABCA1. The gatekeeper for eliminating excess tissue cholesterol. *J Lipid Res.* 2001;42(8):1173–1179.
30. Shvets E, Ludwig A, Nichols BJ. News from the caves: update on the structure and function of caveolae. *Curr Opin Cell Biol.* 2014;29:99–106.
31. Trane AE, Pavlov D, Sharma A, et al. Deciphering the binding of caveolin-1 to client protein endothelial nitric-oxide synthase (eNOS): scaffolding subdomain identification, interaction modeling, and biological significance. *J Biol Chem.* 2014;289(19): 13273–13283. http://dx.doi.org/10.1074/jbc.M113.528695.
32. Barone R, Fiumara A, Jaeken J. Congenital disorders of glycosylation with emphasis on cerebellar involvement. *Semin Neurol.* 2014;34(3):357–366.
33. Freeze HH. Genetic disorders of glycan degration. In: Varki A, Cummings RD, Esko JD, Freeze HH, Stanley P, Bertozzi CR, Hart GW, Etzler ME, eds. *Essentials of Glycobiology*. 2nd ed. Cold Spring Harbor (NY): Cold Spring Harbor Laboratory Press; 2009 [Chapter 41].
34. Schneider C, King RM, Philipson L. Genes specifically expressed at growth arrest of mammalian cells. *Cell.* 1988;54(6):787–793.
35. Ju YT, Chang AC, She BR, et al. gas7: a gene expressed preferentially in growth-arrested fibroblasts and terminally differentiated Purkinje neurons affects neurite formation. *Proc Natl Acad Sci U S A.* 1998;95(19):11423–11428.
36. You JJ, Lin-Chao S. Gas7 functions with N-WASP to regulate the neurite outgrowth of hippocampal neurons. *J Biol Chem.* 2010;285(15):11652–11666.
37. Chao CC, Hung FC, Chao JJ. Gas7 is required for mesenchymal stem cell-derived bone development. *Stem Cells Int.* 2013;2013:137010.
38. Hung FC, Cheng YC, Sun NK, Chao CC. Identification and functional characterization of zebrafish Gas7 gene in early development. *J Neurosci Res.* 2013;91(1):51–61.
39. Sharma S, Burdon KP, Chidlow G, et al. Association of genetic variants in the TMCO1 gene with clinical parameters related to glaucoma and characterization of the protein in the eye. *Invest Ophthalmol Vis Sci.* 2012;53(8):4917–4925.
40. Saleem RA, Banerjee-Basu S, Murphy TC, Baxevanis A, Walter MA. Essential structural and functional determinants within the forkhead domain of FOXC1. *Nucleic Acids Res.* 2004;32(14):4182–4193.
41. Murphy TC, Saleem RA, Footz T, Ritch R, McGillivray B, Walter MA. The wing 2 region of the FOXC1 forkhead domain is necessary for normal DNA-binding and transactivation functions. *Invest Ophthalmol Vis Sci.* 2004;45(8):2531–2538.
42. Ephrussi A, St Johnston D. Seeing is believing: the bicoid morphogen gradient matures. *Cell.* 2004;116(2):143–152.
43. Gage PJ, Camper SA. Pituitary homeobox 2, a novel member of the bicoid-related family of homeobox genes, is a potential regulator of anterior structure formation. *Hum Mol Genet.* 1997;6(3):457–464.
44. Nishina S, Kohsaka S, Yamaguchi Y, et al. PAX6 expression in the developing human eye. *Br J Ophthalmol.* 1999;83(6):723–727.

45. Tsuchiya Y, Nakajima M, Yokoi T. Cytochrome P450-mediated metabolism of estrogens and its regulation in human. *Cancer Lett.* 2005;227:115–124.
46. Choudhary D, Jansson I, Schenkman JB. CYP1B1, a developmental gene with a potential role in glaucoma therapy. *Xenobiotica.* 2009;39(8):606–615.
47. Inoue T, Ohbayashi T, Fujikawa Y, et al. Latent TGF-β binding protein-2 is essential for the development of ciliary zonule microfibrils. *Hum Mol Genet.* 2014;23(21):5672–5682.
48. Jusiak B, Karandikar UC, Kwak SJ, et al. Regulation of Drosophila eye development by the transcription factor Sine oculis. *PLoS One.* 2014;9(2):e89695.
49. Anderson AM, Weasner BM, Weasner BP, Kumar JP. Dual transcriptional activities of SIX proteins define their roles in normal and ectopic eye development. *Development.* 2012;139(5):991–1000.
50. Wiggs JL, Yaspan BL, Hauser MA, et al. Common variants at 9p21 and 8q22 are associated with increased susceptibility to optic nerve degeneration in glaucoma. *PLoS Genet.* 2012;8(4):e1002654.
51. Li N, Zhou Y, Du L, Wei M, Chen X. Overview of Cytochrome P450 1B1 gene mutations in patients with primary congenital glaucoma. *Exp Eye Res.* 2011;93(5): 572–579.
52. Bejjani BA, Lewis RA, Tomey KF, et al. Mutations in CYP1B1, the gene for cytochrome P4501B1, are the predominant cause of primary congenital glaucoma in Saudi Arabia. *Am J Hum Genet.* 1998;62(2):325–333.
53. Stoilov I, Akarsu AN, Alozie I, et al. Sequence analysis and homology modeling suggest that primary congenital glaucoma on 2p21 results from mutations disrupting either the hinge region or the conserved core structures of cytochrome P4501B1. *Am J Hum Genet.* 1998;62(3):573–584.
54. Ali M, McKibbin M, Booth A, et al. Null mutations in LTBP2 cause primary congenital glaucoma. *Am J Hum Genet.* 2009;84(5):664–671.
55. Narooie-Nejad M, Paylakhi SH, Shojaee S, et al. Loss of function mutations in the gene encoding latent transforming growth factor beta binding protein 2, LTBP2, cause primary congenital glaucoma. *Hum Mol Genet.* 2009;18(20):3969–3977.
56. Hjalt TA, Semina EV. Current molecular understanding of Axenfeld-Rieger syndrome. *Expert Rev Mol Med.* 2005;7(25):1–17.
57. Reis LM, Tyler RC, Volkmann Kloss BA, et al. PITX2 and FOXC1 spectrum of mutations in ocular syndromes. *Eur J Hum Genet.* 2012;20(12):1224–1233.
58. D'haene B, Meire F, Claerhout I, et al. Expanding the spectrum of FOXC1 and PITX2 mutations and copy number changes in patients with anterior segment malformations. *Invest Ophthalmol Vis Sci.* 2011;52(1):324–333.
59. Lowry RB, Gould DB, Walter MA, Savage PR. Absence of PITX2, BARX1, and FOXC1 mutations in De Hauwere syndrome (Axenfeld-Rieger anomaly, hydrocephaly, hearing loss): a 25-year follow up. *Am J Med Genet A.* 2007;143A(11):1227–1230.
60. Semina EV, Reiter R, Leysens NJ, et al. Cloning and characterization of a novel bicoid-related homeobox transcription factor gene, RIEG, involved in Rieger syndrome. *Nat Genet.* 1996;14(4):392–399.
61. Alward WL, Semina EV, Kalenak JW, et al. Autosomal dominant iris hypoplasia is caused by a mutation in the Rieger syndrome (RIEG/PITX2) gene. *Am J Ophthalmol.* 1998;125(1):98–100.
62. Kulak SC, Kozlowski K, Semina EV, Pearce WG, Walter MA. Mutation in the RIEG1 gene in patients with iridogoniodysgenesis syndrome. *Hum Mol Genet.* 1998;7(7): 1113–1117.
63. Lee HJ, Colby KA. A review of the clinical and genetic aspects of aniridia. *Semin Ophthalmol.* 2013;28(5-6):306–312.
64. Prosser J, van Heyningen V. PAX6 mutations reviewed. *Hum Mutat.* 1998;11(2): 93–108.

65. Sale MM, Craig JE, Charlesworth JC, et al. Broad phenotypic variability in a single pedigree with a novel 1410delC mutation in the PST domain of the PAX6 gene. *Hum Mutat.* 2002;20(4):322.
66. Désir J, Sznajer Y, Depasse F, et al. LTBP2 null mutations in an autosomal recessive ocular syndrome with megalocornea, spherophakia, and secondary glaucoma. *Eur J Hum Genet.* 2010;18(7):761–767.
67. Kumar A, Duvvari MR, Prabhakaran VC, Shetty JS, Murthy GJ, Blanton SH. A homozygous mutation in LTBP2 causes isolated microspherophakia. *Hum Genet.* 2010;128(4):365–371.
68. Haji-Seyed-Javadi R, Jelodari-Mamaghani S, Paylakhi SH, et al. LTBP2 mutations cause Weill-Marchesani and Weill-Marchesani-like syndrome and affect disruptions in the extracellular matrix. *Hum Mutat.* 2012;33(8):1182–1187.
69. Aldahmesh MA, Khan AO, Hijazi H, Alkuraya FS. Homozygous truncation of SIX6 causes complex microphthalmia in humans. *Clin Genet.* 2013;84(2):198–199.
70. Donegan RK, Hill SE, Freeman DM, et al. Structural basis for misfolding in myocilin-associated glaucoma. *Hum Mol Genet.* 2014;24(8):2111–2124.
71. Allen KF, Gaier E, Wiggs JL. Genetics of primary inherited disorders of the optic nerve: clinical implications. *Cold Spring Harb Perspect Med.* 2015; [In press].
72. Hogewind BF, Mukhopadhyay A, Theelen T, Hollander AI, Hoyng CB. Variable clinical spectrum of the myocilin Gln368X mutation in a Dutch family with primary open angle glaucoma. *Curr Eye Res.* 2010;35(1):31–36.
73. Wiggs JL, Howell GR, Linkroum K, et al. Variations in COL15A1 and COL18A1 influence age of onset of primary open angle glaucoma. *Clin Genet.* 2013;84(2): 167–174.
74. Chen Y, Lin Y, Vithana EN, et al. Common variants near ABCA1 and in PMM2 are associated with primary open-angle glaucoma. *Nat Genet.* 2014;46(10):1115–1119.
75. Thorleifsson G, Walters GB, Hewitt AW, et al. Common variants near CAV1 and CAV2 are associated with primary open-angle glaucoma. *Nat Genet.* 2010;42(10): 906–909.
76. Wiggs JL, Kang JH, Yaspan BL, et al. Common variants near CAV1 and CAV2 are associated with primary open-angle glaucoma in Caucasians from the USA. *Hum Mol Genet.* 2011;20(23):4707–4713.
77. Kato T, Meguro A, Nomura E, et al. Association study of genetic variants on chromosome 7q31 with susceptibility to normal tension glaucoma in a Japanese population. *Eye (Lond).* 2013;27(8):979–983.
78. Ozel AB, Moroi SE, Reed DM, et al. Genome-wide association study and meta-analysis of intraocular pressure. *Hum Genet.* 2014;133(1):41–57.
79. Chen F, Klein AP, Klein BE, et al. Exome array analysis identifies CAV1/CAV2 as a susceptibility locus for intraocular pressure. *Invest Ophthalmol Vis Sci.* 2014;56(1): 544–551.
80. Burdon KP, Macgregor S, Hewitt AW, et al. Genome-wide association study identifies susceptibility loci for open angle glaucoma at TMCO1 and CDKN2B-AS1. *Nat Genet.* 2011;43(6):574–578.
81. Hysi PG, Cheng CY, Springelkamp H, et al. Genome-wide analysis of multi-ancestry cohorts identifies new loci influencing intraocular pressure and susceptibility to glaucoma. *Nat Genet.* 2014;46(10):1126–1130.
82. Lu Y, Vitart V, Burdon KP, et al. Genome-wide association analyses identify multiple loci associated with central corneal thickness and keratoconus. *Nat Genet.* 2013;45(2): 155–163.
83. van Koolwijk LM, Ramdas WD, Ikram MK, et al. Common genetic determinants of intraocular pressure and primary open-angle glaucoma. *PLoS Genet.* 2012;8(5): e1002611.

84. Ramdas WD, van Koolwijk LM, Ikram MK, et al. A genome-wide association study of optic disc parameters. *PLoS Genet.* 2010;6(6):e1000978.
85. Li Z, Allingham RR, Nakano M. A common variant near TGFBR3 is associated with primary open angle glaucoma. *Hum Mol Genet.* 2015 Apr;10. pii: ddv128. [Epub ahead of print].
86. Rezaie T, Child A, Hitchings R, et al. Adult-onset primary open-angle glaucoma caused by mutations in optineurin. *Science.* 2002;295(5557):1077–1079.
87. Aung T, Rezaie T, Okada K, et al. Clinical features and course of patients with glaucoma with the E50K mutation in the optineurin gene. *Invest Ophthalmol Vis Sci.* 2005;46(8):2816–2822.
88. Hauser MA, Sena DF, Flor J, et al. Distribution of optineurin sequence variations in an ethnically diverse population of low-tension glaucoma patients from the United States. *J Glaucoma.* 2006;15(5):358–363.
89. Pasquale LR, Loomis SJ, Kang JH, et al. CDKN2B-AS1 genotype-glaucoma feature correlations in primary open-angle glaucoma patients from the United States. *Am J Ophthalmol.* 2013;155(2):342–353.e5.
90. Burdon KP, Crawford A, Casson RJ, et al. Glaucoma risk alleles at CDKN2B-AS1 are associated with lower intraocular pressure, normal-tension glaucoma, and advanced glaucoma. *Ophthalmology.* 2012;119(8):1539–1545.
91. Thorleifsson G, Magnusson KP, Sulem P, et al. Common sequence variants in the LOXL1 gene confer susceptibility to exfoliation glaucoma. *Science.* 2007;317(5843):1397–1400.
92. Kwartler CS, Chen J, Thakur D, et al. Overexpression of smooth muscle myosin heavy chain leads to activation of the unfolded protein response and autophagic turnover of thick filament-associated proteins in vascular smooth muscle cells. *J Biol Chem.* 2014;289(20):14075–14088.
93. Schröder M, Kaufman RJ. The mammalian unfolded protein response. *Annu Rev Biochem.* 2005;74:739–789.
94. Torres M, Matamala JM, Duran-Aniotz C, Cornejo VH, Foley A, Hetz C. ER stress signaling and neurodegeneration: at the intersection between Alzheimer's disease and prion-related disorders. *Virus Res.* 2014. pii: S0168-1702(14)00523-1. [EPub ahead of print].
95. Papazachariou L, Demosthenous P, Pieri M, et al. Frequency of COL4A3/COL4A4 mutations amongst families segregating glomerular microscopic hematuria and evidence for activation of the unfolded protein response. Focal and segmental glomerulosclerosis is a frequent development during ageing. *PLoS One.* 2014;9(12):e115015.
96. Zhang SX, Ma JH, Bhatta M, Fliesler SJ, Wang JJ. The unfolded protein response in retinal vascular diseases: implications and therapeutic potential beyond protein folding. *Prog Retin Eye Res.* 2015;45:111–131.
97. Wiggs JL, Vollrath D. Molecular and clinical evaluation of a patient hemizygous for TIGR/MYOC. *Arch Ophthalmol.* 2001;119(11):1674–1678.
98. Kim BS, Savinova OV, Reedy MV, et al. Targeted disruption of the Myocilin gene (Myoc) suggests that human glaucoma-causing mutations are gain of function. *Mol Cell Biol.* 2001;21(22):7707–7713.
99. Pang CP, Leung YF, Fan B, et al. TIGR/MYOC gene sequence alterations in individuals with and without primary open-angle glaucoma. *Invest Ophthalmol Vis Sci.* 2002;43(10):3231–3235.
100. Zhou Z, Vollrath D. A cellular assay distinguishes normal and mutant TIGR/myocilin protein. *Hum Mol Genet.* 1999;8(12):2221–2228.
101. Lee B, Rhead W, Diaz GA, et al. Phase 2 comparison of a novel ammonia scavenging agent with sodium phenylbutyrate in patients with urea cycle disorders: safety, pharmacokinetics and ammonia control. *Mol Genet Metab.* 2010;100(3):221–228.

102. Chatterjee A, Villarreal Jr G, Rhee DJ. Matricellular proteins in the trabecular meshwork: review and update. *J Ocul Pharmacol Ther.* 2014;30(6):447–463.
103. Khan AO, Aldahmesh MA, Alkuraya FS. Congenital megalocornea with zonular weakness and childhood lens-related secondary glaucoma—a distinct phenotype caused by recessive LTBP2 mutations. *Mol Vis.* 2011;17:2570–2579 [Epub 2011 Oct 4].
104. Hirani R, Hanssen E, Gibson MA. LTBP-2 specifically interacts with the amino-terminal region of fibrillin-1 and competes with LTBP-1 for binding to this microfibrillar protein. *Matrix Biol.* 2007;26(4):213–223.
105. Ritch R. Perspective on exfoliation syndrome. *J Glaucoma.* 2001;10(5 Suppl 1): S33–S35.
106. Fan BJ, Pasquale LR, Rhee D, Li T, Haines JL, Wiggs JL. LOXL1 promoter haplotypes are associated with exfoliation syndrome in a U.S. Caucasian population. *Invest Ophthalmol Vis Sci.* 2011;52(5):2372–2378.
107. Kim S, Kim Y. Variations in LOXL1 associated with exfoliation glaucoma do not affect amine oxidase activity. *Mol Vis.* 2012;18:265–270.
108. Wiggs JL, Pasquale LR. Expression and regulation of LOXL1 and elastin-related genes in eyes with exfoliation syndrome. *J Glaucoma.* 2014;23(8 Suppl 1):S62–S63.
109. Wiggs JL, Pawlyk B, Connolly E, et al. Disruption of the blood-aqueous barrier and lens abnormalities in mice lacking lysyl oxidase-like 1 (LOXL1). *Invest Ophthalmol Vis Sci.* 2014;55(2):856–864.
110. Fuchshofer R, Tamm ER. The role of TGF-β in the pathogenesis of primary open-angle glaucoma. *Cell Tissue Res.* 2012;347(1):279–290.
111. McDowell CM, Tebow HE, Wordinger RJ, Clark AF. Smad3 is necessary for transforming growth factor-beta2 induced ocular hypertension in mice. *Exp Eye Res.* 2013;116:419–423.
112. Junglas B, Kuespert S, Seleem AA, et al. Connective tissue growth factor causes glaucoma by modifying the actin cytoskeleton of the trabecular meshwork. *Am J Pathol.* 2012;180(6):2386–2403.
113. Wordinger RJ, Sharma T, Clark AF. The role of TGF-β2 and bone morphogenetic proteins in the trabecular meshwork and glaucoma. *J Ocul Pharmacol Ther.* 2014;30(2-3):154–162.
114. Lukas TJ, Miao H, Chen L, et al. Susceptibility to glaucoma: differential comparison of the astrocyte transcriptome from glaucomatous African American and Caucasian American donors. *Genome Biol.* 2008;9(7):R111.
115. Greene LA, Liu DX, Troy CM, Biswas SC. Cell cycle molecules define a pathway required for neuron death in development and disease. *Biochim Biophys Acta.* 2007;1772(4):392–401.
116. Rajaraman P, Melin BS, Wang Z, et al. Genome-wide association study of glioma and meta-analysis. *Hum Genet.* 2012;131(12):1877–1888.
117. Ng SK, Casson RJ, Burdon KP, Craig JE. Chromosome 9p21 primary open-angle glaucoma susceptibility locus: a review. *Clin Experiment Ophthalmol.* 2014;42(1):25–32.
118. Kang JH, Wiggs JL, Pasquale LR. A nested case control study of plasma ICAM-1, E-selectin and TNF receptor 2 levels, and incident primary open-angle glaucoma. *Invest Ophthalmol Vis Sci.* 2013;54(3):1797–1804.
119. Roh M, Zhang Y, Murakami Y, et al. Etanercept, a widely used inhibitor of tumor necrosis factor-α (TNF-α), prevents retinal ganglion cell loss in a rat model of glaucoma. *PLoS One.* 2012;7(7):e40065.
120. Iguchi Y, Katsuno M, Ikenaka K, Ishigaki S, Sobue G. Amyotrophic lateral sclerosis: an update on recent genetic insights. *J Neurol.* 2013;260(11):2917–2927.
121. Ying H, Yue BY. Cellular and molecular biology of optineurin. *Int Rev Cell Mol Biol.* 2012;294:223–258.

122. Yang Z, Klionsky DJ. Mammalian autophagy: core molecular machinery and signaling regulation. *Curr Opin Cell Biol.* 2010;22(2):124–131.
123. Piras A, Gianetto D, Conte D, Bosone A, Vercelli A. Activation of autophagy in a rat model of retinal ischemia following high intraocular pressure. *PLoS One.* 2011;6(7): e22514.
124. Park HY, Kim JH, Park CK. Activation of autophagy induces retinal ganglion cell death in a chronic hypertensive glaucoma model. *Cell Death Dis.* 2012;3:e290.
125. Minegishi Y, Iejima D, Kobayashi H, et al. Enhanced optineurin E50K-TBK1 interaction evokes protein insolubility and initiates familial primary open-angle glaucoma. *Hum Mol Genet.* 2013;22(17):3559–3567.
126. Tucker BA, Solivan-Timpe F, Roos BR, et al. Duplication of TBK1 stimulates autophagy in iPSC-derived retinal cells from a patient with normal tension glaucoma. *J Stem Cell Res Ther.* 2014;3(5):161.
127. Howell GR, Macalinao DG, Sousa GL, et al. Molecular clustering identifies complement and endothelin induction as early events in a mouse model of glaucoma. *J Clin Invest.* 2011;121(4):1429–1444.
128. Yeghiazaryan K, Flammer J, Wunderlich K, Schild HH, Orgul S, Golubnitschaja O. An enhanced expression of ABC 1 transporter in circulating leukocytes as a potential molecular marker for the diagnostics of glaucoma. *Amino Acids.* 2005;28(2):207–211.
129. Stein JD, Newman-Casey PA, Talwar N, Nan B, Richards JE, Musch DC. The relationship between statin use and open-angle glaucoma. *Ophthalmology.* 2012;119(10): 2074–2081.
130. Cavet ME, Vittitow JL, Impagnatiello F, Ongini E, Bastia E. Nitric oxide (NO): an emerging target for the treatment of glaucoma. *Invest Ophthalmol Vis Sci.* 2014;55(8):5005–5015.
131. Dismuke WM, Liang J, Overby DR, Stamer WD. Concentration-related effects of nitric oxide and endothelin-1 on human trabecular meshwork cell contractility. *Exp Eye Res.* 2014;120:28–35.
132. Perkumas KM, Stamer WD. Protein markers and differentiation in culture for Schlemm's canal endothelial cells. *Exp Eye Res.* 2012;96(1):82–87.
133. Toda N, Nakanishi-Toda M. Nitric oxide: ocular blood flow, glaucoma, and diabetic retinopathy. *Prog Retin Eye Res.* 2007;26(3):205–238.
134. Kang JH, Wiggs JL, Haines J, Abdrabou W, Pasquale LR. Reproductive factors and NOS3 variant interactions in primary open-angle glaucoma. *Mol Vis.* 2011;17: 2544–2551.
135. Loomis SJ, Kang JH, Weinreb RN, et al. Association of CAV1/CAV2 genomic variants with primary open-angle glaucoma overall and by gender and pattern of visual field loss. *Ophthalmology.* 2014;121(2):508–516.
136. Vargas-Pinto R, Lai J, Gong H, Ethier CR, Johnson M. Finite element analysis of the pressure-induced deformation of Schlemm's canal endothelial cells. *Biomech Model Mechanobiol.* 2014 [EPub ahead of print].
137. Gotoh A, Hidaka M, Hirose K, Uchida T. Gas7b (growth arrest specific protein 7b) regulates neuronal cell morphology by enhancing microtubule and actin filament assembly. *J Biol Chem.* 2013;288(48):34699–34706.
138. Alanay Y, Ergüner B, Utine E, et al. TMCO1 deficiency causes autosomal recessive cerebrofaciothoracic dysplasia. *Am J Med Genet A.* 2014;164A(2):291–304.
139. Pehlivan D, Karaca E, Aydin H, et al. Whole-exome sequencing links TMCO1 defect syndrome with cerebro-facio-thoracic dysplasia. *Eur J Hum Genet.* 2014;22(9): 1145–1148.
140. Iwamuro S, Saeki M, Kato S. Multi-ubiquitination of a nascent membrane protein produced in a rabbit reticulocyte lysate. *J Biochem.* 1999;126:48–53.

141. Zhang Z, Mo D, Cong P, et al. Molecular cloning, expression patterns and subcellular localization of porcine TMCO1 gene. *Mol Biol Rep*. 2010;37(3):1611–1618. http://dx.doi.org/10.1007/s11033-009-9573-8 [Epub 2009 May 16].
142. Jackson BC, Carpenter C, Nebert DW, Vasiliou V. Update of human and mouse forkhead box (FOX) gene families. *Hum Genomics*. 2010;4(5):345–352.
143. Acharya M, Huang L, Fleisch VC, Allison WT, Walter MA. A complex regulatory network of transcription factors critical for ocular development and disease. *Hum Mol Genet*. 2011;20(8):1610–1624.
144. Berry FB, Lines MA, Oas JM, et al. Functional interactions between FOXC1 and PITX2 underlie the sensitivity to FOXC1 gene dose in Axenfeld-Rieger syndrome and anterior segment dysgenesis. *Hum Mol Genet*. 2006;15(6):905–919.
145. Kelberman D, Islam L, Holder SE, et al. Digenic inheritance of mutations in FOXC1 and PITX2: correlating transcription factor function and Axenfeld-Rieger disease severity. *Hum Mutat*. 2011;32(10):1144–1152. http://dx.doi.org/10.1002/humu.21550 [Epub 2011 Sep 8].
146. Du RF, Huang H, Fan LL, Li XP, Xia K, Xiang R. A novel mutation of FOXC1 (R127L) in an Axenfeld-Rieger syndrome family with glaucoma and multiple congenital heart diseases. *Ophthalmic Genet*. 2014;10:1–5.
147. Drouin J, Lamolet B, Lamonerie T, Lanctôt C, Tremblay JJ. The PTX family of homeodomain transcription factors during pituitary developments. *Mol Cell Endocrinol*. 1998;140(1-2):31–36.
148. Wang J, Bai Y, Li N, et al. Pitx2-microRNA pathway that delimits sinoatrial node development and inhibits predisposition to atrial fibrillation. *Proc Natl Acad Sci USA*. 2014;111(25):9181–9186.
149. Volkmann BA, Zinkevich NS, Mustonen A, et al. Potential novel mechanism for Axenfeld-Rieger syndrome: deletion of a distant region containing regulatory elements of PITX2. *Invest Ophthalmol Vis Sci*. 2011;52(3):1450–1459.
150. Kioussi C, Briata P, Baek SH, et al. Identification of a Wnt/Dvl/beta-Catenin – > Pitx2 pathway mediating cell-type-specific proliferation during development. *Cell*. 2002;111(5):673–685.
151. Smith RS, Zabaleta A, Kume T, et al. Haploinsufficiency of the transcription factors FOXC1 and FOXC2 results in aberrant ocular development. *Hum Mol Genet*. 2000;9(7):1021–1032.
152. Crolla JA, van Heyningen V. Frequent chromosome aberrations revealed by molecular cytogenetic studies in patients with aniridia. *Am J Hum Genet*. 2002;71(5):1138–1149.
153. van Heyningen V, Hoovers JM, de Kraker J, Crolla JA. Raised risk of Wilms tumour in patients with aniridia and submicroscopic WT1 deletion. *J Med Genet*. 2007;44(12): 787–790.
154. Sowers MR, Wilson AL, Kardia SR, Chu J, McConnell DS. CYP1A1 and CYP1B1 polymorphisms and their association with estradiol and estrogen metabolites in women who are premenopausal and perimenopausal. *Am J Med*. 2006;119(9 Suppl 1):S44–S51.
155. Choudhary D, Jansson I, Stoilov I, Sarfarazi M, Schenkman JB. Metabolism of retinoids and arachidonic acid by human and mouse cytochrome P450 1b1. *Drug Metab Dispos*. 2004;32(8):840–847.
156. Pasquale LR, Loomis SJ, Weinreb RN, et al. Estrogen pathway polymorphisms in relation to primary open angle glaucoma: an analysis accounting for gender from the United States. *Mol Vis*. 2013;19:1471–1481.
157. López-Garrido MP, Medina-Trillo C, Morales-Fernandez L, et al. Null CYP1B1 genotypes in primary congenital and nondominant juvenile glaucoma. *Ophthalmology*. 2013;120(4):716–723.
158. Su CC, Liu YF, Li SY, Yang JJ, Yen YC. Mutations in the CYP1B1 gene may contribute to juvenile-onset open-angle glaucoma. *Eye (Lond)*. 2012;26(10):1369–1377.

159. Abu-Amero KK, Morales J, Aljasim LA, Edward DP. CYP1B1 mutations are a major contributor to juvenile-onset open angle glaucoma in Saudi Arabia. *Ophthalmic Genet.* 2013 [Epub ahead of print].
160. Khan AO, Al-Abdi L, Mohamed JY, Aldahmesh MA, Alkuraya FS. Familial juvenile glaucoma with underlying homozygous p.G61E CYP1B1 mutations. *J AAPOS.* 2011;15(2):198–209.
161. López-Garrido MP, Blanco-Marchite C, Sánchez-Sánchez F, et al. Functional analysis of CYP1B1 mutations and association of heterozygous hypomorphic alleles with primary open-angle glaucoma. *Clin Genet.* 2010;77(1):70–78.
162. Wiggs JL, Langguth A, Allen KF. Carrier frequency of CYP1B1 mutations in the United States. *Trans Am Ophthalmol Soc.* 2014;112:94–102.
163. Gallardo ME, Lopez-Rios J, Fernaud-Espinosa I, et al. Genomic cloning and characterization of the human homeobox gene SIX6 reveals a cluster of SIX genes in chromosome 14 and associates SIX6 hemizygosity with bilateral anophthalmia and pituitary anomalies. *Genomics.* 1999;61(1):82–91.
164. Cheng C, Allingham RR, Aung T, et al. Association of common SIX6 polymorphisms with peripapillary retinal nerve fiber layer thickness: the Singapore Chinese eye study. *Invest Ophthalmol Vis Sci.* 2014;56:478–483.
165. Carnes MU, Liu YP, Allingham RR, et al. Discovery and functional annotation of SIX6 variants in primary open-angle glaucoma. *PLoS Genet.* 2014;10(5):e1004372.
166. Iglesias AI, Springelkamp H, van der Linde H, et al. Exome sequencing and functional analyses suggest that SIX6 is a gene involved in an altered proliferation-differentiation balance early in life and optic nerve degeneration at old age. *Hum Mol Genet.* 2014;23(5):1320–1332.
167. Thurman RE, Rynes E, Humbert R, et al. The accessible chromatin landscape of the human genome. *Nature.* 2012;489(7414):75–82.
168. Jonas JB, Wang N. Cerebrospinal fluid pressure and glaucoma. *J Ophthalmic Vis Res.* 2013;8(3):257–263.
169. Ramdas WD, van Koolwijk LM, Cree AJ, et al. Clinical implications of old and new genes for open-angle glaucoma. *Ophthalmology.* 2011;118(12):2389–2397.

CHAPTER TWENTY

Neuroinflammation in Glaucoma and Optic Nerve Damage

Caitlin E. Mac Nair*,†, Robert W. Nickells*,1

*Ophthalmology and Visual Sciences, University of Wisconsin—Madison, Madison, Wisconsin, USA
†Cellular and Molecular Pathology Graduate Program, University of Wisconsin—Madison, Madison, Wisconsin, USA
[1]Corresponding author: e-mail address: nickells@wisc.edu

Contents

Abstract

Glaucoma is a group of optic neuropathies characterized by the degeneration of retinal ganglion cell axons and somas, ultimately preventing light signals in the retina from reaching the brain. Glaucoma is a leading cause of blindness in the world, and treatment options for patients remain limited and minimally efficacious. A number of mechanisms have been linked to glaucomatous pathophysiology. A leading role is now attributed to neuroinflammatory conditions generated by the resident innate immune cells in the optic nerve and retina. Since the eye is immune privileged, the adaptation of these innate immune cells, termed glia, is crucial following trauma. In this chapter, we discuss the mechanisms associated with normal glial function in a healthy eye, and how changes in glial activation can contribute to the process of glaucomatous neurodegeneration in both the optic nerve and retina.

Progress in Molecular Biology and Translational Science, Volume 134
ISSN 1877-1173
http://dx.doi.org/10.1016/bs.pmbts.2015.06.010
© 2015 Elsevier Inc.
All rights reserved.

1. INTRODUCTION

Around 400 BC, the Greek physician Hippocrates first described the ocular deterioration of "glaucosis," in which he noted elderly patients experienced blindness with a characteristic clouding of the pupil. Despite this early recognition, minimal progress was made in preventing and treating this blinding disease until the mid-1800s, when the invention of the ophthalmoscope and tonometer allowed medical professionals to correlate high intraocular pressure (IOP) with the disease described then as "glaucoma."[1] Our knowledge of glaucoma has since been continuously refined, and today this ocular neuropathy encompasses a group of diseases that are characterized by optic nerve damage and retinal ganglion cell (RGC) death, leading to vision loss, and blindness.[2–4] The mechanisms initiating optic nerve damage still remain unclear; however, a major risk factor for developing glaucoma is elevated IOP,[5–7] in which the aqueous humor generated by the ciliary body is unable to drain through the trabecular meshwork.[8,9] This buildup of fluid causes an increase in IOP, and within the closed sphere of the eye, it causes an increase in strain at the optic nerve head (ONH), the site where RGC axons exit the eye and enter the optic nerve.[3,10,11] This strain damages the axons and leads to the deterioration of the optic nerve and ultimately to the neurodegeneration of RGC somas in the retina.[11] Patients with glaucoma, as a result of high IOP, are categorized as having primary open-angle glaucoma and account for 90% of all glaucoma cases. While there are at least 10 additional varieties of glaucoma, everyone remains susceptible, including adults with normal IOP (normal-tension or low-tension glaucoma), and even infants (www.glaucoma.org). Despite research on glaucoma dating back to antiquity, patients today have minimal treatment options and there is no cure, leaving glaucoma as a leading cause of blindness worldwide.[2,12,13] This highlights a critical need for the continuation of basic research to elucidate the mechanisms associated with vision loss and reveal potential targets for therapeutic intervention.

An incredible amount of time and resources have already been dedicated to dissecting the mechanisms associated with glaucomatous damage, which have been aided by the development of *in vitro* and *in vivo* techniques. In addition to culture work that allows for the functional study of retinal cell types, animal models have been developed in zebrafish,[14] pigs,[15] mice,[16–21] rats,[9,22–25] rabbits,[26–28] and nonhuman primates[28–30] using a variety of surgical techniques to mimic glaucomatous neuropathy. Most commonly used are procedures to elevate IOP by laser-induced trabeculoplasty to create ocular hypertension,[7,9,31] the induction of RGC degeneration and death through optic nerve trauma by transection[22,30,32,33] or crush,[17,23,34–36] and animal models that spontaneously develop high IOP accompanied by RGC injury and loss.[6,19,29] Animal models are especially important when studying the complexity of the retinal environment, which is an intricate network of neuronal and support cells that is difficult to replicate *in vitro*. Glaucoma research has revealed changes in a plethora of signaling pathways: from remodeling of the extracellular matrix to the activation of intracellular signaling pathways to epigenetic changes in the nucleus, it is clear that there are many players contributing to the advancement of glaucoma.[4,5,13] One avenue of increasing investigation is the role of the neuroinflammatory response, which has been argued as both beneficial and detrimental to neuronal survival.[4,37] Like many of the molecular signaling networks involved in glaucoma, the immune response is complex, and rather unique compared to the peripheral immune response.

2. IMMUNE PRIVILEGE AND NEUROGLIA

In the 1940s, the British biologist Sir Peter Medawar coined the term "immune privilege" following a series of skin transplantation studies in rabbits.[38–41] Medawar and other researchers noted that transplanted tissue was susceptible to attack by the host immune response, yet skin grafts in the brain and anterior chamber of the eye did not provoke an immune reaction and survived longer.[39] It has since been established that the brain, spinal cord, eyes, testes, and developing fetus display less susceptibility to inflammation.[39,42] This is thought to be a defensive mechanism rooted in evolution, designed to protect these sensitive tissues from irreversible damage and functional loss evoked by a powerful inflammatory response. Initially, this concept was founded on the idea that, without a lymphatic drainage system and the absence of vasculature in the anterior chamber,[42,43] immune cells could not be primed for activation or gain access to these tissues.[39,42] This has since been disproved to an extent, as the anterior chamber does have an active immune response that has been termed anterior chamber-associated immune deviation (ACAID).[40,42,43] Unlike peripheral immune responses, ACAID relies on unique ocular antigen-presenting cells (APCs) that discretely activate T regulatory lymphocytes in the spleen, which migrate to the eye and suppress damaging inflammation.[40,43] This allows visual integrity to be preserved without alerting more powerful inflammatory cells. In addition to these unique APCs, the fluid that fills the front of the eye, called aqueous humor, also exhibits anti-inflammatory potential.[42] In addition to providing nutrients to the iris and cornea, aqueous humor contains factors that prevent natural killer cells from lysing targets, inhibit neutrophil activation, suppress nitric oxide production by macrophages, and interfere with complement activation.[42]

The anterior chamber varies greatly from the posterior chamber and retinal environment at the back of the eye, and while the posterior chamber is also immune privileged, it is facilitated by an entirely different set of cells and is additionally protected by the blood–retinal barrier.[41,44] Like the aqueous humor, the vitreous humor that occupies the posterior chamber appears to exhibit similar anti-inflammatory properties, although these mechanisms are not clearly understood.[45] The blood–retinal barrier is also guarded by retinal pigment epithelial (RPE) cells, which form a layer between the retina and the vascular choroid.[41] These unique cells encompass a number of critical roles for maintaining retinal health, and also express death receptors like FASL and PD-L1 to lyse encroaching peripheral inflammatory cells,[42] and may even initiate inflammatory responses.[41,45] External to the eye, neuroglial cells called oligodendrocytes encase RGC axons as they exit the globe to form the optic nerve and may provide a physical barrier against peripheral immune cell infiltration.[46] In neurodegenerative diseases such as glaucoma, oligodendrocyte loss is attributed to the breakdown of the blood–retinal barrier.[32,46]

With numerous barriers preventing peripheral immune cell infiltration, it is understandable that early researchers concluded that the eye lacked an immune response. However, peripheral immune cells are not the only components of innate immunity, and current research supports that the same cells establishing immune privilege also double as immune responders to injury. With regard to retinal health, RPE cells can participate in both innate and adaptive immunity, express toll-like receptors (TLRs), present antigens through MHC class I and II molecules, and generate cytokines.[41] Additionally, a subset of Mac-2-expressing phagocytic astrocytes in the myelin transition zone monitor axonal health by constitutively internalizing large axonal evulsions, a clearing process that occurs with reduced capacity in glaucoma and allows axonal damage to accumulate.[47] There is also some evidence to indicate that oligodendrocyte progenitor cells are able to proliferate and remyelinate axons after a sublethal injury[32,46,48]; however, these glial cells are not known to play a prominent immunologic role following neurotrauma. Primarily, additional neuroglial subtypes have been shown to be critical mediators of immune privilege and immunological activity following retinal and optic nerve injury.

2.1 Neuroglial Cells

Glial cells were originally thought to be nothing more than scaffolding structures of the nervous system and were named after the German word *glia* for "glue."[10] These cells are better understood today and not only are critical mediators of normal nervous system function but also serve as immune responders following injury.[3,44,49] Glial cells exist in both the peripheral and central nervous system (CNS); however, in the retina, there are three specific glial subtypes of importance: microglia, astrocytes, and Müller cells.[45] Much of our founding knowledge of glial activity has been derived from studies in the brain, particularly in stroke and traumatic injury models.

2.1.1 Microglia

Microglia are much smaller in size than other glial subtypes and are found in all regions of the brain and spinal cord. They are a heterogeneous population of CNS-specific macrophages that are constantly surveying the environment for signs of damage or distress.[50–52] It is believed that these cells enter the CNS during embryonic development as monoctyes and differentiate into resident microglia with cell surface markers' characteristic of macrophages.[53] Resting "ramified" microglia, which exhibit a complex network of processes, migrate through mature CNS tissue and maintain normal physiological function.[45,51,52] Resting microglial cells are unable to mount an immune response, but when they encounter a site of injury they can transform into activated "amoeboid" microglia.[45,51] This transformation is not morphologically uniform, and many intermediates exist with a variety of soma and arbor branch sizes and shapes.[51,53] Activated microglia serve as scavenger cells and respond to infection, inflammation, trauma, ischemia, and neurodegeneration.[53] Their functions include, but are not limited to, phagocytosing cellular debris,[51,53] cytokine production,[44,53–55] and antigen presentation.[44,49] Microglia also upregulate a number of cell surface proteins including TLRs,[44,49] MHC molecules,[44,49] and complement proteins.[44]

2.1.2 Astrocytes

Astrocytes are named after the Greek word *astro* for "star," as they have a characteristic star shape when labeled for filamentous proteins. These cells visually resemble microglia, with an arbor of processes extending from the soma, although astrocytes are much larger and are classified as macroglia.[56] Astrocytes are the most abundant cell type in the CNS,[37] and like microglia take on a variety of heterogeneous morphologies and physiological functions. Resting astrocytes are called "nonreactive," function to maintain homeostasis, regulate blood flow, recycle neurotransmitters, maintain synapses, and participate in neurogenesis.[37] Astrocytes also help maintain the blood–brain barrier and interact closely with vessel capillaries, axon bundles, and neuronal somas.[45] In response to injury, such as trauma, stroke, or neurodegeneration, reactive astrogliosis is a clear hallmark of activated astrocytes and can be visualized as a massive upregulation of the filament protein glial fibrillary acidic protein (GFAP).[37,57] Reactive astrocytes form a barrier to isolate an injury from the surrounding CNS tissue by forming a glial scar,[37] which may prolong survival, but offer little benefit for tissue repair. There is evidence that activated astrocytes can present antigens, generate cytokines such as tumor necrosis factor alpha (TNFα),[58–60] and upregulate a number of different receptors. The downstream consequence of these extracellular signaling molecules, however, is not fully understood.

2.1.3 Müller Glia

Müller cells are a type of radial glia only found in the retina and represent the most abundant retinal glial population.[14,61,62] Like astrocytes, Müller cells are large in size and characterized as macroglial cells.[56] The cell somas are located in the inner nuclear layer of the retina, but unlike other retinal cell types that are confined to only one nuclear layer, Müller cell processes span the entire thickness of the retina,[63] from the outer limiting membrane to the nerve fiber layer, networking with every cell type in the retina.[14,45] This makes Müller cells master regulators of retinal health and function, and they perform a number of critical roles including delivery of neurotrophic factors to neurons, neurotransmitter recycling from the extracellular space, providing a balance to the ionic and pH environments, and maintaining photoreceptor populations.[14,62,64] Unlike microglia and astrocytes, which vary in size and morphology, the Müller cell population is relatively uniform. They are also remarkably resilient to injury and show a minimal change in total population during glaucomatous neurodegeneration.[14] When damage occurs in the retina, Müller cells undergo physiological, morphological, and molecular changes associated with reactive gliosis, similar to astrocytes, including the upregulation of GFAP.[57,61,63] Müller cells have also been found to express TLRs, phagocytose debris, and express cytokines and chemokines.[45]

3. GLAUCOMATOUS NEURODEGENERATION IS COMPARTMENTALIZED

The hallmark feature of glaucomatous neuropathies is the degeneration of RGCs and their axons, which culminate in visual field loss.[2,5,13,52,65] Ganglion cells are composed of distinct structural compartments including the synapse, axon, dendritic tree, and the soma,[29] all of which are affected during RGC degeneration. The insulating myelin sheath is lost as oligodendrocytes die,[31,66] and the axons deteriorate in a die-back pattern from the distal end of the optic nerve to the proximal end.[20] The dendritic arbor field radius and the number of branches decrease in size,[29,67] and the RGC soma and nucleus shrink.[34] Ultimately, BAX-dependent intrinsic apoptosis is activated in the soma and allows for systematic destruction of the critically injured neuron and removal from the retinal environment.[18,34,68] While these degenerative events are predictable in models of glaucoma, an interesting characteristic is that they do not occur synchronously: they are compartmentalized. Nuclear atrophy has been shown to occur within hours of axonal injury before detectable RGC loss,[34] and axonal degeneration precedes somatic loss in the retina.[69] The events of axonal degeneration and somatic loss can also be completely separated by the genetic ablation of the proapoptotic protein BAX. While *Bax*$^{-/-}$ mice still exhibit axonal degeneration like wild-type counterparts,[70] the RGC somas survive indefinitely in BAX-deficient mice after glaucoma onset,[18,70] highlighting a crucial role for BAX in the commitment step of RGC apoptosis. The compartmentalized degeneration of RGCs also occurs in models of optic nerve trauma, such as optic nerve crush, and while optic nerve deterioration varies from glaucoma and instead occurs by Wallerian degeneration,[69] axonal damage also precedes RGC loss, and RGC somas can be protected from crush by genetic deletion of BAX.[18]

In addition to the compartmentalized degeneration of RGCs, the inflammatory responses also differ between compartments. This is not particularly surprising, as the glial populations mediating the immune response are also unique in the axon and the retina. Therefore, the remainder of this chapter will be divided into two sections, with the immune response in the optic nerve and the retina being discussed separately. The subsequent sections will summarize our current understanding of inflammatory responses during glaucomatous neurodegeneration and in animal models of optic nerve trauma, and specifically which immune cell types are thought to be important players during RGC neurodegeneration.

4. IMMUNE RESPONSE IN THE OPTIC NERVE AND ONH

A defining characteristic of glaucoma is the cupping of the ONH and remodeling of the lamina cribrosa through glial cell rearrangement and the deposition of extracellular matrix.[58,71] These changes are thought to injure the optic nerve, which becomes an immunological hot zone populated primarily by astrocytes and microglia as they repair damage and preserve uninjured tissue from further harm. Trauma to the optic nerve also induces a dramatic increase in immune activity, with a focused response at the site of injury.

4.1 Astrocytes

Astrocytes are the most abundant glial cell of the CNS[37] and are arranged over the surface of RGC axon bundles.[45,69] This localization puts them in direct proximity to any injury sustained by the optic nerve, and astrocytes have been shown to undergo changes in gene expression and morphology following optic nerve trauma.[72] Astrocytes are categorized as Type I and Type II, with both types expressing GFAP, while only Type I additionally expresses Connexin 43.[57] Type I is further classified as Ia and Ib, with Ia astrocytes populating the ONH and Ib astrocytes forming the inner limiting membrane in the retina.[57] While both types can become reactive following injury, astrocyte populations in the brain have been shown to respond heterogeneously to injury, with different injuries evoking unique genetic profiles.[72] In a mouse model of glaucoma, astrocytes at the ONH become thicker in animals with moderate to severe disease and exhibited a simplified arbor network.[73] In human glaucoma patients, astrocytes express the cytokine TNFα and its receptor type 1 (TNF-R1) in the ONH.[58,74] Although the consequence of TNFα production is complicated and has been linked to neurodegeneration and neuroprotection,[58,75–81] astrocytes appear to be the primary producer of this cytokine in the ONH.[58] Human glaucomatous eyes have also shown an increase in MHC class II (HLA-DR) labeling in the ONH, suggesting that astrocytes can present antigen and may be particularly sensitive to cytokine signaling when responding to damage.[82]

Optic nerve crush causes axonal swelling and astrocyte degeneration within 1 day of injury, leaving a GFAP-negative void in the optic nerve distal to the crush site.[32,72,83] However, unlike neuronal loss, which is irreversible and permanent, astrocytes repopulate at the crush site, although with processes that appear in an irregular arrangement unlike the ordered array found in uninjured axons.[84,85] Crush also induces a strong and immediate change in inflammatory gene expression in the optic nerve, with an overall decline in astrocyte markers early after injury and a gradual increase in tissue remodeling genes.[72] Astrocyte remodeling allows for mobility by downregulating membrane channels and intercellular connections, which are reestablished as astrocytes repopulate at the injury site.[72] Additionally, the injured neurons downregulate ciliary neurotrophic factor (CNTF) receptor, which may hinder attempts at regeneration,[86] while astrocytes upregulate this receptor, which is thought to contribute to glial scar formation.[72,86] Severe axonal degeneration occurs within 7 days with maximum loss peaking 3 weeks after injury[72]; however, the injury remains permanent and axonal integrity is not restored. Surprisingly, *Gfap* expression at the ONH does not increase following crush, although this is a well-established marker for astrogliosis in the retina.[72] Over time, astrocyte gene expression returns to baseline by 3 months after injury, at which time a glial scar is present at the site of injury.[72] It is important to note that the structure of the mouse ONH differs from that of humans, as it lacks a collagen-rich lamina cribrosa.[72] This may trigger differential astrocytic responses between mice and human glaucoma patients.

4.2 Microglia

Like astrocytes, microglia populate the optic nerve and help maintain normal function, and following injury, quickly become activated.[58] These glial cells show reactive morphology and exhibit retracted processes and large cell bodies, as well as a significant increase in a number of microglial markers.[72,87] In glaucomatous eyes, microglia have been found to produce proinflammatory cytokines, reactive oxygen species, neurotoxic matrix metalloproteinases, and neurotrophic factors.[52,87] Additionally, a small population of microglia have been found to express MHC class II (HLA-DR), TNFR1, and to a lesser extent TNFα. This cytokine is predominantly produced by astrocytes in the optic nerve,[58,80] but microglia appear to be an important source of TNFα in the ONH.[87] In addition to cytokine production, the complement cascade is initiated early after optic nerve injury, which may be triggered by microglial expression of the complement gene *C1qa*.[88] It does not appear that microglia infiltrate into the damage site from the periphery[89]; however, these immune cells have been shown to proliferate as indicated by BrdU incorporation following an injury response.[90] The consequence of microglial activation is still being dissected, although there is evidence to suggest that reducing microglial activation with an antibiotic promotes neuronal survival after injury in mice.[52,87,89]

Fewer studies have examined microgliosis following optic nerve injury, even though microglia appear to be the major proliferating cell type at the site of optic nerve injury.[72] It has been shown that microglia migrate into the injury site within 3 days and remain prominent during the first week after damage. Additional microglia infiltrate proximal and distal to the injury site and maintain a ramified morphology despite a strong reactive molecular profile, although similar to astrocytes, microglial markers return to baseline by 3 weeks post-injury.[72] After optic nerve trauma, it appears that microglia may facilitate RGC damage rather than neuronal survival, as is seen in glaucoma models.[91] It is important to note that markers for identifying microglia, such as AIF1/IBA1, also label macrophages and their monocyte precursors, making it difficult to distinguish between these cell types except by morphology.[72,90]

4.3 Monocytes and Regulatory T-Cells

In a mouse model of glaucoma, activation of the transendothelial migration pathway allows for monocytes to infiltrate the ONH prior to RGC degeneration[89]; it has been shown that monocytes are one of the earliest responders to the ONH. These immune cells are derived in the bone marrow, and splenic cells labeled with the tracer carboxyfluorescein diacetate have allowed for the migration of monocytes to been tracked to various tissues in the body in addition to the ONH. Once monocytes infiltrate a tissue, they differentiate into various types of macrophages[92]; interestingly, the presence of monocytes in the ONH precedes neuronal loss. It is suspected that these immune cells may contribute to RGC pathology, as the infiltration of these cells is prevented in an irradiated mouse eye, which has been correlated with improved neuronal survival. Contrary, neuronal loss can be restored in irradiated eyes by treatment with monocyte-derived endothelin 2 (EDN2).[89] Although these results may be a caveat of this distinct model of glaucoma, these studies do implicate monocytes as players in neuronal damage during glaucoma. However, additional research is required to decipher the role of monocytes in neuronal damage, and the mechanisms of protection induced by irradiation.

Activated T-cells have also been reported to infiltrate and accumulate in the optic nerve, at the site of damage.[93] These cells are principally reactive to components of the myelinated optic nerve tract (i.e., myelin basic protein) and represent a potential autoimmune response to CNS damage. Surprisingly, this response is protective in the case of optic nerve lesions. The protective effect is lost in rats and mice that have previously received a thymectomy, while it is greater in inbred mice that demonstrate greater tolerance to autoimmune encephalomyelitis.[94] The mechanism for this protective autoimmunity is not clear, but may be related to the function of regulatory T-cells that are evoked in response to classical $CD4^+$ T-cells. Regardless, this early finding precipitated several studies to determine if a protective autoimmune response could be generated by a vaccination using nonencephalitogenic peptides, comprising different antigenic regions from the components of myelinated optic nerves. These studies met with moderate success,[95–97] but have not been pursued.

4.4 Impact of Inflammatory Responses on the Optic Nerve

One of the most important cytokines generated during glaucoma and optic nerve trauma is TNFα. This proinflammatory cytokine has been suspected of contributing to oligodendrocyte and axonal pathology in models of ocular hypertension, as their degeneration can be mimicked by a single injection of TNFα, with oligodendrocyte loss occurring as early as 4 days after injection followed by axonal loss at 2 weeks.[31,98] Genetic deletion of *Tnfa* or treatment with the fusion protein etanercept protects the oligodendrocytes and axons from ocular hypertension,[31,87] and this protection has been shown to be mediated through TNFR2.[31] It appears that a primary source of TNFα is the microglia, as mice deficient for CD11b/CD18 (Mac1) show improved optic nerve integrity after the induction of ocular hypertension.[31]

These results link TNFα to optic nerve degeneration, although this is inconsistent with studies examining the effect of this cytokine on RGC survival (see Section 5.5). It appears that the damaging effects of TNFα on the optic nerve occur early after injury and signal directly to the oligodendrocytes and axons, the degeneration of which precede RGC somatic loss in the retina.[27,31] If therapeutics were to be developed against TNFα-mediated injury, treatments would have to be administered quickly to preserve vision given the short window of time before degeneration becomes significant.

5. NEUROINFLAMMATION IN THE RETINA

Unlike the optic nerve, which shows immediate signs of widespread damage following injury, the retinal tissue is relatively resistant to optic nerve trauma. As the RGC axons degenerate and a glial scar forms in the optic nerve, the retina systematically sacrifices the RGC somas while preserving the remainder of the retinal architecture. This process is in no small part controlled by the retinal glia, which recognize the damaged RGCs (possibly through extracellular ATP release by the injured neurons[99–102]) and help remove cellular debris from lost neurons. Glial activation in the retina involves a number of complex signaling pathways and has been linked to both neuroprotective and neurodegenerative pathways.

5.1 Astrocytes

Type Ib astrocytes populate the nerve fiber layer and form the inner limiting membrane, serving as a barrier between the retina and the vitreous.[103] These macroglial cells have not been well studied in models of glaucoma, and much of the current knowledge comes from their behavior in the optic nerve after trauma and from cell culture studies. It is known that the retinal macroglia are particularly resilient to optic nerve damage,[14] and as RGC somas begin to deteriorate quiescent astrocytes enter a reactive state, a transition that has been well documented in glaucoma patients and animal models.[56,63,104] Reactive astrocytes upregulate GFAP and exhibit enlarged somas and thickened processes,[57,72] although they do not migrate and exhibit limited proliferative potential.[103,104] Astrocytes have been shown to become activated by microglial-derived endothelins after crush,[83] and cell culture studies suggest that astrocytes can then regulate microglial activation in a feedback loop by generating additional cytokines such as TNFα, interleukin 6, and colony-stimulating factors.[103] Activated astrocytes have been shown to dramatically upregulate CNTF and its receptor CNTFRα, which may play an important role in regulating GFAP expression and astrocyte morphology during activation.[103] However, the consequence of astrocytic activation on RGC survival remains inadequately understood.

5.2 Microglia

In a normal uninjured retina, the microglial population is the least abundant of the glial cells, although they establish a clear presence in the inner and outer plexiform layers and are localized adjacently to Müller cells and astrocytes.[16] In animal models of glaucoma and following optic nerve injury, the microglial population upregulates AIF1 and expands within the plexiform layers and into the nerve fiber layer and ganglion cell layer.[16,105] Microglial activation has been shown to be an early event in a mouse model of spontaneous glaucoma, preceding somatic loss of RGCs by several months.[16] These glial cells have also been shown to upregulate MHC-II, indicative of antigen presentation.[7] Treatment with the broad-spectrum antibiotic, minocycline, reduced retinal microglial activation[16] and also delayed RGC loss in a spontaneous model of glaucoma,[106] implicating microglia as mediators of RGC damage.

In models of optic nerve injury, microglia have been shown to be critical for phagocytosing debris and secreting trophic and anti-inflammatory factors, although they have also been identified as producers of reactive oxygen species that may additionally contribute to neurotoxicity.[32] Microglia can upregulate endothelin 2 (*Edn2*) in the retina, which has a damaging effect by constricting blood vessels, triggering reactive astrocytosis, and possible by inhibiting axonal transport.[88] Importantly, microglia appear to be a source of TNFα in the retina, which may be a critical autocrine signal for microglial activation.[78,107] Together, microglia are pleiotrophic immune cells and appear to participate in both neuroprotection and neurodegeneration of RGCs.

5.3 Müller Glia

Müller cells are critical for a healthy retinal environment, as genetic ablation of these support cells triggers a number of pathological consequences independent of any injury.[108] The unique distribution of Müller cells through the thickness of the retina allows them to interact with every cell type in the tissue, rendering these glial cells hypersensitive to changes in the retinal environment. These macroglial cells are the most abundant glial cell in the retina and encompass a number of important roles including recycling the neurotransmitter glutamate,[63] an excess of which is toxic to the retina,[109] and in response to glaucoma or damage, Müller glia upregulate the enzyme catalyzing this process, glutamine synthetase.[63,110] Activated Müller cells also upregulate GFAP, which manifests as visible filaments through cross sections of retinal tissue,[63,81] and express growth factors, neurotrophic factors, neurotransmitter transporters, and antioxidant agents that are thought to combat damaging stimuli and protect RGCs.[62] Müller cell activation has been documented during the first week after injury, eventually returning to baseline after several months.[63] It has been shown in culture that Müller cells can generate neurons and glia and that human Müller glia can generate photoreceptors and ganglion cells that may repair damage when transplanted into a rodent retina.[14] The proliferative characteristics of Müller cells have not been replicated *in vivo*, although they suggest that these macroglia may become pluripotent, thus acting as a stem cell population that may contribution to damage repair in the retina.

5.4 Dendritic Cells

An additional type of immune cell has recently been identified as an early responder to retinal injury, but the molecular markers to identify these cells are still being debated. Although not classified as glial cells, dendritic cells (DCs) are bone marrow-derived immune responders that have been shown to participate in immunity and inflammation in the CNS.[111,112] A unique mouse model labeling DCs with GFP allow for their migration from the periphery into the retina to be monitored.[111] In the quiescent retina, DCs exhibit a highly ramified morphology, but are sparse and localized in the ganglion cell layer around blood vessels, and the inner and outer plexiform layers,[111] and although similar to microglia,[22] represent a unique cell population in the retina. Following acute optic nerve injury, a significant increase in DCs was found clustered in the nerve fiber and ganglion cell layers in association with RGC axons and somas and was found to be phagocytosing RGC debris.[112] These cells exhibited an elongated morphology and a radial orientation with MHC class II expression.[111] While it is possible that these cells may serve as an additional source of cytokines and inflammatory factors, additional research is needed to further dissect the signaling mechanisms utilized by DCs.

5.5 Impact of Inflammatory Responses on the Retina

During the mid-1990s, it was determined that the effect of glial activation in the retina was detrimental to ganglion cells, and the theory that developed to explain this relationship was termed secondary degeneration.[113–115] It was proposed that only a subset of neurons were affected during the early stages of glaucoma or following optic nerve trauma and that the cytokines produced by activated retinal glia initiated extrinsic apoptosis of the surviving RGCs, causing a secondary wave of neuronal loss.[30] Support for this theory was founded on the elevation of TNFα in the aqueous humor of glaucoma patients[116] and in rodent models of glaucoma,[117] which correlated with worsening RGC pathology, and that neutralization of TNFα or genetic deletion of the protein protected RGCs from rodent models of ocular hypertension.[31,87] Additional studies weakened the case against TNFα, which showed that genetic deletion of TNFR1 surprisingly offered no protection to RGCs from ocular hypertension,[31] but did protect RGCs from optic nerve crush.[80] Furthermore, a direct intraocular injection of TNFα requires 8 weeks before RGC loss is significant, which is inconsistent with the timeframe of ocular hypertension or crush,[26] suggesting that TNFα is not directly toxic to RGCs.

More recent studies have revealed a protective potential for TNFα in the retina, possibly mediated through TNFR2,[81,118] and contrary to the model of secondary degeneration. In a crush paradigm, when TNFα is delivered prior to injury, RGC survival increased after optic nerve trauma, while TNFα-deficient mice exhibited greater RGC loss after crush, implicating a protective role for this cytokine.[81] Importantly, it appears that the timing of cytokine expression is crucial for determining the long-term outcome, with early expression of TNFα correlating with survival,[81] and long-term expression causing RGC degeneration.[31] The cell type responding to the cytokine may also influence the outcome, as TNFα appears to induce early degeneration of axons, delayed loss of RGC somas, and in Müller cells a rapid upregulation of the transcription factors, nuclear factor kappa B (NFκB) and JUN.[81] It is clear that the retinal immune response is pluripotent and mediated by a complex network of signals and glial subtypes working to preserve retinal tissue while removing critically injured neurons.

6. CONCLUSIONS

There is still much to be learned about the neuroinflammatory responses during RGC degeneration; however, it is becoming clear that the glial response can mediate both protection and degeneration of these neurons and their axons. Future research might allow for these opposing pathways to be untangled and provide a better framework for developing therapeutics and prolonging the quality of life for glaucoma patients.

REFERENCES

1. Grewe R. The history of glaucoma. *Klin Monbl Augenheilkd*. 1986;188:167–169.
2. Almasieh M, Wilson AM, Morquette B, Cueva Vargas JL, Di Polo A. The molecular basis of retinal ganglion cell death in glaucoma. *Prog Retin Eye Res*. 2012;31:152–181.

3. Johnson EC, Morrison JC. Friend or foe? Resolving the impact of glial responses in glaucoma. *J Glaucoma*. 2009;18:341–353.
4. Wax MB, Tezel G. Immunoregulation of retinal ganglion cell fate in glaucoma. *Exp Eye Res*. 2009;88:825–830.
5. Nickells RW, Howell GR, Soto I, John SW. Under pressure: cellular and molecular responses during glaucoma, a common neurodegeneration with axonopathy. *Annu Rev Neurosci*. 2012;35:153–179.
6. McKinnon SJ, Schlamp CL, Nickells RW. Mouse models of retinal ganglion cell death and glaucoma. *Exp Eye Res*. 2009;88:816–824.
7. Gallego BI, Salazar JJ, de Hoz R, et al. IOP induces upregulation of GFAP and MHC-II and microglia reactivity in mice retina contralateral to experimental glaucoma. *J Neuroinflammation*. 2012;9:92.
8. Schlamp CL, Johnson EC, Li Y, Morrison JC, Nickells RW. Changes in Thy1 gene expression associated with damaged retinal ganglion cells. *Mol Vis*. 2001;7:192–201.
9. Levkovitch-Verbin H, Quigley HA, Martin KR, et al. The transcription factor c-jun is activated in retinal ganglion cells in experimental rat glaucoma. *Exp Eye Res*. 2005;80:663–670.
10. Nickells RW. From ocular hypertension to ganglion cell death: a theoretical sequence of events leading to glaucoma. *Can J Ophthalmol*. 2007;42:278–287.
11. Resta V, Novelli E, Vozzi G, et al. Acute retinal ganglion cell injury caused by intraocular pressure spikes is mediated by endogenous extracellular ATP. *Eur J Neurosci*. 2007;25:2741–2754.
12. Gregory MS, Hackett CG, Abernathy EF, et al. Opposing roles for membrane bound and soluble Fas ligand in glaucoma-associated retinal ganglion cell death. *PLoS One*. 2011;6:e17659.
13. Baltmr A, Duggan J, Nizari S, Salt TE, Cordeiro MF. Neuroprotection in glaucoma—is there a future role? *Exp Eye Res*. 2010;91:554–566.
14. Goldman D. Muller glial cell reprogramming and retina regeneration. *Nat Rev Neurosci*. 2014;15:431–442.
15. Ruiz-Ederra J, Garcia M, Martin F, et al. Comparison of three methods of inducing chronic elevation of intraocular pressure in the pig (experimental glaucoma). *Arch Soc Esp Oftalmol*. 2005;80:571–579.
16. Bosco A, Steele MR, Vetter ML. Early microglia activation in a mouse model of chronic glaucoma. *J Comp Neurol*. 2011;519:599–620.
17. Levkovitch-Verbin H, Harris-Cerruti C, Groner Y, Wheeler LA, Schwartz M, Yoles E. RGC death in mice after optic nerve crush injury: oxidative stress and neuroprotection. *Invest Ophthalmol Vis Sci*. 2000;41:4169–4174.
18. Li Y, Schlamp CL, Poulsen KP, Nickells RW. Bax-dependent and independent pathways of retinal ganglion cell death induced by different damaging stimuli. *Exp Eye Res*. 2000;71:209–213.
19. Reichstein D, Ren L, Filippopoulos T, Mittag T, Danias J. Apoptotic retinal ganglion cell death in the DBA/2 mouse model of glaucoma. *Exp Eye Res*. 2007;84:13–21.
20. Schlamp CL, Li Y, Dietz JA, Janssen KT, Nickells RW. Progressive ganglion cell loss and optic nerve degeneration in DBA/2J mice is variable and asymmetric. *BMC Neurosci*. 2006;7:66.
21. Schlamp CL, Montgomery AD, Mac Nair CE, Schuart C, Willmer DJ, Nickells RW. Evaluation of the percentage of ganglion cells in the ganglion cell layer of the rodent retina. *Mol Vis*. 2013;19:1387–1396.
22. Garcia-Valenzuela E, Sharma SC, Pina AL. Multilayered retinal microglial response to optic nerve transection in rats. *Mol Vis*. 2005;11:225–231.
23. Kakurai K, Sugiyama T, Kurimoto T, Oku H, Ikeda T. Involvement of P2X(7) receptors in retinal ganglion cell death after optic nerve crush injury in rats. *Neurosci Lett*. 2013;534:237–241.

24. Nadal-Nicolas FM, Jimenez-Lopez M, Salinas-Navarro M, et al. Whole number, distribution and co-expression of brn3 transcription factors in retinal ganglion cells of adult albino and pigmented rats. *PLoS One*. 2012;7:e49830.
25. Wang X, Fan J, Zhang M, Ni Y, Xu G. Upregulation of SOX9 in Glial (Muller) cells in retinal light damage of rats. *Neurosci Lett*. 2013;556:140–145.
26. Madigan MC, Rao NS, Tenhula WN, Sadun AA. Preliminary morphometric study of tumor necrosis factor-alpha (TNF alpha)-induced rabbit optic neuropathy. *Neurol Res*. 1996;18:233–236.
27. Madigan MC, Sadun AA, Rao NS, Dugel PU, Tenhula WN, Gill PS. Tumor necrosis factor-alpha (TNF-alpha)-induced optic neuropathy in rabbits. *Neurol Res*. 1996;18:176–184.
28. Quigley HA, Nickells RW, Kerrigan LA, Pease ME, Thibault DJ, Zack DJ. Retinal ganglion cell death in experimental glaucoma and after axotomy occurs by apoptosis. *Invest Ophthalmol Vis Sci*. 1995;36:774–786.
29. Nickells RW. Ganglion cell death in glaucoma: from mice to men. *Vet Ophthalmol*. 2007;10(suppl 1):88–94.
30. Levkovitch-Verbin H, Quigley HA, Kerrigan-Baumrind LA, D'Anna SA, Kerrigan D, Pease ME. Optic nerve transection in monkeys may result in secondary degeneration of retinal ganglion cells. *Invest Ophthalmol Vis Sci*. 2001;42:975–982.
31. Nakazawa T, Nakazawa C, Matsubara A, et al. Tumor necrosis factor-alpha mediates oligodendrocyte death and delayed retinal ganglion cell loss in a mouse model of glaucoma. *J Neurosci*. 2006;26:12633–12641.
32. Fitzgerald M, Bartlett CA, Harvey AR, Dunlop SA. Early events of secondary degeneration after partial optic nerve transection: an immunohistochemical study. *J Neurotrauma*. 2010;27:439–452.
33. Rodriguez-Muela N, Germain F, Marino G, Fitze PS, Boya P. Autophagy promotes survival of retinal ganglion cells after optic nerve axotomy in mice. *Cell Death Differ*. 2012;19:162–169.
34. Janssen KT, Mac Nair CE, Dietz JA, Schlamp CL, Nickells RW. Nuclear atrophy of retinal ganglion cells precedes the bax-dependent stage of apoptosis. *Invest Ophthalmol Vis Sci*. 2013;54:1805–1815.
35. Li Y, Semaan SJ, Schlamp CL, Nickells RW. Dominant inheritance of retinal ganglion cell resistance to optic nerve crush in mice. *BMC Neurosci*. 2007;8:19.
36. Templeton JP, Freeman NE, Nickerson JM, et al. Innate immune network in the retina activated by optic nerve crush. *Invest Ophthalmol Vis Sci*. 2013;54:2599–2606.
37. Pekny M, Pekna M. Astrocyte reactivity and reactive astrogliosis: costs and benefits. *Physiol Rev*. 2014;94:1077–1098.
38. Medawar PB. Immunity to homologous grafted skin; the fate of skin homografts transplanted to the brain, to subcutaneous tissue, and to the anterior chamber of the eye. *Br J Exp Pathol*. 1948;29:58–69.
39. Perez VL, Saeed AM, Tan Y, Urbieta M, Cruz-Guilloty F. The eye: a window to the soul of the immune system. *J Autoimmun*. 2013;45:7–14.
40. Hori J, Vega JL, Masli S. Review of ocular immune privilege in the year 2010: modifying the immune privilege of the eye. *Ocul Immunol Inflamm*. 2010;18:325–333.
41. Detrick B, Hooks JJ. Immune regulation in the retina. *Immunol Res*. 2010;47:153–161.
42. Streilein JW. Anterior chamber associated immune deviation: the privilege of immunity in the eye. *Surv Ophthalmol*. 1990;35:67–73.
43. Biros D. Anterior chamber-associated immune deviation. *Vet Clin North Am Small Anim Pract*. 2008;38:309–321, vi–vii.
44. Krizaj D, Ryskamp DA, Tian N, et al. From mechanosensitivity to inflammatory responses: new players in the pathology of glaucoma. *Curr Eye Res*. 2014;39:105–119.

45. Kumar A, Pandey RK, Miller LJ, Singh PK, Kanwar M. Muller glia in retinal innate immunity: a perspective on their roles in endophthalmitis. *Crit Rev Immunol.* 2013;33:119–135.
46. Dewar D, Underhill SM, Goldberg MP. Oligodendrocytes and ischemic brain injury. *J Cereb Blood Flow Metab.* 2003;23:263–274.
47. Nguyen JV, Soto I, Kim KY, et al. Myelination transition zone astrocytes are constitutively phagocytic and have synuclein dependent reactivity in glaucoma. *Proc Natl Acad Sci USA.* 2011;108:1176–1181.
48. Merrill JE. Effects of interleukin-1 and tumor necrosis factor-alpha on astrocytes, microglia, oligodendrocytes, and glial precursors in vitro. *Dev Neurosci.* 1991;13:130–137.
49. Olson JK, Miller SD. Microglia initiate central nervous system innate and adaptive immune responses through multiple TLRs. *J Immunol.* 2004;173:3916–3924.
50. Carson MJ, Doose JM, Melchior B, Schmid CD, Ploix CC. CNS immune privilege: hiding in plain sight. *Immunol Rev.* 2006;213:48–65.
51. Karperien A, Ahammer H, Jelinek HF. Quantitating the subtleties of microglial morphology with fractal analysis. *Front Cell Neurosci.* 2013;7:3.
52. Bosco A, Inman DM, Steele MR, et al. Reduced retina microglial activation and improved optic nerve integrity with minocycline treatment in the DBA/2J mouse model of glaucoma. *Invest Ophthalmol Vis Sci.* 2008;49:1437–1446.
53. Kim SU, de Vellis J. Microglia in health and disease. *J Neurosci Res.* 2005; 81:302–313.
54. Streit WJ. Microglial response to brain injury: a brief synopsis. *Toxicol Pathol.* 2000;28:28–30.
55. Sivakumar V, Foulds WS, Luu CD, Ling EA, Kaur C. Retinal ganglion cell death is induced by microglia derived pro-inflammatory cytokines in the hypoxic neonatal retina. *J Pathol.* 2011;224:245–260.
56. Tezel G, Chauhan BC, LeBlanc RP, Wax MB. Immunohistochemical assessment of the glial mitogen-activated protein kinase activation in glaucoma. *Invest Ophthalmol Vis Sci.* 2003;44:3025–3033.
57. Ganesh BS, Chintala SK. Inhibition of reactive gliosis attenuates excitotoxicity-mediated death of retinal ganglion cells. *PLoS One.* 2011;6:e18305.
58. Yuan L, Neufeld AH. Tumor necrosis factor-alpha: a potentially neurodestructive cytokine produced by glia in the human glaucomatous optic nerve head. *Glia.* 2000;32:42–50.
59. Selmaj KW, Farooq M, Norton WT, Raine CS, Brosnan CF. Proliferation of astrocytes in vitro in response to cytokines. A primary role for tumor necrosis factor. *J Immunol.* 1990;144:129–135.
60. Constantinescu CS, Tani M, Ransohoff RM, et al. Astrocytes as antigen-presenting cells: expression of IL-12/IL-23. *J Neurochem.* 2005;95:331–340.
61. Iandiev I, Biedermann B, Bringmann A, Reichel MB, Reichenbach A, Pannicke T. Atypical gliosis in Muller cells of the slowly degenerating rds mutant mouse retina. *Exp Eye Res.* 2006;82:449–457.
62. Garcia M, Vecino E. Role of Muller glia in neuroprotection and regeneration in the retina. *Histol Histopathol.* 2003;18:1205–1218.
63. Chen H, Weber AJ. Expression of glial fibrillary acidic protein and glutamine synthetase by Muller cells after optic nerve damage and intravitreal application of brain-derived neurotrophic factor. *Glia.* 2002;38:115–125.
64. Toops KA, Berlinicke C, Zack DJ, Nickells RW. Hydrocortisone stimulates neurite outgrowth from mouse retinal explants by modulating macroglial activity. *Invest Ophthalmol Vis Sci.* 2012;53:2046–2061.
65. Li Y, Schlamp CL, Nickells RW. Experimental induction of retinal ganglion cell death in adult mice. *Invest Ophthalmol Vis Sci.* 1999;40:1004–1008.

66. Brambilla R, Dvoriantchikova G, Barakat D, Ivanov D, Bethea JR, Shestopalov VI. Transgenic inhibition of astroglial NF-kappaB protects from optic nerve damage and retinal ganglion cell loss in experimental optic neuritis. *J Neuroinflammation.* 2012;9:213.
67. Morquette JB, Di Polo A. Dendritic and synaptic protection: is it enough to save the retinal ganglion cell body and axon? *J Neuroophthalmol.* 2008;28:144–154.
68. Nickells RW. Apoptosis of retinal ganglion cells in glaucoma: an update of the molecular pathways involved in cell death. *Surv Ophthalmol.* 1999;43(suppl 1): S151–S161.
69. Howell GR, Libby RT, Jakobs TC, et al. Axons of retinal ganglion cells are insulted in the optic nerve early in DBA/2J glaucoma. *J Cell Biol.* 2007;179:1523–1537.
70. Libby RT, Li Y, Savinova OV, et al. Susceptibility to neurodegeneration in a glaucoma is modified by Bax gene dosage. *PLoS Genet.* 2005;1:17–26.
71. Johnson EC, Doser TA, Cepurna WO, et al. Cell proliferation and interleukin-6-type cytokine signaling are implicated by gene expression responses in early optic nerve head injury in rat glaucoma. *Invest Ophthalmol Vis Sci.* 2011;52:504–518.
72. Qu J, Jakobs TC. The time course of gene expression during reactive gliosis in the optic nerve. *PLoS One.* 2013;8:e67094.
73. Lye-Barthel M, Sun D, Jakobs TC. Morphology of astrocytes in a glaucomatous optic nerve. *Invest Ophthalmol Vis Sci.* 2013;54:909–917.
74. Yan X, Tezel G, Wax MB, Edward DP. Matrix metalloproteinases and tumor necrosis factor alpha in glaucomatous optic nerve head. *Arch Ophthalmol.* 2000;118:666–673.
75. Fontaine V, Mohand-Said S, Hanoteau N, Fuchs C, Pfizenmaier K, Eisel U. Neurodegenerative and neuroprotective effects of tumor necrosis factor (TNF) in retinal ischemia: opposite roles of TNF receptor 1 and TNF receptor 2. *J Neurosci.* 2002;22:RC216.
76. Marchetti L, Klein M, Schlett K, Pfizenmaier K, Eisel UL. Tumor necrosis factor (TNF)-mediated neuroprotection against glutamate-induced excitotoxicity is enhanced by N-methyl-D-aspartate receptor activation. Essential role of a TNF receptor 2-mediated phosphatidylinositol 3-kinase-dependent NF-kappa B pathway. *J Biol Chem.* 2004;279:32869–32881.
77. Saha RN, Ghosh A, Palencia CA, Fung YK, Dudek SM, Pahan K. TNF-alpha preconditioning protects neurons via neuron-specific up-regulation of CREB-binding protein. *J Immunol.* 2009;183:2068–2078.
78. Tezel G. TNF-alpha signaling in glaucomatous neurodegeneration. *Prog Brain Res.* 2008;173:409–421.
79. Tezel G, Wax MB. Increased production of tumor necrosis factor-alpha by glial cells exposed to simulated ischemia or elevated hydrostatic pressure induces apoptosis in cocultured retinal ganglion cells. *J Neurosci.* 2000;20:8693–8700.
80. Tezel G, Yang X, Yang J, Wax MB. Role of tumor necrosis factor receptor-1 in the death of retinal ganglion cells following optic nerve crush injury in mice. *Brain Res.* 2004;996:202–212.
81. Mac Nair CE, Fernandes KA, Schlamp CL, Libby RT, Nickells RW. Tumor necrosis factor alpha has an early protective effect on retinal ganglion cells after optic nerve crush. *J Neuroinflammation.* 2014;11:194.
82. Yang J, Yang P, Tezel G, Patil RV, Hernandez MR, Wax MB. Induction of HLA-DR expression in human lamina cribrosa astrocytes by cytokines and simulated ischemia. *Invest Ophthalmol Vis Sci.* 2001;42:365–371.
83. Tonari M, Kurimoto T, Horie T, Sugiyama T, Ikeda T, Oku H. Blocking endothelin-B receptors rescues retinal ganglion cells from optic nerve injury through suppression of neuroinflammation. *Invest Ophthalmol Vis Sci.* 2012;53:3490–3500.
84. Sun D, Lye-Barthel M, Masland RH, Jakobs TC. Structural remodeling of fibrous astrocytes after axonal injury. *J Neurosci.* 2010;30:14008–14019.

85. Sun D, Lye-Barthel M, Masland RH, Jakobs TC. The morphology and spatial arrangement of astrocytes in the optic nerve head of the mouse. *J Comp Neurol*. 2009;516:1–19.
86. Garcia DM, Koke JR. Astrocytes as gate-keepers in optic nerve regeneration—a mini-review. *Comp Biochem Physiol A Mol Integr Physiol*. 2009;152:135–138.
87. Roh M, Zhang Y, Murakami Y, et al. Etanercept, a widely used inhibitor of tumor necrosis factor-alpha (TNF-alpha), prevents retinal ganglion cell loss in a rat model of glaucoma. *PLoS One*. 2012;7:e40065.
88. Howell GR, Macalinao DG, Sousa GL, et al. Molecular clustering identifies complement and endothelin induction as early events in a mouse model of glaucoma. *J Clin Invest*. 2011;121:1429–1444.
89. Howell GR, Soto I, Zhu X, et al. Radiation treatment inhibits monocyte entry into the optic nerve head and prevents neuronal damage in a mouse model of glaucoma. *J Clin Invest*. 2012;122:1246–1261.
90. Wohl SG, Schmeer CW, Witte OW, Isenmann S. Proliferative response of microglia and macrophages in the adult mouse eye after optic nerve lesion. *Invest Ophthalmol Vis Sci*. 2010;51:2686–2696.
91. Liu S, Li ZW, Weinreb RN, et al. Tracking retinal microgliosis in models of retinal ganglion cell damage. *Invest Ophthalmol Vis Sci*. 2012;53:6254–6262.
92. Kigerl KA, Gensel JC, Ankeny DP, Alexander JK, Donnelly DJ, Popovich PG. Identification of two distinct macrophage subsets with divergent effects causing either neurotoxicity or regeneration in the injured mouse spinal cord. *J Neurosci*. 2009;29:13435–13444.
93. Moalem G, Leibowitz-Amit R, Yoles E, Mor F, Cohen IR, Schwartz M. Autoimmune T cells protect neurons from secondary degeneration after central nervous system axotomy. *Nat Med*. 1999;5:49–55.
94. Kipnis J, Yoles E, Schori H, Hauben E, Shaked I, Schwartz M. Neuronal survival after CNS insult is determined by a genetically encoded autoimmune response. *J Neurosci*. 2001;21:4564–4571.
95. Bakalash S, Ben-Shlomo G, Aloni E, et al. T-cell-based vaccination for morphological and functional neuroprotection in a rat model of chronically elevated intraocular pressure. *J Mol Med (Berl)*. 2005;83:904–916.
96. Fisher J, Levkovitch-Verbin H, Schori H, et al. Vaccination for neuroprotection in the mouse optic nerve: implications for optic neuropathies. *J Neurosci*. 2001;21:136–142.
97. Schori H, Kipnis J, Yoles E, et al. Vaccination for protection of retinal ganglion cells against death from glutamate cytotoxicity and ocular hypertension: implications for glaucoma. *Proc Natl Acad Sci USA*. 2001;98:3398–3403.
98. Kitaoka Y, Kitaoka Y, Kwong JM, et al. TNF-alpha-induced optic nerve degeneration and nuclear factor-kappaB p65. *Invest Ophthalmol Vis Sci*. 2006;47:1448–1457.
99. Li A, Zhang X, Zheng D, Ge J, Laties AM, Mitchell CH. Sustained elevation of extracellular ATP in aqueous humor from humans with primary chronic angle-closure glaucoma. *Exp Eye Res*. 2011;93:528–533.
100. Monif M, Reid CA, Powell KL, Smart ML, Williams DA. The P2X7 receptor drives microglial activation and proliferation: a trophic role for P2X7R pore. *J Neurosci*. 2009;29:3781–3791.
101. Reigada D, Lu W, Zhang M, Mitchell CH. Elevated pressure triggers a physiological release of ATP from the retina: possible role for pannexin hemichannels. *Neuroscience*. 2008;157:396–404.
102. Schwiebert EM, Zsembery A. Extracellular ATP as a signaling molecule for epithelial cells. *Biochim Biophys Acta*. 2003;1615:7–32.
103. Yang XT, Huang GH, Feng DF, Chen K. Insight into astrocyte activation after optic nerve injury. *J Neurosci Res*. 2015;93:539–548.

104. Inman DM, Horner PJ. Reactive nonproliferative gliosis predominates in a chronic mouse model of glaucoma. *Glia*. 2007;55:942–953.
105. Baptiste DC, Powell KJ, Jollimore CA, et al. Effects of minocycline and tetracycline on retinal ganglion cell survival after axotomy. *Neuroscience*. 2005;134:575–582.
106. Levkovitch-Verbin H, Kalev-Landoy M, Habot-Wilner Z, Melamed S. Minocycline delays death of retinal ganglion cells in experimental glaucoma and after optic nerve transection. *Arch Ophthalmol*. 2006;124:520–526.
107. Kuno R, Wang J, Kawanokuchi J, Takeuchi H, Mizuno T, Suzumura A. Autocrine activation of microglia by tumor necrosis factor-alpha. *J Neuroimmunol*. 2005;162:89–96.
108. Shen W, Fruttiger M, Zhu L, et al. Conditional Muller cell ablation causes independent neuronal and vascular pathologies in a novel transgenic model. *J Neurosci*. 2012;32:15715–15727.
109. Lebrun-Julien F, Duplan L, Pernet V, et al. Excitotoxic death of retinal neurons in vivo occurs via a non-cell-autonomous mechanism. *J Neurosci*. 2009;29:5536–5545.
110. Carter-Dawson L, Shen F, Harwerth RS, Smith 3rd EL, Crawford ML, Chuang A. Glutamine immunoreactivity in Muller cells of monkey eyes with experimental glaucoma. *Exp Eye Res*. 1998;66:537–545.
111. Lehmann U, Heuss ND, McPherson SW, Roehrich H, Gregerson DS. Dendritic cells are early responders to retinal injury. *Neurobiol Dis*. 2010;40:177–184.
112. Heuss ND, Pierson MJ, Montaniel KR, et al. Retinal dendritic cell recruitment, but not function, was inhibited in MyD88 and TRIF deficient mice. *J Neuroinflammation*. 2014;11:143.
113. Schwartz M, Belkin M, Yoles E, Solomon A. Potential treatment modalities for glaucomatous neuropathy: neuroprotection and neuroregeneration. *J Glaucoma*. 1996;5:427–432.
114. Yoles E, Muller S, Schwartz M. NMDA-receptor antagonist protects neurons from secondary degeneration after partial optic nerve crush. *J Neurotrauma*. 1997;14: 665–675.
115. Yoles E, Schwartz M. Potential neuroprotective therapy for glaucomatous optic neuropathy. *Surv Ophthalmol*. 1998;42:367–372.
116. Tezel G, Li LY, Patil RV, Wax MB. TNF-alpha and TNF-alpha receptor-1 in the retina of normal and glaucomatous eyes. *Invest Ophthalmol Vis Sci*. 2001;42:1787–1794.
117. Bai Y, Shi Z, Zhuo Y, et al. In glaucoma the upregulated truncated TrkC.T1 receptor isoform in glia causes increased TNF-alpha production, leading to retinal ganglion cell death. *Invest Ophthalmol Vis Sci*. 2010;51:6639–6651.
118. Veroni C, Gabriele L, Canini I, et al. Activation of TNF receptor 2 in microglia promotes induction of anti-inflammatory pathways. *Mol Cell Neurosci*. 2010;45:234–244.

CHAPTER TWENTY-ONE

What Animal Models Can Tell Us About Glaucoma

Felix L. Struebing, Eldon E. Geisert[1]
Department of Ophthalmology, Emory University School of Medicine, Atlanta, Georgia, USA
[1]Corresponding author: e-mail address: egeiser@emory.edu

Contents

Abstract

Well defined animal models facilitate the study of ocular diseases. Each model brings a unique perspective to the understanding of the disease process, and in some cases, the models are critical to the development of therapeutic approaches for treatments. This is especially the case for glaucoma. Glaucoma is a family of diseases that can be caused by very different biological processes. The one thing in common is the end result, the loss of retinal ganglion cells and blindness. In this review, we will attempt to relate the findings from a number of animal models to specific types of glaucoma, emphasizing the contributions that each of the models makes to our overall understanding of the complex collection of diseases we call glaucoma.

1. INTRODUCTION

Animal models are one of the most viable tools for researchers to study diseases, especially complex diseases with multiple genetic and environmental risk factors. In general, each model brings a unique benefit to the understanding of the disease process, and in specific cases, the animal models are critical to the development of therapeutics and treatments. This is especially the case for glaucoma. Glaucoma is a family of diseases affecting multiple

Progress in Molecular Biology and Translational Science, Volume 134
ISSN 1877-1173
http://dx.doi.org/10.1016/bs.pmbts.2015.06.003
© 2015 Elsevier Inc.
All rights reserved.

biological systems, all of which result in the loss of retinal ganglion cells (RGCs) and, if left untreated, blindness. In this review, we attempt to relate the findings from a number of animal models to specific types of glaucoma, emphasizing the contributions that each of the models makes to our overall understanding of the complex collection of diseases we call glaucoma.

2. ANIMALS IN GLAUCOMA RESEARCH

Traditionally, glaucoma in humans is classified as primary open-angle glaucoma, primary closed-angle glaucoma, congenital glaucoma, exfoliation glaucoma, pigmentary glaucoma, traumatic glaucoma, and normotensive glaucoma.[1] The one common feature of all of these forms of glaucoma is the axonal damage at the optic nerve head with the final common denominator of RGC death. The most common risk factor for developing glaucoma is an elevation in intraocular pressure (IOP). Recent evidence suggests that this represents a differential pressure between the intraocular and cerebral spinal fluid pressure at the lamina cribosa, implying that a large pressure gradient between the two compartments would increase the risk of developing glaucoma.[2] The complexity of this group of diseases means that a single animal model will only be able to define limited aspects of the disease process or a specific subtype of glaucoma. Furthermore, the genetics of glaucoma are not simple. There are many forms that follow Mendelian heritability. In general, these are congenital in juvenile forms of glaucoma and include autosomal recessive mutations in genes such as *CYP1B1* and *LTBP2*, and autosomal dominant mutations in *MYOC*, *OPTN*, or *WDR36*.[3] Glaucoma in adults has a complex inheritance with multiple genetic risk factors potentially interacting with environmental factors that all contribute to a complex genetic disease. These complex genetic risk factors are the focus of recent genome-wide association studies that are not only defining genetic risk factors but also aiding in our understanding of the putative causes of glaucoma.[4–6] These would include the identification of genes affecting cerebral spinal fluid pressure[7] and ultimately the differential pressure observed across the optic nerve head.[8] Recent findings suggest that RGC number is also a risk factor for glaucoma. Several polymorphisms in the genes *SIX6* and *ATOH7* were identified to be responsible for lower RGC number, suggesting that humans with a lower count of RGCs are at increased risk of developing glaucoma.[9–11] To unravel the mechanisms responsible for the complex series of diseases called glaucoma requires research on a variety of different animal models.

There are many different types of glaucoma animal models ranging from large animals like cattle, dogs, and cats to small animals like rats, mice, and zebrafish. There are a number of excellent review articles discussing these animal models and their use.[12–16] In general, large animals offer the advantage of giving a better access to the eye due to its larger size. This is especially the case for nonhuman primates, which have been used in glaucoma research for decades. The benefits of using nonhuman primate models center around their resemblance in ocular anatomy, as compared to humans. In rhesus monkeys, elevated IOP was induced by laser coagulation of the trabecular meshwork.[17,18] There are several models of inherited glaucoma in dogs and cats providing for a naturally occurring model without experimental manipulation.[19,20] Topical application or injection of steroids in sheep, cow, and rabbit eyes has proven to be valuable glaucoma models by inducing elevated IOP.[21–23] In this review, we focus on rodent models of glaucoma, emphasizing the contributions of specific models to our overall understanding of glaucoma in humans.

3. RODENT MODELS

Rodent models of glaucoma have gained favor in the research community due to their ease of handling and the lower costs associated with acquisition and care. In particular, the mouse provides a number of useful genetic approaches to create models and to test specific molecular interactions associated with the disease process. Furthermore, the mouse genome is relatively conserved compared to the human genome. One of the informative mouse models is the DBA/2J mouse that is widely used in glaucoma research by many different groups.[24–29] The DBA/2J mouse carries two mutations ($Tyrp1^{b}$ and $Gpnmb^{R150X}$), which together cause an inherited disease similar to pigmentary dispersion glaucoma in humans. Even though the DBA/2J mouse may not be a perfect model of human pigment dispersion syndrome, it allows researchers to study the effects of elevated IOP on the retina and optic nerve head revealing fundamental discoveries about the disease process in glaucoma.

There are several inducible models of glaucoma that either damage the ganglion cell axons directly[30–35] or through experimentally elevated IOP[36,37] (Table 1). The advantage of optic nerve crush, where the optic nerve is crushed using a self-closing fine forceps, is a dramatic and temporally synchronized injury to the RGC axons.[46] This allows the research scientist to look at the effects of injury in a relatively large number of cells simultaneously. The disadvantage is that this does not represent the type of continuous nonsynchronized cell death that is observed in either acute glaucoma or chronic glaucoma. For this type of response, it is best to elevate IOP and let the

Table 1 Summary of Genetically Engineered or Naturally Occurring Models and Experimentally Inducible Models

Genetic Models

Model		Affected Gene(s)	Mechanism of RGC Death	Onset of RGC Loss (Age)	References
Naturally occurring models	Nee mutation	$B10\text{-}Sh3pxd2b^{nee}$	Inhibited podosome development of the iridocorneal angle, elevated IOP	3 months	38
	DBA/2J mouse strain	$Gpnmb^{R150X}$, $Tyrp1^{b}$	Pigment dispersion and blockage of outflow pathways, elevated IOP	8–9 months	25
Inducible/engineered models	Myocilin-mutation	$Myoc^{Tyr437His}$	Reduced exocytosis of myocilin and accumulation thereof in TM endothelial cells, elevated IOP	18 months	39
	Collagen type 1-mutation	$Col1a1^{r/r}$	Collagen accumulation in outflow pathways through inability of cleavage by matrix metalloproteinases, elevated IOP	18–24 weeks	40
	Vav2/Vav3-deficient mice	$Vav2^{-/-}$ $Vav3^{-/-}$	Molecular mechanism unknown, but progressive iridocorneal angle pathology leads to elevated IOP	<6 months	41

Experimentally Inducible Models

Model		Mechanism of RGC Death	References
Injection	Latex/polystyrene microbeads	Blocking of TM outflow facility, elevated IOP	37
	Magnetic microbeads	Blocking of TM outflow facility, elevated IOP	36
	Hypertonic saline	Sclerosing outflow facility vessels, elevated IOP	42
Cauterization/ coagulation	Episcleral vein cauterization	Sclerosing outflow facility vessels, elevated IOP	43
	Episcleral and limbal vein argon laser photocoagulation	Sclerosing outflow facility vessels, elevated IOP	44
Other models	Topical application of steroids	Elevated IOP as a side effect	45
	Optic nerve crush	Direct damage to the optic nerve and subsequent degeneration of RGCs	31, 33, 34, 46–48

increased pressure cause the injury at the optic nerve head.[27] There are models that induce elevated IOP, either through direct or indirect blockage of the trabecular meshwork. Examples include injections of polystyrene or magnetic microspheres into the anterior chamber to block the trabecular meshwork[36,37]; other models mimic glaucoma by injection of hypertonic saline into the episcleral veins[42] or cauterization of the episcleral vein,[43] which both block the outflow facility further distally, leading to increased IOP. Episcleral and limbal veins can also be photocoagulated with argon laser, leading to the same effect.[44] All of these inducible models cause IOP elevation within a few days after the procedure, and in most models, IOP stays elevated for a couple of weeks, which is enough to compromise RGCs.

4. MODELS INDUCED BY GENETIC MANIPULATION

Genetic models have been developed that placed the mutated human myocilin gene into the mouse genome. The MYOC Tyr437His mutation is known to cause an autosomal dominant form of severe glaucoma in humans with juvenile onset.[49] Mice with the mutated form of the gene show elevated IOP at 18 months of age and a 20% decrease in RGC number.[39,50] In another model, the alpha subunit of fibrillar collagen type 1 (*Col1a1*) was targeted to induce a mutation that abrogates matrix metalloproteinase-related cleavage of collagen, which in turn is necessary for extracellular matrix turnover in the trabecular meshwork. The accumulating collagen then eventually leads to IOP elevation and RGC death at about 24 weeks of age.[40] Another example is *Vav2*/*Vav3*-deficient mice that exhibit a phenotype resembling human glaucoma; however, the exact molecular pathomechanism is unclear.[41] These mice develop early-onset buphthalmia and progressive iridocorneal angle changes at less than 6 months of age. Interestingly, *Vav2* and *Vav3* have also been found to be candidate genes for a Japanese subgroup of open-angle glaucoma in a genome-wide association study.[51] Another recently developed model includes the nee-mutant mouse, which carries a spontaneously arising mutation in *Sh3pxd2b*.[38] This mutation has been associated with a rare syndrome in humans called Frank–ter Haar syndrome, which apart from skeletal and cardiovascular abnormalities also includes early formation of glaucoma.[52] In the mouse model, the normal development of podosomes is impaired in the trabecular meshwork and animals develop severe iridocorneal adhesions as a consequence of aqueous humor outflow blockage and high IOP. At 3–4 months of age, these mice show signs of progressive RGC loss.

5. MECHANISMS UNDERLYING RGC DEATH IN GLAUCOMA

The final common feature of virtually all types of glaucoma is the death of the RGCs. This is one of the most complex cell types in the mammalian retina, and the mechanisms by which RGCs die are equally complex.[14,53,54] Rodent models have offered significant insights into the biology of RGCs and the mechanisms involved in their death. The RGC is composed of distinct anatomical compartments, and there is clear evidence that different compartments die by different molecular mechanisms.[55,56] If the RGC axon is severed, the portion distal to the site of injury degenerates by a process called Wallerian degeneration,[57] and the portion of the axon proximal to the injury site is eliminated by a process termed dying back.[58] The RGC soma dies by apoptosis.[59–61] Taking these findings together, it now appears that molecularly distinct pathways are responsible for RGC axon and RGC soma death after injury.

Early on, many investigators believed that Wallerian degeneration occurred when the axon was cut off from its nutritive center, the neuronal soma. Then, a mouse mutation was discovered that significantly slowed down axon degeneration, the Wallerian degeneration slow (WldS) mouse. In this mouse, degenerative axon changes only take place after weeks rather than the usual 2 days,[62] which was discovered to be due to a spontaneous mutation resulting in a tandem triplication, the fusion of two genes, *Nmnat1* and *Ube4b*.[63] A recent series of studies defining the molecular mechanisms involved in Wallerian Degeneration has used the WldS mouse.[14,64] The mutation protects axons by maintaining mitochondrial functions[65,66] as well as axonal transport.[67] Associated with this protection is an attenuation of calcium influx that is normally found when axons degenerate.[68,69] These data point to the potential for a specific molecular mechanism modulating axon degeneration. Several mouse models were generated to study the effect of a specific gene or protein on axon degeneration following injury. Knocking out *Bax*, a proapoptotic gene, promotes the survival of RGC bodies but does not rescue axons.[56] There is clear evidence that the JNK (c-Jun, N-terminal kinase) pathway is involved in both axonal and soma degradation following optic nerve injury, since deficiency in both *Jnk2* and *Jnk3* results in protection from RGC death after optic nerve crush.[31] Likewise, knocking out *Dlk* (dual leucine zipper kinase) caused an attenuated cell death in the ganglion cell soma but did not affect the degeneration of the axon, demonstrating that the activation of JNK in the RGC soma is DLK dependent. This suggests that distinct upstream kinases regulate the activation of JNK in different cellular compartments.[55]

6. INNATE IMMUNE NETWORK IN GLAUCOMA

Our group has made extensive use of the BXD recombinant inbred (RI) strain of mice to define the genomic basis of complex traits and genetic networks associated with the injury of RGCs.[33,70,71] In BXD strains, which are derived by crossbreeding C57BL/6 and DBA/2J mice, the segregating pattern of maternal and paternal alleles generates a natural range of gene and protein expression. Because the genomes of both parental strains are fully sequenced and all BXD strains are fully mapped, the BXD RI panel is well suited to study the complex genetic contributions in glaucoma. Two comprehensive and complementary transcriptome datasets were created, the normal retina and the retina 2 days after optic nerve crush (both datasets are presented on www.genenetwork.org). By examining the changes that occur in the gene expression profiles between these conditions, our group discovered an innate immune network that is rapidly activated by optic nerve crush.[33] This added to the previous work of others that showed members of the complement cascade to be involved in retinal injury and glaucoma.[72–75] The importance of the complement cascade in glaucoma is revealed by knocking out *C1qa* on a DBA/2J background. In these animals, both pigmentary dispersion glaucoma and elevated IOP occur, yet the expected loss of axons in the optic nerve is dramatically mitigated.[74] Thus, C1QA and potentially the complement cascade play a pivotal role in the degeneration of axons within the optic nerve of DBA/2J glaucoma mouse model.

To illustrate the relationships of the genes in the innate immunity network, we selected a small group of genes (*App*, *C21orf91*, *Cd74*, *Icam1*, *C3*, *Serping1*, *H2-T23*, *C4b*, *Ifitm1*, *Tnfrsf21*, *Ifitm3*) and generated network diagrams representing their mutual correlation of expression (Fig. 1). In the normal retina, the selected genes form a genetic network but the correlation is rather low (Table 2). However after optic nerve crush, these genes increase in expression and form a highly correlated network (Fig. 1; Table 2). This suggests that injury to the optic nerve results in activation of this innate immunity network by upregulating gene expression and their mutual correlation.[70]

The tightly correlated expression of genes in this network indicates that there are common regulatory mechanisms in the retina that are activated by injury. We performed genome-wide association mapping to define

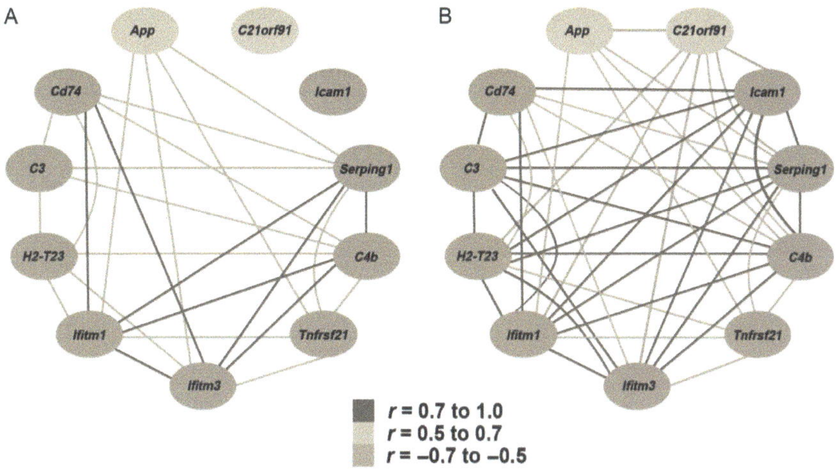

Figure 1 Genes of the innate immunity were selected from www.genenetwork.org, and their correlation in expression in two different conditions was drawn in the form of a network diagram. The correlation among those genes in the Normal Retina Database is shown in (A). After optic nerve crush, these genes are upregulated in expression, and their correlation increases dramatically (B). The scale indicates the Pearson correlation coefficient *r* between genes. Note that as the only member of this subset of genes, the *C21orf91* gene is negatively correlated with the rest of the network.

quantitative trait loci (QTLs) that could modulate this network and found a significant QTL on Chromosome 16 from 80 to 95 Mb that correlates to the expression of *C4b* in the normal retina.[70] This locus is potentially modulating the changes in gene expression we observed in the innate immune network occurring after axon injury. Within this region, *cis*-acting QTLs are putative candidate genes modulating the innate immunity network.[76] We found two genes with significant *cis*-QTLs, *Adamts5* (disintegrin-like metallopeptidase) and *Cxard* (Coxsackievirus and adenovirus receptor). In addition, this interval harbors five genes with nonsynonymous single nucleotide polymorphisms: *C21orf91* (*D16Ertd472e*), *Chodl* (chondrolectin), *App* (amyloid precursor protein), *Adamts1* (disintegrin-like metallopeptidase type 1), and *Adamts5* (disintegrin-like metallopeptidase type 5) (Table 3). Of these seven genes, the best candidate genes are the ones that show an expression pattern across the whole BXD RI strain set similar to the genes in the innate immune network. Another way of explaining this is, if a gene is a modulator of a network, its expression should correlate with the genes that form it. Two of the seven candidate genes (*App* and *C21orf91*) have a relatively high correlation with genes in the innate immune network and can thus be understood as putative upstream modulators. Notice that the

Table 2 Gene Expression and Correlation Changes in the Selected Subgroup of Innate Immunity Genes from Fig. 1

		Expression		Correlation to *C4b*	
Symbol	**Description**	**Control Retina**	**ONC**	**Control Retina**	**ONC**
C4b	Complement component 4B	8.75	10.23	1	1
Serping1	Serine peptidase inhibitor, clade G, member 1	8.48	10.62	0.768	0.811
C3	Complement component 3	8.14	8.54	0.636	0.779
Icam1	Intercellular adhesion molecule 1	7.66	7.50	0.231	0.749
Tnfrsf21	Tumor necrosis factor receptor subfamily, member 21	10.14	10.38	0.565	0.655
H2-T23	Histocompatibility 2, T region locus 23	9.14	9.63	0.588	0.771
Ifitm1	Interferon-induced transmembrane protein 1	10.77	11.51	0.735	0.821
Ifitm3	Interferon-induced transmembrane protein 3	10.77	11.58	0.752	0.777
C21orf91	Chromosome 21 open reading frame 91	14.05	14.17	−0.071	−0.555
Cd74	CD74 antigen	6.22	7.66	0.721	0.657
App	Amyloid beta (A4) precursor protein	13.23	13.14	0.483	0.553
Averaged		9.75	10.45	0.58	0.63

Expression was averaged by computing *Z*-scores for the raw data values and then adding an offset of 8 units. The goal of this transformation was to produce a set of *Z*-like scores for each probe that had a mean of 8, a variance of 4, and a standard deviation of 2. The advantage of this was that a twofold difference in expression corresponded to approximately 1 unit. Correlation is expressed by means of Pearson's coefficient.

upstream modulators are correlated with many different genes in the network and even each other after optic nerve crush but not before the insult. Interestingly, APP is believed to be a proinflammatory activator of the innate immune system in the retina[77,78] and the brain.[79] Our results confirm the role of *App* as an upstream modulator of the innate immune system in the retina.

Within the retina as in the rest of the central nervous system, cells are protected from outside influences by the blood–retina and –brain barriers.

Table 3 List of Candidate Genes at the Significant QTL on Chromosome 16
Candidate Genes on Chromosome 16, 78–87 Mb

	Gene Symbol	Gene Description
cis-QTLs	*Adamts5*	Disintegrin-like metallopeptidase type 5
	Cxadr	Coxsackievirus and adenovirus receptor
Nonsynonymous SNPs	*D16Ertd472e*	C21orf91
	Chodl	Chondrolectin
	App	Amyloid precursor protein
	Adamts1	Disintegrin-like metallopeptidase type 1
	Adamts5	Disintegrin-like metallopeptidase type 5

In this locus, *Admats5* and *Cxadr* are *cis*-QTLs, meaning that these genes are regulating their own expression.

This creates an immunologically privileged environment where peripheral blood components do not enter the parenchyma of the retina or optic nerve. We demonstrated that many components of the innate immune system are expressed intrinsically by retinal cells (Retinal Database, www.genenetwork.org[80]). Others find that components of the complement cascade are expressed in human RPE.[81,82] There is also a very strong association of several members of the complement system and age-related macular degeneration (AMD). Many of the specific allelic variants of the components of the complement system are risk factors for developing AMD.[83,84] Furthermore, complement proteins are found within drusen deposits that typify AMD.[25,85–87] In models of pigment dispersion glaucoma,[74] changes in the innate immune system appear to play a very important role in the initial phases of injury to the retina. The identification of the innate immune network was made possible due to the BXD mouse strain set.[76] The contribution of this type of defined mouse system cannot be underestimated. Furthermore, the same can be said of the DBA/2J model, which has allowed researchers to define the early changes in glaucoma[74] as well as test the role of the innate immune system in the progression of glaucomatous degeneration.[74,88]

ACKNOWLEDGMENTS

This work was supported by the National Eye Institute R01EY017841, P30EY006360, an unrestricted grant to the Department of Ophthalmology at Emory University from Research to Prevent Blindness, Inc., DOD grant W81XWH-12-1-0255.

REFERENCES

1. Quigley HA. Open-angle glaucoma. *N Engl J Med.* 1993;328(15):1097–1106. http://dx.doi.org/10.1056/NEJM199304153281507.
2. Berdahl JP, Allingham RR, Johnson DH. Cerebrospinal fluid pressure is decreased in primary open-angle glaucoma. *Ophthalmology.* 2008;115(5):763–768. http://dx.doi.org/10.1016/j.ophtha.2008.01.013.
3. Khan AO. Genetics of primary glaucoma. *Curr Opin Ophthalmol.* 2011;22(5):347–355. http://dx.doi.org/10.1097/ICU.0b013e32834922d2.
4. Ozel AB, Moroi SE, Reed DM, et al. Genome-wide association study and meta-analysis of intraocular pressure. *Hum Genet.* 2014;133(1):41–57. http://dx.doi.org/10.1007/s00439-013-1349-5.
5. Wang R, Wiggs JL. Common and rare genetic risk factors for glaucoma. *Cold Spring Harb Perspect Med.* 2014;4(12):a017244. http://dx.doi.org/10.1101/cshperspect.a017244.
6. Wiggs JL, Hauser MA, Abdrabou W, et al. The NEIGHBOR consortium primary open-angle glaucoma genome-wide association study: rationale, study design, and clinical variables. *J Glaucoma.* 2013;22(7):517–525. http://dx.doi.org/10.1097/IJG.0b013e31824d4fd8.
7. Pasquale LR, Loomis SJ, Kang JH, et al. CDKN2B-AS1 genotype-glaucoma feature correlations in primary open-angle glaucoma patients from the United States. *Am J Ophthalmol.* 2013;155(2):342–353. http://dx.doi.org/10.1016/j.ajo.2012.07.023, e345.
8. Fleischman D, Berdahl J, Stinnett SS, Fautsch MP, Allingham RR. Cerebrospinal fluid pressure trends in diseases associated with primary open-angle glaucoma. *Acta Ophthalmol.* 2014;93(3):e234–e236. http://dx.doi.org/10.1111/aos.12551.
9. Carnes MU, Liu YP, Allingham RR, et al. Discovery and functional annotation of SIX6 variants in primary open-angle glaucoma. *PLoS Genet.* 2014;10(5):e1004372. http://dx.doi.org/10.1371/journal.pgen.1004372.
10. Macgregor S, Hewitt AW, Hysi PG, et al. Genome-wide association identifies ATOH7 as a major gene determining human optic disc size. *Hum Mol Genet.* 2010;19(13):2716–2724. http://dx.doi.org/10.1093/hmg/ddq144.
11. Prasov L, Nagy M, Rudolph DD, Glaser T. Math5 (Atoh7) gene dosage limits retinal ganglion cell genesis. *Neuroreport.* 2012;23(10):631–634. http://dx.doi.org/10.1097/WNR.0b013e328355f260.
12. Bouhenni RA, Dunmire J, Sewell A, Edward DP. Animal models of glaucoma. *J Biomed Biotechnol.* 2012;2012:692609. http://dx.doi.org/10.1155/2012/692609.
13. Johnson TV, Tomarev SI. Rodent models of glaucoma. *Brain Res Bull.* 2010;81(2–3):349–358. http://dx.doi.org/10.1016/j.brainresbull.2009.04.004.
14. Nickells RW, Howell GR, Soto I, John SW. Under pressure: cellular and molecular responses during glaucoma, a common neurodegeneration with axonopathy. *Annu Rev Neurosci.* 2012;35:153–179. http://dx.doi.org/10.1146/annurev.neuro.051508.135728.
15. Qu J, Wang D, Grosskreutz CL. Mechanisms of retinal ganglion cell injury and defense in glaucoma. *Exp Eye Res.* 2010;91(1):48–53. http://dx.doi.org/10.1016/j.exer.2010.04.002.
16. Vecino E, Sharma S. Glaucoma animal models. In: Rumelt S, ed. *Glaucoma—Basic and Clinical Concepts.* InTech; 2011. ISBN: 978-953-307-591-4. http://dx.doi.org/10.5772/18498. Available from: http://www.intechopen.com/books/glaucoma-basic-and-clinical-concepts/glaucoma-animal-models.
17. Gaasterland D, Kupfer C. Experimental glaucoma in the rhesus monkey. *Invest Ophthalmol.* 1974;13(6):455–457.
18. Quigley HA, Hohman RM. Laser energy levels for trabecular meshwork damage in the primate eye. *Invest Ophthalmol Vis Sci.* 1983;24(9):1305–1307.

19. Wyman M, Ketring K. Congenital glaucoma in the basset hound: a biologic model. *Trans Sect Ophthalmol Am Acad Ophthalmol Otolaryngol*. 1976;81(4 Pt 1):OP645–OP652.
20. Zhan GL, Miranda OC, Bito LZ. Steroid glaucoma: corticosteroid-induced ocular hypertension in cats. *Exp Eye Res*. 1992;54(2):211–218.
21. Bonomi L, Perfetti S, Noya E, Bellucci R, Tomazzoli L. Experimental corticosteroid ocular hypertension in the rabbit. *Albrecht Von Graefes Arch Klin Exp Ophthalmol*. 1978;209(2):73–82.
22. Gerometta R, Podos SM, Candia OA, et al. Steroid-induced ocular hypertension in normal cattle. *Arch Ophthalmol*. 2004;122(10):1492–1497. http://dx.doi.org/10.1001/archopht.122.10.1492.
23. Gerometta R, Podos SM, Danias J, Candia OA. Steroid-induced ocular hypertension in normal sheep. *Invest Ophthalmol Vis Sci*. 2009;50(2):669–673. http://dx.doi.org/10.1167/iovs.08-2410.
24. Anderson MG, Libby RT, Gould DB, Smith RS, John SW. High-dose radiation with bone marrow transfer prevents neurodegeneration in an inherited glaucoma. *Proc Natl Acad Sci USA*. 2005;102(12):4566–4571. http://dx.doi.org/10.1073/pnas.0407357102.
25. Anderson MG, Smith RS, Hawes NL, et al. Mutations in genes encoding melanosomal proteins cause pigmentary glaucoma in DBA/2J mice. *Nat Genet*. 2002;30(1):81–85. http://dx.doi.org/10.1038/ng794.
26. Dengler-Crish CM, Smith MA, Inman DM, Wilson GN, Young JW, Crish SD. Anterograde transport blockade precedes deficits in retrograde transport in the visual projection of the DBA/2J mouse model of glaucoma. *Front Neurosci*. 2014;8:290. http://dx.doi.org/10.3389/fnins.2014.00290.
27. Howell GR, Libby RT, Jakobs TC, et al. Axons of retinal ganglion cells are insulted in the optic nerve early in DBA/2J glaucoma. *J Cell Biol*. 2007;179(7):1523–1537. http://dx.doi.org/10.1083/jcb.200706181.
28. Inman DM, Sappington RM, Horner PJ, Calkins DJ. Quantitative correlation of optic nerve pathology with ocular pressure and corneal thickness in the DBA/2 mouse model of glaucoma. *Invest Ophthalmol Vis Sci*. 2006;47(3):986–996. http://dx.doi.org/10.1167/iovs.05-0925.
29. Ju WK, Kim KY, Angert M, et al. Memantine blocks mitochondrial OPA1 and cytochrome c release and subsequent apoptotic cell death in glaucomatous retina. *Invest Ophthalmol Vis Sci*. 2009;50(2):707–716. http://dx.doi.org/10.1167/iovs.08-2499.
30. Dietz JA, Maes ME, Huang S, et al. Spink2 modulates apoptotic susceptibility and is a candidate gene in the Rgcs1 QTL that affects retinal ganglion cell death after optic nerve damage. *PLoS One*. 2014;9(4):e93564. http://dx.doi.org/10.1371/journal.pone.0093564.
31. Fernandes KA, Harder JM, Kim J, Libby RT. JUN regulates early transcriptional responses to axonal injury in retinal ganglion cells. *Exp Eye Res*. 2013;112:106–117. http://dx.doi.org/10.1016/j.exer.2013.04.021.
32. Li Y, Semaan SJ, Schlamp CL, Nickells RW. Dominant inheritance of retinal ganglion cell resistance to optic nerve crush in mice. *BMC Neurosci*. 2007;8:19. http://dx.doi.org/10.1186/1471-2202-8-19.
33. Templeton JP, Freeman NE, Nickerson JM, et al. Innate immune network in the retina activated by optic nerve crush. *Invest Ophthalmol Vis Sci*. 2013;54(4):2599–2606. http://dx.doi.org/10.1167/iovs.12-11175.
34. Templeton JP, Geisert EE. A practical approach to optic nerve crush in the mouse. *Mol Vis*. 2012;18:2147–2152.
35. Templeton JP, Nassr M, Vazquez-Chona F, et al. Differential response of C57BL/6J mouse and DBA/2J mouse to optic nerve crush. *BMC Neurosci*. 2009;10:90. http://dx.doi.org/10.1186/1471-2202-10-90.

36. Samsel PA, Kisiswa L, Erichsen JT, Cross SD, Morgan JE. A novel method for the induction of experimental glaucoma using magnetic microspheres. *Invest Ophthalmol Vis Sci.* 2011;52(3):1671–1675. http://dx.doi.org/10.1167/iovs.09-3921.
37. Sappington RM, Carlson BJ, Crish SD, Calkins DJ. The microbead occlusion model: a paradigm for induced ocular hypertension in rats and mice. *Invest Ophthalmol Vis Sci.* 2010;51(1):207–216. http://dx.doi.org/10.1167/iovs.09-3947.
38. Mao M, Hedberg-Buenz A, Koehn D, John SW, Anderson MG. Anterior segment dysgenesis and early-onset glaucoma in nee mice with mutation of Sh3pxd2b. *Invest Ophthalmol Vis Sci.* 2011;52(5):2679–2688. http://dx.doi.org/10.1167/iovs.10-5993.
39. Senatorov V, Malyukova I, Fariss R, et al. Expression of mutated mouse myocilin induces open-angle glaucoma in transgenic mice. *J Neurosci.* 2006;26(46): 11903–11914. http://dx.doi.org/10.1523/JNEUROSCI.3020-06.2006.
40. Aihara M, Lindsey JD, Weinreb RN. Ocular hypertension in mice with a targeted type I collagen mutation. *Invest Ophthalmol Vis Sci.* 2003;44(4):1581–1585.
41. Fujikawa K, Iwata T, Inoue K, et al. VAV2 and VAV3 as candidate disease genes for spontaneous glaucoma in mice and humans. *PLoS One.* 2010;5(2):e9050. http://dx.doi.org/10.1371/journal.pone.0009050.
42. Morrison JC, Moore CG, Deppmeier LM, Gold BG, Meshul CK, Johnson EC. A rat model of chronic pressure-induced optic nerve damage. *Exp Eye Res.* 1997;64(1):85–96. http://dx.doi.org/10.1006/exer.1996.0184.
43. Ruiz-Ederra J, Verkman AS. Mouse model of sustained elevation in intraocular pressure produced by episcleral vein occlusion. *Exp Eye Res.* 2006;82(5):879–884. http://dx.doi.org/10.1016/j.exer.2005.10.019.
44. Gross RL, Ji J, Chang P, et al. A mouse model of elevated intraocular pressure: retina and optic nerve findings. *Trans Am Ophthalmol Soc.* 2003;101:163–169, discussion 169–171.
45. Sawaguchi K, Nakamura Y, Nakamura Y, Sakai H, Sawaguchi S. Myocilin gene expression in the trabecular meshwork of rats in a steroid-induced ocular hypertension model. *Ophthalmic Res.* 2005;37(5):235–242. http://dx.doi.org/10.1159/000086946.
46. Allcutt D, Berry M, Sievers J. A quantitative comparison of the reactions of retinal ganglion cells to optic nerve crush in neonatal and adult mice. *Brain Res.* 1984;318(2):219–230.
47. Li Y, Schlamp CL, Nickells RW. Experimental induction of retinal ganglion cell death in adult mice. *Invest Ophthalmol Vis Sci.* 1999;40(5):1004–1008.
48. Liu Y, McDowell CM, Zhang Z, Tebow HE, Wordinger RJ, Clark AF. Monitoring retinal morphologic and functional changes in mice following optic nerve crush. *Invest Ophthalmol Vis Sci.* 2014;55(6):3766–3774. http://dx.doi.org/10.1167/iovs.14-13895.
49. Fingert JH, Stone EM, Sheffield VC, Alward WL. Myocilin glaucoma. *Surv Ophthalmol.* 2002;47(6):547–561.
50. Zhou Y, Grinchuk O, Tomarev SI. Transgenic mice expressing the Tyr437His mutant of human myocilin protein develop glaucoma. *Invest Ophthalmol Vis Sci.* 2008;49(5):1932–1939. http://dx.doi.org/10.1167/iovs.07-1339.
51. Shi D, Takano Y, Nakazawa T, et al. Molecular genetic analysis of primary open-angle glaucoma, normal tension glaucoma, and developmental glaucoma for the VAV2 and VAV3 gene variants in Japanese subjects. *Biochem Biophys Res Commun.* 2013;432(3): 509–512. http://dx.doi.org/10.1016/j.bbrc.2013.02.003.
52. Iqbal Z, Cejudo-Martin P, de Brouwer A, et al. Disruption of the podosome adaptor protein TKS4 (SH3PXD2B) causes the skeletal dysplasia, eye, and cardiac abnormalities of Frank-Ter Haar Syndrome. *Am J Hum Genet.* 2010;86(2):254–261. http://dx.doi.org/10.1016/j.ajhg.2010.01.009.
53. Calkins DJ. Age-related changes in the visual pathways: blame it on the axon. *Invest Ophthalmol Vis Sci.* 2013;54(14):ORSF37–ORSF41. http://dx.doi.org/10.1167/iovs.13-12784.

54. Whitmore AV, Libby RT, John SW. Glaucoma: thinking in new ways-a role for autonomous axonal self-destruction and other compartmentalised processes? *Prog Retin Eye Res*. 2005;24(6):639–662. http://dx.doi.org/10.1016/j.preteyeres.2005.04.004.
55. Fernandes KA, Harder JM, John SW, Shrager P, Libby RT. DLK-dependent signaling is important for somal but not axonal degeneration of retinal ganglion cells following axonal injury. *Neurobiol Dis*. 2014;69:108–116. http://dx.doi.org/10.1016/j.nbd.2014.05.015.
56. Libby RT, Li Y, Savinova OV, et al. Susceptibility to neurodegeneration in a glaucoma is modified by Bax gene dosage. *PLoS Genet*. 2005;1(1):17–26. http://dx.doi.org/10.1371/journal.pgen.0010004.
57. Waller A. Experiments on the section of the glossopharyngeal and hypoglossal nerves of the frog, and observations of the alterations produced thereby in the structure of their primitive fibres. *Philos Trans R Soc Lond*. 1850;140:423–429. http://dx.doi.org/10.1098/rstl.1850.0021.
58. Cavanagh JB. The significance of the "dying back" process in experimental and human neurological disease. *Int Rev Exp Pathol*. 1964;3:219–267.
59. Libby RT, Anderson MG, Pang IH, et al. Inherited glaucoma in DBA/2J mice: pertinent disease features for studying the neurodegeneration. *Vis Neurosci*. 2005;22(5):637–648. http://dx.doi.org/10.1017/S0952523805225130.
60. Nickells RW, Zack DJ. Apoptosis in ocular disease: a molecular overview. *Ophthalmic Genet*. 1996;17(4):145–165. http://dx.doi.org/10.3109/13816819609057889.
61. Quigley HA, Nickells RW, Kerrigan LA, Pease ME, Thibault DJ, Zack DJ. Retinal ganglion cell death in experimental glaucoma and after axotomy occurs by apoptosis. *Invest Ophthalmol Vis Sci*. 1995;36(5):774–786.
62. Lunn ER, Perry VH, Brown MC, Rosen H, Gordon S. Absence of Wallerian degeneration does not hinder regeneration in peripheral nerve. *Eur J Neurosci*. 1989;1(1):27–33.
63. Coleman MP, Conforti L, Buckmaster EA, et al. An 85-kb tandem triplication in the slow Wallerian degeneration (Wlds) mouse. *Proc Natl Acad Sci USA*. 1998;95(17):9985–9990.
64. Freeman MR. Signaling mechanisms regulating Wallerian degeneration. *Curr Opin Neurobiol*. 2014;27:224–231. http://dx.doi.org/10.1016/j.conb.2014.05.001.
65. Ikegami K, Koike T. Non-apoptotic neurite degeneration in apoptotic neuronal death: pivotal role of mitochondrial function in neurites. *Neuroscience*. 2003;122(3): 617–626.
66. O'Donnell KC, Vargas ME, Sagasti A. WldS and PGC-1alpha regulate mitochondrial transport and oxidation state after axonal injury. *J Neurosci*. 2013;33(37):14778–14790. http://dx.doi.org/10.1523/JNEUROSCI.1331-13.2013.
67. Fang C, Decker H, Banker G. Axonal transport plays a crucial role in mediating the axon-protective effects of NmNAT. *Neurobiol Dis*. 2014;68:78–90. http://dx.doi.org/10.1016/j.nbd.2014.04.013.
68. Barrientos SA, Martinez NW, Yoo S, et al. Axonal degeneration is mediated by the mitochondrial permeability transition pore. *J Neurosci*. 2011;31(3):966–978. http://dx.doi.org/10.1523/JNEUROSCI.4065-10.2011.
69. Sunio A, Bittner GD. Cyclosporin A retards the Wallerian degeneration of peripheral mammalian axons. *Exp Neurol*. 1997;146(1):46–56. http://dx.doi.org/10.1006/exnr.1997.6484.
70. Templeton JP, Wang X, Freeman NE, et al. A crystallin gene network in the mouse retina. *Exp Eye Res*. 2013;116:129–140. http://dx.doi.org/10.1016/j.exer.2013.08.001.
71. Vazquez-Chona FR, Geisert EE. Networks modulating the retinal response to injury: insights from microarrays, expression genetics, and bioinformatics. *Adv Exp Med Biol*. 2012;723:649–656. http://dx.doi.org/10.1007/978-1-4614-0631-0_82.
72. Ahmed F, Brown KM, Stephan DA, et al. Microarray analysis of changes in mRNA levels in the rat retina after experimental elevation of intraocular pressure. *Invest Ophthalmol Vis Sci*. 2004;45(4):1247–1258.

73. Fan W, Li X, Wang W, Mo JS, Kaplan H, Cooper NG. Early involvement of immune/inflammatory response genes in retinal degeneration in DBA/2J mice. *Ophthalmol Eye Dis*. 2010;1:23–41.
74. Howell GR, Macalinao DG, Sousa GL, et al. Molecular clustering identifies complement and endothelin induction as early events in a mouse model of glaucoma. *J Clin Invest*. 2011;121(4):1429–1444. http://dx.doi.org/10.1172/JCI44646.
75. Steele MR, Inman DM, Calkins DJ, Horner PJ, Vetter ML. Microarray analysis of retinal gene expression in the DBA/2J model of glaucoma. *Invest Ophthalmol Vis Sci*. 2006;47(3):977–985. http://dx.doi.org/10.1167/iovs.05-0865.
76. Geisert EE, Lu L, Freeman-Anderson NE, et al. Gene expression in the mouse eye: an online resource for genetics using 103 strains of mice. *Mol Vis*. 2009;15:1730–1763.
77. Ding JD, Johnson LV, Herrmann R, et al. Anti-amyloid therapy protects against retinal pigmented epithelium damage and vision loss in a model of age-related macular degeneration. *Proc Natl Acad Sci USA*. 2011;108(28):E279–E287. http://dx.doi.org/10.1073/pnas.1100901108.
78. Loffler KU, Edward DP, Tso MO. Immunoreactivity against tau, amyloid precursor protein, and beta-amyloid in the human retina. *Invest Ophthalmol Vis Sci*. 1995;36(1):24–31.
79. Bradt BM, Kolb WP, Cooper NR. Complement-dependent proinflammatory properties of the Alzheimer's disease beta-peptide. *J Exp Med*. 1998;188(3):431–438.
80. Hadziahmetovic M, Kumar U, Song Y, et al. Microarray analysis of murine retinal light damage reveals changes in iron regulatory, complement, and antioxidant genes in the neurosensory retina and isolated RPE. *Invest Ophthalmol Vis Sci*. 2012;53(9):5231–5241. http://dx.doi.org/10.1167/iovs.12-10204.
81. Anderson DH, Radeke MJ, Gallo NB, et al. The pivotal role of the complement system in aging and age-related macular degeneration: hypothesis re-visited. *Prog Retin Eye Res*. 2010;29(2):95–112. http://dx.doi.org/10.1016/j.preteyeres.2009.11.003.
82. Sparrow JR, Ueda K, Zhou J. Complement dysregulation in AMD: RPE-Bruch's membrane-choroid. *Mol Aspects Med*. 2012;33(4):436–445. http://dx.doi.org/10.1016/j.mam.2012.03.007.
83. Hageman GS, Anderson DH, Johnson LV, et al. A common haplotype in the complement regulatory gene factor H (HF1/CFH) predisposes individuals to age-related macular degeneration. *Proc Natl Acad Sci USA*. 2005;102(20):7227–7232. http://dx.doi.org/10.1073/pnas.0501536102.
84. Klein RJ, Zeiss C, Chew EY, et al. Complement factor H polymorphism in age-related macular degeneration. *Science*. 2005;308(5720):385–389. http://dx.doi.org/10.1126/science.1109557.
85. Crabb JW, Miyagi M, Gu X, et al. Drusen proteome analysis: an approach to the etiology of age-related macular degeneration. *Proc Natl Acad Sci USA*. 2002;99(23):14682–14687. http://dx.doi.org/10.1073/pnas.222551899.
86. Johnson PT, Betts KE, Radeke MJ, Hageman GS, Anderson DH, Johnson LV. Individuals homozygous for the age-related macular degeneration risk-conferring variant of complement factor H have elevated levels of CRP in the choroid. *Proc Natl Acad Sci USA*. 2006;103(46):17456–17461. http://dx.doi.org/10.1073/pnas.0606234103.
87. Zhou J, Jang YP, Kim SR, Sparrow JR. Complement activation by photooxidation products of A2E, a lipofuscin constituent of the retinal pigment epithelium. *Proc Natl Acad Sci USA*. 2006;103(44):16182–16187. http://dx.doi.org/10.1073/pnas.0604255103.
88. Howell GR, Soto I, Ryan M, Graham LC, Smith RS, John SW. Deficiency of complement component 5 ameliorates glaucoma in DBA/2J mice. *J Neuroinflammation*. 2013;10(1):76. http://dx.doi.org/10.1186/1742-2094-10-76.

SECTION VI

Retina

CHAPTER TWENTY-TWO

Introduction to the Retina

Hans E. Grossniklaus[1], Eldon E. Geisert, John M. Nickerson
Department of Ophthalmology, Emory University School of Medicine, Atlanta, Georgia, USA
[1]Corresponding author: e-mail address: ophtheg@emory.edu

Contents

Abstract

Here, we briefly introduce and survey the retina including terminology and naming conventions. We consider anatomical structures of the retina including gross and microscopic anatomy of the fundus, layers of the retina, retinal cell types, and neuronal wiring of the retina. We discuss briefly biochemistry of the visual transduction cascade and the Vitamin A cycle. We introduce physiological processes including outer segment disk shedding and origins of electrical signals detected by the electroretinogram. This sets the stage for a more in-depth look at specific aspects of retinal structure and function in subsequent sections in this chapter.

1. INTRODUCTION

Shakespeare said, "The eyes are the window to your soul." Little did he realize the truth of his epigram. When Romeo gazed into Juliet's eyes, he may have seen the love that was in her soul; however, an eye doctor sees much more. Gazing through the pupil to the back of the eye reveals a wondrous tissue, the retina, which takes patterned light and turns it into neural signals. This chapter will explore the retina in all its glory. This introduction

Progress in Molecular Biology and Translational Science, Volume 134
ISSN 1877-1173
http://dx.doi.org/10.1016/bs.pmbts.2015.06.001
© 2015 Elsevier Inc.
All rights reserved.

will provide a 30,000-foot brief overview of this complex tissue. The introduction will include a survey of the fundus, the layers of the retina, the retinal cell types, the visual transduction cascade, the Vitamin A cycle, the neuronal wiring of the retina, and the Nobel-level achievements that relied on the unique structure of retina. This will set the stage for a more in-depth look at specific aspects of retinal structure and function in subsequent sections of this chapter.

2. THE FUNDUS

During the normal eye examination, the vision health expert uses simple optical instruments such as the direct ophthalmoscope to literally look at living neural tissue the retina. At the back of the eye, the image of the fundus is a view of the central nervous system. Developmentally, the retina is an out-pouching of the developing brain that in combination with other tissues forms the eye. The retina is a diaphanous tissue, like a sheer curtain. When one looks at the fundus, the retinal pigment epithelium, which underlies the neurosensory retina, is seen through the neurosensory retina as an orange (dark gray in the print version) background. The fundus image includes the neurosensory retina, its vascular supply, and the underlying retinal pigment epithelium.

Figure 1 is a color photo taken of a fundus illuminated with white light. This image is taken with a fundus camera, reveals the health of the retina, the

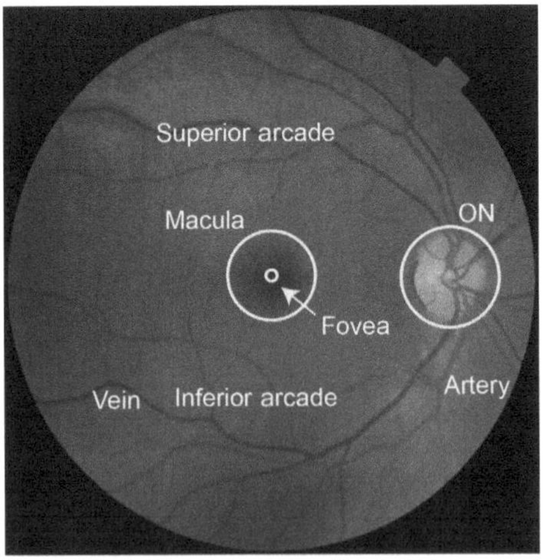

Figure 1 The fundus. Major structure of the normal fundus of the right eye is indicated. Wikimedia Commons. CC0 1.0. The overlays were created in Adobe Illustrator. *The underlying image of the fundus is the work of Mikael Häggström*[1].

eye, and to some extent the body. Noninvasive visualization through the optics of the eye itself provides a clear image and enables an assessment of the health of many components of the eye including the cornea, lens, and vitreous body. The orange (dark gray in the print version) color is due to the underlying retinal pigment epithelium. The fundus photo in Fig. 1 is of the right eye, because the ON lies to the left of the macula.

The center of the fundus lies on the optical axis; this is the fovea, which collects the best-resolved images, and it is usually associated with small yellow dot, the macula lutea. The anatomic and clinical foveola, fovea, and macula are indicated on the diagram. The major vascular supply of the retina forms from the superior and inferior arcade of blood vessels. The retinal area between the superior and inferior arcade is known as the area centralis or posterior pole. The center of this posterior pole contains the macula, which is redder (dark gray in the print version) and denser in color than the surrounding retina. This is due to more photoreceptors packed at high densities and more pigments behind the photoreceptor cells. The macula lutea refers to yellow xanthophyll pigment within the retina in the center of the macula. The very center of the macula is referred to as the fovea, which is a 500-μm in diameter avascular area that essentially is composed of inner limiting membrane and specialized cone photoreceptor cells, known as the bundle of Rochenne Duverney. The major vessels visible in this color fundus photo lie in the superficial retina. Many fine vessels penetrate deeper into the inner layers of the retina (see Fig. 2) and provide the vascular supply to inner retina

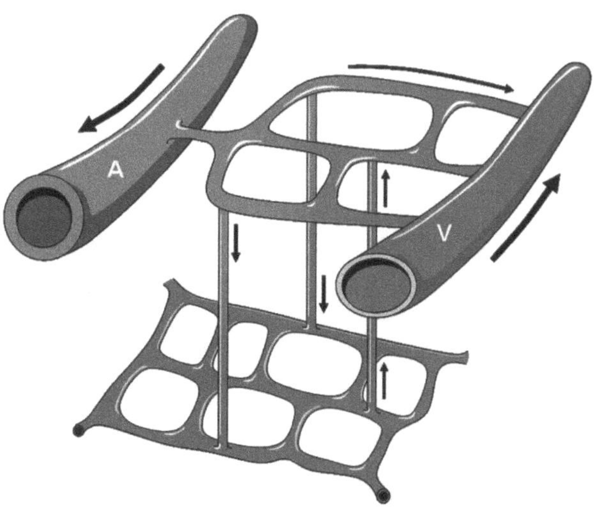

Figure 2 Blood vessels in the inner layers of the retina of the mouse. A diagram of blood flow. *With permission of the authors[2] who hold the copyrights, CCL3.0 BY-AT.*

out to the outer 1/3 of the outer nuclear layer. Special properties of the retina can be observed with numerous imaging modalities that may represent natural fluorescence of eye constituents, or fluorescence from dyes injected into the patient or reflectance of specific wavelengths of light. However, fundus exam imaging with white light is the best method to clinically examine the retina.

3. THE STRUCTURE OF THE RETINA IN CROSS-SECTION

Recent advances in optics using adaptive optics, interference, and tomography to create high-resolution images through the multiple layers of the retina (~250 μm thick). These technical breakthroughs advance our abilities to image the retina with high resolution. This noninvasive imaging can be repeated with excellent alignment using key anatomical landmarks, making it possible to follow specific features of any retina over a time. The time series of images is particularly useful in monitoring disease progression and the effects of surgical intervention or treatment efficacy. The layers of the retina cut in cross-section are shown in Fig. 3. Cells associated with each layer are indicated and the relationship of the images created

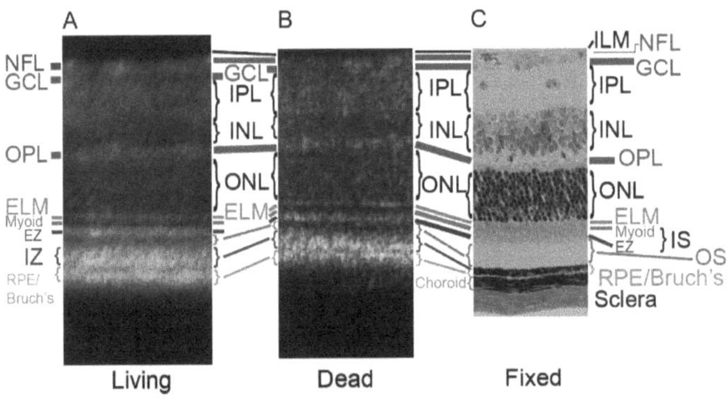

Figure 3 Structure of the retina in cross-section. (A) Spectral domain imaging of a living mouse taken with a Micron 4 (Phoenix Research Labs, Pleasanton, CA). (B) Spectral domain imaging of the same mouse moments after death taken with the same Micron 4. (C) Histological preparation of the mouse eye, hematoxylin, and eosin staining. Abbreviations: ILM, inner limiting membrane; NFL, nerve fiber layer; GCL, ganglion cell layer; IPL, inner plexiform layer; INL, inner nuclear layer; OPL, outer plexiform layer; ONL, outer nuclear layer; ELM, external limiting membrane; Myoid, myoid component of the inner segment; EZ, ellipsoid zone of the inner segment; IZ; interdigitization zone of the outer segments and apical processes of the RPE; IS, inner segment; OS, outer segment; RPE/Bruch's, retinal pigmented epithelium and Bruch's membrane; Choroid; combined layers of the choroid; Sclera; sclera through Tenon's capsule.

by interference methods (spectral domain optical coherence tomography; SD-OCT) is compared to the retinal histology. Historic studies by Ramon y Cajal discovered the connections among specific cell types of the retina that led to the neuron doctrine discussed below in Section 9. One of his diagrams from the 1890s is included in Fig. 4. He was able to resolve and identify RPE, photoreceptor cells, horizontal cells, bipolar cells, amacrine cells, ganglion cells, and Müller cells in the retina. He classified many subtypes of each cell and identified the major circuitry of one cell to the next in the retina, illustrated in Fig. 4.

4. THE BLOOD SUPPLY OF THE RETINA

Blood vessels are needed to supply large amounts of oxygen to the inner and outer parts of the retina. The choroid (Fig. 5) supplies nourishment and oxygen to the RPE outer retina up to the outer two-thirds of the outer nuclear layer, while the retinal vessels, readily visible by fundus exam, supply the inner layers to the inner one-third of the outer nuclear layer (a trypsin digestion preparation of the retina illustrating the retinal vessels and capillary bed) (Fig. 6). The inner portion of the choroid, the choriocapillaris, which consists of a network of capillaries intimately associated with the outer aspect of Bruch's membrane, supplies oxygen and nutrition to the RPE and outer retinal layers. The RPE overlies the inner aspect of Bruch's membrane. Photoreceptor outer segments dangle posteriorly and are closely associated with microvilli on the inner aspect of the RPE.

5. *EN FACE* IMAGING AND PATTERNS IN THE RETINA

En face, the cells of the retina exhibit to a degree a regularity of spacing across the surface layers in the retina. Cell soma and nuclei of one subtype of cells are spaced about the same distance apart from one neighbor to each of its immediately adjacent neighbors. For example in Fig. 7D, the panel illustrates a regular spacing of cell bodies of homotypic cell types. The dendrites of one cell may block the intrusion of dendrites from another cell. Other cell types exhibit regular spacing of cell bodies yet have large overlaps of dendrites over long ranges. Finally, a uniform cell cross-sectional area and high-density packing can generate cell mosaics by the physics of packing (this occurs in the foveola with cone outer segments). Reese[3] said, "...retinal mosaics are commonly associated with the phenomenon of dendritic tiling, by which each cell in a mosaic extends its dendritic

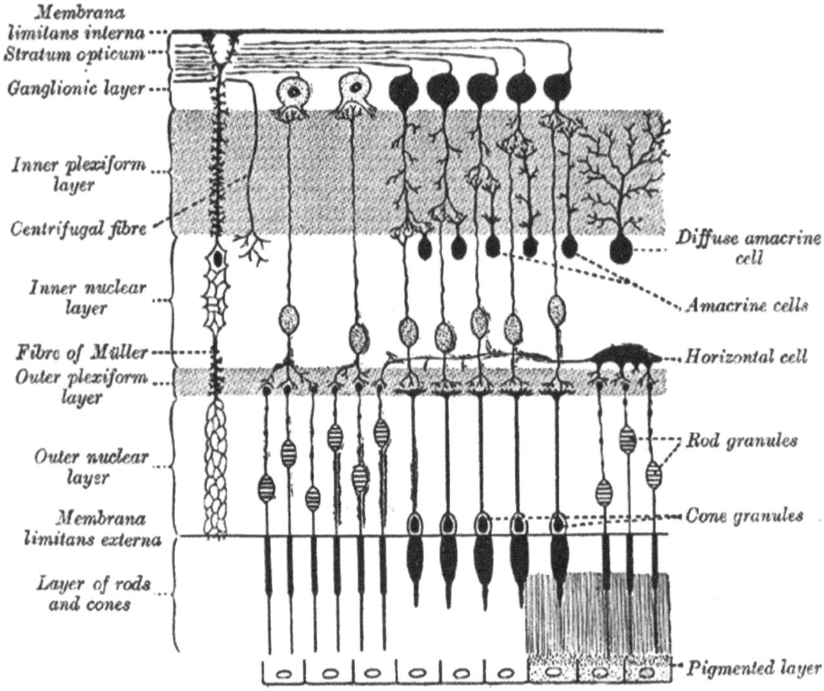
Membrana limitans interna
Stratum opticum
Ganglionic layer
Inner plexiform layer
Centrifugal fibre
Inner nuclear layer
Fibre of Müller
Outer plexiform layer
Outer nuclear layer
Membrana limitans externa
Layer of rods and cones
Diffuse amacrine cell
Amacrine cells
Horizontal cell
Rod granules
Cone granules
Pigmented layer

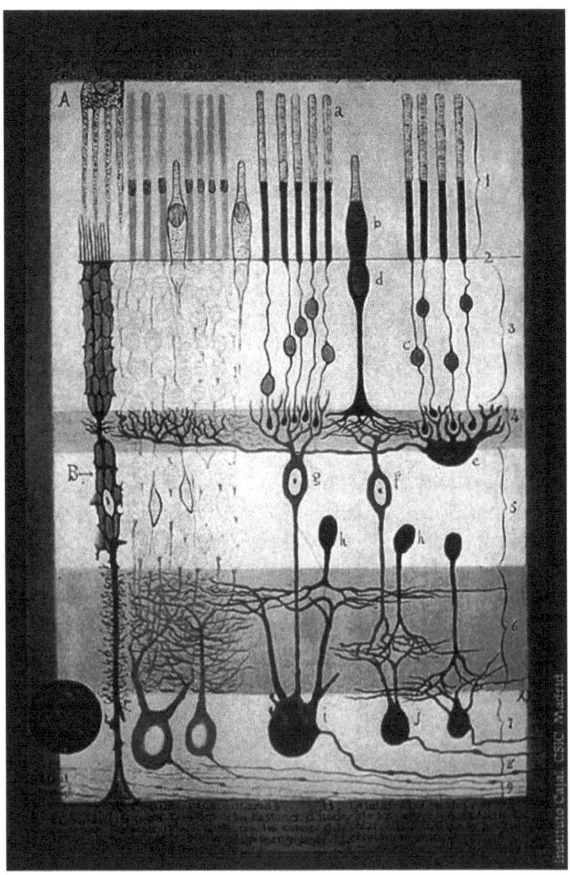
Instituto Cajal. CSIC. Madrid

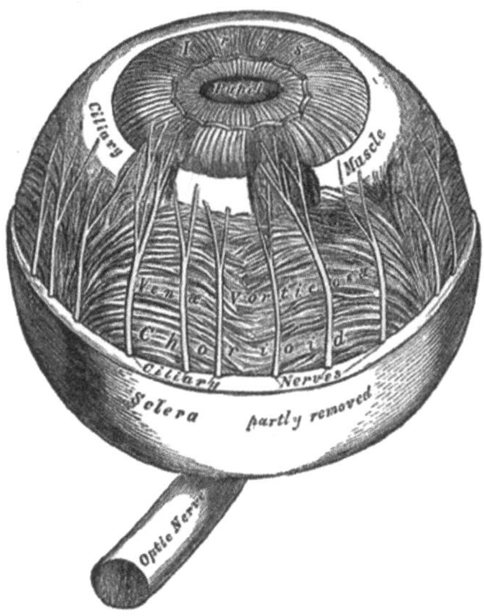

Figure 5 The choroid. *Image from Gray's Anatomy (1916) in the public domain.*

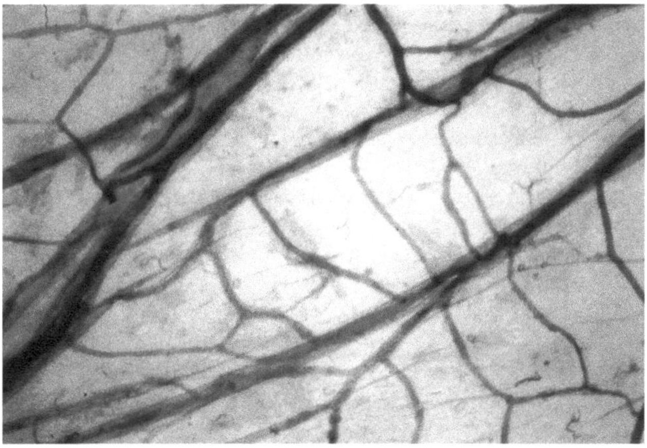

Figure 6 Trypsin digestion preparation of the retina illustrating the retinal vessels and capillary bed of the inner retinal vessels.

Figure 4 (A) Organization of connections among the cells of the retina. (B) Diagram from about 1890 of Ramon y Cajal images of the organization of retinal cells of the same class and type in the public domain. *Panel (A): Based on Ramon y Cajal's work, image from Gray's Anatomy (1916) in the public domain.*

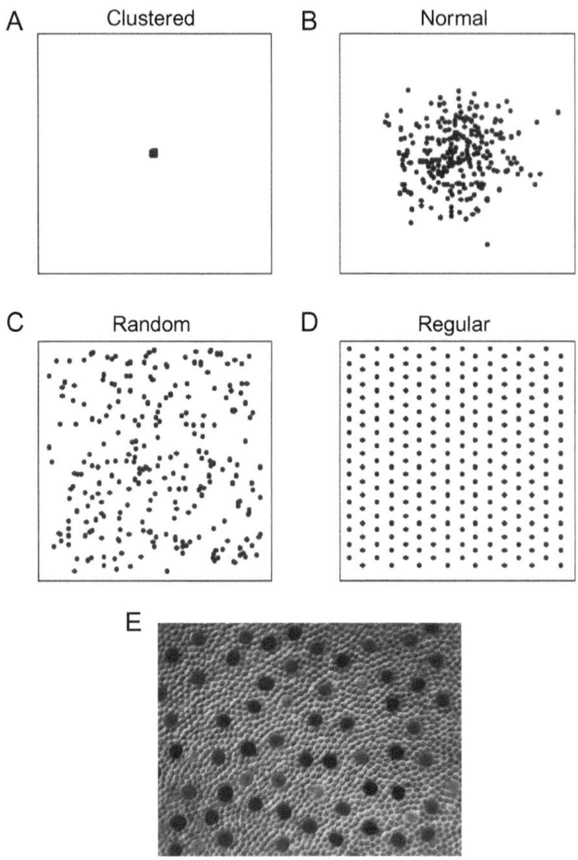

Figure 7 Spatial point patterns. The retina typically exhibits a regular pattern of nuclear spacing of cells of the same type often referred to as homotypic interactions, exclusion zones, when the retina is viewed *en face*. Shown here are four types of classic spatial point patterns. (A) Clustered spacing of 256 points. (B) Normally distributed pattern of 256 points. (C) Random distribution of 256 points. (D) Regular patterning of 256 points such as that observed in many inner retina cells. (E) Cone mosaics, demonstrating equal distance among cones of the same color. *Panels (A–D): Courtesy of "Ldecola," Wikimedia name. CCL3.0 AT-SA. Panel (E): Courtesy of Dr. Nicolas Cuenca, Universidad de Alicante, Alicante, Spain.*

processes to "tessellate" the retinal surface by limiting further growth at the boundaries with homotypic dendritic fields."

6. THE VISUAL TRANSDUCTION CASCADE

The retina converts photons of light to signals describing visual information and sends them to the brain. This process takes place in the outer segments of the photoreceptors, including in the cell membranes and

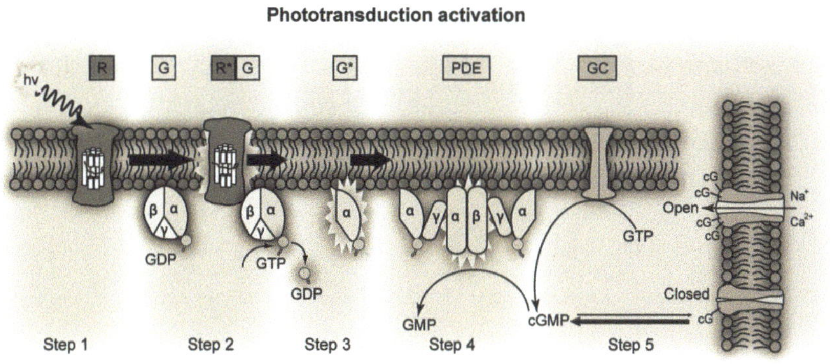

Figure 8 The visual transduction cascade. *Based on the image of Jason J. Corneveaux, CCL 3.0 AT-SA and redrawn after Leskov* et al.[4].

pigment disk membranes of the outer segments. The steps in the absorption of a photon of light and conversion to useful visual information are called the visual transduction cascade (Fig. 8). In the first step, a photon strikes 11-*cis*-retinal linked to an opsin by a Schiff's base linkage, converting this retinoid to the all-*trans* form. This results in subtle shape changes to the opsin that activates the transduction cascade starting first with Transducin. The opsins are part of the large family of G-protein-mediated receptors, and Transducin is a G-protein. Transducin in turn activates a cGMP-phosphodiesterase (PDE6). This enzyme selectively and specifically converts cGMP to GMP. The cGMP is an allosteric effector of the sodium channel. The drop in levels of cGMP, normally at about 10 μM (in the dark), to about 1 μM (in bright light) changes a Na channel from open to closed. This results in a reduction in the flow of sodium ions and less current flow (known as the Dark Current). In a dose–response manner, the reduced current flow reduces the release of the neurotransmitter glutamate in the photoreceptor synapse. In the absence of light, the visual transduction cascade can be reset to the dark state. An accumulation of cGMP results from the activity of guanylate cyclase (mediated by GCAPs) that converts GTP to cGMP. This resets the Na channels to their open state restoring the Dark Current. Rhodopsin is restored with 11-*cis*-retinal. Transducin is restored by the combined actions of RGS9 and recoverin reuniting all the subunits. The PDE6 is reset by combining with a gamma subunit.

7. OUTER SEGMENT DISK GENESIS AND SHEDDING

Since phototransduction takes place in the cell membranes and disk membranes of the outer segments of the rods and cones, and the opsins

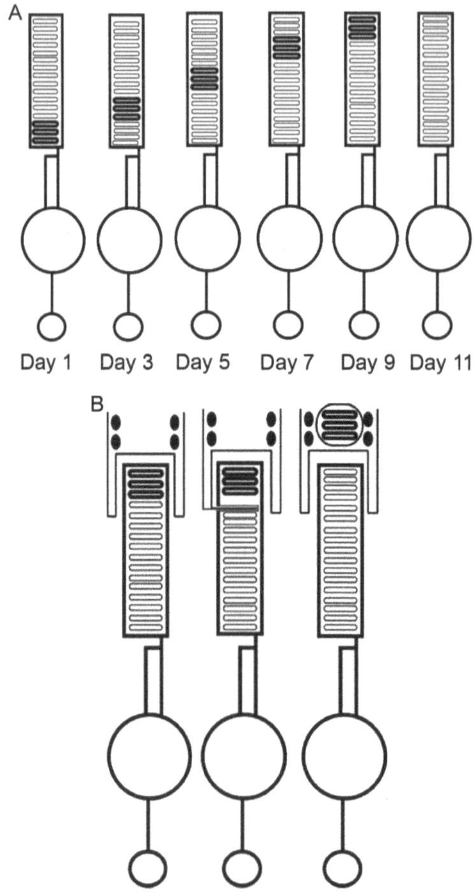

Figure 9 Outer segment disk genesis and shedding. (A) The movement of outer segment disks in the outer segment of rods. Daily, a group of about 50–100 new outer segment disks are synthesized at the base of the rod outer segment, and they migrate upward daily, pushed upward by the next new group of disks. (B) Correspondingly, the oldest 50–100 disks (now 10 days old) at the distal tip of the outer segment are phagocytosed by the RPE, resulting in a uniform length of 500–1000 disks per outer segment. Quoting Young and Bok from 1969[5], "Repeated formation of new disks at the base of the outer segment progressively displaces older disks away from the site of assembly, and toward the apical end of the cell."

continually need to be replenished, including with Vitamin A provided by the systemic circulation via the choriocapillaris, there is a continual production and shedding of these outer membrane structures. This remarkable process rejuvenates the outer segment structure providing for fresh photosensitive pigments on a continuous basis within the cells (Fig. 9).

The tips of outer segments are shed daily and phagocytosed by the RPE cells. The new outer segment disks are generated next to the ciliary process of the photoreceptors. Briefly, new outer segment membrane disks are assembled at the proximal tip of an outer segment and pushed like a penny roll toward the distal tip over the course of approximately 10 days. Roughly, 10% of the outer segment is removed each, with outer segments of the rods shedding at dawn and outer of the cones segments shedding at dusk. The RPE performs many other notable functions. The RPE is the major blood–retina barrier. It absorbs stray light. It specifically and selectively transports metabolites. It makes and regulates numerous growth and signaling factors. It stores and metabolizes retinoids and produces the critical visual chromophore 11-*cis*-retinal. Of note, the Vitamin A cycle is split among the photoreceptor cells, the RPE, and intriguingly for rapid cone pigment regeneration the Müller cells.

8. RETINAL CIRCUITRY

The photoreceptors transduce light into a neuronal signal and this visual information will eventually pass through the retinal ganglion cell to the brain. Before the signal from the photoreceptor is sent to the ganglion cell, there are a number of neuronal processes that transform the simple point absorption of the photoreceptor signal into the center-surround receptive field of the ganglion cell. These two cells do not directly connect. Intermediaries in the inner layers of the retinas refine the signals that are integrated and processed in many ways to detect changes in lighting patterns with position and time, so that information about contrast and intensity are presented to the ganglion cells. The bipolar cells connect photoreceptors to the ganglion cells providing for a vertical integration through the retina that is responsible for the center response of the ganglion cells. The horizontal and amacrine cells process information orthogonally to the bipolar cells and provide input that contributes to the antagonistic surround. Each ganglion cell with its center-surround organization is made up of bits of information that is derived from the integration of many photoreceptor cells. A basic wiring diagram for the photopic to scotopic shift retinal circuit is illustrated in Fig. 10.

The retinal ganglion cells further process the signals and then send them out of the eye in nerve fibers through the optic nerve head. The ganglion cell bodies and axonal processes (the nerve fiber layer) are illustrated *en face* in Fig. 11.

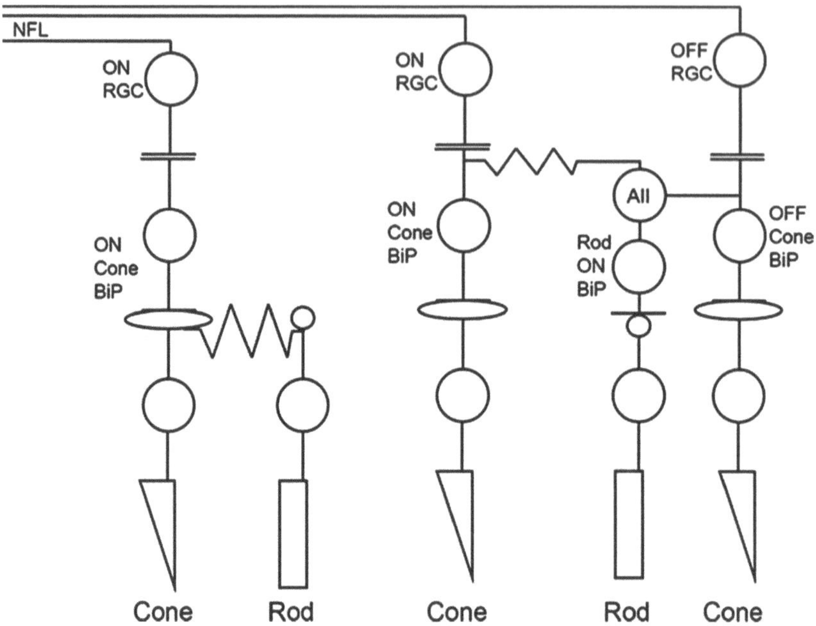

Figure 10 The scotopic to photopic circuits of the retina.

Figure 11 The ganglion cell bodies and axonal processes (the nerve fiber layer) is illustrated *en face*. Axonal processes initiate at the RGC cell body and migrate toward the optic nerve head in a centripetal manner.

9. NOBEL PRIZES BASED ON UNDERSTANDING THE RETINA

In closing this introduction, we remind the reader that the retina has served as a springboard for remarkable conceptual advances in our understanding of biology and medicine. These are reflected in Nobel Prizes to Ramon y Cajal and Golgi in 1906, and Hartline, Granit, and Wald in 1967. We briefly summarize their accomplishments below.

The Neuron Doctrine developed by Ramón y Cajal is that the nervous system is built from discrete neurons, each of which is a single cell with a nucleus and neuronal processes. Two neurons are connected by a synapse where signals transit an extracellular space between cells. As mentioned earlier, Ramón y Cajal used Golgi's silver staining methods to discover many different types and classes of cells in the retina and the basic wiring of the retina (Fig. 11 and most sections of this chapter). His hand drawings of microscopic images continue to be extremely accurate, precise, and demonstrate a level of artistic beauty. He knew that the eye transmitted signals to the brain and that the retinal ganglion cells were only cells that sent a process (the axon) through the optic nerve to the brain. This gave him the ability to not only form the neuron doctrine but to ascribe function to different components of the neuron. The dendrites were for receiving signals and the axon for transmitting the signal.

Hartline discovered that photoreceptor cells are coupled so that if one is excited, others nearby are inhibited. This results in increased contrast and perception of shapes. This provided a framework for lateral inhibition seen in many sensory systems to increase the signal-to-noise ratio and to better localize sensory input.

Wald built on the work on Kuehn and Boll to discover that photoreceptors require the 11-*cis* form of retinal to renew functional visual pigments. He discovered that the Vitamin A cycle is the visual cycle and proposed that an enzyme, retinol isomerase, resides in the RPE to regenerate 11-*cis*-retinal (see chapter, "A history of the classical visual cycle" by Wright).

Granit discovered light-dependent excitatory and inhibitory signals in the optic nerve. Also, he developed the electroretinogram (ERG) and used it to understand visual pigments and sensitivity to levels and hues of light. The ERG remains an essential tool in vision sciences and clinical practice.

In conclusion, Plato said, "Beauty lies in the eyes of the beholder." Through the lessons presented in the following sections of this chapter, it is our hope that the reader will see beauty also lies in the wondrous structure of the eye and retina.

ACKNOWLEDGMENTS

This work was supported by the National Eye Institute (R01EY016470 (J.N.), R01EY021592 (J.N.), R01EY017841 (E.G.)), P30EY006360, an unrestricted grant to the

Department of Ophthalmology at Emory University from Research to Prevent Blindness, Inc., DOD grant W81XWH-12-1-0255 (E.G.), and the National Cancer Institute R01CA176001 (H.E.G.).

REFERENCES

1. Häggström M. Medical gallery of Mikael Häggström 2014. *Wiver J Med.* 2014;1(2). http://dx.doi.org/10.15347/wjm/2014.008.
2. Ramos D, Navarro M, Mendes-Jorge L, et al. Confocal laser microscopy – Principles and applications in medicine, biology, and the food sciences. In: Lagali N, ed. Biochemistry, Genetics and Molecular Biology, 2013:19–37. http://dx.doi.org/10.5772/56131.
3. Reese BE. Retinal mosaics: pattern formation driven by local interactions between homotypic neighbors. *Front Neural Circuits.* 2012;6:24.
4. Leskov IB, Klenchin VA, Handy JW, et al. The gain of rod phototransduction: reconciliation of biochemical and electrophysiological measurements. *Neuron.* 2000;27(3):525–537.
5. Young RW, Bok D. Participation of the retinal pigment epithelium in the rod outer segment renewal process. *J Cell Biol.* 1969;42(2):392–403.

CHAPTER TWENTY-THREE

Development of the Vertebrate Eye and Retina

Deborah L. Stenkamp[1]
Department of Biological Sciences, University of Idaho, Moscow, Idaho, USA
[1]Corresponding author: e-mail address: dstenkam@uidaho.edu

Contents

Abstract

The mature, functional, and healthy eye is generated by the coordinated regulatory interaction of numerous and diverse developing tissues. The neural retina of the eye must undergo the neurogenesis of multiple retinal cell types in the correct ratios and spatial patterns. This chapter provides an overview of retinal development, and includes a summary of the process of eye organogenesis, a discussion of major principles of retinal neurogenesis, and describes some of the key molecular factors critical for retinal development. Defects in many of these factors underlie diseases of the eye, and an understanding of the process of retinal development will be critical for successful future applications of regenerative therapies for eye disease.

1. INTRODUCTION

The vertebrate retina contains seven major cell types organized into three layers of cell bodies termed nuclear layers, separated by two synaptic, or plexiform layers (Fig. 1A and B). The outermost layer (outer nuclear layer; ONL) contains the cell bodies of photoreceptors (rods and cones), which project their inner and outer segments toward the adjacent retinal

Progress in Molecular Biology and Translational Science, Volume 134
ISSN 1877-1173
http://dx.doi.org/10.1016/bs.pmbts.2015.06.006
© 2015 Elsevier Inc.
All rights reserved.

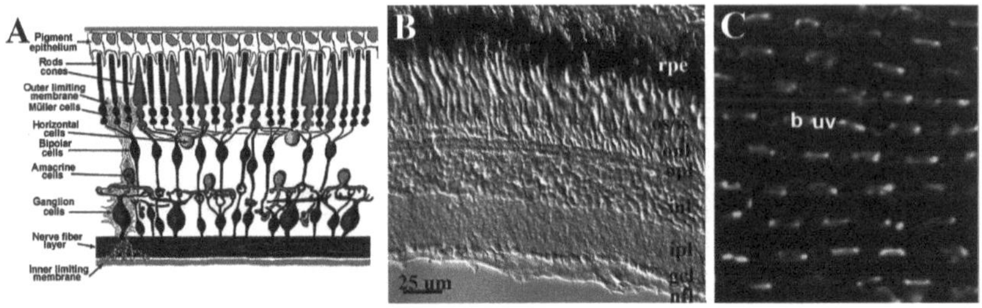

Figure 1 Histology and cell arrangements in the mature vertebrate retina. (A) Cartoon showing the major cell types of the retina. (B) Nomarski (DIC) image of a cryosection of the adult zebrafish retina; RPE, retinal pigmented epithelium; OS/IS, outer segments and inner segments of photoreceptors; ONL, outer nuclear layer; OPL, outer plexiform layer; INL, inner nuclear layer; IPL, inner plexiform layer; GCL, ganglion cell layer; NFL, nerve fiber layer. (C) Dual *in situ* hybridization of zebrafish retinal wholemount with cRNA probes complementary to mRNA encoding SWS2 (blue-sensitive; b) opsin and SWS1 (UV-sensitive; UV) opsin, showing geometrically precise mosaic. *Figure panels reproduced from Ref. 1 with permission.*

pigmented epithelium (RPE). Rods and cones detect photons and convey this information by chemical signaling. The innermost layer of the retina (retinal ganglion cell layer; GCL) contains cell bodies of retinal ganglion cells (RGCs), which are the output neurons of the retina, projecting axons via the optic nerve and tract to specific targets in the brain. Bipolar cells, horizontal cells, and amacrine cells, which have cell bodies located in the inner nuclear layer (INL) between the photoreceptors and ganglion cells, process and transmit information from photoreceptors to the ganglion cells, allowing for the extraction of features from the visual field including luminance contrast, color contrast, and movement. The final major cell type derived from retinal progenitors is the Müller glial cell (MG). MGs have cell bodies which reside in the INL, and processes that span the retina, providing structure to the retina as well as supporting homeostasis. MGs are known to respond to retinal injury or disease by reentering the cell cycle, and in some nonmammalian vertebrates can generate new neurons that functionally integrate into the retina.[2–5] In addition to cells generated from retinal progenitor cells, the retina also contains astrocytes, microglia, and cells of the retinal vasculature. Each of the six major retinal neuronal categories (rod, cone, horizontal cell, bipolar cell, amacrine cell, RGC) includes numerous specific morphological and functional subtypes, and many of these subtypes are distributed across the retinal sheet in regularly spaced mosaics. In teleost fish, the cone photoreceptor mosaics may be geometrically precise; Fig. 2C.[1]

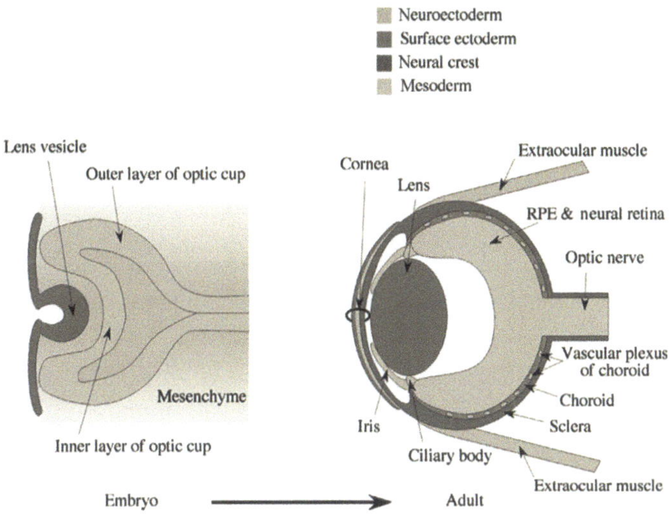

Figure 2 Formation of the major tissues of the eye. Forebrain neuroectoderm (blue (light gray in the print version)) contributes to RPE, iris, and neural retina; surface ectoderm contributes to lens and cornea (green (gray in the print version)); neural crest contributes to cornea, choroid, and sclera (pink (dark gray in the print version)); and mesoderm contributes to extraocular muscles and cornea (yellow (light gray in the print version)). *Figure from Ref. 6 with permission.*

This regularity of two-dimensional neuronal pattern is likely functionally important for representative sampling of the visual field.

The mature vertebrate retina, with its stereotypical laminar and mosaic organizations, is derived from an apparently uniform, single-layered retinal neuroepithelium during development. Our understanding of the process and molecular mechanisms that result in the generation, differentiation, patterning, and subsequent wiring of numerous neuronal types continues to grow, yet remains incomplete. Great insights into these mechanisms have been revealed recently through the use of model organisms, particularly mouse, chick, and zebrafish. Mouse is a favored model because of the ability of transgenic resources. Chick offers an easy to manipulate system that is well developed upon hatching. The zebrafish model offers advantages in genetic manipulation, direct visualization, and rapid experimentation. Zebrafish generate large clutches of embryos that develop externally and remain transparent through the period of retinal neurogenesis, and a functional retina develops by 72 h postfertilization (hpf).[1,7] The rest of this chapter will emphasize the advancements in our understanding of retinal neurogenesis that have benefited from the use of the zebrafish and other model organisms.

2. OVERVIEW OF EYE MORPHOGENESIS

The retinal neuroepithelium is itself derived from the embryonic neural tube, from a region that expresses several eye field transcription factors (EFTFs); these homeodomain-containing factors include *pax6*, one or more of the *rx/rax* genes, *six6/optx2*, *six3*, *lhx2*, and *otx2* (reviewed by Graw[8]). Mutations or knockdown of expression of these genes results in severe developmental defects of the eye; for example, null mutations of the *rx3* gene in zebrafish results in the absence of eyes.[9] Similarly, mutations in *rx* in humans are associated with anophthalmia.[10] The EFTF-expressing region of the neural tube expands bilaterally to form a pair of optic vesicles. Mutations or knockdown of expression of the sonic hedgehog (*Shh*) gene, which is expressed at the ventral midline of the neural tube, results in cyclopia in zebrafish[11–13] and in humans.[14] Shh is a secreted signaling molecule that acts to downregulate *pax6* expression, thereby restricting *pax6* expression laterally, and thus the embryonic region capable of forming an eye, to the more distal regions of the optic vesicles, and this activity results in the formation of bilateral eyes. Midline Shh signaling continues to regulate the timing of retinal neurogenesis.[15] The most distal portion of the optic vesicle interacts with overlying surface ectoderm and surrounding periocular mesenchyme to result in the induction of the nonneural tissues of the eye: the lens and cornea, derived from surface ectoderm, and iris and ciliary body, derived from the margins of the optic cup and periocular mesenchyme (reviewed by Graw and Harada *et al.*[6,8]) (Fig. 2).

3. PRINCIPLES OF RETINAL NEUROGENESIS

At 20 h postfertilization (hpf) in zebrafish, the optic vesicle is collapsed into two layers, with the distal portion fated to become the neural retina, and the more proximal portion fated to become the RPE.[16] The future neural retina is at this point a proliferative neuroepithelium, still expressing the EFTFs, which act as positive regulators of proliferation.[17] This proliferative neuroepithelium then generates the neurons and glia of the retina in a process that illustrates several principles of retinal neurogenesis observed in all vertebrates. The first principle is that specific types of neurons are generated in sequential but overlapping waves. In the zebrafish this sequence is RGCs, followed by cells of the INL, followed by photoreceptor cells[18] (Fig. 3A). Variations on this sequence are seen in other vertebrates,

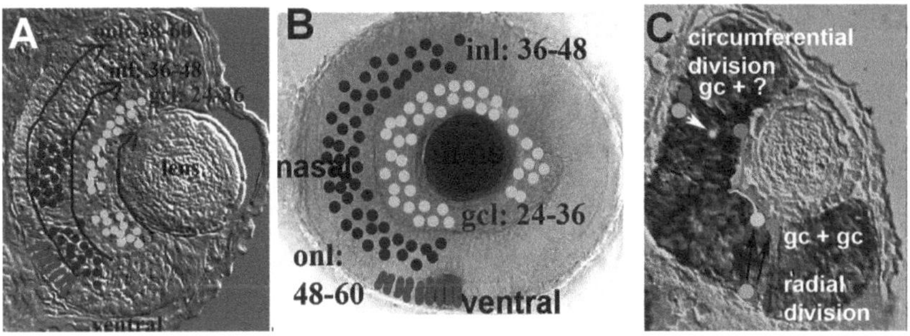

Figure 3 Patterns and principles of retinal neurogenesis. (A) Nomarski (DIC) image of a cryosection of the embryonic zebrafish retina, with patterns of neurogenesis of cells of the ganglion cell layer (GCL, in green (light gray in the print version)), inner nuclear layer (INL, in blue (black in the print version)), and outer nuclear layer (ONL, in red (dark gray in the print version)). (B) Wholemounted embryonic zebrafish eye, with patterns of neurogenesis as in A. (C) Nomarski (DIC) image of a cryosection of an embryonic zebrafish retina earlier in neurogenesis, with orientation of terminal mitotic division predictive of symmetric versus asymmetric fates; GC, ganglion cell. *Figure panels reproduced from Ref. 1 with permission.*

but the "RGCs first, rod photoreceptors and Müller glia last" order of neuron generation is well conserved (reviewed in Refs. 19,20).

The second principle is that neurons of each cell population are generated in a stereotyped spatiotemporal pattern. In the zebrafish, neurogenesis is initiated in a "ventral patch" immediately adjacent to the optic stalk and future position of the optic nerve head, and progresses nasally, and then dorsally and finally temporally, a pattern that has been likened to the opening of a fan[18,21] (Fig. 3B). In other vertebrates the spatiotemporal pattern is manifested as a less complex central-to-peripheral gradient, with the first cells of each cell type to be generated located near the future site of the optic nerve head.[22]

A third principle of retinal neurogenesis is one that has recently been visualized directly in the zebrafish embryo: interkinetic nuclear migration. Nuclei of proliferating progenitors can be found at various apico-basal locations during most phases of the cell cycle. However, immediately prior to M-phase, nuclei migrate rapidly to the apical side of the neuroepithelium and undergo mitosis at that location.[23] The orientation of the mitotic plane of cleavage may be either circumferential or radial with respect to the retinal hemisphere. As RGCs are generated, radial divisions are more likely to generate a pair of RGCs that migrate to the appropriate layer and produce axons. Circumferential divisions are more likely to generate asymmetric fates (e.g., RGC plus a progenitor or RGC plus some other cell type;

Fig. 3C). This process has not been directly visualized in any other vertebrate, so it is not known if this is a conserved mechanism. However, factors known to establish and maintain apico-basal polarity of retinal progenitors in zebrafish—such as Crumbs, nok, and par[24–26]—are required for normal retinal neurogenesis and cell differentiation in the fish, and mutations in apico-basal polarity genes are also associated with human retinopathies.[27]

The stereotyped birth order of retinal neurons supported a model for the generation of neuronal diversity that is referred to as the "competence model."[28] This model states that progenitor cells pass through sequential competence states, and while in each state progenitors are capable of generating a progressively restricted set of cell types. Live imaging in the zebrafish embryo has again informed this model, revealing that, while birth order biases the fates of retinal progenitor cells, it does not restrict fates as strictly as predicted by the model.[29,30]

4. RETINAL NEURONAL DIVERSITY: REGULATION BY INTRINSIC AND EXTRINSIC FACTORS

Both intrinsic factors (specific transcription factors) and extrinsic factors (extracellular signals) endogenously regulate the production of specific retinal cell types in the appropriate numbers. The major intrinsic factors involved in retinal neurogenesis are transcription factors in the homeodomain-containing gene family, the basic helix-loop-helix (bHLH) family, and the nuclear hormone receptor family. Selected intrinsic factors involved in retinal neurogenesis are shown in Table 1. Some of these are required to specify lineage; for example, in zebrafish the thyroid hormone receptor, TRβ2 is expressed in cells that give rise to horizontal cells and long-wavelength-sensitive cones.[66] However, it is thought that intrinsic factors act in a combinatorial manner to restrict or permit progenitor cell competence to specific fates; for example, the homeodomain-containing *vsx2/chx10* is required for bipolar cell development, but promotes this fate together with the bHLH genes *ascl1* and *ath3*.[54]

Extrinsic factors known to regulate retinal neurogenesis and retinal cell differentiation include Shh, Fibroblast growth factors (Fgfs), Wnts, Bone morphogenetic proteins (Bmps/Gdfs/TGFb), Notch, and retinoic acid (RA) (Table 2). The major emphases in the field of retinal development have been centered on understanding factors regulating the neurogenesis of RGCs and of photoreceptor cells. These cell types are lost or damaged—and not replaced, resulting in permanent loss of visual function—in

Table 1 Selected Intrinsic Factors Regulating Retinal Neurogenesis[a]

Family	Factor	Regulatory Role: Eye Morphogenesis	Regulatory Role: Retinal Progenitor Cell Cycle Control	Regulatory Role: Retinal Cell Determination and Differentiation
Homeodomain-containing	Pax6	11,31	17	32,33
	Otx2	34,35		
	Six3	36	37	37
	Six6	38		
	Lhx2	39		
	Pax2	11,40		
	Vax	41		42
	Chx20/Vsx		37,43,44	44
	Crx		45	45–47
	Rx/Rax	9,48,49	50	50,51
bHLH	Atoh7		30	12,13,15,30,52
	Ascl1			53–55
	Math3			56
	NeuroD		57	58,59
Nuclear hormone receptor or orphan receptor	RARα			60,61
	RXRγ			62
	PNR/Nr2e3			63,64
	TRβ2			65,66
	RORα			67
Other	Brn3b			68
	Tbx2b			69
	Nrl			70
Chromatin remodeling	Brg1			71,72
	Add		73	

[a]Selected citations emphasize literature pertaining to the zebrafish/teleost fish models.

Table 2 Selected Extrinsic Factors Regulating Retinal Neurogenesis[a]

	Regulatory Role		
Factor	**Eye Morphogenesis**	**Retinal Progenitor Cell Cycle Control**	**Retinal Cell Determination and Differentiation**
IGFs		74–76	
FGFs	77,78		75,79–81
Wnts	82	73,83	84
Hh	11,13,41	83,85,86	13,83,87–89
RA	90–92		60,84,93–95
T3			96,97
Notch		73	98,99

[a]Selected citations emphasize literature pertaining to the zebrafish/teleost fish models.

glaucoma (RGCs), hereditary retinal degenerations such as retinitis pigmentosa (photoreceptors), and in age-related macular degeneration (photoreceptors). These retinal disorders collectively affect millions of people in the United States alone. A deep understanding of the mechanisms underlying the development of these cell populations is likely to lead to treatments for these disorders that promote the survival and/or regeneration of these cells.

4.1 RGC Neurogenesis: Intrinsic and Extrinsic Factors

An important intrinsic factor for RGC neurogenesis is the bHLH transcription factor *atonal7* (*atoh7*, also referred to as *ath5*). In all vertebrates, *atoh7* is expressed immediately prior to the neurogenesis of RGCs, in a spatiotemporal gradient that anticipates the pattern of RGC genesis.[12] Zebrafish with null mutations in *atoh7* fail to develop RGCs, indicating that atoh7 is required for the development of this cell type.[52] *Atoh7* has been shown to regulate a network of RGC-specific genes; for example, in mice *Atoh7* regulates the transcription factor *Brn3b*, a factor that is required for the development of specific subpopulations of RGCs.[100] Using an *atoh7* reporter system (*atoh7:GFP*), the fates of *atoh7*-expressing progenitor cells were tracked in living zebrafish embryos. GFP+ cells underwent a single cell division which generated one daughter cell that migrated to the RGC layer and differentiated as an RGC—the fate of the other daughter cell remains unknown but may include a photoreceptor fate.[30]

Key extrinsic factors that regulate RGC neurogenesis include the extracellular signaling proteins FGF and Shh. FGF has been shown to be produced in the most distal region of the optic stalk, immediately adjacent to the retinal region where neurogenesis is initiated. Pharmacological inhibition of FGF signaling in zebrafish embryos prevents the expression of *atoh7*, and hence the production of RGCs.[78] Mutants lacking functional Fgf8 and Fgf3 genes (double mutants) display a similar phenotype.[79] Shh signaling has similarly been investigated in zebrafish using both gain- and loss-of-function approaches. Unlike FGF, Shh is expressed by the RGCs themselves, as well as their *atoh7+* progenitors, and there is evidence for two specific roles for Shh signaling in regulating RGC production. The first role is in propagating the wave of RGC neurogenesis; the zebrafish *shh* mutant displays arrested RGC differentiation, and reduced cell proliferation.[13,87] The second role is in regulating RGC number; in other animal models, defects in Shh signaling result in overproduction of RGCs, suggesting that Shh acts as a negative feedback regulator of RGC neurogenesis.[101] Interestingly, reduced Shh signaling during retinal regeneration in zebrafish also results in overproduction of neurons.[5] Shh signals are likely integrated with other signals (such as FGF and Notch) via intracellular signaling cascades to influence cell cycle progression and fate.[19] Retinal progenitor cells express *ptch2*,[89,102] a transmembrane component of the Shh receptor complex, and likely integrate this information.

4.2 Photoreceptor Neurogenesis: Intrinsic and Extrinsic Factors

Numerous specific transcription factors have been demonstrated as important for the neurogenesis of photoreceptors, including those that regulate rod versus cone fate. This "photoreceptor transcription factor network" includes the homeobox genes *crx* and *rx* and the bHLH gene *NeuroD*, all of which are essential for photoreceptor development (reviewed by Swaroop *et al.*[103]). Mutations or knockdown of these genes result in failure of rod and cone development in animal models,[45,46,50] and human mutations in *crx* are associated with hereditary retinal degenerations.[104] The orphan nuclear receptors *Nr2e3* and *RORa,* and the leucine zipper-containing transcription factor *nrl* are all key regulators of rod fate and rod-specific gene expression.[103] When these genes are mutated in mice, retinas contain no rods and instead show an abundance of photoreceptors with an unusual cone-like phenotype most similar to short-wavelength-sensitive (S; blue) cones.[70,105] In the zebrafish, the T-box transcription factor *tbx2b* is required for the genesis of ultraviolet-sensitive (UV) cones (the zebrafish UV opsin is orthologous to the mouse S opsin), and functional nulls instead have supernumerary rods.[69] We have recently observed in the zebrafish that the nuclear hormone receptor RARαb is needed for a retinoic acid-mediated cone-to-rod fate change of progenitor cells[106]. In addition, the nuclear hormone receptors RXRγ and TRβ2 promote the production of cones expressing a medium or long-wavelength-sensitive opsin rather than S or UV opsin.[62,65,66]

A large number of extrinsic factors have been shown regulate photoreceptor neurogenesis and differentiation, including Fgfs, Wnts, Shh, RA, and thyroid hormone (T3). The latter two are particularly interesting because they act directly upon the nuclear hormone receptors (RARs, RXRs, and TRs) mentioned above. Gradients in the concentration of these lipophilic signaling molecules are associated with gradients in photoreceptor-specific opsin expression,[60,65,107] and while the patterns of rod and cone photoreceptors are distinct in specific animal models, the use of these gradients appears to be a conserved mechanism for regulating photoreceptor fate decisions and therefore photoreceptor diversity and ratios.

5. PERSISTENT RETINAL NEUROGENESIS AND REGENERATION

The field of retinal development aspires not only to understand the developmental processes and mechanisms underlying the generation of cell diversity in the retina but also to translate this knowledge into treatments for retinal disorders. Potential treatments include gene therapy, cell transplantation, and regenerative therapies that utilize endogenous sources of new neurons. Regenerative therapies have become of high interest as the National Eye Institute has named as a priority area (the "Audacious Goal") the regeneration of retinal neurons and their connections to restore vision in retinal disorders. To accomplish this goal, several laboratories are taking advantage of the zebrafish animal model using approaches other than the direct study of embryonic retinal development. This is possible because zebrafish, like other teleost fishes, display persistent retinal neurogenesis as well as the capacity for retinal regeneration.[1] Persistent neurogenesis in zebrafish includes the postembryonic addition of new retina at the periphery, generated by a population of stem cells known as the ciliary marginal zone or circumferential germinal zone (CGZ; Fig. 4). This persistent neurogenesis allows the fish retina to grow in concert with body size. CGZ-like

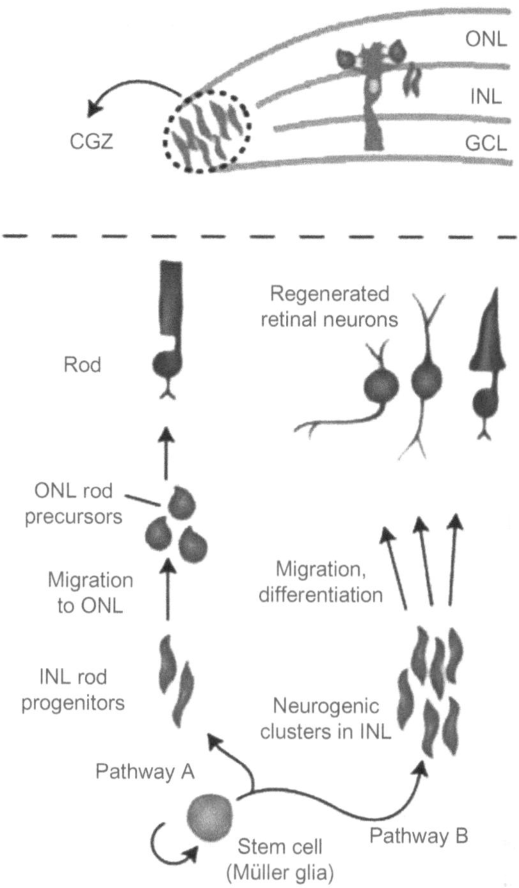

Figure 4 Persistent neurogenesis and retinal regeneration in adult teleost fish. Top: Stem cells of the circumferential germinal zone (CGZ) generate all retinal cells except rod photoreceptors. Bottom, left: Müller glia are stem cells of the rod lineage (Pathway A). Bottom, right: Following damage, Müller glia became multipotent, and contribute to the neurogenesis of all retinal neuronal types (Pathway B). *Figure reproduced from Ref. 106 with permission.*

characteristics have been described to exist at the retinal margin in mammals,[108] although their location in the periphery makes them inconvenient as potential sources of new neurons in treatment of any retinal disease that affects central vision. A second aspect of persistent neurogenesis in zebrafish is the continuous generation of new rod photoreceptors by a dedicated lineage of cells known as the rod lineage.[106] At the apex of this lineage—the stem cell—is the MG, which generates progenitors that proliferate within the INL (a transit-amplifying population), migrate to the

ONL, then undergo terminal mitosis and differentiate as rods[2] (Fig. 4). MG in mammals have very limited potential for cell cycle reentry, and do not contribute to a rod lineage. However, an understanding of this lineage and its regulation in zebrafish may lead to rod replacement treatments in humans. Finally, in cases of retinal damage, the MG of zebrafish not only contribute to the rod lineage but also contribute to the replacement of all retinal cell types[2,109–111] (Fig. 4), resulting in the restoration of vision.[4,5] Normally, MG proliferation in mammals gives rise to gliosis; however, it has been observed that in special circumstances and with specific treatments the MG of mammals can reenter the cell cycle and generate at least some new neurons.[112,113] These are exciting findings that the vision research community plans to build upon to develop regenerative therapies for retinal disorders.[114]

6. SUMMARY

This chapter has provided an overview of retinal development, summarizing the process of organogenesis, principles of retinal neurogenesis, and describing some of the major molecular factors critical for retinal development. The field is actively engaged in deciphering the mechanisms through which intrinsic and extrinsic factors are integrated to regulate progenitor cell competence, identifying additional important players such as chromatin remodeling factors and micro RNAs, and understanding the mechanisms underlying retinal regeneration. This information can be applied for treatment of developmental disorders of the eye, and the development of stem cell and regenerative therapies for retinal degenerative disease and damage.

ACKNOWLEDGMENTS

I thank Drs. Diana Mitchell and Peter Fuerst for constructive comments on the manuscript. Work in this laboratory is supported by National Institutes of Health Grant R01 EY012146, the NIH-INBRE program (P20 RR016454), the UI Office of Research and Economic Development, the NSF-BEACON consortium, and the Malcolm Renfrew Faculty Fellowship.

REFERENCES

1. Stenkamp DL. Neurogenesis in the fish retina. *Int Rev Cytol.* 2007;259:173–224.
2. Bernardos RL, Barthel LK, Meyers JR, Raymond PA. Late-stage neuronal progenitors in the retina are radial Muller glia that function as retinal stem cells. *J Neurosci.* 2007;27(26):7028–7040.
3. Fausett BV, Goldman D. A role for alpha1 tubulin-expressing Muller glia in regeneration of the injured zebrafish retina. *J Neurosci.* 2006;26(23):6303–6313.

4. Sherpa T, Fimbel SM, Mallory DE, et al. Ganglion cell regeneration following whole-retina destruction in zebrafish. *Dev Neurobiol.* 2008;68(2):166–181.
5. Sherpa T, Lankford T, McGinn TE, et al. Retinal regeneration is facilitated by the presence of surviving neurons. *Dev Neurobiol.* 2014;74(9):851–876. http://dx.doi.org/10.1002/dneu.22167.
6. Harada T, Harada C, Parada LF. Molecular regulation of visual system development: more than meets the eye. *Genes Dev.* 2007;21(4):367–378. http://dx.doi.org/10.1101/gad.1504307.
7. Easter Jr SS, Malicki JJ. The zebrafish eye: developmental and genetic analysis. *Results Probl Cell Differ.* 2002;40:346–370.
8. Graw J. Eye development. *Curr Top Dev Biol.* 2010;90:343–386. http://dx.doi.org/10.1016/S0070-2153(10)90010-0.
9. Loosli F, Staub W, Finger-Baier KC, et al. Loss of eyes in zebrafish caused by mutation of chokh/rx3. *EMBO Rep.* 2003;4(9):894–899. http://dx.doi.org/10.1038/sj.embor.embor919.
10. Voronina VA, Kozhemyakina EA, O'Kernick CM, et al. Mutations in the human RAX homeobox gene in a patient with anophthalmia and sclerocornea. *Hum Mol Genet.* 2004;13(3):315–322. http://dx.doi.org/10.1093/hmg/ddh025.
11. Macdonald R, Barth KA, Xu Q, Holder N, Mikkola I, Wilson SW. Midline signalling is required for Pax gene regulation and patterning of the eyes. *Development.* 1995;121(10):3267–3278.
12. Masai I, Stemple DL, Okamoto H, Wilson SW. Midline signals regulate retinal neurogenesis in zebrafish. *Neuron.* 2000;27(2):251–263.
13. Stenkamp DL, Frey RA. Extraretinal and retinal hedgehog signaling sequentially regulate retinal differentiation in zebrafish. *Dev Biol.* 2003;258(2):349–363.
14. Maity T, Fuse N, Beachy PA. Molecular mechanisms of Sonic hedgehog mutant effects in holoprosencephaly. *Proc Natl Acad Sci U S A.* 2005;102(47):17026–17031. http://dx.doi.org/10.1073/pnas.0507848102.
15. Kay JN, Link BA, Baier H. Staggered cell-intrinsic timing of ath5 expression underlies the wave of ganglion cell neurogenesis in the zebrafish retina. *Development.* 2005;132(11):2573–2585.
16. Schmitt EA, Dowling JE. Early eye morphogenesis in the zebrafish, Brachydanio rerio. *J Comp Neurol.* 1994;344(4):532–542. http://dx.doi.org/10.1002/cne.903440404.
17. Hsieh YW, Yang XJ. Dynamic Pax6 expression during the neurogenic cell cycle influences proliferation and cell fate choices of retinal progenitors. *Neural Dev.* 2009;4:32. http://dx.doi.org/10.1186/1749-8104-4-32.
18. Hu M, Easter SS. Retinal neurogenesis: the formation of the initial central patch of postmitotic cells. *Dev Biol.* 1999;207(2):309–321.
19. Bassett EA, Wallace VA. Cell fate determination in the vertebrate retina. *Trends Neurosci.* 2012;35(9):565–573. http://dx.doi.org/10.1016/j.tins.2012.05.004.
20. Centanin L, Wittbrodt J. Retinal neurogenesis. *Development.* 2014;141(2):241–244. http://dx.doi.org/10.1242/dev.083642.
21. Raymond PA, Barthel LK, Curran GA. Developmental patterning of rod and cone photoreceptors in embryonic zebrafish. *J Comp Neurol.* 1995;359(4):537–550.
22. McCabe KL, Gunther EC, Reh TA. The development of the pattern of retinal ganglion cells in the chick retina: mechanisms that control differentiation. *Development.* 1999;126(24):5713–5724.
23. Baye LM, Link BA. Interkinetic nuclear migration and the selection of neurogenic cell divisions during vertebrate retinogenesis. *J Neurosci.* 2007;27(38):10143–10152.
24. Krock BL, Perkins BD. The Par-PrkC polarity complex is required for cilia growth in zebrafish photoreceptors. *PLoS One.* 2014;9(8):e104661. http://dx.doi.org/10.1371/journal.pone.0104661.

25. Yamaguchi M, Imai F, Tonou-Fujimori N, Masai I. Mutations in N-cadherin and a Stardust homolog, Nagie oko, affect cell-cycle exit in zebrafish retina. *Mech Dev.* 2010;127(5-6):247–264. http://dx.doi.org/10.1016/j.mod.2010.03.004.
26. Zou J, Wen Y, Yang X, Wei X. Spatial-temporal expressions of Crumbs and Nagie oko and their interdependence in zebrafish central nervous system during early development. *Int J Dev Neurosci.* 2013;31(8):770–782. http://dx.doi.org/10.1016/j.ijdevneu.2013.09.005.
27. Richard M, Roepman R, Aartsen WM, et al. Towards understanding CRUMBS function in retinal dystrophies. *Hum Mol Genet.* 2006;15(Spec No 2):R235–R243. http://dx.doi.org/10.1093/hmg/ddl195.
28. Cepko CL, Austin CP, Yang X, Alexiades M, Ezzeddine D. Cell fate determination in the vertebrate retina. *Proc Natl Acad Sci U S A.* 1996;93(2):589–595.
29. He J, Zhang G, Almeida AD, Cayouette M, Simons BD, Harris WA. How variable clones build an invariant retina. *Neuron.* 2012;75(5):786–798. http://dx.doi.org/10.1016/j.neuron.2012.06.033.
30. Poggi L, Vitorino M, Masai I, Harris WA. Influences on neural lineage and mode of division in the zebrafish retina in vivo. *J Cell Biol.* 2005;171(6):991–999.
31. Nornes S, Clarkson M, Mikkola I, et al. Zebrafish contains two pax6 genes involved in eye development. *Mech Dev.* 1998;77(2):185–196.
32. Hitchcock PF, Macdonald RE, VanDeRyt JT, Wilson SW. Antibodies against Pax6 immunostain amacrine and ganglion cells and neuronal progenitors, but not rod precursors, in the normal and regenerating retina of the goldfish. *J Neurobiol.* 1996;29(3):399–413.
33. Otteson DC, D'Costa AR, Hitchcock PF. Putative stem cells and the lineage of rod photoreceptors in the mature retina of the goldfish. *Dev Biol.* 2001;232(1):62–76.
34. Bernard C, Kim HT, Torero Ibad R, et al. Graded Otx2 activities demonstrate dose-sensitive eye and retina phenotypes. *Hum Mol Genet.* 2014;23(7):1742–1753. http://dx.doi.org/10.1093/hmg/ddt562.
35. Martinez-Morales JR, Signore M, Acampora D, Simeone A, Bovolenta P. Otx genes are required for tissue specification in the developing eye. *Development.* 2001;128(11):2019–2030.
36. Loosli F, Winkler S, Wittbrodt J. Six3 overexpression initiates the formation of ectopic retina. *Genes Dev.* 1999;13(6):649–654.
37. Dyer MA. Regulation of proliferation, cell fate specification and differentiation by the homeodomain proteins Prox1, Six3, and Chx10 in the developing retina. *Cell Cycle.* 2003;2(4):350–357.
38. Bernier G, Panitz F, Zhou X, Hollemann T, Gruss P, Pieler T. Expanded retina territory by midbrain transformation upon overexpression of Six6 (Optx2) in Xenopus embryos. *Mech Dev.* 2000;93(1-2):59–69.
39. Tetreault N, Champagne MP, Bernier G. The LIM homeobox transcription factor Lhx2 is required to specify the retina field and synergistically cooperates with Pax6 for Six6 trans-activation. *Dev Biol.* 2009;327(2):541–550. http://dx.doi.org/10.1016/j.ydbio.2008.12.022.
40. Macdonald R, Scholes J, Strahle U, et al. The Pax protein Noi is required for commissural axon pathway formation in the rostral forebrain. *Development.* 1997;124(12):2397–2408.
41. Take-uchi M, Clarke JD, Wilson SW. Hedgehog signalling maintains the optic stalk-retinal interface through the regulation of Vax gene activity. *Development.* 2003;130(5):955–968.
42. Vitorino M, Jusuf PR, Maurus D, Kimura Y, Higashijima S, Harris WA. Vsx2 in the zebrafish retina: restricted lineages through derepression. *Neural Dev.* 2009;4:14.

43. Barabino SM, Spada F, Cotelli F, Boncinelli E. Inactivation of the zebrafish homologue of Chx10 by antisense oligonucleotides causes eye malformations similar to the ocular retardation phenotype. *Mech Dev.* 1997;63(2):133–143.
44. Passini MA, Raymond PA, Schechter N. Vsx-2, a gene encoding a paired-type homeodomain, is expressed in the retina, hindbrain, and spinal cord during goldfish embryogenesis. *Brain Res Dev Brain Res.* 1998;109(2):129–135.
45. Shen YC, Raymond PA. Zebrafish cone-rod (crx) homeobox gene promotes retinogenesis. *Dev Biol.* 2004;269(1):237–251.
46. Hennig AK, Peng GH, Chen S. Regulation of photoreceptor gene expression by Crx-associated transcription factor network. *Brain Res.* 2008;1192:114–133.
47. Liu Y, Shen Y, Rest JS, Raymond PA, Zack DJ. Isolation and characterization of a zebrafish homologue of the cone rod homeobox gene. *Invest Ophthalmol Vis Sci.* 2001;42(2):481–487.
48. Chuang JC, Raymond PA. Zebrafish genes rx1 and rx2 help define the region of forebrain that gives rise to retina. *Dev Biol.* 2001;231(1):13–30.
49. Rojas-Munoz A, Dahm R, Nusslein-Volhard C. chokh/rx3 specifies the retinal pigment epithelium fate independently of eye morphogenesis. *Dev Biol.* 2005;288(2): 348–362.
50. Nelson SM, Park L, Stenkamp DL. Retinal homeobox 1 is required for retinal neurogenesis and photoreceptor differentiation in embryonic zebrafish. *Dev Biol.* 2009;328(1):24–39.
51. Chuang JC, Mathers PH, Raymond PA. Expression of three Rx homeobox genes in embryonic and adult zebrafish. *Mech Dev.* 1999;84(1-2):195–198.
52. Kay JN, Finger-Baier KC, Roeser T, Staub W, Baier H. Retinal ganglion cell genesis requires lakritz, a Zebrafish atonal Homolog. *Neuron.* 2001;30(3):725–736.
53. Brzezinski 4th JA, Kim EJ, Johnson JE, Reh TA. Ascl1 expression defines a subpopulation of lineage-restricted progenitors in the mammalian retina. *Development.* 2011;138(16):3519–3531. http://dx.doi.org/10.1242/dev.064006.
54. Hatakeyama J, Tomita K, Inoue T, Kageyama R. Roles of homeobox and bHLH genes in specification of a retinal cell type. *Development.* 2001;128(8):1313–1322.
55. Mazurier N, Parain K, Parlier D, et al. Ascl1 as a novel player in the Ptf1a transcriptional network for GABAergic cell specification in the retina. *PLoS One.* 2014;9(3):e92113. http://dx.doi.org/10.1371/journal.pone.0092113.
56. Inoue T, Hojo M, Bessho Y, Tano Y, Lee JE, Kageyama R. Math3 and NeuroD regulate amacrine cell fate specification in the retina. *Development.* 2002;129(4):831–842.
57. Ochocinska MJ, Hitchcock PF. NeuroD regulates proliferation of photoreceptor progenitors in the retina of the zebrafish. *Mech Dev.* 2009;126(3-4):128–141.
58. Hitchcock P, Kakuk-Atkins L. The basic helix-loop-helix transcription factor neuroD is expressed in the rod lineage of the teleost retina. *J Comp Neurol.* 2004;477(1):108–117.
59. Korzh V, Sleptsova I, Liao J, He J, Gong Z. Expression of zebrafish bHLH genes ngn1 and nrd defines distinct stages of neural differentiation. *Dev Dyn.* 1998;213(1):92–104.
60. Stevens CB, Cameron DA, Stenkamp DL. Plasticity of photoreceptor-generating retinal progenitors revealed by prolonged retinoic acid exposure. *BMC Dev Biol.* 2011;11:51. http://dx.doi.org/10.1186/1471-213X-11-51.
61. Tallafuss A, Hale LA, Yan YL, Dudley L, Eisen JS, Postlethwait JH. Characterization of retinoid-X receptor genes rxra, rxrba, rxrbb and rxrg during zebrafish development. *Gene Expr Patterns.* 2006;6(5):556–565.
62. Roberts MR, Hendrickson A, McGuire CR, Reh TA. Retinoid X receptor (gamma) is necessary to establish the S-opsin gradient in cone photoreceptors of the developing mouse retina. *Invest Ophthalmol Vis Sci.* 2005;46(8):2897–2904. http://dx.doi.org/10.1167/iovs.05-0093.

63. Chen J, Rattner A, Nathans J. The rod photoreceptor-specific nuclear receptor Nr2e3 represses transcription of multiple cone-specific genes. *J Neurosci*. 2005;25(1):118–129.
64. Nelson SM, Frey RA, Wardwell SL, Stenkamp DL. The developmental sequence of gene expression within the rod photoreceptor lineage in embryonic zebrafish. *Dev Dyn*. 2008;237(10):2903–2917.
65. Roberts MR, Srinivas M, Forrest D, Morreale de Escobar G, Reh TA. Making the gradient: thyroid hormone regulates cone opsin expression in the developing mouse retina. *Proc Natl Acad Sci U S A*. 2006;103(16):6218–6223. http://dx.doi.org/10.1073/pnas.0509981103.
66. Suzuki SC, Bleckert A, Williams PR, Takechi M, Kawamura S, Wong RO. Cone photoreceptor types in zebrafish are generated by symmetric terminal divisions of dedicated precursors. *Proc Natl Acad Sci U S A*. 2013;110(37):15109–15114. http://dx.doi.org/10.1073/pnas.1303551110.
67. Forrest D, Swaroop A. Minireview: the role of nuclear receptors in photoreceptor differentiation and disease. *Mol Endocrinol*. 2012;26(6):905–915. http://dx.doi.org/10.1210/me.2012-1010.
68. DeCarvalho AC, Cappendijk SL, Fadool JM. Developmental expression of the POU domain transcription factor Brn-3b (Pou4f2) in the lateral line and visual system of zebrafish. *Dev Dyn*. 2004;229(4):869–876.
69. Alvarez-Delfin K, Morris AC, Snelson CD, et al. Tbx2b is required for ultraviolet photoreceptor cell specification during zebrafish retinal development. *Proc Natl Acad Sci U S A*. 2009;106(6):2023–2028.
70. Mears AJ, Kondo M, Swain PK, et al. Nrl is required for rod photoreceptor development. *Nat Genet*. 2001;29(4):447–452. http://dx.doi.org/10.1038/ng774.
71. Gregg RG, Willer GB, Fadool JM, Dowling JE, Link BA. Positional cloning of the young mutation identifies an essential role for the Brahma chromatin remodeling complex in mediating retinal cell differentiation. *Proc Natl Acad Sci U S A*. 2003;100(11):6535–6540.
72. Link BA, Fadool JM, Malicki J, Dowling JE. The zebrafish young mutation acts non-cell-autonomously to uncouple differentiation from specification for all retinal cells. *Development*. 2000;127(10):2177–2188.
73. Yamaguchi M, Tonou-Fujimori N, Komori A, et al. Histone deacetylase 1 regulates retinal neurogenesis in zebrafish by suppressing Wnt and Notch signaling pathways. *Development*. 2005;132(13):3027–3043.
74. Boucher SE, Hitchcock PF. Insulin-like growth factor-I binds in the inner plexiform layer and circumferential germinal zone in the retina of the goldfish. *J Comp Neurol*. 1998;394(3):395–401.
75. Mack AF, Fernald RD. Regulation of cell division and rod differentiation in the teleost retina. *Brain Res Dev Brain Res*. 1993;76(2):183–187.
76. Otteson DC, Cirenza PF, Hitchcock PF. Persistent neurogenesis in the teleost retina: evidence for regulation by the growth-hormone/insulin-like growth factor-I axis. *Mech Dev*. 2002;117(1-2):137–149.
77. Heisenberg CP, Brennan C, Wilson SW. Zebrafish aussicht mutant embryos exhibit widespread overexpression of ace (fgf8) and coincident defects in CNS development. *Development*. 1999;126(10):2129–2140.
78. Picker A, Brand M. Fgf signals from a novel signaling center determine axial patterning of the prospective neural retina. *Development*. 2005;132(22):4951–4962. http://dx.doi.org/10.1242/dev.02071.
79. Martinez-Morales JR, Del Bene F, Nica G, Hammerschmidt M, Bovolenta P, Wittbrodt J. Differentiation of the vertebrate retina is coordinated by an FGF signaling center. *Dev Cell*. 2005;8(4):565–574. http://dx.doi.org/10.1016/j.devcel.2005.01.022.

80. Hochmann S, Kaslin J, Hans S, et al. Fgf signaling is required for photoreceptor maintenance in the adult zebrafish retina. *PLoS One*. 2012;7(1):e30365.
81. Qin Z, Kidd 3rd AR, Thomas JL, et al. FGF signaling regulates rod photoreceptor cell maintenance and regeneration in zebrafish. *Exp Eye Res*. 2011;93(5):726–734.
82. Veien ES, Rosenthal JS, Kruse-Bend RC, Chien CB, Dorsky RI. Canonical Wnt signaling is required for the maintenance of dorsal retinal identity. *Development*. 2008;135(24):4101–4111. http://dx.doi.org/10.1242/dev.027367.
83. Masai I, Yamaguchi M, Tonou-Fujimori N, Komori A, Okamoto H. The hedgehog-PKA pathway regulates two distinct steps of the differentiation of retinal ganglion cells: the cell-cycle exit of retinoblasts and their neuronal maturation. *Development*. 2005;132(7):1539–1553.
84. Meyers JR, Hu L, Moses A, Kaboli K, Papandrea A, Raymond PA. beta-catenin/Wnt signaling controls progenitor fate in the developing and regenerating zebrafish retina. *Neural Dev*. 2012;7:30. http://dx.doi.org/10.1186/1749-8104-7-30.
85. Shkumatava A, Neumann CJ. Shh directs cell-cycle exit by activating p57Kip2 in the zebrafish retina. *EMBO Rep*. 2005;6(6):563–569.
86. Stenkamp DL, Frey RA, Mallory DE, Shupe EE. Embryonic retinal gene expression in sonic-you mutant zebrafish. *Dev Dyn*. 2002;225(3):344–350.
87. Neumann CJ, Nuesslein-Volhard C. Patterning of the zebrafish retina by a wave of sonic hedgehog activity. *Science*. 2000;289(5487):2137–2139.
88. Shkumatava A, Fischer S, Muller F, Strahle U, Neumann CJ. Sonic hedgehog, secreted by amacrine cells, acts as a short-range signal to direct differentiation and lamination in the zebrafish retina. *Development*. 2004;131(16):3849–3858.
89. Stenkamp DL, Frey RA, Prabhudesai SN, Raymond PA. Function for Hedgehog genes in zebrafish retinal development. *Dev Biol*. 2000;220(2):238–252.
90. Hyatt GA, Schmitt EA, Marsh-Armstrong N, McCaffery P, Drager UC, Dowling JE. Retinoic acid establishes ventral retinal characteristics. *Development*. 1996;122(1):195–204.
91. Hyatt GA, Schmitt EA, Marsh-Armstrong NR, Dowling JE. Retinoic acid-induced duplication of the zebrafish retina. *Proc Natl Acad Sci U S A*. 1992;89(17):8293–8297.
92. Marsh-Armstrong N, McCaffery P, Gilbert W, Dowling JE, Drager UC. Retinoic acid is necessary for development of the ventral retina in zebrafish. *Proc Natl Acad Sci U S A*. 1994;91(15):7286–7290.
93. Biehlmaier O, Lampert JM, von Lintig J, Kohler K. Photoreceptor morphology is severely affected in the beta, beta-carotene-15,15'-oxygenase (bcox) zebrafish morphant. *Eur J Neurosci*. 2005;21(1):59–68. http://dx.doi.org/10.1111/j.1460-9568.2004.03830.x.
94. Hyatt GA, Schmitt EA, Fadool JM, Dowling JE. Retinoic acid alters photoreceptor development in vivo. *Proc Natl Acad Sci U S A*. 1996;93(23):13298–13303.
95. Prabhudesai SN, Cameron DA, Stenkamp DL. Targeted effects of retinoic acid signaling upon photoreceptor development in zebrafish. *Dev Biol*. 2005;287(1):157–167.
96. Pessoa CN, Santiago LA, Santiago DA, et al. Thyroid hormone action is required for normal cone opsin expression during mouse retinal development. *Invest Ophthalmol Vis Sci*. 2008;49(5):2039–2045. http://dx.doi.org/10.1167/iovs.07-0908.
97. Suliman T, Novales Flamarique I. Visual pigments and opsin expression in the juveniles of three species of fish (rainbow trout, zebrafish, and killifish) following prolonged exposure to thyroid hormone or retinoic acid. *J Comp Neurol*. 2014;522(1):98–117. http://dx.doi.org/10.1002/cne.23391.
98. Bernardos RL, Lentz SI, Wolfe MS, Raymond PA. Notch-Delta signaling is required for spatial patterning and Muller glia differentiation in the zebrafish retina. *Dev Biol*. 2005;278(2):381–395.

99. Scheer N, Groth A, Hans S, Campos-Ortega JA. An instructive function for Notch in promoting gliogenesis in the zebrafish retina. *Development*. 2001;128(7):1099–1107.
100. Liu W, Mo Z, Xiang M. The Ath5 proneural genes function upstream of Brn3 POU domain transcription factor genes to promote retinal ganglion cell development. *Proc Natl Acad Sci U S A*. 2001;98(4):1649–1654. http://dx.doi.org/10.1073/pnas.98.4.1649.
101. Zhang XM, Yang XJ. Regulation of retinal ganglion cell production by Sonic hedgehog. *Development*. 2001;128(6):943–957.
102. Bibliowicz J, Gross JM. Ectopic proliferation contributes to retinal dysplasia in the juvenile zebrafish patched2 mutant eye. *Invest Ophthalmol Vis Sci*. 2011;52(12):8868–8877.
103. Swaroop A, Kim D, Forrest D. Transcriptional regulation of photoreceptor development and homeostasis in the mammalian retina. *Nat Rev Neurosci*. 2010;11(8):563–576. http://dx.doi.org/10.1038/nrn2880.
104. Huang L, Xiao X, Li S, et al. CRX variants in cone-rod dystrophy and mutation overview. *Biochem Biophys Res Commun*. 2012;426(4):498–503. http://dx.doi.org/10.1016/j.bbrc.2012.08.110.
105. Webber AL, Hodor P, Thut CJ, et al. Dual role of Nr2e3 in photoreceptor development and maintenance. *Exp Eye Res*. 2008;87(1):35–48. http://dx.doi.org/10.1016/j.exer.2008.04.006.
106. Stenkamp DL. The rod photoreceptor lineage of teleost fish. *Prog Retin Eye Res*. 2011;30(6):395–404.
107. Mey J, McCaffery P, Klemeit M. Sources and sink of retinoic acid in the embryonic chick retina: distribution of aldehyde dehydrogenase activities, CRABP-I, and sites of retinoic acid inactivation. *Brain Res Dev Brain Res*. 2001;127(2):135–148.
108. Kubota R, Hokoc JN, Moshiri A, McGuire C, Reh TA. A comparative study of neurogenesis in the retinal ciliary marginal zone of homeothermic vertebrates. *Brain Res Dev Brain Res*. 2002;134(1-2):31–41.
109. Qin Z, Barthel LK, Raymond PA. Genetic evidence for shared mechanisms of epimorphic regeneration in zebrafish. *Proc Natl Acad Sci U S A*. 2009;106(23):9310–9315.
110. Ramachandran R, Reifler A, Parent JM, Goldman D. Conditional gene expression and lineage tracing of tuba1a expressing cells during zebrafish development and retina regeneration. *J Comp Neurol*. 2010;518(20):4196–4212. http://dx.doi.org/10.1002/cne.22448.
111. Yurco P, Cameron DA. Responses of Muller glia to retinal injury in adult zebrafish. *Vision Res*. 2005;45(8):991–1002.
112. Karl MO, Hayes S, Nelson BR, Tan K, Buckingham B, Reh TA. Stimulation of neural regeneration in the mouse retina. *Proc Natl Acad Sci U S A*. 2008;105(49):19508–19513.
113. Pollak J, Wilken MS, Ueki Y, et al. ASCL1 reprograms mouse Muller glia into neurogenic retinal progenitors. *Development*. 2013;140(12):2619–2631. http://dx.doi.org/10.1242/dev.091355.
114. Fischer AJ, Bongini R. Turning Muller glia into neural progenitors in the retina. *Mol Neurobiol*. 2011;42(3):199–209.

CHAPTER TWENTY-FOUR

Insights into the Molecular Properties of ABCA4 and Its Role in the Visual Cycle and Stargardt Disease

Robert S. Molday[1]
Department of Biochemistry and Molecular Biology, Centre for Macular Research, University of British Columbia, Vancouver, British Columbia, Canada
[1]Corresponding author: e-mail address: molday@mail.ubc.ca

Contents

Abstract

ABCA4 is a member of the A-subfamily of ATP-binding cassette (ABC) transporters localized in rod and cone outer segment disc membranes. Over 800 mutations in ABCA4 are now known to cause Stargardt macular degeneration and related retinal degenerative diseases. Biochemical studies have shown that ABCA4 transports or flips the 11-*cis* and all-*trans* isomers of *N*-retinylidene-phosphatidylethanolamine across disc membranes, thereby facilitating the removal of retinal from disc membranes through the visual cycle and preventing the accumulation of potentially toxic bisretinoid compounds in photoreceptor and retinal pigment epithelial cells. This chapter summarizes studies which have led to our understanding of the role of ABCA4 in the visual cycle and Stargardt disease.

Progress in Molecular Biology and Translational Science, Volume 134
ISSN 1877-1173
http://dx.doi.org/10.1016/bs.pmbts.2015.06.008
© 2015 Elsevier Inc.
All rights reserved.

1. INTRODUCTION

Stargardt disease is an autosomal recessive early onset macular dystrophy first described by the German ophthalmologist Karl Stargardt in 1909.[1] It accounts for up to 7% of the retinopathies with a prevalence of 1 in 8000–10,000 individuals. Although there is a considerable variation in the age of onset and disease severity, most Stargardt patients experience significant reduction in visual acuity associated with loss in central vision in their first or second decade of life.[2–4] Progressive reduction in visual acuity typically occurs throughout life with values reaching 20/200 or greater in the advanced stage of the disease.[5–7] Affected individuals also experience some loss in color vision.[8,9] Ophthalmic examination generally reveals bilateral atrophic changes in the macula associated with the loss of photoreceptors and RPE cells, the appearance of yellow–white flecks around the midperiphery of the retina, variable decrease in rod and cone electroretinograms (ERGs), and a reduction in dark adaptation.[9–11]

The gene responsible for autosomal recessive Stargardt disease was first identified in 1997 by Allikmets and colleagues[12] and shown to encode an ATP-binding cassette (ABC) transporter, previously described as an abundant high molecular weight glycoprotein called the Rim protein and now generally known as ABCA4.[13,14] Since this discovery, molecular geneticists have identified over 800 different mutations in ABCA4 associated with Stargardt disease.[3,6,15–19] Mutations in the gene encoding ABCA4 are also responsible for other related but clinically defined diseases including fundus flavimaculatus and a subset of autosomal recessive cone-rod dystrophy, retinitis pigmentosa, and age-related macular degeneration.[6,11,16,20,21] A wide range of different types of disease-associated mutations occurs throughout the gene. These include frame shifts, small deletions and insertions, and splice-site, nonsense, and missense mutations with missense mutations being the most common.

Since the identification of the gene associated with Stargardt disease in 1997, substantial progress has been made in determining the molecular properties of ABCA4 and defining its role in photoreceptor cell biology and Stargardt disease. This chapter provides an overview of the structural and functional properties of ABCA4 and insights into its role in the visual cycle. Molecular and cellular mechanisms responsible for Stargardt disease linked to mutations in ABCA4 are also briefly discussed.

2. ABC TRANSPORTERS

ATP-binding cassette or ABC transporters are a large superfamily of proteins found in all living organisms including bacteria, archaea, and eukarya.[22] These proteins typically transport a variety of compounds across cell membranes including drugs, ions, metabolites, lipids, vitamins, and organic compounds among others. ABC transporters have a core structure consisting of two transmembrane domains (TMDs) and two ABCs or nucleotide binding domains (NBDs). In prokaryotes, these domains can reside on individual polypeptides or fused onto larger polypeptides with the individual subunits assembling to form the core functional transporter. In eukaryotes, ABC transporters exist as either full transporters with all four domains residing on a single polypeptide or half transporters in which two polypeptides each having a TMD and NBD associate to form a homo- or heterodimer. The two core TMDs generally consisting of 12 membrane-spanning segments form the substrate binding site and its translocation pathway across the membrane, while the NBDs provide energy for substrate transport through the binding and hydrolysis of ATP. The NBDs contain a number of conserved structural elements including a Walker A motif (GXXGXGK(S/T)), a Walker B motif (hhhhD where h is a hydrophobic residue), a Q-loop (Q), an H-motif (H), and the ABC signature motif (LSGGQ) which is only found in ABC transporters.[23]

ABC transporters can translocate their substrates either in an outward direction from the cytoplasmic to the exocytoplasmic (extracellular/lumen) side of membranes and are called exporters or in an inner direction from the exocytoplasmic (lumen/extracellular) side to the cytoplasmic side of membranes and are referred to as importers. Prokaryote ABC transporters can be either importers or exporters, whereas most, but not all, eukaryotic ABC transporters are exporters.

3. ABCA SUBFAMILY

The human genome codes for 49 ABC transporters.[24,25] These have been organized into seven subfamilies (ABCA–ABCG) based on phylogenetic similarities. The ABCA subfamily consists of 12 full transporters which show a high degree of similarity in sequence and structural organization. All ABCA transporters exist as nonidentical tandem halves organized as TMD1–NBD1–TMD2–NBD2. On closer inspection, the first membrane-spanning segment of TMD1 and TMD2 is separated from the remaining five membrane-spanning segments by large glycosylated extracellular domains, ECD1 in the N-half and ECD2 in the C-half, respectively.[26] ABCA transporters have been further subdivided into two groups.[27] One group (ABCA5, ABCA6, ABCA8, ABCA9, and ABCA10) is encoded by a cluster of genes that map to chromosome 17q24.2–q24.3. These transporters consist of 1543–1642 amino acids and display a 53–78% similarity in amino acid sequence. Another group (ABCA1, ABCA2, ABCA3, ABCA4, ABCA7, ABCA12, and ABCA13) are encoded by genes on various chromosomes.

To date, mutations in five human ABCA transporters have been linked to monogenic diseases. Mutations in ABCA1 cause Tangier disease associated with a loss in cholesterol efflux from cells and a deficiency in high-density lipoprotein formation.[28–30] Mutations in ABCA3 cause neonatal surfactant deficiency and pediatric interstitial lung disease associated with defective surfactant secretion into the lungs and abnormal lamellar body formation.[31,32] ABCA4 has been linked to autosomal recessive retinal degenerative diseases including Stargardt disease as discussed below. Mutations in ABCA12 are associated with harlequin and lamellar ichthyosis, related skin diseases caused by defective lipid transport.[33,34] Most recently, mutations in ABCA5 have been reported to cause autosomal recessive congenital generalized hypertrichosis terminalis characterized by excessive hair growth and other congenital defects.[35] Several other ABCA transporters including ABCA2, ABCA5, and ABCA7 have been associated with the etiology of Alzheimer's disease.[36,37]

Substrates transported by the various ABCA transporters have been identified in only a few cases. ABCA1 and ABCA7 function as exporters transporting or flipping phosphatidylcholine and phosphatidylserine from the cytoplasmic to the exocytoplasmic leaflet of membranes or to bound ApoA1 or ApoE, thereby facilitating the efflux of cholesterol from cells.[38–40] In contrast, ABCA4 functions as an importer flipping *N*-retinylidene-phosphatidylethanolamine (*N*-ret-PE) and phosphatidylethanolamine (PE) from the lumen to the cytoplasmic side of photoreceptor disc membranes.[41] Genetic and cell-based studies have implicated other ABCA proteins in the transport of lipids across membranes, although the lipid substrates and direction of transport remain to be determined at a biochemical level.

4. ABCA4

4.1 Structural Features and Localization

ABCA4 consists of 50 exons and maps to position 1p21–p22.1 on chromosome 1.[12,17,42] The human gene encodes a protein of 2273 amino acids with an approximate molecular weight of 250 kDa. ABCA4 is highly conserved among vertebrates with the human protein being 88% identical in sequence to the bovine and mouse proteins and 66% identical to the *Xenopus lavis* ortholog.[43] ABCA4 shares the highest sequence identity with ABCA1 (52%) and ABCA7 (49%).

Like other ABCA proteins, ABCA4 is organized in two nonidentical tandem halves with each half containing a transmembrane segment followed by a large glycosylated ECD, five transmembrane segments, and a NBD (Fig. 1A).[13,26] ECD1 in the N-half contains over 600 amino acids, whereas ECD2 in the C-half is smaller with 289 amino acids. The ECDs have numerous cysteine residues, most of which form disulfide bonds within and between the ECDs and N-linked glycosylation sites.[26,44] NBD1 and NBD2 consisting of over 200 amino acids share 37% identity in sequence with both domains containing the structural features found in other ABC transporters.[45] The TMDs have not been studied in detail, although they likely pack to form the lipid substrate binding site and translocation pathway. Recent structural studies using single particle analysis for electron microscopy have generated a global structure of ABCA4 at 18.1 Å resolution.[46] ABCA4 has an elongated asymmetrical profile with a length of 20 nm and a shape resembling an upside down ice-cream cone with a "handle" and "head" connected by a thin linker region. ConA binding studies indicate that the glycosylated ECD domains are present in the "handle" and "linker." The binding of a nonhydrolyzable ATP analogue induces a conformational change in the cytoplasmic and transmembrane regions of ABCA4. To date, P-glycoprotein is the only eukaryotic ABC transporter for which there is a high-resolution structure.[47] The structures of numerous prokaryotic ABC transporters, however, have been determined in various states.[23,48]

Gene expression studies have shown that *ABCA4* is highly expressed in the retina.[12] At the protein level, ABCA4 has been localized to rod and cone photoreceptor outer segments by immunofluorescence microscopy.[49,50] At a higher resolution, ABCA4 has been localized to the rim region of photoreceptor disc membranes by immunoelectron microscopy.[13,51] The

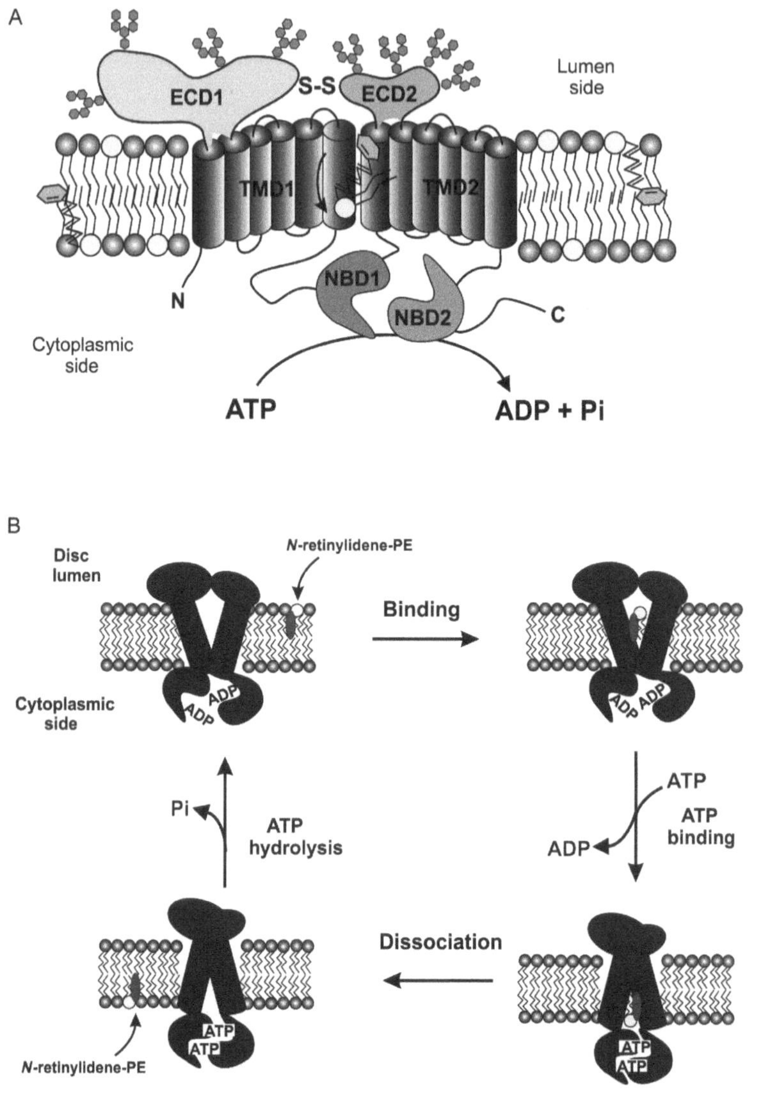

Figure 1 Structural features and proposed reaction cycle of ABCA4. (A) Topological organization of ABCA4 in disc membranes. ABCA4 is a full transporter consisting of two transmembrane domains (TMD1 and TMD2), two exocytoplasmic domains (ECD1 and ECD2) linked by one or more disulfide bonds, and two nucleotide binding domains (NBD1 and NBD2) on a single polypeptide of 2273 amino acids (human protein). (B) Proposed alternating access model for the active transport of *N*-ret-PE across the disc membrane. ABCA4, possibly containing ADP bound to its NBDs, binds its substrate *N*-ret-PE with high affinity on the lumen side of the disc membrane (outward face). ATP binds to one or both NBDs resulting in a protein conformational change which converts the high-affinity substrate binding site to a low-affinity site accessible on the cytoplasmic side of the disc membrane (inward face). *N*-ret-PE dissociates from ABCA4 and ATP is hydrolyzed to return ABCA4 to its initial state.

functional significance of rim localization remains to be determined; however, sequence analysis and recent structural studies indicate that the tightly packed ECDs are too large to fit within the lumen space of the flattened regions of disc membranes, but can be accommodated within the hairpin rim region.[45,46]

4.2 *N*-Ret-PE as the Substrate for ABCA4

The first indication that ABCA4 may function as a retinoid transporter arose from biochemical characterization of ABCA4 and analysis of *Abca4* knockout mice. In the former study, ABCA4 purified by immunoaffinity chromatography and reconstituted into liposomes displayed ATPase activity that was activated two- to five-fold by all-*trans* and 11-*cis* retinal, but not all-*trans* retinol or retinylesters.[52] The basal- and retinal-activated ATPase activity required PE.[53] Since retinal is known to reversibly react with PE, activation of the ATPase activity by retinal could be due to free retinal or *N*-ret-PE, the Schiff base adduct of retinal and PE. Previous studies of other ABC transporters including P-glycoprotein showed that compounds that activated its ATPase activity also served as transport substrates.[54] From these studies, it was concluded that ABCA4 may function as a retinoid transporter.[52] In another approach, *Abca4* knockout mice were generated and shown to display a light-dependent elevation in all-*trans* retinal, protonated *N*-ret-PE, and PE compared to age-matched WT mice.[55] In addition, these mice were reported to accumulate lipofuscin deposits and the bisretinoid pyridinium compound A2E in their RPE cells in a light-dependent manner. This led to the suggestion that ABCA4 may function in the transport of *N*-ret-PE across disc membranes and deficiency in *N*-ret-PE transport by ABCA4 results in the accumulation of retinoids and PE in the photoreceptors.[55]

Subsequent biochemical studies confirmed the direct interaction of the all-*trans* isomer of *N*-ret-PE with ABCA4.[56] When all-*trans* retinal was added to ABCA4 in the presence of PE, ABCA4 bound all-*trans*-*N*-ret-PE with high affinity. Addition of ATP resulted in the release of bound all-*trans*-*N*-ret-PE. These studies are consistent with high-affinity binding of *N*-ret-PE to ABCA4 in the absence of ATP. The addition of ATP results in a conformational change in the NBDs and TMDs leading to the loss in high-affinity substrate binding and dissociation from ABCA4, a sequence of steps that likely comprise part of the substrate transport cycle as illustrated in Fig. 1B.[45]

These studies taken together led to the conclusion that ABCA4 plays a crucial role in the removal of retinoids from disc membranes following photoexcitation through its ability to actively transport all-*trans*-*N*-ret-PE from the lumen to the cytoplasmic side of disc membranes.

4.3 ABCA4 Functions as an *N*-Ret-PE and PE Importer

Although the studies described above provide indirect support for the all-*trans* isomer of *N*-ret-PE (all-*trans*-*N*-ret-PE) as the physiological substrate transported by ABCA4, direct substrate transport or flipping was not measured in the studies described above and the direction of transport across disc membranes was not defined. The latter issue has been a particular concern since all previously characterized mammalian ABC transporters and most eukaryotic ABC transporters have been shown to be exporters transporting their substrates from the cytoplasmic to the exocytoplasmic side of cells, a direction that is opposite of that proposed for ABCA4.

To address these issues, an assay that measures ATP-dependent transport was recently developed.[41] In this assay, ATP-dependent transfer of radiolabeled all-*trans* retinal from proteoliposome containing ABCA4 to liposomes was measured. This process requires ATP as an energy source for ABCA4 to flip all-*trans*-*N*-ret-PE from the inner leaflet of the proteoliposomes (equivalent to the lumen side of disc membranes) to the outer leaflet (equivalent to the cytoplasmic side of disc membranes). ATP-dependent all-*trans*-*N*-ret-PE transport activity of ABCA4 was also measured in disc membranes by replacing the proteoliposomes with isolated disc membranes containing endogenous ABCA4.[41] These studies provided the first direct evidence for the ABCA4-mediated transport of *N*-ret-PE across membranes and importantly showed that the direction of transport was indeed that of an importer.

On the basis of these studies, a reaction cycle based on the alternating access model has been proposed (Fig. 1B). In this model, ABCA4 binds its substrate all-*trans*-*N*-ret-PE with high affinity on the lumen side of the disc membrane (outward face). The binding of ATP to one or possibly both NBDs induces a conformational change which brings the two NBDs in close proximity to each other. This dimerization is coupled to a conformational change in the TMDs of ABCA4 resulting in a low-affinity substrate binding site exposed on the cytoplasmic side of the disc membrane (inward face). All-*trans*-*N*-ret-PE is then released from ABCA4 on the cytoplasmic side of the disc membrane and ATP hydrolysis returns ABCA4 to its initial conformation enabling it to bind another substrate on the lumen side to continue the reaction cycle.[45]

In related studies, ABCA4 was also shown to actively flip PE from the exocytoplasmic (lumen) to the cytoplasmic leaflet in the absence of retinal.[40,41] The basal ATPase activity of ABCA4 observed after purification and reconstitution in PE-containing liposomes supplies the energy needed for this transport function. The import direction of transport or flipping is in contrast to the export direction of phospholipid flipping observed for purified and reconstituted ABCA1 and ABCA7.[40] The physiological relevance of PE transport by ABCA4 is unclear as disc membranes appear to have a symmetrical distribution of PE and other phospholipids due to the phospholipid scramblase activity of rhodopsin in disc membranes.[57–59]

5. ROLE OF ABCA4 IN THE VISUAL CYCLE

The visual cycle also known as the retinoid cycle is a series of enzyme-catalyzed reactions which converts all-*trans* retinal derived from the photobleaching of rhodopsin and cone opsin back to 11-*cis* retinal for the regeneration of photopigments (Fig. 2).[60–62] In the classical rod visual cycle, all-*trans* retinal released from rhodopsin is reduced to all-*trans* retinol by retinol dehydrogenases (RDHs), nominally RDH8 present in photoreceptor disc membranes. All-*trans* retinol is then shuttled to RPE cells through its interaction with the interphotoreceptor retinoid binding protein (IRBP). In RPE cells, all-*trans* retinol is esterified by lecithin: retinol acyltransferase (LRAT). Retinylesters can be stored in a cellular compartment known as retinosomes or directly isomerized to 11-*cis* retinol by RPE65 isomerohydrolase. Subsequently, 11-*cis* retinol is oxidized to 11-*cis* retinal by 11-*cis*-RDHs and returned to photoreceptors with the aid of IRBP for regeneration of rhodopsin. Cone photoreceptors use a similar mechanism but also rely on another cone visual cycle which utilizes a series of enzymes present in Mueller cells.[62]

The classical visual cycle has been modified to include the proposed role of ABCA4 in the complete removal of all-*trans* retinal from disc membranes following photoexcitation (Fig. 2). In this scheme, a significant fraction of all-*trans* retinal released from rhodopsin or cone opsin following photoexcitation is directly reduced to all-*trans* retinol by RDH8 as proposed in the classical visual cycle. However, up to 24% of all-*trans* retinal reversibly reacts with PE in disc membranes to form all-*trans*-*N*-ret-PE, a portion of which is trapped on the lumen leaflet of disc membranes.[55] ABCA4 actively

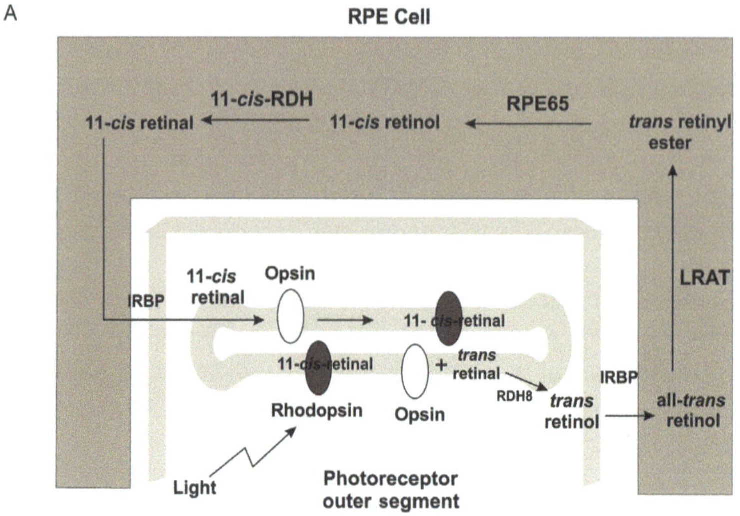

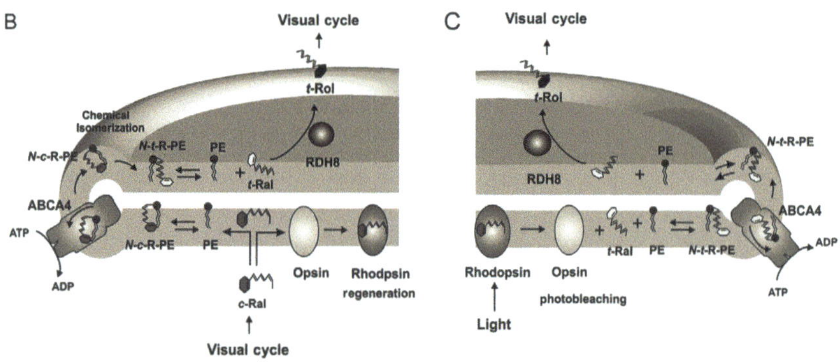

Figure 2 The visual cycle and ABCA4. (A) Series of reactions in which all-*trans* retinal generated from the bleaching of rhodopsin is converted to 11-*cis* retinal for the regeneration of rhodopsin. In the simple scheme, all-*trans* retinal is reduced to all-*trans* retinol by retinol dehydrogenases (RDH8). All-*trans* retinol is then shuttled to RPE cells through interphotoreceptor retinoid binding protein (IRBP) and esterified to *trans* retinylester by lecithin:retinol acyltransferase (LRAT). RPE65 isomerohydrolase catalyzes the isomerization to 11-*cis* retinol and subsequently oxidized by 11-*cis* retinol dehydrogenases to 11-*cis* retinal which is then shuttled back to photoreceptors via IRBP to regenerate rhodopsin. (B) The role of ABCA4 in the removal of excess 11-*cis* retinal from disc membranes as part of the visual cycle. 11-*cis* Retinal entering outer segments from the visual cycle preferentially combines with opsin to regenerate rhodopsin. Excess 11-*cis* retinal reacts reversibly with phosphatidylethanolamine (PE) to form *N*-*cis*-retinylidene-PE (*N*-*c*-R-PE), a portion of which is trapped on the lumen side of the disc membrane. ABCA4 transports *N*-*c*-R-PE to the cytoplasmic side of the disc membrane

transports or flips all-*trans*-*N*-ret-PE across the disc membrane.[41] All-*trans* retinal produced upon hydrolysis of all-*trans*-ret-PE is then reduced to all-*trans* retinol by RDH8 as part of the visual cycle.

Recent studies by Boyer *et al.*,[63] however, have indicated that ABCA4 may play a more important role in the removal of the excess 11-*cis* retinal that is not needed for the regeneration of rhodopsin or cone opsins. They showed that *Abca4* knockout mice generate substantially more lipofuscin and A2E than age-matched wild-type mice, but in contrast to the earlier studies of Travis and colleagues,[55] *Abca4* knockout mice reared in the dark were found to generate as much lipofuscin and A2E as knockout mice reared under cyclic lighting conditions.[63] These and other studies led to the premise that in the absence of ABCA4, 11-*cis* retinal entering photoreceptor disc membranes accumulates in disc membranes where it can form potentially toxic bisretinoid compounds independent of photoexcitation. The mechanisms for how ABCA4 may function in the removal of excess 11-*cis* retinal, however, were not resolved in these studies.

To address the role of ABCA4 in the removal of excess 11-*cis* retinal from disc membranes, we assessed the ability of ABCA4 to transport 11-*cis*-*N*-ret-PE across membranes. PE-containing proteoliposomes reconstituted with ABCA4 were shown to transport 11-*cis*-*N*-ret-PE in an ATP-dependent reaction to the same extent as all-*trans* *N*-ret-PE and in the import direction.[64] Chemical isomerization of 11-*cis*-*N*-ret-PE to all-*trans* *N*-ret-PE before or after ABCA4-mediated flipping of *N*-ret-PE together with its hydrolysis to all-*trans* retinal and PE enables RDH8 to reduce all-*trans* retinal to all-*trans* retinol for entry back into the visual cycle.[64] The chemical isomerization step is important since previous studies have shown that RDH8 reduces all-*trans* retinal, but not 11-*cis* retinal.[65] The chemical isomerization reaction requires PE and the rate of isomerization is

where it undergoes chemical isomerization to *N*-*trans*-retinylidene-PE (N-*t*-R-PE). Hydrolysis yields all-*trans* retinal (*t*-Ral) which can be reduced to all-*trans* retinol (*t*-Rol) by retinol dehydrogenase 8 (RDH8) as part of the visual cycle. Chemical hydrolysis can also occur on the lumen side. In this case, ABCA4 transports N-*t*-R-PE to the cytoplasmic side of the disc membrane. (C) The role of ABCA4 in the removal of all-*trans* retinal following photoexcitation. All-*trans* retinal (*t*-Ral) from photobleaching of rhodopsin can be directly reduced to all-*trans* retinol (*t*-Rol) for entry into the visual cycle (not shown). A fraction of *t*-Ral reacts with PE to form *N*-*t*-R-PE, some of which is trapped on the lumen side of the disc membrane. ABCA4 transports *N*-*t*-R-PE to the cytoplasmic side of the disc membrane where upon hydrolysis *t*-Ral can be reduced to *t*-Rol by RDH8 as part of the visual cycle.

accelerated for PE containing a docosahexaenoic acid (DHA) fatty acyl chain.[64,66] This explains in part the high PE and DHA content of photoreceptor outer segment disc membranes.[67]

These studies provide a general mechanism for the role of ABCA4 in the visual cycle under dark as well as light conditions.[64] Briefly, 11-*cis* retinal enters the photoreceptor disc membranes from the visual cycle in the dark as well as in the light and combines with opsin to regenerate rhodopsin and cone opsin in rod and cone outer segments, respectively (Fig. 2). Excess 11-*cis* retinal not required for the photopigment regeneration reacts reversibly with PE to form 11-*cis*-*N*-ret-PE, a portion of which becomes trapped in the lumen leaflet of disc membranes. ABCA4 effectively transports 11-*cis*-*N*-ret-PE to the cytoplasmic surface of disc membranes where it can undergo chemical isomerization to all-*trans*-*N*-Ret-PE. Alternatively, chemical isomerization of 11-*cis*-ret-PE to all-*trans*-ret-PE can occur in the lumen leaflet. In this case, ABCA4 flips all-*trans*-*N*-ret-PE to the cytoplasmic leaflet of disc membranes. In either case, all-*trans*-retinal generated through hydrolysis of all-*trans*-*N*-ret-PE can be reduced to all-*trans* retinol by RDH8 for entry into the visual cycle. ABCA4 can also function in the transport of all-*trans*-*N*-ret-PE after photoexcitation as discussed above (Fig. 2).

Retinal is known to diffuse between membranes.[68,69] Some 11-*cis* and all-*trans* retinal in disc membranes can diffuse out of the photoreceptor outer segments. RDHs present in other cells and photoreceptor subcellular organelles including RDH12 in the inner segments protect these cellular compartments against the toxic effects of retinal via reduction to retinol.

6. LOSS OF ABCA4 FUNCTION AND STARGARDT DISEASE

Biochemical studies have shown that disease-causing mutations in ABCA4 cause reduced or complete loss in retinal-activated ATPase activity and/or *N*-ret-PE flipping activity.[40,41,70,71] Disease-causing mutations can also cause ABCA4 mislocalization in photoreceptor cells and transfected culture cells due to protein misfolding.[40,71,72] The loss in retinoid transport activity and/or mislocalization of ABCA4 can prevent the efficient removal of retinal from disc membranes resulting in the accumulation retinoids in photoreceptor outer segment discs. If retinal and *N*-ret-PE are not efficiently removed from photoreceptor outer segments, they can react non-enzymatically to form dihydropyridinium compounds, which subsequently undergoes autoxidation to form the phosphatidyl-pyridinium bisretinoid

A2PE in photoreceptor membranes.[73–75] Upon phagocytosis, A2PE is hydrolyzed by phospholipase D to form A2E which accumulates as a component of lipofuscin in RPE cells. A2E and other bisretinal compounds can undergo photooxidation to generate products which can cause cell death.[76,77] It remains to be determined which bisretinoid compounds are responsible for RPE cell death.

7. CONCLUSIONS

Significant progress has been made in understanding the role of ABCA4 in photoreceptor cells and the molecular basis for Stargardt disease and related retinal degenerative diseases arising from loss in function mutations in the ABCA4. Retinal transport assays indicate that ABCA4 functions as an importer flipping all-*trans* and 11-*cis* isomers of *N*-ret-PE across disc membranes. This is the first known mammalian ABC transporter that transports its substrate in the "inward" or import direction. This function together with chemical isomerization and enzymatic reduction of retinal to retinol facilitates the complete removal of excess 11-*cis* and all-*trans* retinal from discs preventing the buildup of toxic bisretinoid compounds. Disease-associated mutations which cause a loss in function and/or mislocalization of ABCA4 lead to a progressive accumulation of retinal and its bisretinoid products, a hallmark of Stargardt disease. Understanding the molecular basis of Stargardt disease provides a foundation in which to develop cell-, gene-, or drug-based therapies for Stargardt disease.

ACKNOWLEDGMENTS

This work was supported by grants from NIH (EY002422), CIHR (MOP-106667), and the Macula Vision Research Foundation. Thanks is extended to the past and current members of the Molday lab involved in the molecular and cellular characterization of ABCA4 and its role in the visual cycle and Stargardt disease.

REFERENCES

1. Stargardt K. Uber familiare, progressive degeenration under makulagegend des augen. *Albrecht Von Graefes Arch Ophthalmol.* 1909;71:534–550.
2. Gelisken O, De Laey JJ. A clinical review of Stargardt's disease and/or fundus flavimaculatus with follow-up. *Int Ophthalmol.* 1985;8(4):225–235.
3. Webster AR, Heon E, Lotery AJ, et al. An analysis of allelic variation in the ABCA4 gene. *Invest Ophthalmol Vis Sci.* 2001;42(6):1179–1189.
4. Weleber RG. Stargardt's macular dystrophy. *Arch Ophthalmol.* 1994;112(6):752–754.
5. Fishman GA, Farber M, Patel BS, Derlacki DJ. Visual acuity loss in patients with Stargardt's macular dystrophy. *Ophthalmology.* 1987;94(7):809–814.

6. Rivera A, White K, Stohr H, et al. A comprehensive survey of sequence variation in the ABCA4 (ABCR) gene in Stargardt disease and age-related macular degeneration. *Am J Hum Genet.* 2000;67(4):800–813.
7. Rotenstreich Y, Fishman GA, Anderson RJ. Visual acuity loss and clinical observations in a large series of patients with Stargardt disease. *Ophthalmology.* 2003;110(6): 1151–1158.
8. Fishman GA, Stone EM, Grover S, Derlacki DJ, Haines HL, Hockey RR. Variation of clinical expression in patients with Stargardt dystrophy and sequence variations in the ABCR gene. *Arch Ophthalmol.* 1999;117(4):504–510.
9. Mantyjarvi M, Tuppurainen K. Color vision in Stargardt's disease. *Int Ophthalmol.* 1992;16(6):423–428.
10. Fishman GA, Farbman JS, Alexander KR. Delayed rod dark adaptation in patients with Stargardt's disease. *Ophthalmology.* 1991;98(6):957–962.
11. Lois N, Holder GE, Bunce C, Fitzke FW, Bird AC. Phenotypic subtypes of Stargardt macular dystrophy-fundus flavimaculatus. *Arch Ophthalmol.* 2001;119(3):359–369.
12. Allikmets R, Singh N, Sun H, et al. A photoreceptor cell-specific ATP-binding transporter gene (ABCR) is mutated in recessive Stargardt macular dystrophy [see comments]. *Nat Genet.* 1997;15(3):236–246.
13. Illing M, Molday LL, Molday RS. The 220-kDa rim protein of retinal rod outer segments is a member of the ABC transporter superfamily. *J Biol Chem.* 1997;272(15):10303–10310.
14. Papermaster DS, Schneider BG, Zorn MA, Kraehenbuhl JP. Immunocytochemical localization of a large intrinsic membrane protein to the incisures and margins of frog rod outer segment disks. *J Cell Biol.* 1978;78(2):415–425.
15. Allikmets R. Simple and complex ABCR: genetic predisposition to retinal disease. *Am J Hum Genet.* 2000;67(4):793–799.
16. Martinez-Mir A, Paloma E, Allikmets R, et al. Retinitis pigmentosa caused by a homozygous mutation in the Stargardt disease gene ABCR. *Nat Genet.* 1998;18(1):11–12.
17. Nasonkin I, Illing M, Koehler MR, Schmid M, Molday RS, Weber BH. Mapping of the rod photoreceptor ABC transporter (ABCR) to 1p21-p22.1 and identification of novel mutations in Stargardt's disease. *Hum Genet.* 1998;102(1):21–26.
18. Stone EM, Webster AR, Vandenburgh K, et al. Allelic variation in ABCR associated with Stargardt disease but not age-related macular degeneration. *Nat Genet.* 1998;20(4):328–329.
19. Zernant J, Schubert C, Im KM, et al. Analysis of the ABCA4 gene by next-generation sequencing. *Invest Ophthalmol Vis Sci.* 2014;52(11):8479–8487.
20. Allikmets R, Shroyer NF, Singh N, et al. Mutation of the Stargardt disease gene (ABCR) in age-related macular degeneration. *Science.* 1997;277(5333):1805–1807.
21. Maugeri A, Klevering BJ, Rohrschneider K, et al. Mutations in the ABCA4 (ABCR) gene are the major cause of autosomal recessive cone-Rod dystrophy. *Am J Hum Genet.* 2000;67(4):960–966.
22. Higgins CF. ABC transporters: from microorganisms to man. *Annu Rev Cell Biol.* 1992;8:67–113.
23. Rees DC, Johnson E, Lewinson O. ABC transporters: the power to change. *Nat Rev Mol Cell Biol.* 2009;10(3):218–227.
24. Dean M, Allikmets R. Complete characterization of the human ABC gene family. *J Bioenerg Biomembr.* 2001;33(6):475–479.
25. Dean M, Annilo T. Evolution of the ATP-binding cassette (ABC) transporter superfamily in vertebrates. *Annu Rev Genomics Hum Genet.* 2005;6:123–142.
26. Bungert S, Molday LL, Molday RS. Membrane topology of the ATP binding cassette transporter ABCR and its relationship to ABC1 and related ABCA transporters: identification of N-linked glycosylation sites. *J Biol Chem.* 2001;276(26):23539–23546.

27. Kaminski WE, Piehler A, Wenzel JJ. ABC A-subfamily transporters: structure, function and disease. *Biochim Biophys Acta.* 2006;1762(5):510–524.
28. Bodzioch M, Orso E, Klucken J, et al. The gene encoding ATP-binding cassette transporter 1 is mutated in Tangier disease. *Nat Genet.* 1999;22(4):347–351.
29. Brooks-Wilson A, Marcil M, Clee SM, et al. Mutations in ABC1 in Tangier disease and familial high-density lipoprotein deficiency. *Nat Genet.* 1999;22(4):336–345.
30. Rust S, Rosier M, Funke H, et al. Tangier disease is caused by mutations in the gene encoding ATP-binding cassette transporter 1. *Nat Genet.* 1999;22(4):352–355.
31. Ban N, Matsumura Y, Sakai H, et al. ABCA3 as a lipid transporter in pulmonary surfactant biogenesis. *J Biol Chem.* 2007;282(13):9628–9634.
32. Shulenin S, Nogee LM, Annilo T, Wert SE, Whitsett JA, Dean M. ABCA3 gene mutations in newborns with fatal surfactant deficiency. *N Engl J Med.* 2004;350(13):1296–1303.
33. Akiyama M, Sugiyama-Nakagiri Y, Sakai K, et al. Mutations in lipid transporter ABCA12 in harlequin ichthyosis and functional recovery by corrective gene transfer. *J Clin Invest.* 2005;115(7):1777–1784.
34. Lefevre C, Audebert S, Jobard F, et al. Mutations in the transporter ABCA12 are associated with lamellar ichthyosis type 2. *Hum Mol Genet.* 2003;12(18):2369–2378.
35. DeStefano GM, Kurban M, Anyane-Yeboa K, et al. Mutations in the cholesterol transporter gene ABCA5 are associated with excessive hair overgrowth. *PLoS Genet.* 2014;10(5):e1004333.
36. Fu Y, Hsiao JH, Paxinos G, Halliday GM, Kim WS. ABCA5 regulates amyloid-beta peptide production and is associated with Alzheimer's disease neuropathology. *J Alzheimers Dis.* 2015;43(3):857–869.
37. Reitz C, Jun G, Naj A, et al. Variants in the ATP-binding cassette transporter (ABCA7), apolipoprotein E 4, and the risk of late-onset Alzheimer disease in African Americans. *JAMA.* 2013;309(14):1483–1492.
38. Coleman JA, Quazi F, Molday RS. Mammalian P4-ATPases and ABC transporters and their role in phospholipid transport. *Biochim Biophys Acta.* 2013;1831(3):555–574.
39. Phillips MC. Molecular mechanisms of cellular cholesterol efflux. *J Biol Chem.* 2014;289(35):24020–24029.
40. Quazi F, Molday RS. Differential phospholipid substrates and directional transport by ATP-binding cassette proteins ABCA1, ABCA7, and ABCA4 and disease-causing mutants. *J Biol Chem.* 2013;288(48):34414–34426.
41. Quazi F, Lenevich S, Molday RS. ABCA4 is an N-retinylidene-phosphatidylethanolamine and phosphatidylethanolamine importer. *Nat Commun.* 2012;3:925.
42. Azarian SM, Megarity CF, Weng J, Horvath DH, Travis GH. The human photoreceptor rim protein gene (ABCR): genomic structure and primer set information for mutation analysis. *Hum Genet.* 1998;102(6):699–705.
43. Yatsenko AN, Wiszniewski W, Zaremba CM, Jamrich M, Lupski JR. Evolution of ABCA4 proteins in vertebrates. *J Mol Evol.* 2005;60(1):72–80.
44. Tsybovsky Y, Wang B, Quazi F, Molday RS, Palczewski K. Posttranslational modifications of the photoreceptor-specific ABC transporter ABCA4. *Biochemistry.* 2011;50(32):6855–6866.
45. Molday RS, Zhong M, Quazi F. The role of the photoreceptor ABC transporter ABCA4 in lipid transport and Stargardt macular degeneration. *Biochim Biophys Acta.* 2009;1791(7):573–583.
46. Tsybovsky Y, Orban T, Molday RS, Taylor D, Palczewski K. Molecular organization and ATP-induced conformational changes of ABCA4, the photoreceptor-specific ABC transporter. *Structure.* 2013;21(5):854–860.

47. Aller SG, Yu J, Ward A, et al. Structure of P-glycoprotein reveals a molecular basis for poly-specific drug binding. *Science*. 2009;323(5922):1718–1722.
48. Hollenstein K, Frei DC, Locher KP. Structure of an ABC transporter in complex with its binding protein. *Nature*. 2007;446(7132):213–216.
49. Molday LL, Rabin AR, Molday RS. ABCR expression in foveal cone photoreceptors and its role in Stargardt macular dystrophy. *Nat Genet*. 2000;25(3):257–258.
50. Sun H, Nathans J. Stargardt's ABCR is localized to the disc membrane of retinal rod outer segments [letter]. *Nat Genet*. 1997;17(1):15–16.
51. Papermaster DS, Reilly P, Schneider BG. Cone lamellae and red and green rod outer segment disks contain a large intrinsic membrane protein on their margins: an ultrastructural immunocytochemical study of frog retinas. *Vision Res*. 1982;22(12):1417–1428.
52. Sun H, Molday RS, Nathans J. Retinal stimulates ATP hydrolysis by purified and reconstituted ABCR, the photoreceptor-specific ATP-binding cassette transporter responsible for Stargardt disease. *J Biol Chem*. 1999;274(12):8269–8281.
53. Ahn J, Wong JT, Molday RS. The effect of lipid environment and retinoids on the ATPase activity of ABCR, the photoreceptor ABC transporter responsible for Stargardt macular dystrophy. *J Biol Chem*. 2000;275(27):20399–20405.
54. Shapiro AB, Ling V. ATPase activity of purified and reconstituted P-glycoprotein from Chinese hamster ovary cells. *J Biol Chem*. 1994;269(5):3745–3754.
55. Weng J, Mata NL, Azarian SM, Tzekov RT, Birch DG, Travis GH. Insights into the function of rim protein in photoreceptors and etiology of Stargardt's Disease from the phenotype in *abcr* knockout mice. *Cell*. 1999;98:13–23.
56. Beharry S, Zhong M, Molday RS. N-retinylidene-phosphatidylethanolamine is the preferred retinoid substrate for the photoreceptor-specific ABC transporter ABCA4 (ABCR). *J Biol Chem*. 2004;279(52):53972–53979.
57. Hessel E, Muller P, Herrmann A, Hofmann KP. Light-induced reorganization of phospholipids in rod disc membranes. *J Biol Chem*. 2001;276(4):2538–2543.
58. Menon I, Huber T, Sanyal S, et al. Opsin is a phospholipid flippase. *Curr Biol*. 2011;21(2):149–153.
59. Wu G, Hubbell WL. Phospholipid asymmetry and transmembrane diffusion in photoreceptor disc membranes. *Biochemistry*. 1993;32(3):879–888.
60. Kiser PD, Golczak M, Palczewski K. Chemistry of the retinoid (visual) cycle. *Chem Rev*. 2013;114(1):194–232.
61. Saari JC. Vitamin A metabolism in rod and cone visual cycles. *Annu Rev Nutr*. 2012;32:125–145.
62. Travis GH, Golczak M, Moise AR, Palczewski K. Diseases caused by defects in the visual cycle: retinoids as potential therapeutic agents. *Annu Rev Pharmacol Toxicol*. 2007;47:469–512.
63. Boyer NP, Higbee D, Currin MB, et al. Lipofuscin and N-retinylidene-N-retinylethanolamine (A2E) accumulate in retinal pigment epithelium in absence of light exposure: their origin is 11-cis-retinal. *J Biol Chem*. 2012;287(26):22276–22286.
64. Quazi F, Molday RS. ATP-binding cassette transporter ABCA4 and chemical isomerization protect photoreceptor cells from the toxic accumulation of excess 11-cis-retinal. *Proc Natl Acad Sci USA*. 2014;111(13):5024–5029.
65. Palczewski K, Jager S, Buczylko J, et al. Rod outer segment retinol dehydrogenase: substrate specificity and role in phototransduction. *Biochemistry*. 1994;33(46):13741–13750.
66. Groenendijk GW, Jacobs CW, Bonting SL, Daemen FJ. Dark isomerization of retinals in the presence of phosphatidylethanolamine. *Eur J Biochem*. 1980;106(1):119–128.
67. Fliesler SJ, Anderson RE. Chemistry and metabolism of lipids in the vertebrate retina. *Prog Lipid Res*. 1983;22(2):79–131.
68. Noy N, Xu ZJ. Kinetic parameters of the interactions of retinol with lipid bilayers. *Biochemistry*. 1990;29(16):3883–3888.

69. Rando RR, Bangerter FW. The rapid intermembraneous transfer of retinoids. *Biochem Biophys Res Commun*. 1982;104(2):430–436.
70. Sun H, Smallwood PM, Nathans J. Biochemical defects in ABCR protein variants associated with human retinopathies. *Nat Genet*. 2000;26(2):242–246.
71. Zhong M, Molday LL, Molday RS. Role of the C terminus of the photoreceptor ABCA4 transporter in protein folding, function, and retinal degenerative diseases. *J Biol Chem*. 2009;284(6):3640–3649.
72. Wiszniewski W, Zaremba CM, Yatsenko AN, et al. ABCA4 mutations causing mislocalization are found frequently in patients with severe retinal dystrophies. *Hum Mol Genet*. 2005;14(19):2769–2778.
73. Ben-Shabat S, Parish CA, Vollmer HR, et al. Biosynthetic studies of A2E, a major fluorophore of retinal pigment epithelial lipofuscin. *J Biol Chem*. 2002;277(9):7183–7190.
74. Mata NL, Weng J, Travis GH. Biosynthesis of a major lipofuscin fluorophore in mice and humans with ABCR-mediated retinal and macular degeneration. *Proc Natl Acad Sci USA*. 2000;97(13):7154–7159.
75. Sparrow JR, Boulton M. RPE lipofuscin and its role in retinal pathobiology. *Exp Eye Res*. 2005;80(5):595–606.
76. Radu RA, Mata NL, Bagla A, Travis GH. Light exposure stimulates formation of A2E oxiranes in a mouse model of Stargardt's macular degeneration. *Proc Natl Acad Sci USA*. 2004;101(16):5928–5933.
77. Sparrow JR, Nakanishi K, Parish CA. The lipofuscin fluorophore A2E mediates blue light-induced damage to retinal pigmented epithelial cells. *Invest Ophthalmol Vis Sci*. 2000;41(7):1981–1989.

CHAPTER TWENTY-FIVE

A History of the Classical Visual Cycle

Charles B. Wright*[,1], T. Michael Redmond[†], John M. Nickerson[‡]
*Department of Ophthalmology and Visual Sciences, University of Kentucky, Lexington, Kentucky, USA
[†]Laboratory of Retinal Cell and Molecular Biology, National Eye Institute, NIH, Bethesda, Maryland, USA
[‡]Department of Ophthalmology, Emory University School of Medicine, Atlanta, Georgia, USA
[1]Corresponding author: e-mail address: wright.charles@uky.edu

Contents

Abstract

The visual cycle, the biochemical process by which the light-sensitive isomer of vitamin A is continually recycled, is crucial to vision in a healthy eye. More than 150 years of research into this remarkable biochemical process has given invaluable understanding in debilitating visual diseases that impact thousands of individuals worldwide, many of them children. The visual cycle spans photoreceptor cells in the retina and the underlying retinal pigment epithelium (RPE) and requires a protein called RPE65 for its function. In many ways, RPE65 is the capstone to the cyclical processing of vitamin A in the eye, and the discovery of this retinol isomerase helped fill a critical gap in the understanding of retinoid processing in vision. This chapter will focus on the history of visual cycle research, from the first experiments well over a century ago to the discovery of RPE65. Because of the undeniable importance of RPE65 in the visual cycle, this chapter will also focus on the protein structure and mechanism by which it converts light-insensitive all-*trans*-vitamin A to light-sensitive 11-*cis*-vitamin A for continued visual function. Finally, this chapter will briefly discuss RPE65 and its known disease associations in the clinical setting. Thanks to the efforts of researchers for well over a century in studying the visual cycle, the medical community is now poised to make significant gains in the treatment of blindness.

Progress in Molecular Biology and Translational Science, Volume 134
ISSN 1877-1173
http://dx.doi.org/10.1016/bs.pmbts.2015.06.009
© 2015 Elsevier Inc.
All rights reserved.

1. INTRODUCTION

The modern understanding of the rod visual cycle and its components is the result of more than 137 years of research. The recognition of RPE65 and its role in the visual cycle is a relatively recent addition to the field's knowledge of proteins required for visual function. RPE65 has only been recently identified, and even then, its significance in the visual cycle was not fully appreciated for almost another 10 years. Since then, the field's understanding and knowledge of RPE65 and its various structural and functional characteristics has greatly increased to the point where RPE65 and Leber congenital amaurosis, type 2 (LCA2) are partially understood.

2. HISTORY OF VISUAL CYCLE RESEARCH

The investigations of the visual cycle occurred in 1876 with the work of Boll, who observed and described a light-sensitive part of the eye that changed color when exposed to light and turned back into its original color when left in the dark. This became known as Boll's phenomenon.[1,2] In 1878, Kühne reported that visual purple (dark gray in the print version) (now known as light-sensitive rhodopsin which absorbs light in the green-yellow (light gray in the print version) portion of the visual spectrum),[3] when dissected in the dark and exposed to light, was bleached to form visual yellow (light gray in the print version) (now known to be bleached free opsin without chromophore).[3] Visual yellow (light gray in the print version) could again become the light-sensitive visual purple (dark gray in the print version) if the retina was left on top of the retinal epithelium (now called the retinal pigment epithelium (RPE)); this recovery of visual purple (dark gray in the print version), however, would not occur if the retina was dissected away from the RPE[4] (Fig. 1). The recovery of visual purple (dark gray in the print version) in the presence of the RPE[4] would form the foundation of the modern understanding of the visual cycle and the critical role the RPE plays in its maintenance.

Despite the early contributions by Boll and Kühne in the late nineteenth century, the advance of the modern molecular understanding of the visual cycle would not occur until 1935. When Wald exposed the visual purple (dark gray in the print version) to light, it allowed vitamin A (which he referred to as retinene) to be freely released from the retina; likewise, when visual yellow (light gray in the print version) was restored to visual purple

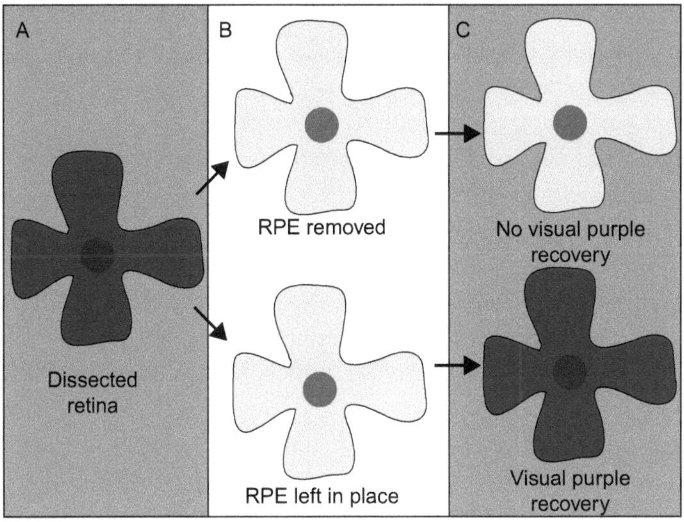

Figure 1 The retina only recovers sensitivity to light if the RPE is left attached. The initial importance of the RPE to visual sensitivity and the visual cycle was first described by Boll and Kühne in the nineteenth century. (A) If the retina, RPE, and choroid are dissected in the dark, the retina will appear a reddish-purple color (dark gray in the print version) because of visual purple (dark gray in the print version) (i.e., rhodopsin). (B) If exposed to light, visual purple (dark gray in the print version) will bleach to form visual yellow (light gray in the print version) (i.e., bleached rhodopsin). Visual yellow will form even if the RPE is not attached. (C) If the RPE is dissected away from the retina, though, visual purple (dark gray in the print version) will not reappear if the retina is placed back in the dark. The RPE is required for recovery of visual sensitivity to light.

(dark gray in the print version), retinene was lost.[5] Wald concluded that vitamin A was a prosthetic group on visual purple (dark gray in the print version), and that its continuous replacement was necessary for continued vision.[5] In 1937, Wald used the concept of the visual cycle to explain dark and light adaptation.[6] By 1953, Wald established the chief importance of the visual cycle to be the conversion of all-*trans*-vitamin A (retinal) to 11-*cis*-vitamin A for the regeneration of visual pigment.[7] Wald's recognition in 1953 of the importance of regeneration of the 11-*cis* double bond for the light sensitivity of visual pigment resulted in the molecular description of bleaching 10 years later.[8] By 1963, Wald's laboratory had discovered 11-*cis*-retinal (RAL) was bound to opsin by a Schiff base to form light-sensitive rhodopsin; after exposure to light, all-*trans*-RAL was discovered to dissociate from rhodopsin to leave free opsin apoprotein, and the necessary recycling of the 11-*cis* double bond did not occur directly on the protein.[8] Four years later, the importance and value of Wald's contributions to the molecular understanding of vision were recognized with the 1967 Nobel Prize in Physiology or Medicine.[9] A full understanding of the visual cycle still lacked the roles the RPE fulfilled.

The understanding of the RPE's importance in the visual cycle (Fig. 2) built upon Kühne's and Boll's qualitative observations concerning visual purple (dark gray in the print version) and the retinal epithelium.[4] All-*trans*-RAL could be converted to all-*trans*-ROL by the retinol dehydrogenases RDH8 and/or RDH12[10–12] in the rod OS. Further work on all-*trans*-retinoids indicated all-*trans*-ROL was esterified in the RPE (now known to be performed by lecithin retinol acyltransferase).[13,14] Bernstein and Rando discovered that isomerase activity was localized to RPE membranes, and this isomerase activity was abolished if the RPE was dissociated with detergents. However, the direct substrate for the isomerase reaction was still not understood.[15]

Initially, the importance of vitamin A esterification was not well understood in the context of the visual cycle because 11-*cis*-RAL,[16] not all-*trans*-retinyl ester (RE) or 11-*cis*-RE, could restore functional rhodopsin.[17] By 1990, though, all-*trans*-RE was recognized as being a critical intermediate to the reformation of 11-*cis*-vitamin A in the RPE.[18] Eventually, it was

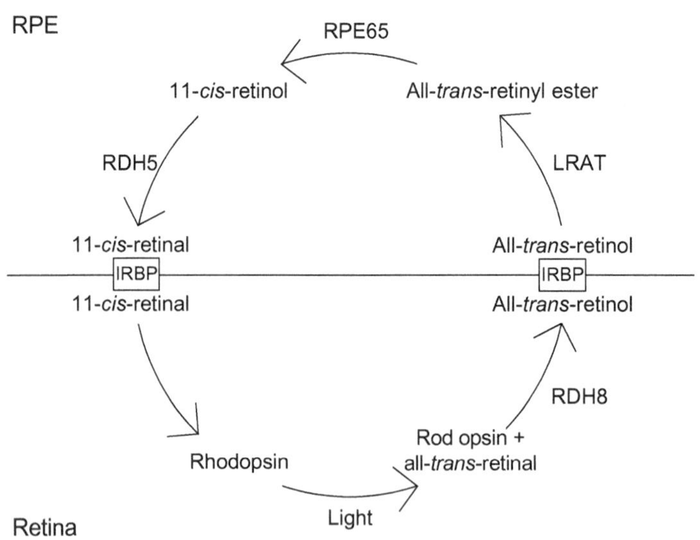

Figure 2 Biochemistry of the classical visual cycle. The retina and RPE contain components necessary for the classical visual cycle, which recycles all-*trans*-retinol to 11-*cis*-retinal to continuously replenish the supply of functional, light-sensitive opsin proteins in the photoreceptors.

widely understood an isomerase-like enzyme performed a one-step conversion of all-*trans*-RE to 11-*cis*-ROL (later identified as RPE65). 11-*cis*-ROL would subsequently be enzymatically converted to 11-*cis*-RAL by RDH5 in the RPE.[19] Wiggert *et al.* in 1977 described retinol receptors in both the outer retina and RPE, suggesting the existence of a protein that was capable of shuttling retinoids in the subretinal space (the subretinal or interphotoreceptor space is the small boundary between the retina and RPE).[20] This protein, interphotoreceptor retinol-binding protein (IRBP), was isolated in 1983.[21] IRBP was found to be localized in the interphotoreceptor space and binds retinoids in the subretinal space.[21,22] At this point, it was clear the isomerase was in the RPE, but there was a gap in understanding between the RE substrate and the 11-*cis*-ROL product.

Before the retinol isomerase was identified, it was known the isomerase activity was localized in the RPE membranes.[15,23] A possible candidate protein for this isomerase-like activity would not emerge until 1991, when a monoclonal antibody (MAb) to chicken RPE cell antigens identified a protein that had a molecular weight of about 63 kDa.[24] Another group using MAb to human RPE membranes identified a protein in bovine and other RPE[25] and cloned it[26] naming it RPE65 because it appeared to be RPE specific. Subsequent work indicated this protein's localization coincided with the expected location of retinoid isomerization.[15,23,24] In 1998, gene deletion of the mouse *Rpe65* gene resulted in a blind mouse model lacking 11-*cis*-RAL in the photoreceptors but with overaccumulation of all-*trans*-RE, the putative isomerase substrate, in the RPE, providing prima facie evidence that RPE65 was the visual cycle isomerase.[27] RPE65 was finally biochemically confirmed as the long-sought retinal isomerase in 2005 when cell culture experiments demonstrated all-*trans*-RE could be directly converted to 11-*cis*-ROL upon introduction of RPE65 gene expression.[28–30]

3. RPE65: STRUCTURE, FUNCTION, AND BIOCHEMICAL MECHANISM

RPE65 was first described in 1991 as a 63-kDa RPE-specific protein[24,31] and was first successfully cloned in 1993 identifying a protein of 533 amino acids in length with a computed molecular weight of 61 kDa, and with an mRNA of approximately 3.2 kb.[26] In this study, it was noted the mRNA was relatively abundant in primary cultured RPE while the protein was not translated. They speculated that the 3′-untranslated region (UTR) of the mRNA might contain elements key to translational regulation,[26] an observation that was later supported by increased protein levels after deletion of the 3′-UTR region in the mRNA.[32] Although the protein was shown to localize to the RPE microsomal membrane,[24,26] the amino acid sequence of the protein suggested RPE65 had no transmembrane domains, suggesting that it was associated with the lipid membrane instead of being embedded within it.[26]

RPE65 was initially reported to be expressed in an RPE-specific manner,[26,33] but more recent evidence suggests it may be expressed at relatively lower levels in a wider variety of cell types. Aberrant *RPE65* expression has been reported in cancer patients with transformed kidney cells[34] and in human embryonic kidney 293 cells,[34,35] but it is important to note *RPE65* expression was undetectable in human kidneys without renal tumors.[34] Furthermore, RPE65 may be expressed in normal cone photoreceptor cells in mammals as well. Although many older reports show amphibians with a cone-dominant retina express RPE65 in cone photoreceptors,[36] RPE65 expression in rod-dominant mammalian retina tissue[37,38] is just now being identified. At this time, RPE65's role is only understood in the context of the RPE. Its exact function in cone photoreceptors and kidney cancers is still being investigated.

As previously mentioned, RPE65's primary structure shows no evidence of any transmembrane domains.[26] RPE65 belongs to a family of carotenoid oxygenase enzymes that also includes the human enzyme beta-carotene 15,15′-monooxygenase 1 (BCMO1; Fig. 3).[39] Because BCMO1 was known to convert carotenoids into vitamin A and had a similar protein sequence (40% identity between BCMO1 and RPE65; Fig. 3), it corroborated the hypothesis that RPE65 was the retinoid isomerase in the visual cycle.[39] As previously mentioned, this hypothesis would later be supported.

The structure of RPE65 has been resolved to a 1.9 Å resolution by X-ray crystallography. In terms of secondary structure, RPE65 consists mostly of beta strands arranged in an antiparallel fashion to form beta sheets (each beta sheet containing at least four beta strands), which make up seven beta-propeller blades, and connecting loops.[40,41] To form the tertiary structure (Fig. 4), the seven beta-propeller blades of RPE65, in addition to an iron cofactor, fold together to form a structure with a mostly hydrophilic exterior.[40,41] One face of RPE65, though, is hydrophobic, and it is this portion of the protein that is thought to associate with the microsomal membrane in the RPE.[41] Upon associating with the microsomal membrane, the protein undergoes a conformational change that is required for enzymatic activity.[42]

Score	Expect	Method	Identities	Positives	Gaps
407 bits (1047)	1e-139	Compositional matrix adjust.	206/519 (40%)	321/519 (61%)	19/519 (3%)

```
RPE65  20   EELSSPLTAHVTGRIPLWLTGSLLRCGPGLFEVGSEPFYHLFDGQALLHKFDFKEGHVTY  79
            +E   P+ A VTG+IP WL G+LLR GPG+  VG   + H FDG ALLH F  ++G V Y
BCMO1  10   KEQLEPVRAKVTGKIPAWLQGTLLRNGPGMHTVGESRYNHWFDGLALLHSFTIRDGEVYY  69

RPE65  80   HRRFIRTDAYVRAMTEKRIVITEFGTCAFPDPCKNIFSRFFSYFRGV--EVTDNALVNVY  137
              +++R+D Y   +   RIV++EFGT A+PDPCKNIFS+ FSY      + TDN L+N+
BCMO1  70   RSKYLRSDTYNTNIEANRIVVSEFGTMAYPDPCKNIFSKAFSYLSHTIPDFTDNCLINIM  129

RPE65  138  PVGEDYYACTETNFITKINPETLETIKQVDLCNYVSVNGATAHPHIENDGTVYNIGNCFG  197
              GED+YA +ETN+I KINP+TLET+++VD   YV+VN AT+HPH +  G V N+G
BCMO1  130  KCGEDFYATSETNYIRKINPQTLETLEKVDYRKYVAVNLATSHPHYDEAGNVLNMGTSIV  189

RPE65  198  KNFSIAYNIVKIPPL----QADKEDPISKSEIVVQFPCSDRFKPSYVHSFGLTPNYIVFV  253
            +     Y I KIP      +   + P   +E+    P      PSY HSFG+T NY++F+
BCMO1  190  EKGKTKYVIFKIPATVPEGKKQGKSPWKHTEVFCSIPSRSLLSPSYYHSFGVTENYVIFL  249

RPE65  254  ETPVKINLFKFLSSWSLWGANYMDCFESNETMGVWLHIADKKRKKYLNNKYRTSPFNLFH  313
            E P ++++ K  +++ +   ++  C   +     ++HI D++ ++ + K+ T    +FH
BCMO1  250  EQPFRLDILKMATAY-IRRMSWASCLAFHREEKTYIHIIDQRTRQPVQTKFYTDAMVVFH  308

RPE65  314  HINTYEDNGFLIVDLCCWKGFEFVYNYLYLANLRENWEEVKKNARKAPQPEVRRYVLPLN  373
            H+N YE++G ++ D+  ++    +Y   YLANL ++++E   N+R    P +RR+ +PL+
BCMO1  309  HVNAYEEDGCIVFDVIAYED-NSLYQLFYLANLNQDFKE---NSRLTSVPTLRRFAVPLH  364

RPE65  374  IDK-ADTGKNLVTLPNTTATAILCSDETIWLEPEVLFSGPRQAFEFPQINYQKYCGKPYT  432
            +DK A+ G NL+ + +TTATA+   D  ++ +PE L+ G     E P++NY  + GK Y
BCMO1  365  VDKNAEVGTNLIKVASTTATALKEEDGQVYCQPEFLYEG----LELPRVNY-AHNGKQYR  419

RPE65  433  YAYGLGLN-HFVPDRLCKLNVKTKETWVWQEPDSYPSEPIFVSHPDALEEDDGVVLSVVV  491
            Y +  G+    +P ++ K ++ TK +  W+E D +P+EP+FV  P A +EDDGV+LS +V
BCMO1  420  YVFATGVQWSPIPTKIIKYDILTKSSLKWREDDCWPAEPLFVPAPGAKDEDDGVILSAIV  479

RPE65  492  SPGAGQKPAYLLILNAKDLSEVARAEVEINIPVTFHGLF  530
            S    QK  +LLIL+AK  +E+ARA V++++ +  HGLF
BCMO1  480  STDP-QKLPFLLILDAKSFTELARASVDVDMHMDLHGLF  517
```

Figure 3 Alignment of human RPE65 and BCMO1 protein sequences. RPE65 shares significant sequence identity (40%) with BCMO1, a protein that cleaves beta-carotene to yield two vitamin A molecules. The significant identity of RPE65 with carotenoid oxygenases combined with observations in KO mice that 11-*cis*-retinoid (isomerase product) levels were not detectable and all-*trans*-RE (retinoid substrate) levels were elevated indicated that RPE65 might be the retinol isomerase in the visual cycle (because the absence of RPE65 produced a block in the visual cycle). The mouse nonsense R44X mutation and missense R91W and F229S mutations are highlighted in red (gray in the print version) on the protein sequence.

From the hydrophobic face, a hydrophobic tunnel leads to the interior of the protein where the iron cofactor-containing active site resides.[40,41] The knowledge of the quaternary structure of the protein is still largely theoretical, but RPE65 is capable of forming homodimers when crystallized,[41] and it is predicted that RPE65 may form a multiprotein complex with 11-*cis*-retinol dehydrogenase (RDH5) and retinal G protein-coupled receptor at the microsomal membrane in the RPE.[41] It should be noted that RPE65 has only been observed to form homodimers in crystals; this has never been

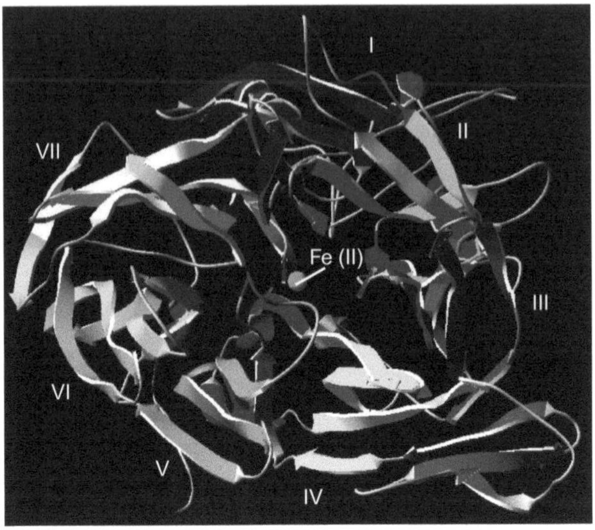

Figure 4 Tertiary structure of RPE65 with iron (II) cofactor labeled. RPE65 has a hydrophobic tunnel that leads to the interior of the protein for the isomerization of all-*trans*-RE substrate to 11-*cis*-ROL product. This one tunnel is the only channel that leads from the exterior of the protein to the interior. The tunnel opens on the RPE membrane but is closed on the other (where the active site is located). For another molecule of substrate to enter, a molecule of product must leave first. The seven beta-propeller blade structures (I–VII) and active site-bound Fe(II) cofactor are labeled.

reported *in vivo*. These and other protein interactions have been proposed, but as of yet, no direct protein interactions have been confirmed. These structural data have aided in the mechanistic understanding of RPE65 isomerase activity.

Experimental data have suggested that posttranslational modification is necessary for RPE65 association with the microsomal membrane,[43] and it is now known that RPE65 requires association with a lipid membrane for its enzymatic activity.[42] Initial work suggested palmitoylation might be required for RPE65 association with the plasma membrane because mutation of Cys residues (which can be sites for palmitoylation) C231, C329, and C330 of the human protein resulted in a decrease in RPE65 activity.[44] More recent data, though, suggest that although membrane association is necessary for RPE65 activity,[45] palmitoylation of those Cys residues is not necessary for membrane association.[41,46,47] C112 palmitoylation, though, does appear to be necessary for appropriate membrane association and enzymatic activity.[48] These data suggest that proper protein folding alone is not sufficient for RPE65 function.

In addition to these posttranslational modifications, the iron cofactor previously mentioned is critical for the isomerase activity of RPE65; chelation of iron with either 2,2′-bipyridine or 1,10-phenanthroline abolishes all enzymatic activity.[40,49] The iron cofactor sits in the active site of RPE65 and is held in an iron coordination motif by four His residues.[42] Mutation of any one of the His-binding residues H180, H241, H417, or H527 in bovine RPE65 abolishes enzymatic activity.[30,50] In the active site, it is currently thought that the iron cofactor plays an important role in the initial catalysis of the isomerization by acting as a Lewis acid,[42] implying the Fe(II) cofactor is temporarily oxidized to Fe(III) during the conversion of all-*trans*-RE to 11-*cis*-ROL.[42]

The actual mechanism of RPE65 isomerase activity was unresolved for several years as two different mechanisms for all-*trans* to 11-*cis* isomerization seemed possible. The first mechanism proposed that an S_N2' mechanism of nucleophilic addition to the C_{11} carbon could result in 11-*cis* production.[51] The second mechanism proposed all-*trans* to 11-*cis* isomerization proceeded through the formation of a carbocation intermediate that results from protonation of the substrate and subsequent bond rearrangement because of electron delocalization.[52] Eventual resolution of the isomerization mechanism came in 2010 and relied on predictions concerning 11-*cis*-retinoid production from each mechanistic proposal. If the isomerization proceeded through the proposed S_N2' nucleophilic substitution, then RPE65 would produce only 11-*cis*-ROL without any 13-*cis*-ROL by-product formed.[53] If, on the other hand, the isomerization proceeded through formation of a carbocation intermediate, it would be expected some 13-*cis*-ROL would be produced by RPE65 as well.[53] When the canine RPE65 residues Thr147 and Phe103, which both reside in the active site but are not involved in iron coordination, were mutated, RPE65 specificity for 11-*cis* and 13-*cis*-ROL production is altered.[53] Tyr338 and Phe 526 also play a similar role in accommodating the substrate and/or intermediate.[54] So, while it appears that isomerization proceeds through the formation of a carbocation intermediate, it is not possible, however, to exclude the possibility that the intermediate might be a radial cation.[54,55]

The possible interactions of RPE65 with other visual cycle proteins at the microsomal membrane of the RPE[41] may allow for efficient processing of hydrophobic retinoids in an assembly line-like fashion. Even if an "assembly line" of retinoid processing does exist in the RPE, the rate of 11-*cis*-ROL synthesis is too slow to account for the rate of regeneration needed for light-adapted vision.[56] In order to operate in a bright environment, cone photoreceptors require a more rapid synthesis of chromophore to compensate for the higher demand.[56] Given the structure of RPE65 and the fact that substrate and product must proceed deep into the protein structure to bind to the active site, RPE65 may not be able to synthesize the product fast enough for cone photoreceptor demand. Other work, which is beyond the scope of this chapter to explore in necessary detail, suggests a cone visual cycle (with a Müller cell retinoid isomerase) may supplement chromophore generation in the retina for light-adapted vision.[56]

4. RPE65 AND ITS KNOWN DISEASE ASSOCIATIONS

Before the Rpe65 knockout was made[27] and it was biochemically established that RPE65 was the isomerase in the visual cycle,[28–30] it became apparent that *RPE65* mutations could have profound impacts on vision. In 1997, mutations in RPE65 were found to be associated with early onset blindness[57,58]) called Leber congenital amaurosis, type 2 (LCA2), or autosomal recessive childhood-onset severe retinal dystrophy,[57] autosomal recessive retinitis pigmentosa,[59] and fundus albipunctatus.[60] LCA is a disease originally (and sometimes still) considered a severe, congenital form of retinitis pigmentosa when it was first formally described by Leber in 1869,[61] with four criteria that must be met by patients for the diagnosis of LCA: (1) severe, early childhood vision loss; (2) nystagmus; (3) severely reduced pupillary responses; and (4) nearly complete abolishment of ERG response.[61] Collectively, LCA is an inherited genetic disease that affects approximately 1 in 80,000 people[62] and 20% of students in schools for blind children.[63] In dogs, *RPE65* mutations are linked to LCA2,[64] congenital stationary night blindness,[65] and autosomal recessive retinal dystrophy.[64] Multiple *RPE65* mutations have been identified in humans,[57,58,66–72] and all inherit in an autosomal recessive fashion, except one dominant-negative mutant, *RPE65*^{D477G}, identified in 2011.[73] It is not yet understood how the RPE65^{D477G} mutation induces a dominant-negative effect on vision or even if protein–protein interactions are involved in the dominantly inheriting negative effects on vision.

Pathogenic *RPE65* mutations are found throughout the entire gene in all its exons and introns; unlike other proteins, which may have mutations that cluster near catalytic active sites or other single functional domain, there is no single region in *RPE65* that appears particularly enriched in pathogenic mutations (Fig. 5A). As illustrated in Fig. 5B, which highlights amino acid sequence stretches that form important structural features of RPE65 along

A

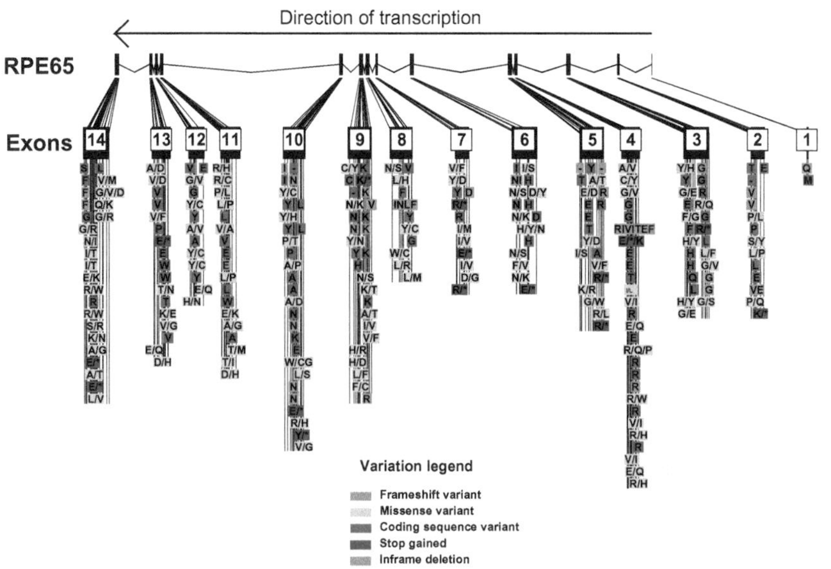

B

1- MSIQVEHPAGGYKKLFETVEELSSPLTAHVTGRIPLWLTGSLLRCGPGLF
51- EVGSEPFYHLFDGQALLHKFDFKEGHVTYHRRFIRTDAYVRAMTEKRIVI
101- TEFGTCAFPDPCKNIFSRFFSYFRGVEVTDNALVNVYPVGEDYYACTETN
151- FITKINPETLETIKQVDLCNYVSVNGATAHPHIENDGTVYNIGNCFGKNF
201- SIAYNIVKIPPLQADKEDPISKSEIVVQFPCSDRFKPSYVHSFGLTPNYI
251- VFVETPVKINLFKFLSSWSLWGANYMDCFESNETMGVWLHIADKKRKKYL
301- NNKYRTSPFNLFHHINTYEDNGFLIVDLCCWKGFEFVYNYLYLANLRENW
351- EEVKKNARKAPQPEVRRYVLPLNIDKADTGKNLVTLPNTTATAILCSDET
401- IWLEPEVLFSGPRQAFEFPQINYQKYCGKPYTYAYGLGLNHFVPDRLCKL
451- NVKTKETWVWQEPDSYPSEPIFVSHPDALEEDDGVVLSVVVSPGAGQKPA
501- YLLILNAKDLSEVARAEVEINIPVTFHGLFKKS

Figure 5 Known human mutations in RPE65. (A) The human RPE65 gene is shown above with all known coding, missense, stop, frameshift, and deletion mutations. (B) Human RPE65 protein with known structural motifs in the protein. Alpha helices are highlighted in green (gray in the print version), beta-propeller I is highlighted in cyan (light gray in the print version), beta-propeller II is highlighted in magenta (gray in the print version), beta-propeller III is highlighted in blue (dark gray in the print version), beta-propeller IV is highlighted in teal (dark gray in the print version), beta-propeller V is highlighted in green (dark gray in the print version), beta-propeller VI is highlighted in yellow (light gray in the print version), and beta-propeller VII is highlighted in red (dark gray in the print version).[40] Iron-coordinating histidine residues, which form the active site of the protein, are highlighted in black with white font. The residues that form the active site are H180, H241, H313, and H527.[30]

with the histidine residues that are responsible for iron (II) cofactor coordination, the wide distribution of pathogenic mutations in the protein may indicate many of these mutations may disrupt protein folding rather than directly affecting enzymatic activity via the disruption of the active site. More work would be needed to confirm this, however.

5. CONCLUSIONS

The collected knowledge of almost 150 years of research has enabled a comprehensive understanding of retinoid processing and the biochemical mechanisms that are necessary for visual function in the healthy eye. With the advent of novel gene delivery methods and other treatment strategies, for the first time in history, patients with heretofore untreatable visual diseases stand a real chance of maintaining visual function or regaining lost visual function. Future work will still need to be done to fill in the gaps in the visual cycle, and despite early successes in small-scale clinical trials for therapies based on the molecular understanding of the visual cycle, the medical and research communities are just beginning to apply the wealth of knowledge to the clinic. These early successes will nonetheless pave the way for future therapies, making the future of visual cycle research even more exciting than what has come to date.

Conflicts of Interest

The authors declare no conflicts of interest.

REFERENCES

1. Baumann C. Boll's phenomenon. *Vision Res.* 1977;17(11–12):1325–1327.
2. Marmor MF, Martin LJ. 100 years of the visual cycle. *Surv Ophthalmol.* 1978;22(4):279–285.
3. Kühne W. On the stable colours of the retina. *J Physiol.* 1878;1(2–3):109–212, 105.
4. Kühne W, Sewall H. On the physiology of the retinal epithelium. *J Physiol.* 1880;3(1):88–92.
5. Wald G. Carotenoids and the visual cycle. *J Gen Physiol.* 1935;19(2):351–371.
6. Wald G, Clark AB. Visual adaptation and chemistry of the rods. *J Gen Physiol.* 1937;21(1):93–105.
7. Wald G, Brown PK. The molar extinction of rhodopsin. *J Gen Physiol.* 1953;37(2):189–200.
8. Matthews RG, Hubbard R, Brown PK, Wald G. Tautomeric forms of metarhodopsin. *J Gen Physiol.* 1963;47:215–240.
9. Wald G. The molecular basis of visual excitation. *Nature.* 1968;219(5156):800–807.
10. Parker RO, Crouch RK. Retinol dehydrogenases (RDHs) in the visual cycle. *Exp Eye Res.* 2010;91(6):788–792.
11. Maeda A, Maeda T, Imanishi Y, et al. Role of photoreceptor-specific retinol dehydrogenase in the retinoid cycle in vivo. *J Biol Chem.* 2005;280(19):18822–18832.

12. Janecke AR, Thompson DA, Utermann G, et al. Mutations in RDH12 encoding a photoreceptor cell retinol dehydrogenase cause childhood-onset severe retinal dystrophy. *Nat Genet.* 2004;36(8):850–854.
13. Andrews JS, Futterman S. Metabolism of the retina. V. The role of microsomes in vitamin A esterification in the visual cycle. *J Biol Chem.* 1964;239:4073–4076.
14. Zimmerman WF. The distribution and proportions of vitamin A compounds during the visual cycle in the rat. *Vision Res.* 1974;14(9):795–802.
15. Bernstein PS, Law WC, Rando RR. Isomerization of all-trans-retinoids to 11-cis-retinoids in vitro. *Proc Natl Acad Sci USA.* 1987;84(7):1849–1853.
16. Perlman JI, Nodes BR, Pepperberg DR. Utilization of retinoids in the bullfrog retina. *J Gen Physiol.* 1982;80(6):885–913.
17. Canada FJ, Law WC, Rando RR, Yamamoto T, Derguini F, Nakanishi K. Substrate specificities and mechanism in the enzymatic processing of vitamin A into 11-cis-retinol. *Biochemistry.* 1990;29(41):9690–9697.
18. Trehan A, Canada FJ, Rando RR. Inhibitors of retinyl ester formation also prevent the biosynthesis of 11-cis-retinol. *Biochemistry.* 1990;29(2):309–312.
19. Yamamoto H, Simon A, Eriksson U, Harris E, Berson EL, Dryja TP. Mutations in the gene encoding 11-cis retinol dehydrogenase cause delayed dark adaptation and fundus albipunctatus. *Nat Genet.* 1999;22(2):188–191.
20. Abe T, Wiggert B, Bergsma DR, Chader GJ. Vitamin A receptors. I. Comparison of retinol binding to serum retinol-binding protein and to tissue receptors in chick retina and pigment epithelium. *Biochim Biophys Acta.* 1977;498(1):355–365.
21. Adler AJ, Evans CD. Rapid isolation of bovine interphotoreceptor retinol-binding protein. *Biochim Biophys Acta.* 1983;761(3):217–222.
22. Bridges CD, Alvarez RA, Fong SL, Gonzalez-Fernandez F, Lam DM, Liou GI. Visual cycle in the mammalian eye. Retinoid-binding proteins and the distribution of 11-cis retinoids. *Vision Res.* 1984;24(11):1581–1594.
23. Bernstein PS, Law WC, Rando RR. Biochemical characterization of the retinoid isomerase system of the eye. *J Biol Chem.* 1987;262(35):16848–16857.
24. Sagara H, Hirosawa K. Monoclonal antibodies which recognize endoplasmic reticulum in the retinal pigment epithelium. *Exp Eye Res.* 1991;53(6):765–771.
25. Hamel CP, Tsilou E, Harris E, et al. A developmentally regulated microsomal protein specific for the pigment epithelium of the vertebrate retina. *J Neurosci Res.* 1993;34(4):414–425.
26. Hamel CP, Tsilou E, Pfeffer BA, Hooks JJ, Detrick B, Redmond TM. Molecular cloning and expression of RPE65, a novel retinal pigment epithelium-specific microsomal protein that is post-transcriptionally regulated in vitro. *J Biol Chem.* 1993;268(21):15751–15757.
27. Redmond TM, Yu S, Lee E, et al. Rpe65 is necessary for production of 11-cis-vitamin A in the retinal visual cycle. *Nat Genet.* 1998;20(4):344–351.
28. Moiseyev G, Chen Y, Takahashi Y, Wu BX, Ma JX. RPE65 is the isomerohydrolase in the retinoid visual cycle. *Proc Natl Acad Sci USA.* 2005;102(35):12413–12418.
29. Jin M, Li S, Moghrabi WN, Sun H, Travis GH. Rpe65 is the retinoid isomerase in bovine retinal pigment epithelium. *Cell.* 2005;122(3):449–459.
30. Redmond TM, Poliakov E, Yu S, Tsai J-Y, Lu Z, Gentleman S. Mutation of key residues of RPE65 abolishes its enzymatic role as isomerohydrolase in the visual cycle. *Proc Natl Acad Sci USA.* 2005;102(38):13658–13663.
31. Kiser PD, Palczewski K. Membrane-binding and enzymatic properties of RPE65. *Prog Retin Eye Res.* 2010;29(5):428–442.
32. Liu SY, Redmond TM. Role of the 3′-untranslated region of RPE65 mRNA in the translational regulation of the RPE65 gene: identification of a specific translation inhibitory element. *Arch Biochem Biophys.* 1998;357(1):37–44.

33. Nicoletti A, Wong DJ, Kawase K, et al. Molecular characterization of the human gene encoding an abundant 61 kDa protein specific to the retinal pigment epithelium. *Hum Mol Genet.* 1995;4(4):641–649.
34. Ma JX, Zhang D, Laser M, et al. Identification of RPE65 in transformed kidney cells. *FEBS Lett.* 1999;452(3):199–204.
35. Chen Y, Moiseyev G, Wu BX, Ma JX, Crouch RK. Visual cycle retinoid processing proteins are present in HEK293S cells. *Vision Res.* 2003;43(28):3037–3044.
36. Ma J, Xu L, Othersen DK, Redmond TM, Crouch RK. Cloning and localization of RPE65 mRNA in salamander cone photoreceptor cells. *Biochim Biophys Acta.* 1998;1443(1–2):255–261.
37. Znoiko SL, Crouch RK, Moiseyev G, Ma J-X. Identification of the RPE65 protein in mammalian cone photoreceptors. *Invest Ophthalmol Vis Sci.* 2002;43(5):1604–1609.
38. Tang PH, Buhusi MC, Ma J-X, Crouch RK. RPE65 is present in human green/red cones and promotes photopigment regeneration in an in vitro cone cell model. *J Neurosci.* 2011;31(50):18618–18626.
39. Redmond TM, Gentleman S, Duncan T, et al. Identification, expression, and substrate specificity of a mammalian beta-carotene 15,15′-dioxygenase. *J Biol Chem.* 2001;276(9):6560–6565.
40. Kiser PD, Golczak M, Lodowski DT, Chance MR, Palczewski K. Crystal structure of native RPE65, the retinoid isomerase of the visual cycle. *Proc Natl Acad Sci USA.* 2009;106(41):17325–17330.
41. Golczak M, Kiser PD, Lodowski DT, Maeda A, Palczewski K. Importance of membrane structural integrity for RPE65 retinoid isomerization activity. *J Biol Chem.* 2010;285(13):9667–9682.
42. Kiser PD, Farquhar ER, Shi W, Sui X, Chance MR, Palczewski K. Structure of RPE65 isomerase in a lipidic matrix reveals roles for phospholipids and iron in catalysis. *Proc Natl Acad Sci USA.* 2012;109(41):E2747–E2756.
43. Ma J, Zhang J, Othersen KL, et al. Expression, purification, and MALDI analysis of RPE65. *Invest Ophthalmol Vis Sci.* 2001;42(7):1429–1435.
44. Takahashi Y, Moiseyev G, Chen Y, Ma JX. The roles of three palmitoylation sites of RPE65 in its membrane association and isomerohydrolase activity. *Invest Ophthalmol Vis Sci.* 2006;47(12):5191–5196.
45. Nikolaeva O, Moiseyev G, Rodgers K, Ma J-X. Binding to lipid membrane induces conformational changes in RPE65: implications for its isomerohydrolase activity. *Biochem J.* 2011;436(3):591–597.
46. Trudel E, Beaufils S, Renault A, Breton R, Salesse C. Binding of RPE65 fragments to lipid monolayers and identification of its partners by glutathione S-transferase pull-down assays. *Biochemistry.* 2006;45(10):3337–3347.
47. Jin M, Yuan Q, Li S, Travis GH. Role of LRAT on the retinoid isomerase activity and membrane association of Rpe65. *J Biol Chem.* 2007;282(29):20915–20924.
48. Takahashi Y, Moiseyev G, Ablonczy Z, Chen Y, Crouch RK, Ma JX. Identification of a novel palmitoylation site essential for membrane association and isomerohydrolase activity of RPE65. *J Biol Chem.* 2009;284(5):3211–3218.
49. Moiseyev G, Takahashi Y, Chen Y, et al. RPE65 is an iron(II)-dependent isomerohydrolase in the retinoid visual cycle. *J Biol Chem.* 2006;281(5):2835–2840.
50. Takahashi Y, Moiseyev G, Chen Y, Ma JX. Identification of conserved histidines and glutamic acid as key residues for isomerohydrolase activity of RPE65, an enzyme of the visual cycle in the retinal pigment epithelium. *FEBS Lett.* 2005;579(24):5414–5418.
51. Deigner PS, Law WC, Canada FJ, Rando RR. Membranes as the energy source in the endergonic transformation of vitamin A to 11-cis-retinol. *Science.* 1989;244(4907):968–971.

52. McBee JK, Kuksa V, Alvarez R, et al. Isomerization of all-trans-retinol to cis-retinols in bovine retinal pigment epithelial cells: dependence on the specificity of retinoid-binding proteins. *Biochemistry*. 2000;39(37):11370–11380.
53. Redmond TM, Poliakov E, Kuo S, Chander P, Gentleman S. RPE65, visual cycle retinol isomerase, is not inherently 11-cis-specific: support for a carbocation mechanism of retinol isomerization. *J Biol Chem*. 2010;285(3):1919–1927.
54. Chander P, Gentleman S, Poliakov E, Redmond TM. Aromatic residues in the substrate cleft of RPE65 protein govern retinol isomerization and modulate its progression. *J Biol Chem*. 2012;287(36):30552–30559.
55. Poliakov E, Parikh T, Ayele M, et al. Aromatic lipophilic spin traps effectively inhibit RPE65 isomerohydrolase activity. *Biochemistry*. 2011;50(32):6739–6741.
56. Wang J-S, Kefalov VJ. The cone-specific visual cycle. *Prog Retin Eye Res*. 2011;30(2):115–128.
57. Gu SM, Thompson DA, Srikumari CR, et al. Mutations in RPE65 cause autosomal recessive childhood-onset severe retinal dystrophy. *Nat Genet*. 1997;17(2):194–197.
58. Marlhens F, Bareil C, Griffoin JM, et al. Mutations in RPE65 cause Leber's congenital amaurosis. *Nat Genet*. 1997;17(2):139–141.
59. Poehner WJ, Fossarello M, Rapoport AL, et al. A homozygous deletion in RPE65 in a small Sardinian family with autosomal recessive retinal dystrophy. *Mol Vis*. 2000;6:192–198.
60. Schatz P, Preising M, Lorenz B, Sander B, Larsen M, Rosenberg T. Fundus albipunctatus associated with compound heterozygous mutations in RPE65. *Ophthalmology*. 2011;118(5):888–894.
61. den Hollander AI, Roepman R, Koenekoop RK, Cremers FP. Leber congenital amaurosis: genes, proteins and disease mechanisms. *Prog Retin Eye Res*. 2008;27(4):391–419.
62. Cideciyan AV. Leber congenital amaurosis due to RPE65 mutations and its treatment with gene therapy. *Prog Retin Eye Res*. 2010;29(5):398–427.
63. Ahmed E, Loewenstein J. Leber congenital amaurosis: disease, genetics and therapy. *Semin Ophthalmol*. 2008;23(1):39–43.
64. Veske A, Nilsson SE, Narfstrom K, Gal A. Retinal dystrophy of Swedish briard/briard-beagle dogs is due to a 4-bp deletion in RPE65. *Genomics*. 1999;57(1):57–61.
65. Aguirre GD, Baldwin V, Pearce-Kelling S, Narfstrom K, Ray K, Acland GM. Congenital stationary night blindness in the dog: common mutation in the RPE65 gene indicates founder effect. *Mol Vis*. 1998;4:23.
66. Chen Y, Moiseyev G, Takahashi Y, Ma JX. Impacts of two point mutations of RPE65 from Leber's congenital amaurosis on the stability, subcellular localization and isomerohydrolase activity of RPE65. *FEBS Lett*. 2006;580(17):4200–4204.
67. Hanein S, Perrault I, Gerber S, et al. Leber congenital amaurosis: comprehensive survey of the genetic heterogeneity, refinement of the clinical definition, and genotype-phenotype correlations as a strategy for molecular diagnosis. *Hum Mutat*. 2004;23(4):306–317.
68. Morimura H, Fishman GA, Grover SA, Fulton AB, Berson EL, Dryja TP. Mutations in the RPE65 gene in patients with autosomal recessive retinitis pigmentosa or Leber congenital amaurosis. *Proc Natl Acad Sci USA*. 1998;95(6):3088–3093.
69. Philp AR, Jin M, Li S, et al. Predicting the pathogenicity of RPE65 mutations. *Hum Mutat*. 2009;30(8):1183–1188.
70. Stone EM. Leber congenital amaurosis—a model for efficient genetic testing of heterogeneous disorders: LXIV Edward Jackson Memorial Lecture. *Am J Ophthalmol*. 2007;144(6):791–811.

71. Sundaresan P, Vijayalakshmi P, Thompson S, Ko AC, Fingert JH, Stone EM. Mutations that are a common cause of Leber congenital amaurosis in northern America are rare in southern India. *Mol Vis*. 2009;15:1781–1787.
72. Xu F, Dong Q, Liu L, et al. Novel RPE65 mutations associated with Leber congenital amaurosis in Chinese patients. *Mol Vis*. 2012;18:744–750.
73. Bowne SJ, Humphries MM, Sullivan LS, et al. A dominant mutation in RPE65 identified by whole-exome sequencing causes retinitis pigmentosa with choroidal involvement. *Eur J Hum Genet*. 2011;19(10):1074–1081.

CHAPTER TWENTY-SIX

A2E and Lipofuscin

Rosalie K. Crouch*,[1,2], Yiannis Koutalos*, Masahiro Kono*, Kevin Schey†, Zsolt Ablonczy*

*Department of Ophthalmology, Albert Florens Storm Eye Institute, Medical University of South Carolina, Charleston, South Carolina, USA

†Department of Biochemistry, Vanderbilt University, Nashville, Tennessee, USA

[1]Corresponding author: e-mail address: crouchrk@musc.edu

Contents

Abstract

Lipofuscin is highly fluorescent material, formed in several tissues but best studied in the eye. The accumulation of lipofuscin in the retinal pigment epithelium (RPE) is a hallmark of aging in the eye and has been implicated in various retinal degenerations, including age-related macular degeneration. The *bis*-retinoid *N*-retinyl-*N*-retinylidene ethanolamine (A2E), formed from retinal, has been identified as a byproduct of the visual cycle, and numerous *in vitro* studies have found toxicity associated with this compound. The compound is known to accumulate in the RPE with age and was the first identified compound extracted from lipofuscin. Our studies have correlated the distribution of lipofuscin and A2E across the human and mouse RPE. Lipofuscin fluorescence was imaged in the RPE from human donors of various ages and from assorted mouse models. The spatial distribution of A2E was determined using matrix-assisted laser desorption–ionization imaging mass spectrometry on both flat-mounted and transversally sectioned RPE tissue. Our data support the clinical observations in humans of strong RPE fluorescence, increasing with age, in the central area of the RPE. However, there was no correlation between the distribution of A2E and lipofuscin, as the levels of A2E were highest in the far periphery and decreased toward the central region. Interestingly, in all the mouse models, A2E distribution and lipofuscin fluorescence correlate

[2] Senior author.

Progress in Molecular Biology and Translational Science, Volume 134
ISSN 1877-1173
http://dx.doi.org/10.1016/bs.pmbts.2015.06.005
© 2015 Elsevier Inc.
All rights reserved.

well. These data demonstrate that the accumulation of A2E is not responsible for the increase in lipofuscin fluorescence observed in the central RPE with aging in humans.

ABBREVIATIONS

A2E *N*-retinyl-*N*-retinylidene ethanolamine
AMD age-related macular degeneration
MALDI matrix-assisted laser desorption–ionization
RPE retinal pigment epithelium

1. INTRODUCTION

Age-related macular degeneration (AMD) is the leading cause of blindness in individuals over 55 in Western countries. This number is expected to double in the next 15 years as the population ages. Approximately 90% of AMD patients have the slowly progressing form of the disease, known as dry AMD; however, the disease can transform into a more severe form, known as wet AMD which accounts for the majority of severe vision loss in AMD. The cellular and molecular events responsible for the development of the disease and the transition from the dry to the wet form are not known. Although AMD is a multifactorial disease, inflammation seems to be involved in the degenerative process. Understanding the molecular mechanisms responsible for the pathogenesis of AMD is one of the most pressing issues of current vision research.

Accumulation of lipofuscin (Fig. 1A) with age is well documented, and the first compound to be isolated from lipofuscin was the *bis*-retinoid *N*-retinyl-*N*-retinylidene ethanolamine (A2E; Fig. 1B).[1] This compound can be

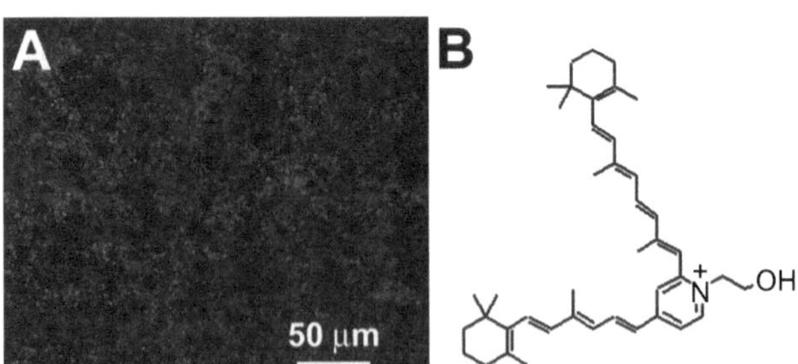

Figure 1 Lipofuscin fluorescence and A2E. Image of the fluorescence (excitation: 450–490 nm) emitted by the RPE of a cyclic-light-reared 12-month-old 129/sv wild-type mouse (A). Structure of A2E (B).

synthesized[2] and there have been extensive *in vitro* studies addressing the possible involvement of A2E in the destructive processes that occur in the retina with aging. However, our research has shown that the spatial distribution of A2E in humans does not correlate with lipofuscin distribution, unlike in mouse models. We discuss here our findings which indicate that the mouse is not a good model of the human for the purpose of understanding lipofuscin composition and its potential role in retinal degeneration and further emphasize the necessity of understanding the composition of lipofuscin in humans.

2. LIPOFUSCIN

Lipofuscin (sometimes referred to as "aging pigment") is a term given to intracellular, fluorescent pigment granules that mainly accumulate in postmitotic cells such as neurons, cardiac muscle, and the retinal pigment epithelium (RPE). This material is distinct from ceroids, the lipopigments found in mitotic cells in many tissues that accumulate as a result of a pathological condition (for a review, see Ref. 3). Lipofuscin accumulates slowly with age and is a complex heterogeneous mixture of intracellular compounds. In general, it has been reported to be a mixture of oxidized proteins, lipids, carbohydrates, metals (predominately iron), and other bioorganic molecules, but this is very variable depending on the tissue of origin.[4,5] This mixture has fluorescence at approximately 600 nm, but the spectral range again depends on the tissue from which it is isolated. The composition is not really well described for any tissue. The role of lipofuscin on cellular function in general is not clear, but there is general agreement that accumulation of this material increases vulnerability to oxidative stress and decreases lysosomal degradation.[6]

Lipofuscin formed in the RPE is the best studied (Fig. 1A). The major source of this lipofuscin is thought to be the photoreceptor outer segments phagocytosed by the RPE cells. Interestingly, the proteomic study by Ng *et al.*,[7] showed that the lipofuscin granules themselves contained little protein (about 2%). The lipofuscin in the RPE is highly fluorescent (Fig. 1A). The fluorophores in mouse lipofuscin require a functioning retinoid visual cycle as shown from studies on vitamin A-deprived animals[8] as well as on the RPE65 knockout mouse, which is missing the protein RPE65 known to be essential for the production of 11-*cis* retinal.[9]

Traditional methods for ocular lipofuscin analysis involve the collection of granules from homogenized RPE tissue, which eliminates any spatial resolution. The homogenate is then extracted with chloroform/methanol and the extract separated by HPLC.[10] The isolated lipofuscin compounds have been evaluated by various analytical methods. The chemistry and the synthetic techniques involved are quite complex due to the extensive conjugation of these systems. Conclusive identification is often only possible by the synthesis and simultaneous analysis of authentic standards. Moreover, the chloroform-soluble fraction only represents a small portion of the lipofuscin components, although it contains the predominant fluorophores, at least in mouse lipofuscin (e.g., [11]). To date, about 25 *bis*-retinoid components of lipofuscin have been identified.[12]

Ocular lipofuscin is highly photoreactive, generating reactive oxygen species upon irradiation with visible light which has been proposed to affect lysomal function and possibly be a cause of AMD.[13,14] The retina is vulnerable to damage, due to its exposure to visible light, high oxygen tension, and the high composition of unsaturated fatty acids in the photoreceptors. In addition, there is the presence of retinoid-based chromophores. In humans, the RPE shows increased lipofuscin fluorescence in the center of the eye, particularly around the perifoveal ring (for a recent review of geographic atrophy, see Ref. 15). Interestingly, a recent study by Rudolf *et al.* [16] on human eyes with geographic atrophy suggested that the florescence usually attributed to lipofuscin may be arising from vertically superimposed cells and cellular fragments rather than lipofuscin itself.

3. A2E AND OTHER *BIS*-RETINOIDS

The visual cycle involves the processing of the chromophore, 11-*cis* retinal, for regeneration of the photosensitivity of rod and cone visual pigments. The critical role of vitamin A in this process was established by Wald and his colleagues, work for which they received the Nobel prize in 1967.[17] Interestingly, although night blindness, which results from the rod photoreceptors lacking their chromophore, is the first sign of vitamin A deficiency[18]; it is the cone photoreceptors that deteriorate most rapidly, due to mislocalization of the opsins, the protein portion of the visual pigments.[19] This deficiency can either arise due to the lack of dietary intake or to the mutation or deficiency of one of the various enzymes or transport proteins involved in the visual cycle.

As a result of the absorption of a photon by the visual pigments (both rod and cone), all-*trans* retinal is produced (see Fig. 2). This product has been shown to produce A2PE in the rod photoreceptors.[20] More recently, it

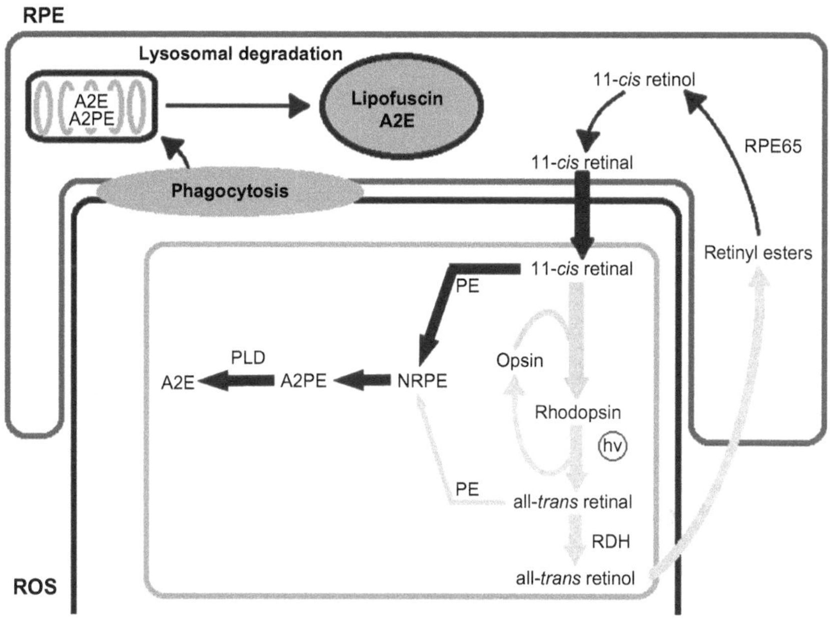

Figure 2 Visual cycle and A2E. Lipofuscin precursors, such as A2PE, are formed primarily by 11-*cis* retinal in the photoreceptor outer segment membranes. The majority of the all-*trans* retinal generated by light exposure is reduced to retinol and is not a major source of lipofuscin and A2E. Lack of the Rpe65 protein stops the generation of 11-*cis* retinal and eliminates formation of lipofuscin precursors regardless of light exposure. NRPE, *N*-retinylidene-phosphatidylethanolamine; PLD, phospholipase D; RDH, retinol dehydrogenase. *This figure is modified from Boyer* et al.[21]

has also been found that A2PE can also form in rod photoreceptors from excess 11-*cis* retinal.[21,22] It is proposed that A2E (*N*-retinylidene-*N*-retinylethanolamine) is then generated after phagocytosis of the photoreceptors in the RPE by cleavage of the phosphatidyl group. A2E was the first compound to be identified in extracts of ocular lipofuscin,[1] and it has been shown that the compound increases with age, both in murine models and in humans, and in certain retinal degenerations.

A number of *bis*-retinoids have now been isolated (25 to date)[23]; this includes various *cis* isomers, oxidized forms, and various intermediates. In addition, there are modified *bis*-retinoids with varying fatty acids on the phosphatidyl group. All of these compounds are reported to be present in human lipofuscin,[23] although the spatial localization for most is not known. All the known *bis*-retinoids contain a central pyridinium ring with amine nitrogen carrying a positive charge and multiple conjugated double bonds.

Careful chemical structural evaluation and comparison to synthetic standards support the proposed structures, mainly conducted by the Sparrow group (for recent reviews, see Refs. 12,23).

The *bis*-retinoid components of lipofuscin have received considerable attention as they have been shown to produce RPE damage *in vitro*. A2E can be synthesized[2] and thus has been used extensively for studies *in vitro*, with cultured RPE cells and in animal models. There are multiple effects that have been reported, including impairment of RPE phagocytosis,[24] induction of detergent-like action on lysosomal membranes,[25] and inhibition of lysosomal degradation of lipids.[26] In addition, A2E can act as a photosensitizer on exposure to light[27]; but the extent of its reactivity has been questioned as it is less active than all-*trans* retinal itself.[28] In this respect, some reports suggest that the role of A2E may be to detoxify the more reactive all-*trans* retinal.[11,29,30]

4. MATRIX-ASSISTED LASER DESORPTION–IONIZATION TISSUE-IMAGING MASS SPECTROMETRY

Matrix-assisted laser desorption–ionization (MALDI) tissue imaging[31] allows the collection of spatial abundance information of individual molecules. It has already been successfully applied to direct studies of brain,[32] kidney,[33] pancreas,[34] prostate,[35] and embryonic[36] and ocular lens tissues[37]. The most well-known aspect of this state-of-the-art approach is its ability to identify protein signatures unique to different normal or diseased tissues.[38] However, the methodology can be extended to peptides,[39] lipids,[40] and small-molecule metabolites[41]. We recently have shown that this technique can also be applied to cross sections of the retina as well as flat-mounted tissue and have used cross sections to study the spatial distribution of lipids as well as retinoid metabolites in mouse models.[42]

MALDI imaging technology is an extension of MALDI mass spectrometry, using whole tissue instead of extracted protein/peptide/compound solutions. In the imaging experiment, tissue sections or thin flat-mounted tissues are coated with a matrix solution, and a laser scans the tissues at a specified step size. This process allows the acquisition of mass spectra (MALDI profile) at each point where the laser hits the matrix, resulting in a large two-dimensional profile data set. From this dataset, particular molecular ions (m/z value) can be displayed as spatial intensities (see Fig. 3). The spatial resolution depends on the laser spot size, the matrix system, and the quality of the matrix coating. It is normally between 50 and 200 μm; therefore,

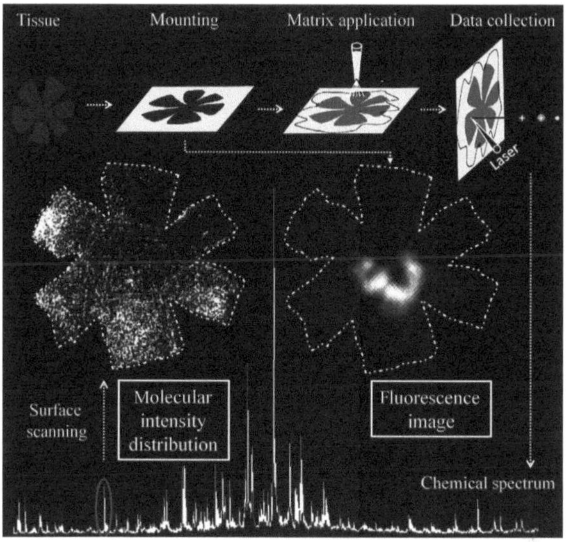

Figure 3 MALDI imaging process. The multimodal MALDI imaging workflow allows the comparison of the molecular distribution of any ion in the three-dimensional imaging database to images obtained by traditional imaging methods, such as fluorescence or histology. The inset figures show images from an 86-year right human eye: (left) the image of the molecular ion (*m/z* 592) corresponding to A2E in the MALDI imaging database and (right) lipofuscin fluorescence recorded in the same tissue before MALDI matrix application. The images represent the respective intensity distributions on a false color heat map (red (light gray in the print version)—high intensity, blue (dark gray in the print version)—low intensity). *The inset panels were modified from Ablonczy* et al.[51]

subcellular structures are not visible by MALDI imaging, but single-cell mass spectra have already been acquired.[43] In contrast to traditional imaging methods, however, MALDI tissue imaging allows the simultaneous collection of hundreds of molecular signals in a single experiment. This abundance of signals is then utilized to determine regions of interesting pathology or differentiation. Alternatively, regions of particular anatomy or pathology can also be targeted with MALDI imaging for the identification of novel metabolites or markers with high spatial resolution. Moreover, as the tissue is analyzed with minimal manipulation, it is expected that previously unobserved molecules, which were lost during traditional sample preparation, might be detected. However, conclusive identification of these compounds depends on the acquisition of the structurally specific fragmentation pattern of the molecule by tandem mass spectrometry. In Fig. 4, the comparison of the mass spectrum of A2E detected from the mouse RPE is shown with that of a synthetic sample, demonstrating that the identification is correct.

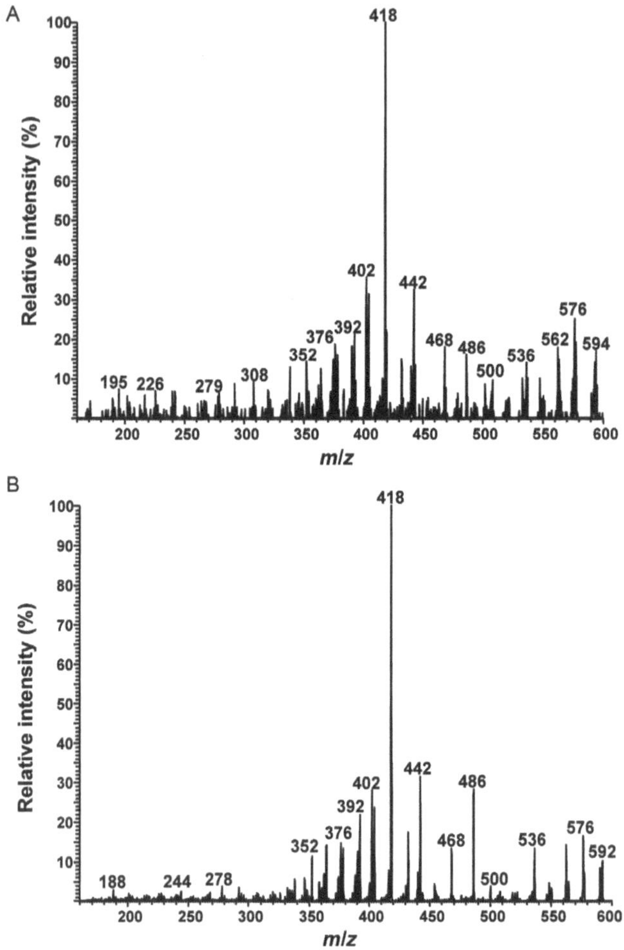

Figure 4 Tandem mass spectra of A2E. Recorded from a sample of synthetic A2E (A) and directly from matrix-coated 6-month-old *Abca4*$^{-/-}$ mouse RPE tissue (B). The selected precursor ion for the MS/MS experiment was *m/z* 592. *This figure is modified from Grey et al.*[46]

The quality of images obtained with MALDI imaging is determined mainly by the sample preparation. Different tissues, or the same tissue from different species, may require different sample preparation protocols and instrument settings for optimal mass spectrometric imaging. Tissue sections can be washed prior to matrix deposition, and matrices based on different solvents and solutes can be used. The utilized washing/matrix systems have a significant impact on the quality and number of the detected mass spectrometric hits and the quality of the resulting tissue images. Spray coating, spot coating, or matrix sublimation can be utilized to apply matrices. Homogeneous matrix crystal formation is essential, as this will lead to reproducible MALDI images from a given tissue. The method of optimization allows overcoming the most significant limitation of the MALDI imaging method, the detection of only those molecules which are abundantly present. This is a powerful technique particularly for the spatial localization of molecules which cannot be localized by traditional immune-localization methods.

5. A2E AND LIPOFUSCIN IN MICE

For our studies, we have employed a multimodal imaging strategy on the same tissue by the simultaneous application of fluorescence and MALDI imaging, which is an added advantage of the technology. This strategy allows the comparison of A2E with lipofuscin distribution. This technique allows good spatial resolution; however, the individual components cannot be identified.[44] One of the main advantages of MALDI imaging is that the images provided by this technique can be compared to fluorescence measurements to identify whether the fluorescent signal originates from a particular compound.

We have examined both wild-type mice and several mouse models of retinal degenerations. Lipofuscin fluorescence was measured in a Leica confocal microscope, with 488 nm excitation. The resulting fluorescence emission spectra are in good agreement with those described for lipofuscin.[45]

The tissue was then examined by MALDI, and the distribution of A2E was mapped. In every mouse model that we have examined, the distribution of A2E correlates well with the fluorescence intensity as seen in Fig. 5.[44,46] In addition, the majority of lipofuscin components correlating with lipofuscin intensity were found to be retinoid based.[47] From these mouse images and the large amount of data from extractions of RPE homogenates, a reasonable conclusion would be that the fluorescence in mouse lipofuscin is indeed due mainly to A2E and/or its derivatives. Moreover, with the strong *in vitro* data showing toxicity of A2E to RPE cells, the hypothesis that A2E is the culprit in various retinal degenerations is not unreasonable.

6. A2E AND LIPOFUSCIN IN HUMANS

We have used these same techniques of MALDI imaging and fluorescence measurements discussed above to examine the distribution of A2E and

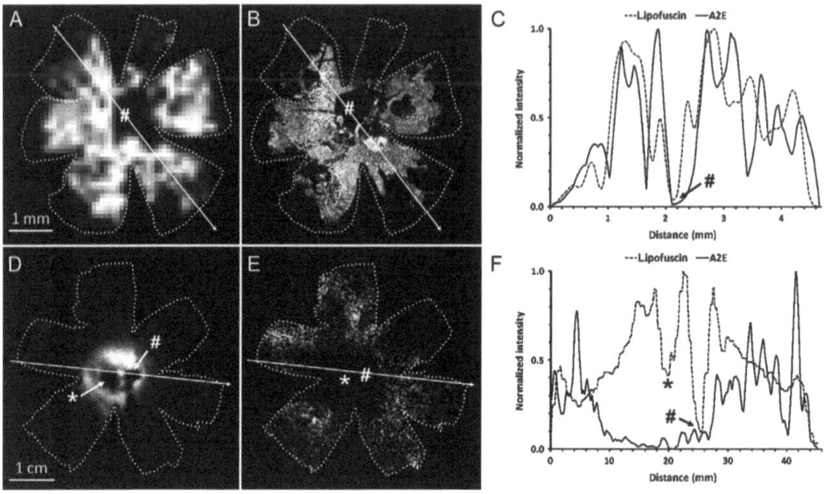

Figure 5 Comparison of A2E and lipofuscin in the mouse and human. The spatial distribution of lipofuscin fluorescence (A) and of A2E (B) in the right eye of a 6-month-old *Abca4*$^{-/-}$ mouse. Cross-sectional profiles along the arrow indicated in the previous panels show good correlation between lipofuscin and A2E intensities (C). The fluorescence image was collaged from individual fields taken with a fluorescent microscope (λ_{exc} = 488 nm). The MALDI image of A2E was collected at 150 mm resolution. Bar = 1 mm. The spatial distributions of lipofuscin fluorescence (D) and of A2E (E) in a 94-year right human eye. Cross-sectional profiles along the arrow indicated in the previous panels show that lipofuscin (which is central) does not correlate with A2E (which is peripheral) (F). The fluorescence image was acquired in a Xenogen IVIS 200 bioluminescence imaging system (λ_{exc} = 450–490 nm, λ_{em} = 575–650 nm) at a final resolution of 500 μm. The MALDI image of A2E was collected at 300 μm resolution. Bar = 1 cm. The intensity is coded by the indicated false color heat map. The images are oriented: dorsal—top; ventral—bottom; nasal—left; and temporal—right. * indicates the macula, and # is for the optic nerve head.

lipofuscin in human tissue. A previous report based on extraction of A2E from punches of human RPE had suggested that the quantity of A2E observed in the center tissue was less than that observed in the peripheral samples.[48] Our data clearly demonstrated that in humans across all ages, A2E is concentrated in the peripheral tissue.[49] A similar distribution has been observed for the several A2E metabolites studied. Our fluorescence images agree with those reported by numerous groups (e.g.,[50] showing that fluorescence is highest in the center of the human RPE; Fig. 6). These data were unexpected and are in sharp contrast to the data obtained from mice.

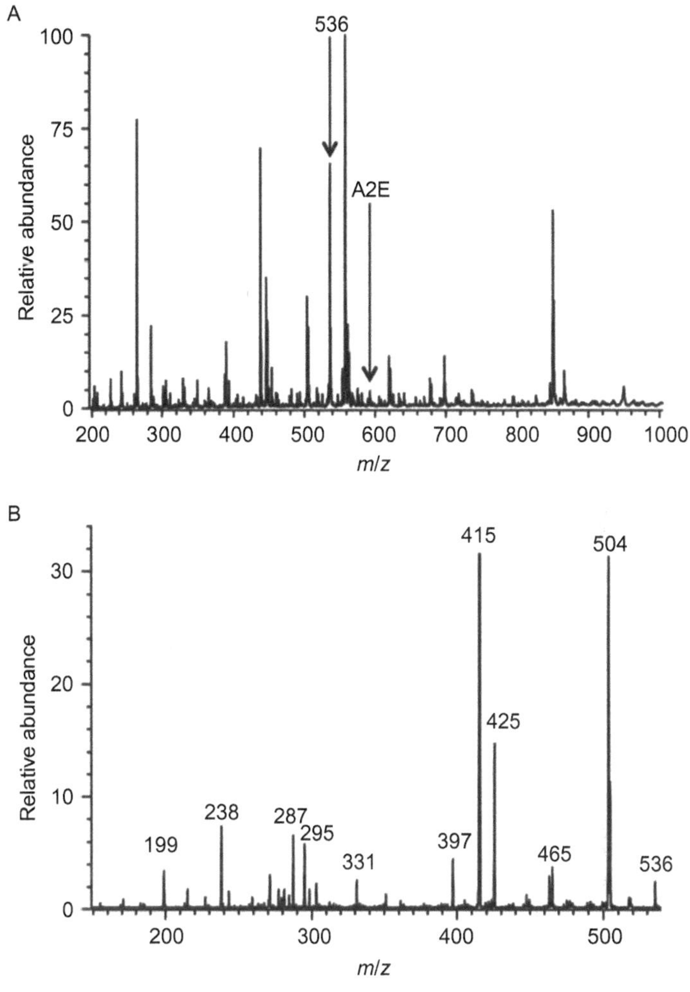

Figure 6 Example of a mass spectrometric analysis of sample fluorescent compound from human lipofuscin. A polar extract of human lipofuscin was chromatographed on silica thin-layer plates and an orange fluorescent thin-layer chromatography band separated. The analysis was performed following solubilizing the compounds from this band in methanol and direct injection of the extract into an LTQ ion trap mass spectrometer. The mass spectrum indicates the presence of numerous compounds (A). A peak corresponding to the molecular weight of A2E (*m/z* 592) as well as its first two oxidized species (*m/z* 608 and *m/z* 624) was detected; however, the most abundant peaks did not match previously identified *bis*-retinoids. The peak for *m/z* 536, a prominent peak in the spectrum, is indicated (B). Tandem mass spectrum corresponding to the *m/z* 536 peak shows a new chemically unidentified compound.

7. FUTURE DIRECTIONS

The surprising data obtained from the MALDI imaging studies are clearly not in agreement with the proposals that inhibition of A2E formation will prevent AMD. From the results on mice, both the extracts and the MALDI imaging, and from the *in vitro* results on A2E toxicity, this proposal was certainly not an unreasonable assumption. However, the results on human tissue indicate that for issues of retinal degeneration, the mouse is not a good model likely due to the differences in anatomy between the two species. The relative lack of A2E in the center of the RPE suggests that either cones, which are concentrated in the central area, have a different mechanism of handling the excess retinal or that the central RPE in the human has specific mechanisms for dealing with retinoid by-products. From our data, clearly the strong fluorescence in the center of the human RPE cannot be due to A2E. Our paucity of knowledge as to the actual content of lipofuscin in humans suggests this is a potential direction for future research.

Our group has therefore begun to examine the molecules in human lipofuscin which correlated in abundance with the fluorescence patterns in human lipofuscin, i.e., are strongest in abundance in the center of the RPE and weakest at the periphery. Our approach is to extract fluorescent compounds from human lipofuscin, determine the molecular weight and fragmentation pattern, and then map the spatial distribution of these compounds in the human RPE. If the distribution is in agreement with the lipofuscin pattern, the compounds will be structurally identified by structural chemical techniques. An example of the mass spectrum of one such compound (mass 536), which was isolated from lipofuscin and which does have the distribution pattern of human lipofuscin, is shown in Fig. 6. To our knowledge, this compound and indeed many of the most abundant molecules in human lipofuscin are not retinoid based (Z. Ablonczy, unpublished data). Now that it is established that the strong fluorescence of human lipofuscin likely does not come from A2E itself, it is critical to understand the source(s) of this fluorescence and to identify the compounds that are responsible so that new therapeutic approaches can be developed.

ACKNOWLEDGMENTS

This work was supported in part by National Institutes of Health Grants R01 EY004939 (R.K.C.), R01 EY019065 (Z.A.), R01 EY014850(Y.K.), R21EY020661 (Z.A., R.K.C.),

GM103391 (Vanderbilt University), and CO6 RR015455 (Medical University of South Carolina); Foundation Fighting Blindness, Inc. (R.K.C.); and an unrestricted grant (Department of Ophthalmology, Medical University of South Carolina) from Research to Prevent Blindness (RPB); RPB Senior Scientific Investigator Award (R.K.C.).

REFERENCES

1. Eldred GE, Katz ML. Fluorophores of the human retinal pigment epithelium: separation and spectral characterization. *Exp Eye Res.* 1988;47:71–86.
2. Parish CA, Hashimoto M, Nakanishi K, Dillon J, Sparrow J. Isolation and one-step preparation of A2E and iso-A2E, fluorophores from human retinal pigment epithelium. *Proc Natl Acad Sci USA.* 1998;95:14609–14613.
3. Seehafer SS, Pearce DA. You say lipofuscin, we say ceroid: defining autofluorescent storage material. *Neurobiol Aging.* 2006;27:576–588.
4. Brunk UT, Terman A. Lipofuscin: mechanisms of age-related accumulation and influence on cell function. *Free Radic Biol Med.* 2002;33:611–619.
5. Terman A, Brunk UT. Lipofuscin. *Int J Biochem Cell Biol.* 2004;36:1400–1404.
6. Yin D. Biochemical basis of lipofuscin, ceroid, and age pigment-like fluorophores. *Free Radic Biol Med.* 1996;21:871–888.
7. Ng KP, Gugiu B, Renganathan K, et al. Retinal pigment epithelium lipofuscin proteomics. *Mol Cell Proteomics.* 2008;7:1397–1405.
8. Katz ML, Gao CL. Vitamin A incorporation into lipofuscin-like inclusions in the retinal pigment epithelium. *Mech Ageing Dev.* 1995;84:29–38.
9. Katz ML, Redmond TM. Effect of Rpe65 knockout on accumulation of lipofuscin fluorophores in the retinal pigment epithelium. *Invest Ophthalmol Vis Sci.* 2001;42: 3023–3030.
10. Boulton M, Rozanowska M, Rozanowski B, Wess T. The photoreactivity of ocular lipofuscin. *Photochem Photobiol Sci.* 2004;3:759–764.
11. Sparrow JR, Nakanishi K, Parish CA. The lipofuscin fluorophore A2E mediates blue light-induced damage to retinal pigmented epithelial cells. *Invest Ophthalmol Vis Sci.* 2000;41:1981–1989.
12. Sparrow JR, Yamamoto K. The bisretinoids of RPE lipofuscin: a complex mixture. *Adv Exp Med Biol.* 2012;723:761–767.
13. Wassell J, Davies S, Bardsley W, Boulton M. The photoreactivity of the retinal age pigment lipofuscin. *J Biol Chem.* 1999;274:23828–23832.
14. Davies S, Elliott MH, Floor E, et al. Photocytotoxicity of lipofuscin in human retinal pigment epithelial cells. *Free Radic Biol Med.* 2001;31:256–265.
15. Holz FG, Strauss EC, Schmitz-Valckenberg S, van Lookeren CM. Geographic atrophy: clinical features and potential therapeutic approaches. *Ophthalmology.* 2014;121: 1079–1091.
16. Rudolf M, Vogt SD, Curcio CA, et al. Histologic basis of variations in retinal pigment epithelium autofluorescence in eyes with geographic atrophy. *Ophthalmology.* 2013;120:821–828.
17. Wald G. The molecular basis of visual excitation. *Nature.* 1968;219:800–807.
18. Dowling JE, Wald G. Vitamin A deficiency and night blindness. *Proc Natl Acad Sci USA.* 1958;44:648–661.
19. Rohrer B, Lohr HR, Humphries P, Redmond TM, Seeliger MW, Crouch RK. Cone opsin mislocalization in $RPE65^{-/-}$ mice: a defect that can be corrected by 11-cis retinal. *Invest Ophthalmol Vis Sci.* 2005;46:3876–3882.
20. Liu J, Itagaki Y, Ben-Shabat S, Nakanishi K, Sparrow JR. The biosynthesis of A2E, a fluorophore of aging retina, involves the formation of the precursor, A2-PE, in the photoreceptor outer segment membrane. *J Biol Chem.* 2000;275:29354–29360.

21. Boyer NP, Higbee D, Currin MB, et al. Lipofuscin and N-retinylidene-N-retinylethanolamine (A2E) accumulate in retinal pigment epithelium in absence of light exposure: their origin is 11-cis-retinal. *J Biol Chem*. 2012;287:22276–22286.
22. Quazi F, Molday RS. ATP-binding cassette transporter ABCA4 and chemical isomerization protect photoreceptor cells from the toxic accumulation of excess 11-cis-retinal. *Proc Natl Acad Sci USA*. 2014;111:5024–5029.
23. Sparrow JR, Gregory-Roberts E, Yamamoto K, et al. The bisretinoids of retinal pigment epithelium. *Prog Retin Eye Res*. 2012;31:121–135.
24. Vives-Bauza C, Anand M, Shiraz AK, et al. The age lipid A2E and mitochondrial dysfunction synergistically impair phagocytosis by retinal pigment epithelial cells. *J Biol Chem*. 2008;283:24770–24780.
25. Schutt F, Bergmann M, Kopitz J. Holz FG [Detergent-like effects of the lipofuscin retinoid component A2-E in retinal pigment epithelial cells]. *Ophthalmologe*. 2002;99: 861–865.
26. Lakkaraju A, Finnemann SC, Rodriguez-Boulan E. The lipofuscin fluorophore A2E perturbs cholesterol metabolism in retinal pigment epithelial cells. *Proc Natl Acad Sci USA*. 2007;104:11026–11031.
27. Wielgus AR, Chignell CF, Ceger P, Roberts JE. Comparison of A2E cytotoxicity and phototoxicity with all-trans-retinal in human retinal pigment epithelial cells. *Photochem Photobiol*. 2010;86:781–791.
28. Roberts JE, Kukielczak BM, Hu DN, et al. The role of A2E in prevention or enhancement of light damage in human retinal pigment epithelial cells. *Photochem Photobiol*. 2002;75:184–190.
29. Schutt F, Davies S, Kopitz J, Holz FG, Boulton ME. Photodamage to human RPE cells by A2-E, a retinoid component of lipofuscin. *Invest Ophthalmol Vis Sci*. 2000;41:2303–2308.
30. Suter M, Reme C, Grimm C, et al. Age-related macular degeneration. The lipofusion component N-retinyl-N-retinylidene ethanolamine detaches proapoptotic proteins from mitochondria and induces apoptosis in mammalian retinal pigment epithelial cells. *J Biol Chem*. 2000;275:39625–39630.
31. Caprioli RM, Farmer TB, Gile J. Molecular imaging of biological samples: localization of peptides and proteins using MALDI-TOF MS. *Anal Chem*. 1997;69:4751–4760.
32. Chan K, Lanthier P, Liu X, Sandhu JK, Stanimirovic D, Li J. MALDI mass spectrometry imaging of gangliosides in mouse brain using ionic liquid matrix. *Anal Chim Acta*. 2009;639:57–61.
33. Herring KD, Oppenheimer SR, Caprioli RM. Direct tissue analysis by matrix-assisted laser desorption ionization mass spectrometry: application to kidney biology. *Semin Nephrol*. 2007;27:597–608.
34. Minerva L, Clerens S, Baggerman G, Arckens L. Direct profiling and identification of peptide expression differences in the pancreas of control and ob/ob mice by imaging mass spectrometry. *Proteomics*. 2008;8:3763–3774.
35. Schwamborn K, Krieg RC, Reska M, Jakse G, Knuechel R, Wellmann A. Identifying prostate carcinoma by MALDI-imaging. *Int J Mol Med*. 2007;20:155–159.
36. Burnum KE, Tranguch S, Mi D, Daikoku T, Dey SK, Caprioli RM. Imaging mass spectrometry reveals unique protein profiles during embryo implantation. *Endocrinology*. 2008;149:3274–3278.
37. Grey AC, Chaurand P, Caprioli RM, Schey KL. MALDI imaging mass spectrometry of integral membrane proteins from ocular lens and retinal tissue. *J Proteome Res*. 2009;8(7):3278–3283.
38. Chaurand P, Sanders ME, Jensen RA, Caprioli RM. Proteomics in diagnostic pathology: profiling and imaging proteins directly in tissue sections. *Am J Pathol*. 2004;165:1057–1068.

39. Goodwin RJ, Pennington SR, Pitt AR. Protein and peptides in pictures: imaging with MALDI mass spectrometry. *Proteomics*. 2008;8:3785–3800.
40. Murphy RC, Hankin JA, Barkley RM. Imaging of lipid species by MALDI mass spectrometry. *J Lipid Res*. 2009;50(Suppl):S317–S322.
41. Drexler DM, Garrett TJ, Cantone JL, et al. Utility of imaging mass spectrometry (IMS) by matrix-assisted laser desorption ionization (MALDI) on an ion trap mass spectrometer in the analysis of drugs and metabolites in biological tissues. *J Pharmacol Toxicol Methods*. 2007;55:279–288.
42. Anderson DM, Ablonczy Z, Koutalos Y, et al. High resolution MALDI imaging mass spectrometry of retinal tissue lipids. *J Am Soc Mass Spectrom*. 2014;25:1394–1403.
43. Rubakhin SS, Sweedler JV. Characterizing peptides in individual mammalian cells using mass spectrometry. *Nat Protoc*. 2007;2:1987–1997.
44. Gutierrez DB, Blakeley L, Goletz PW, et al. Mass spectrometry provides accurate and sensitive quantitation of A2E. *Photochem Photobiol Sci*. 2010;9:1513–1519.
45. Bui TV, Han Y, Radu RA, Travis GH, Mata NL. Characterization of native retinal fluorophores involved in biosynthesis of A2E and lipofuscin-associated retinopathies. *J Biol Chem*. 2006;281:18112–18119.
46. Grey AC, Crouch RK, Koutalos Y, Schey KL, Ablonczy Z. Spatial localization of A2E in the retinal pigment epithelium. *Invest Ophthalmol Vis Sci*. 2011;52:3926–3933.
47. Ablonczy Z, Smith N, Anderson DM, et al. The utilization of fluorescence to identify the components of lipofuscin by imaging mass spectrometry. *Proteomics*. 2014;14:936–944.
48. Bhosale P, Serban B, Bernstein PS. Retinal carotenoids can attenuate formation of A2E in the retinal pigment epithelium. *Arch Biochem Biophys*. 2009;483:175–181.
49. Ablonczy Z, Higbee D, Anderson DM, et al. Lack of correlation between the spatial distribution of A2E and lipofuscin fluorescence in the human retinal pigment epithelium. *Invest Ophthalmol Vis Sci*. 2013;54:5535–5542.
50. Wing GL, Blanchard GC, Weiter JJ. The topography and age relationship of lipofuscin concentration in the retinal pigment epithelium. *Invest Ophthalmol Vis Sci*. 1978;17:601–607.
51. Ablonczy Z, Higbee D, Grey AC, Koutalos Y, Schey KL, Crouch RK. Similar molecules spatially correlate with lipofuscin and N-retinylidene-N-retinylethanolamine in the mouse but not in the human retinal pigment epithelium. *Arch Biochem Biophys*. 2013;539:196–202.

CHAPTER TWENTY-SEVEN

Cone Health and Retinoids

Masahiro Kono[1]
Department of Ophthalmology, Albert Florens Storm Eye Institute, Medical University of South Carolina, Charleston, South Carolina, USA
[1]Corresponding author: e-mail address: konom@musc.edu

Contents

Abstract

Cones are photoreceptor cells used for bright light and color vision. Retinoids are vitamin A derivatives, one of which is the 11-*cis* aldehyde form that serves as the chromophore for both cone and rod visual pigments. In the visual disease, Type 2 Leber congenital amaurosis (LCA2), 11-*cis*-retinal generation is inhibited or abolished. Work by others has shown that patients with LCA2 have symptoms consistent with degenerating cones. In mouse models for LCA2, early cone degeneration is readily apparent: cone opsins and other proteins associated with the outer segment are delocalized and cell numbers decline rapidly within the first month. Rods would appear normal morphologically and functionally, if not for the absence of chromophore. Supplementation of mouse models of LCA2 with *cis*-retinoids has been shown to slow loss of cone photoreceptor cells if mice were maintained in darkness. Thus, 11-*cis*-retinal appears not only to have a role in the light response reaction but also to promote proper trafficking of the cone opsins and maintain viable cones.

ABBREVIATIONS

DES-1 dihydroceramide desaturase
ER endoplasmic reticulum
LCA Leber congenital amaurosis
LRAT lecithin–retinol acyltransferase
M/LWS middle-/long-wavelength-sensitive
RDH retinol dehydrogenase
RPE retinal pigment epithelium
SWS1 class 1 short-wavelength-sensitive

Progress in Molecular Biology and Translational Science, Volume 134
ISSN 1877-1173
http://dx.doi.org/10.1016/bs.pmbts.2015.06.002
© 2015 Elsevier Inc.
All rights reserved.

1. INTRODUCTION

The light reaction in vision occurs in rod and cone photoreceptor cells. In the outer segments of rods and cones are photopigments that consist of a protein component called opsin and a retinoid-based chromophore, 11-*cis*-retinal (Fig. 1). This chromophore is the 11-*cis* aldehyde form of

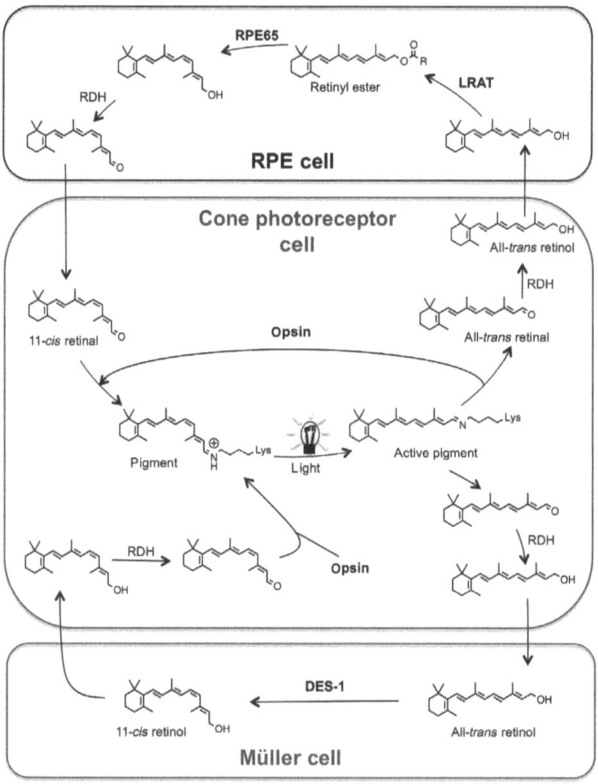

Figure 1 Schematic of the visual cycles associated with cone photoreceptor cells. Cones have two sources of 11-*cis*-retinoids to regeneratate cone visual pigments after photoactivation of the cone pigments. After photoactivtion of the pigment, the isomerized chromophore is released as all-*trans*-retinal, which is reduced to all-*trans*-retinol (vitamin A). This can then be transported to either the RPE or Müller cells. Isomerization from all-*trans*-retinol to 11-*cis*-retinal in the RPE uses LRAT and RPE65 in the process. 11-*cis*-retinol can be generated directly from all-*trans*-retinol in the Müller cells using DES-1[1] and converted to the useful 11-*cis*-retinal only in cones.[2,3] What occurs in the RPE, photoreceptor, and Müller cells are illustrated and represented by the separate boxes labeled accordingly.

vitamin A and common to all visual pigments. Because 11-*cis*-retinal is the compound that absorbs light, its interaction with the different opsins results in tuning their spectral sensitivity and functional properties. In visual pigments, 11-*cis*-retinal is covalently bound to a strictly conserved Lys within the seventh transmembrane helix via a Schiff-base linkage. Rods are utilized in dim light vision and cones used for bright light and color vision. All visual pigment proteins are highly homologous with about 40% identity at the amino acid level and members of the superfamily of G protein-coupled receptors. Furthermore, sequence analysis indicates that human pigments fall in three families[4]: Rh1 for rhodopsin, class 1 short-wavelength-sensitive (SWS1) for the blue cone pigment, and middle-/long-wavelength-sensitive (M/LWS) for both the green and red cone pigments.

When light is absorbed by the visual pigment, the bound retinoid isomerizes from the 11-*cis* to all-*trans* form. This photoisomerization results in a conformational change of the protein such that it is able to activate its G protein, transducin, to initiate the visual signaling cascade. During this process, the pigment is destroyed, and in order to reestablish light sensitivity, new 11-*cis*-retinal must be supplied. Thus, 11-*cis*-retinal is required for the light response and without it, vision is not possible. The generation of 11-*cis*-retinoids occurs through a process known as the visual cycle.[5]

In a disease where 11-*cis*-retinal generation is greatly inhibited such as with Type 2 Leber congenital amaurosis (LCA2), patients display early loss of visual acuity and degradation of the fovea consistent with loss of cones. In mouse models for LCA2, cones are clearly degenerating rapidly and supplementation with 11-*cis*-retinal improves cone function and survival indicating a role for 11-*cis*-retinal beyond imparting photosensitivity.

2. CONE PIGMENTS AND OPSINS

Although the general features of visual signaling components between cones and rods are similar, there are differences, in particular with how the opsins interact with retinoids. Single-cell electrophysiological studies have shown faster response and recovery kinetics with cones[6,7] than in rods.[8,9] *In vitro* studies of cone and rod visual pigments demonstrated retinal release after photoactivation is also faster in cone pigments than rhodopsin.[10,11] The bound chromophore is accessible to small molecules such as hydroxylamine in the dark with cone pigments but not with rhodopsin suggesting a more open structure with cone pigments.[12,13] It should be noted that salamander and mouse SWS1 cone pigments appear to be quite stable in the presence of hydroxylamine in the dark which may be due to a highly protected environment or unprotonated Schiff-base linkage.[11,14] Further demonstration of the looser hold on 11-*cis*-retinal in cone pigments was demonstrated by the ability of cone pigments to exchange chromophore in the dark.[15,16] Similarly, purified cone pigments are able to release retinal in the dark when excess chromophore is not present.[17]

Another role of 11-*cis*-retinal is to serve as an inverse agonist to the opsin.[18] In biochemical assays, both rod and cone opsins have been shown to be constitutively active in that they are able to activate transducin, the G protein in phototransduction, in the absence of ligand and light.[18–24] Depending on the opsin type, the degree of activity can differ. Our data indicate that SWS1 opsins, which include the human and mouse blue cone opsin, are more active than M/LWS cone opsins, which include the human red and green and mouse green cone opsins, and both classes of opsins are more active than the rod opsin.[2]

3. TWO VISUAL CYCLES FOR CONES

As mentioned in Section 1, the visual cycle describes the process that converts all-*trans*-retinol to an 11-*cis*-retinoid such that it can be used to regenerate visual pigment (Fig. 1). Classically, this process occurs in the retinal pigment epithelium (RPE). Its main features were described by Hubbard and Wald.[5] After the visual pigment is photobleached in the outer segments of photoreceptor cells (rods and cones), the chromophore dissociates from the opsin as all-*trans*-retinal, which is reduced to all-*trans*-retinol (Vitamin A). The retinol is then transported to the RPE where it is acylated (usually a palmitate group) via an ester linkage by lecithin–retinol acyltransferase (LRAT).[25,26] This retinyl ester is then isomerized to the 11-*cis* form and hydrolyzed by the protein RPE65[27–29] to form 11-*cis*-retinol. This is oxidized to 11-*cis*-retinal by retinol dehydrogenase 5 (RDH5),[30] although RDH11 could also contribute.[31] 11-*cis*-retinal can then be transported to the rod and cone outer segments to combine with the opsins to reform rhodopsin and cone visual pigments.

A second source for 11-*cis*-retinoids has recently been described for the Müller cells and referred to as the cone cycle. Das *et al.*[32] described results where cultured chicken Müller cells could take in all-*trans*-retinol and release 11-*cis*-retinol into the media. Mata *et al.*[33] further described that chicken Müller cells had an isomerase that converted all-*trans*-retinol to 11-*cis*-retinol directly. Coupled with earlier findings from Cornwall and coworkers,[3,34] a cone-specific visual cycle emerged. Jones *et al.*[3]

demonstrated that bleached cones, but not bleached rods, could regain sensitivity when administered 11-*cis*-retinol. They noted that 11-cis-retinol had a toxic effect on the rods. We later interpreted that as 11-*cis*-retinol serving as an agonist to rod opsins.[2,35] Cones, however, are able to oxidize 11-*cis*-retinol to 11-*cis*-retinal. Furthermore, Jin *et al.*[34] found that if only the cell bodies of cones were exposed to 11-*cis*-retinal, cones regained sensitivity; whereas, rods could not. These results suggested that 11-*cis*-retinoids have access to cones from the cell body region, which coincides with the extent to which Müller cells reach within the retina. Kefalov's group showed using retinas that an intraretinal cycle was needed to fully dark-adapt cones rapidly.[36] Kaylor *et al.*[1] has identified dihydroceramide desaturase-1 (DES-1) as the isomerase within Müller cells through an extensive screening process.

4. LEBER CONGENITAL AMAUROSIS

LCA2 is a cone-rod disease that severely affects vision early in childhood and is associated with mutations to the *RPE65* gene, which result in reduced or abolished levels of 11-*cis*-retinal.[37,38] Jacobson *et al.*[39] report that patients with LCA have thin foveas that worsen with age and no detectable blue vision, which is mediated by SWS1 cones. These are consistent with degenerating cones with those containing the more active opsins being more severely affected. Recently, there have been successful results with *RPE65* gene therapy trials to LCA2 patients.[40–42] Better results have occurred with younger patients,[43] presumably because fewer cones have been permanently lost. While vision has improved with these patients, a 3-year update of the *RPE65* gene therapy trials indicated that photoreceptors continued to degenerate.[44] LCA14 has a similar phenotype as LCA2 but mapped to mutations to LRAT.[45] Thus, mutations to enzymes critical to the generation of 11-*cis*-retinal in the visual cycle lead to visual defects.

There has been a recent study where an oral form of a 9-*cis*-retinoid was used to treat patients with LCA2 and LCA14.[46] Visual acuity was reported to be improved in 6 of the 14 patients with mutations to either RPE65 or LRAT, and 10 patients had improvements in their visual fields. The rationale for retinoid supplementation arose from animal model studies described in Section 5.

5. MOUSE MODELS FOR LCA

Redmond and coworkers[38] successfully knocked out the *Rpe65* gene in mice and demonstrated that this enzyme was necessary for generating

11-*cis*-retinal. LRAT was subsequently knocked out from mice[47] and found to have similar phenotype to the RPE65 knockout.[48] In both knockout models, there is a rapid degeneration of cones, in particular those predominantly populated with the blue cone opsins.[48,49] As the cones degenerate, Rohrer *et al.*[49] first noted that the opsins no longer were localized to the outer segments but rather distributed throughout the cells from the synapse to the outer segment, but cone morphology and function could be preserved if administered 11-*cis*-retinal in the dark (Fig. 2). 9-*cis*-retinoid was also used to prevent rapid cone death[51] as 9-*cis*-retinal can also function as a chromophore to rod and cone opsins.[52] If pigment formation was an important component in proper trafficking of cone opsins to the outer segment, we wondered if the presence of light would destroy the benefits of chromophore supplementation. Indeed, if mice lacking RPE65 were treated with

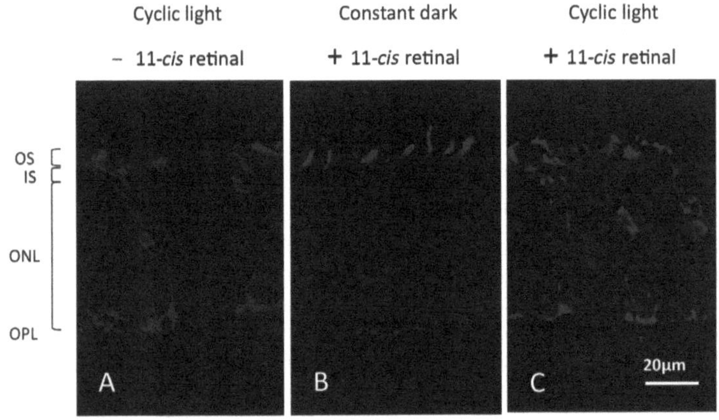

Figure 2 Mouse blue cone opsin localization in cone photoreceptors lacking endogenous 11-*cis*-retinal. Cross-sections of $Rpe65^{-/-}Rho^{-/-}$ mouse retinas at postnatal day 21 are shown. (A) Untreated mice, (B) 11-*cis*-retinal-treated mice maintained in the dark, and (C) 11-*cis*-retinal-treated mice maintained in 12 h light/12 h dark conditions. The sections were probed for the blue cone opsin (red, light gray in the print version) with a polyclonal antibody and nuclei (blue, dark gray in the print version) with DRAQ5. 11-*cis*-retinal-treated mice displayed improved opsin localization in the outer segment if maintained in the dark; whereas, under cyclic light conditions, there is still a pronounced level of mislocalized opsin in the cell body and down to the synapse. The double knockout (RPE65 and rhodopsin) was used to eliminate interference from the rod opsin in the study. The locations of the photoreceptor outer segment (OS), photoreceptor inner segment (IS), outer nuclear layer (ONL), and outer plexiform layer (OPL) are labeled beside the first image. *The figure has been adapted from Fan* et al.[50] *(the Association for Research in Vision and Ophthalmology is the copyright holder of the original figure).*

11-*cis*-retinal at around postnatal day 10 and kept in the dark, cross-sections of the retina 10 days later or counting cones 20 days later showed marked improvements over sham-treated mice.[50] However, if the same treatment was conducted on mice which were maintained on 12 h light/12 h dark schedule, cone opsin mislocalization was apparent and no apparent preservation of functional cones (Fig. 2).

6. ROLE OF RETINOIDS IN MAINTAINING HEALTHY CONES

It is quite clear that the absence of 11-*cis*-retinal results in cone degeneration in both animal models and in humans. Supplementation of *cis*-retinoids can preserve cones in mouse models[48,49,51,53] and in some patients with LCA.[46] At this point, we can only speculate on the mechanisms for degeneration and preservation through this treatment protocol. Fu and coworkers have approached this problem from an endoplasmic reticulum (ER) stress perspective.[54] Further support for ER stress contribution to cell death was shown by treating LRAT knockout mice with tauroursodeoxycholic acid, which can act as a chemical chaperone to relieve ER stress and thereby prevent cell death.[55] They did find that more cones remained in the retina, but the underlying problem of cone opsin trafficking was not solved. The same group subsequently showed that genetically deleting the gene for the blue cone opsin in an LCA mouse model prevented the rapid cone loss found in the single knockout.[56] Thus, there is truth to their assertion that the presence of cone opsins contributes to the rate of cone degeneration.[54,56] Recall that in LCA2 patients, blue vision is absent.[39] Furthermore, in LCA mouse models, the blue cone degeneration is much more severe than green cone degeneration,[48,51,53] and now the absence of mouse blue opsins prevented rapid cone cell death.[56]

If opsin is mediating photoreceptor degeneration, then how does 11-*cis*-retinal fit in? Our working hypothesis for this is that cones without 11-*cis*-retinal degenerate similarly to how rods degenerate with certain retinitis pigmentosa (RP) rhodopsin mutants. A rhodopsin mutant in which Lys296 has been substituted was identified as being associated with RP.[57] This is the lysine to which 11-*cis*-retinal forms a Schiff-base linkage. *In vitro* biochemical studies found such mutants to be highly constitutively active.[58] However, *in vivo* transgenic *Xenopus*[59] and mouse[60,61] studies demonstrated that the mutant opsins are inactive in rods. Instead, rod opsins are phosphorylated at their C-termini, associated with arrestin, and are no longer localized in the outer segments.[60,61] This should not be surprising as Rim *et al.*[62] reported that both rhodopsin kinase, which phosphorylates the C-terminal tail of rhodopsin, and transducin recognize the same active conformation of rhodopsin. Work pioneered by Deretic has further demonstrated that the C-terminus of rhodopsin is critical to the assembly of a complex of proteins into a vesicle that targets the cilium,[63–66] and its modification due to the properties of the mutant may disrupt this normal behavior.

The fate of cones under conditions of chromophore depletion may be similar to rods containing K296E rhodopsin. This active rhodopsin mutant cannot bind 11-*cis*-retinal and remains an opsin. Cone opsins are more constitutively active than wild-type rod opsin.[2] In mice that do not generate 11-*cis*-retinal, cone opsins are mislocalized and cone cells subsequently degenerate.[48,49,67] Cone opsins in the absence of 11-*cis*-retinal could be phosphorylated, associated with arrestin, leading to photoreceptor degeneration as opsins cannot all be packaged into a transport vesicle heading to the cilium and outer segment. Thus, targeting cone opsins to the membrane could be 11-*cis*-retinal dependent. Such ligand-dependent improvement of processing and transport of nascent mutant receptor proteins from the ER to the plasma membrane has been reported[68,69] and includes the P23H rhodopsin RP mutant in COS and HEK293 cells with an inverse agonist,[70] mutants of the V2 vasopressin receptor associated with nephrogenic diabetes insipidus with antagonists,[71] and the mu and delta opioid receptor with agonists.[72]

If true, then the cone visual cycle is well positioned to supply 11-*cis*-retinoids exclusively to cones via the cell body. If 11-*cis*-retinol oxidation were to occur there, then pigment could be formed as the cone opsins are being expressed and folded. Binding 11-*cis*-retinal would result in an opsin conformation that is not a good substrate for kinase and thus could be packaged into a vesicle targeted to the cilium and then the outer segment. Light would destroy the bound 11-*cis*-retinoid and result either in a photoactivated pigment or constitutively active opsin and prevent proper opsin localization.[50] One question that might arise is why the clinical trials on humans with 9-*cis*-retinoids would work. The answer could lie in what the actual genotype of the disease is. Koenekoop *et al.*[46] only list whether the mutation of the patient was in RPE65 or LRAT. Samardzija *et al.*[73,74] reported that when a mouse with a point mutation (R91W) in RPE65 was knocked-in, cones degenerated more slowly than if the gene had been completely knocked out. With the single mutant present,

11-*cis*-retinal was still being generated but at much lower levels than the wild type. Thus, depending on the specific mutation in the patient, the benefits of a retinoid supplement could vary such that those with milder conditions would likely have the better results. This could be why a portion of the patients improved significantly, while several patients saw no improvements.[46] With *cis*-retinoids coming from both the visual cycle and the external source, the patient with milder forms of LCA might be able to overcome the day-to-day ambient light conditions.

ACKNOWLEDGMENTS

This work was supported in part by National Institutes of Health Grant R01 EY019515 and an unrestricted grant from Research to Prevent Blindness to the Department of Ophthalmology at the Medical University of South Carolina.

REFERENCES

1. Kaylor JJ, Yuan Q, Cook J, et al. Identification of DES1 as a vitamin A isomerase in Müller glial cells of the retina. *Nat Chem Biol.* 2013;9:30–36.
2. Ala-Laurila P, Cornwall MC, Crouch RK, Kono M. The action of 11-*cis*-retinol on cone opsins and intact cone photoreceptors. *J Biol Chem.* 2009;284:16492–16500.
3. Jones GJ, Crouch RK, Wiggert B, Cornwall MC, Chader GJ. Retinoid requirements for recovery of sensitivity after visual-pigment bleaching in isolated photoreceptors. *Proc Natl Acad Sci USA.* 1989;86:9606–9610.
4. Ebrey T, Koutalos Y. Vertebrate photoreceptors. *Prog Retin Eye Res.* 2001;20:49–94.
5. Hubbard R, Wald G. Cis-trans isomers of vitamin A and retinene in the rhodopsin system. *J Gen Physiol.* 1952;36:269–315.
6. Perry RJ, McNaughton PA. Response properties of cones from the retina of the tiger salamander. *J Physiol.* 1991;433:561–587.
7. Schnapf JL, Nunn BJ, Meister M, Baylor DA. Visual transduction in cones of the monkey *Macaca fascicularis*. *J Physiol.* 1990;427:681–713.
8. Baylor DA, Lamb TD, Yau K-W. The membrane current of single rod outer segments. *J Physiol.* 1979;288:589–611.
9. Baylor DA, Nunn BJ, Schnapf JL. The photocurrent, noise and spectral sensitivity of rods of the monkey *Macaca fascicularis*. *J Physiol.* 1984;357:575–607.
10. Chen M-H, Kuemmel C, Birge RR, Knox BE. Rapid retinal release from a cone visual pigment following photoactivation. *Biochemistry.* 2012;51:4117–4125.
11. Das J, Crouch RK, Ma JX, Oprian DD, Kono M. Role of the 9-methyl group of retinal in cone visual pigments. *Biochemistry.* 2004;43:5532–5538.
12. Fukada Y, Okano T, Shichida Y, et al. Comparative study on the chromophore binding sites of rod and red-sensitive cone visual pigments by use of synthetic retinal isomers and analogs. *Biochemistry.* 1990;29:3133–3140.
13. Wald G, Brown PK, Smith PH. Iodopsin. *J Gen Physiol.* 1955;38:623–681.
14. Tsutsui K, Imai H, Shichida Y. Photoisomerization efficiency in UV-absorbing visual pigments: protein-directed isomerization of an unprotonated retinal Schiff base. *Biochemistry.* 2007;46:6437–6445.
15. Crescitelli F. The gecko visual pigment: the dark exchange of chromophore. *Vision Res.* 1984;24:1551–1553.

16. Matsumoto H, Tokunaga F, Yoshizawa T. Accessibility of the iodopsin chromophore. *Biochim Biophys Acta*. 1975;404:300–308.
17. Kefalov VJ, Estevez ME, Kono M, et al. Breaking the covalent bond—a pigment property that contributes to desensitization in cones. *Neuron*. 2005;46:879–890.
18. Cohen GB, Yang T, Robinson PR, Oprian DD. Constitutive activation of opsin: influence of charge at position 134 and size at position 296. *Biochemistry*. 1993;32: 6111–6115.
19. Surya A, Foster KW, Knox BE. Transducin activation by the bovine opsin apoprotein. *J Biol Chem*. 1995;270:5024–5031.
20. Melia Jr TJ, Cowan CW, Angleson JK, Wensel TG. A comparison of the efficiency of G protein activation by ligand-free and light-activated forms of rhodopsin. *Biophys J*. 1997;73:3182–3191.
21. Palczewski K, Jäger S, Buczylko J, et al. Rod outer segment retinol dehydrogenase: substrate specificity and role in phototransduction. *Biochemistry*. 1994;33:13741–13750.
22. Buczylko J, Saari JC, Crouch RK, Palczewski K. Mechanisms of opsin activation. *J Biol Chem*. 1996;271:20621–20630.
23. Kono M. Constitutive activity of a UV cone opsin. *FEBS Lett*. 2006;580:229–232.
24. Isayama T, Chen Y, Kono M, et al. Differences in the pharmacological activation of visual opsins. *Vis Neurosci*. 2006;23:899–908.
25. Deigner PS, Law WC, Cañada FJ, Rando RR. Membranes as the energy source in the endergonic transformation of vitamin A to 11-*cis*-retinol. *Science*. 1989;244:968–971.
26. Trehan A, Cañada FJ, Rando RR. Inhibitors of retinyl ester formation also prevent the biosynthesis of 11-*cis*-retinol. *Biochemistry*. 1989;29:309–312.
27. Jin M, Li S, Moghrabi WN, Sun H, Travis GH. Rpe65 is the retinoid isomerase in bovine pigment epithelium. *Cell*. 2005;122:449–459.
28. Moiseyev G, Chen Y, Takahashi Y, Wu BX, Ma J-x. RPE65 is the isomerohydrolase in the retinoid visual cycle. *Proc Natl Acad Sci USA*. 2005;102:12413–12418.
29. Redmond TM, Poliakov E, Yu S, Tsai J-Y, Lu Z, Gentleman S. Mutation of key residues of RPE65 abolishes its enzymatic role as isomerohydrolase in the visual cycle. *Proc Natl Acad Sci USA*. 2005;102:13658–13663.
30. Yamamoto H, Simon A, Eriksson U, Harris E, Berson EL, Dryja TP. Mutations in the gene encoding 11-*cis* retinol dehydrogenase cause delayed dark adaptation and fundus albipunctatus. *Nat Genet*. 1999;22:188–191.
31. Kim TS, Maeda A, Maeda T, et al. Delayed dark adaptation in 11-*cis*-retinol dehydrogenase-deficient mice: a role of RDH11 in visual processes *in vivo*. *J Biol Chem*. 2005;280:8694–8704.
32. Das SR, Bhardwaj N, Kjeldbye H, Gouras P. Muller cells of chicken retina synthesize 11-cis-retinol. *Biochem J*. 1992;285:907–913.
33. Mata NL, Ruiz A, Radu RA, Bui TV, Travis GH. Chicken retinas contain a retinoid isomerase activity that catalyzes the direct conversion of all-*trans*-retinol to 11-*cis*-retinol. *Biochemistry*. 2005;44:11715–11721.
34. Jin J, Jones GJ, Cornwall MC. Movement of retinal along cone and rod photoreceptors. *Vis Neurosci*. 1994;11:389–399.
35. Kono M, Goletz PW, Crouch RK. 11-*cis* and all-*trans* retinols can activate rod opsin: rational design of the visual cycle. *Biochemistry*. 2008;47:7567–7571.
36. Wang J-S, Estevez ME, Cornwall MC, Kefalov VJ. Intra-retinal visual cycle required for rapid and complete cone dark adaptation. *Nat Neurosci*. 2009;12:295–302.
37. Marlhens F, Bareil C, Griffoin J-M, et al. Mutations in *RPE65* cause Leber's congenital amaurosis. *Nat Genet*. 1997;17:139–141.
38. Redmond TM, Yu S, Lee E, et al. *Rpe65* is necessary for production of 11-*cis*-vitamin A in the retinal visual cycle. *Nat Genet*. 1998;20:344–451.
39. Jacobson SG, Aleman TS, Cideciyan AV, et al. Human cone photoreceptor dependence on RPE65 isomerase. *Proc Natl Acad Sci USA*. 2007;104:15123–15128.

40. Maguire AM, Simonelli F, Pierce EA, et al. Safety and efficacy of gene transfer for Leber's congenital amaurosis. *N Engl J Med.* 2008;358:2240–2248.
41. Bainbridge JWB, Smith AJ, Barker SS, et al. Effect of gene therapy on visual function in Leber's congenital amaurosis. *N Engl J Med.* 2008;358:2231–2239.
42. Cideciyan AV, Aleman TS, Boye SL, et al. Human gene therapy for RPE65 isomerase deficiency activates the retinoid cycle but with slow rod kinetics. *Proc Natl Acad Sci USA.* 2008;105:15112–15117.
43. Maguire AM, High KA, Auricchio A, et al. Age-dependent effects of *RPE65* gene therapy for Leber's congenital amaurosis: a phase 1 dose-escalation trial. *Lancet.* 2009;374:1597–1605.
44. Cideciyan AV, Jacobson SG, Beltran WA, et al. Human retinal gene therapy for Leber congenital amaurosis shows advancing retinal degeneration despite enduring visual improvement. *Proc Natl Acad Sci USA.* 2013;110:E517–E525.
45. Thompson DA, Li Y, McHenry CL, et al. Mutations in the gene encoding lecithin retinol acyltransferase are associated with early-onset severe retinal dystrophy. *Nat Genet.* 2001;28:123–124.
46. Koenekoop RK, Sui R, Sallum J, et al. Oral 9-cis retinoid for childhood blindness due to Leber congenital amaurosis caused by RPE65 or LRAT mutations: an open-label phase 1b trial. *Lancet.* 2014;384:1513–1520.
47. Batten ML, Imanishi Y, Maeda T, et al. Lecithin-retinol acyltransferase is essential for accumulation of all-*trans*-retinyl esters in the eye and in the liver. *J Biol Chem.* 2004;279:10422–10432.
48. Fan J, Rohrer B, Frederick JM, Baehr W, Crouch RK. Rpe65$^{-/-}$ and Lrat$^{-/-}$ mice: comparable models of Leber congenital amaurosis. *Invest Ophthalmol Vis Sci.* 2008;49:2384–2389.
49. Rohrer B, Lohr HR, Humphries P, Redmond TM, Seeliger MW, Crouch RK. Cone opsin mislocalization in Rpe65$^{-/-}$ mice: a defect that can be corrected by 11-*cis* retinal. *Invest Ophthalmol Vis Sci.* 2005;46:3876–3882.
50. Fan J, Crouch RK, Kono M. Light prevents exogenous 11-*cis* retinal from maintaining cone photoreceptors in chromophore-deficient mice. *Invest Ophthalmol Vis Sci.* 2011;52:2412–2416.
51. Maeda T, Cideciyan AV, Maeda A, et al. Loss of cone photoreceptors caused by chromophore depletion is partially prevented by the artificial chromophore pro-drug, 9-*cis*-retinyl acetate. *Hum Mol Genet.* 2009;18:2277–2287.
52. Hubbard R, Kropf A. The action of light on rhodopsin. *Proc Natl Acad Sci USA.* 1958;44:130–139.
53. Zhang H, Fan J, Li S, et al. Trafficking of membrane-associated proteins to cone photoreceptor outer segments requires the chromophore 11-*cis*-retinal. *J Neurosci.* 2008;28:4008–4014.
54. Zhang T, Zhang N, Baehr W, Fu Y. Cone opsin determines the time course of cone photoreceptor degeneration in Leber congenital amaurosis. *Proc Natl Acad Sci USA.* 2011;108:8879–8884.
55. Zhang T, Baehr W, Fu Y. Chemical chaperone TUDCA preserves cone photoreceptors in a mouse model of Leber congenital amaurosis. *Invest Ophthalmol Vis Sci.* 2012;53:3349–3356.
56. Zhang T, Enemchukwu NO, Jones A, et al. Genetic deletion of S-opsin prevents rapid cone degeneration in a mouse model of Leber congenital amaurosis. *Hum Mol Genet.* 2014;24:1755–1763.
57. Keen TJ, Inglehearn CF, Lester DH, et al. Autosomal dominant retinitis pigmentosa: four new mutations in rhodopsin, one of them in the retinal attachment site. *Genomics.* 1991;11:199–205.
58. Zhukovsky EA, Robinson PR, Oprian DD. Transducin activation by rhodopsin without a covalent bond to the 11-*cis*-retinal chromophore. *Science.* 1991;251:558–560.

59. Tam BM, Xie G, Oprian DD, Moritz OL. Mislocalized rhodopsin does not require activation to cause retinal degeneration and neurite outgrowth in *Xenopus laevis*. *J Neurosci*. 2006;26:203–209.
60. Li T, Franson WK, Gordon JW, Berson EL, Dryja TP. Constitutive activation of phototransduction by K296E opsin is not a cause of photoreceptor degeneration. *Proc Natl Acad Sci USA*. 1995;92:3551–3555.
61. Chen J, Shi G, Concepcion F, Xie G, Oprian D, Chen J. Stable rhodopsin/arrestin complex leads to retinal degeneration in a transgenic mouse model of autosomal dominant retinitis pigmentosa. *J Neurosci*. 2006;26:11929–11937.
62. Rim J, Oprian DD. Constitutive activation of opsin: interaction of mutants with rhodopsin kinase and arrestin. *Biochemistry*. 1995;34:11938–11945.
63. Deretic D, Puleo-Scheppke B, Trippe C. Cytoplasmic domain of rhodopsin is essential for post-Golgi vesicle formation in a retinal cell-free system. *J Biol Chem*. 1996;271:2279–2286.
64. Deretic D, Schmerl S, Hargrave PA, Arendt A, McDowell JH. Regulation of sorting and post-Golgi trafficking of rhodopsin by its C-terminal sequence QVS(A)PA. *Proc Natl Acad Sci USA*. 1998;95:10620–10625.
65. Deretic D, Williams AH, Ransom N, Morel V, Hargrave PA, Arendt A. Rhodopsin C terminus, the site of mutations causing retinal disease, regulates trafficking by binding to ADP-ribosylation factor 4 (ARF4). *Proc Natl Acad Sci USA*. 2005;102:3301–3306.
66. Mazelova J, Astuto-Gribble L, Inoue H, et al. Ciliary targeting motif VxPx directs assembly of a trafficking module through Arf4. *EMBO J*. 2009;28:183–192.
67. Znoiko SL, Rohrer B, Lu K, Lohr HR, Crouch RK, Ma J-x. Downregulation of cone-specific gene expression and degeneration of cone photoreceptors in the *rpe65*$^{-/-}$ mouse at early ages. *Invest Ophthalmol Vis Sci*. 2005;46:1473–1479.
68. Conn PM, Ulloa-Aguirre A, Ito J, Janovick JA. G protein-coupled receptor trafficking in health and disease: lessons learned to prepare for therapeutic mutant rescue in vivo. *Pharmacol Rev*. 2007;59:225–250.
69. Drake MT, Shenoy SK, Lefkowitz RJ. Trafficking of G protein-coupled receptors. *Circ Res*. 2006;99:570–582.
70. Noorwez SM, Kuksa V, Imanishi Y, et al. Pharmacological chaperone-mediated in vivo folding and stabilization of the P23H-opsin mutant associated with autosomal dominant retinitis pigmentosa. *J Biol Chem*. 2003;278:14442–14450.
71. Morello J-P, Salahpour A, Laperrière A, et al. Pharmacological chaperones rescue cell-surface expression and function of misfolded V2 vasopressin receptor mutants. *J Clin Invest*. 2000;105:887–895.
72. Hasbi A, Nguyen T, Fan T, et al. Trafficking of preassembled opioid m-d heteroligomer-Gz signaling complexes to the plasma membrane: coregulation by agonists. *Biochemistry*. 2007;46:12997–13009.
73. Samardzija M, von Lintig J, Tanimoto N, et al. R91W mutation in Rpe65 leads to milder early-onset retinal dystrophy due to the generation of low levels of 11-*cis*-retinal. *Hum Mol Genet*. 2008;17:281–292.
74. Samardzija M, Tanimoto N, Kostic C, et al. In conditions of limited chromophore supply rods entrap 11-cis-retinal leading to loss of cone function and cell death. *Hum Mol Genet*. 2009;18:1266–1275.

CHAPTER TWENTY-EIGHT

Retinoid Processing in Induced Pluripotent Stem Cell-Derived Retinal Pigment Epithelium Cultures

Mark A. Fields[1,2], Hannah E. Bowrey, Jie Gong, Zsolt Ablonczy, Lucian V. Del Priore

Department of Ophthalmology, Albert Florens Storm Eye Institute, Medical University of South Carolina, Charleston, South Carolina, USA

[1]Corresponding author: e-mail address: fieldm@musc.edu

Contents

Abstract

Stem cell therapy for retinal degenerative diseases such as age-related macular degeneration is a promising clinical option for the replacement of photoreceptors and retinal pigment epithelium (RPE). Induced pluripotent stem cell technology has emerged as a viable potential source of cells for transplantation in retinal degenerative disorders. Induced pluripotent stem cells have been used to derive RPE and have been tested for their functional behavior. These cells have the ability to express RPE-specific proteins and morphologically resemble native RPE. Induced pluripotent stem cell-derived RPE are also able to contribute to the visual cycle by their ability to metabolize all-*trans* retinol, a critical function of RPE in maintaining visual function. Advances in induced

[2] Senior author.

Progress in Molecular Biology and Translational Science, Volume 134
ISSN 1877-1173
http://dx.doi.org/10.1016/bs.pmbts.2015.06.004
© 2015 Elsevier Inc.
All rights reserved.

pluripotent stem cell technology will contribute to the development of clinical therapies for retinal degenerative diseases as well as provide a tool to understand the pathology of these disorders.

ABBREVIATIONS

AMD age-related macular degeneration
CRALBP cellular retinaldehyde-binding protein
ESC embryonic stem cell
GA geographic atrophy
HPLC high-performance liquid chromatography
iPSC induced pluripotent stem cell
iPSC-RPE induced pluripotent stem cell-derived retinal pigment epithelium
LRAT lecithin retinol acetyltransferase
PC phosphatidylcholine
RPE retinal pigment epithelium
RPE65 retinal pigment epithelium-specific 65
TER transepithelial resistance
ZO-1 zona occluden-1

1. INTRODUCTION

There has been well-founded optimism that stem cell therapy may provide a novel avenue for the treatment of disease by replacing dying or malignant cells with healthy ones. To mention every disease treatment being investigated by stem cells is beyond the scope of this chapter. Suffice it to say that the medical applications of stem cells are broad, and there is optimism that stem cell treatment may become available for diseases such as autism, epilepsy, Parkinson, and Alzheimer. Pertinently, stem cell treatments to restore sight from blinding diseases are currently being developed.

In this chapter, we present a review of the current state of stem cells for treatments of blinding diseases, such as age-related macular degeneration (AMD). Additionally, the merit of stem cell-derived models of normal visual function, including retinoid processing, will be discussed.

2. STEM CELL DEFINITIONS AND TYPES

The term "stem cell" refers—not to a single cell type—but to two groups of different cell types: adult stem cells and embryonic stem cells (ESCs). Adult stem cells (or tissue cells) replenish tissues throughout the life of the adult. For example, blood cells make blood; fat cells make fat. Thus, adult stem cells are dedicated to repair and tissue maintenance throughout life. Embryonic (or pluripotent) stem cells are present in early development, before tissue is formed. Stem cells, whether adult or embryonic, are defined by two properties: self-renewal and differentiation. Self-renewal refers to the ability to complete many cycles of cell division while still retaining the undifferentiated state. Differentiation means these cells can become different types of cells.

The greater the number of cell types these stem cells can become, the greater their potency. Stem cells are categorized into five levels by their potency: totipotent, pluripotent, multipotent, oligopotent, bipotent, and unipotent. Totipotent stem cells have the ability to differentiate into embryonic cell types. These cells are produced by the fusion of the ovum and sperm cell. Additionally, the first several divisions of the zygote are totipotent. Pluripotent stem cells descend from totipotent stem cells and can potentially take on many cell fates derived from any of the three germ layers.[1] Multipotent stem cells can differentiate into several cell types, but fewer than those of pluripotent stem cells. Oligopotent, bipotent, and unipotent stem cells can develop into a few, two, or a single cell type.[2,3] Only embryonic cells are naturally pluripotent.

Despite the value of embryonic cells as a resource for basic and translational research, their use presents some major challenges for research. In order to utilize these cells, embryos must be destroyed. Groups citing concerns pertaining primarily to the moral, ethical, and religious implications of embryo research have, in various countries, led to the restriction of such study. Additionally, as with any transplantation procedure, graft rejection can occur, and has serious health consequences for the patient.

3. INDUCED PLURIPOTENT STEM CELLS

Induced pluripotent stem cells (iPSCs) were initially generated by the Yamanaka group in 2006.[4] The group used an adenovirus-based vector to introduce transcriptional regulators SOX2, OCT3/4, Klf4, and c-MYC to reprogram mouse fibroblasts, which became iPSCs after 2–3 weeks. This combination of transgenes has since been used to reprogram somatic cells of a number of species, including the rat,[5] dog,[6] rhesus monkey,[7] and human[8]. After reprogramming, these iPSCs are capable of differentiating into the three germ layer cell types (mesoderm, ectoderm, and endoderm), similar to that of embryonic cells.[4,8–10] Valuably, just like embryonic cells, these induced stem cells are pluripotent and thus can take on many cell fates. Therefore, this significant advancement of the stem cell research field through iPSCs has allowed researchers to generate stem cells without the controversial use of embryos or the serious health consequences of graft rejection.

The applications for such iPSCs are vast, but of particular importance are two application classes: disease modeling and cell-based therapy. iPSCs offer a platform to study the mechanism of disease as the cells are patient specific. By generating iPSCs from patients with disease, models that simulate disease phenotypes can be used to better understand the pathophysiology of these diseases. Such models have been used for understanding diseases such as Down syndrome and Type 1 diabetes (for an in-depth review, see Ref. 11). Importantly, models can also be developed to better understand inherited diseases given their clinical and genetic heterogeneity.[12] Additionally, cells derived from a particular patient can be used for patient-specific testing of drug discovery and/or toxicity, providing a new standard for personalized medicine.[13] These applications have already had an impact on the understanding and treatment of blinding disorders such as AMD and retinitis pigmentosa, and are discussed below.

4. RETINAL PIGMENT EPITHELIUM, THE VISUAL CYCLE, AND AGE-RELATED MACULAR DEGENERATION

Retinoid metabolism is a principal function of the visual system, and the retinal pigment epithelium (RPE) plays an important role in this process. The uptake and processing of retinoids via the visual cycle is a critical aspect of phototransduction and there is a flow of these molecules between the photoreceptors and RPE.[14,15] As depicted in Fig. 1, this process involves the shuttling of 11-*cis*-retinal to the rod photoreceptor cell from the RPE, where it binds to a protein called opsin to form the visual pigment, rhodopsin.[14,16,17] Absorption of a photon catalyzes the isomerization of 11-*cis*-retinal to all-*trans*-retinal and results in its release.[18,19] This isomerization triggers a cascade of events, leading to the generation of an electrical signal to the optic nerve resulting in vision. All-*trans* retinal is reduced to all-*trans*-retinol within the photoreceptors and then transported to the RPE, where it is converted to an all-*trans* retinyl ester. All-*trans* retinyl ester is then isomerized by the protein retinal pigment epithelium-specific 65 (RPE65) and hydrolyzed to release 11-*cis* retinol. 11-*cis* retinol is then oxidized by

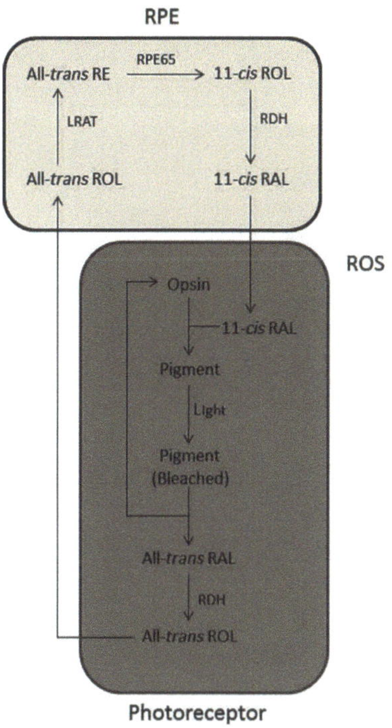

Figure 1 Classic visual cycle. Retinoids are metabolized and shuttled between the RPE (beige; light gray in the print version) and both rod and cone OS (outer segments; depicted generically in orange; dark gray in the print version). RAL, retinal; ROL, retinol; RDH, retinol dehydrogenase; LRAT, lecithin retinol acetyltransferase; RE, retinyl-ester. *With the full permission of all authors of the original publication, Fig. 4 of Ref. 14 has been modified for inclusion here.*

11-*cis* retinol dehydrogenase into 11-*cis*-retinal and then transported back to the photoreceptors as mentioned above. This loop is critical for light perception and eventual vision. The Ablonczy lab has demonstrated this process *in vitro* in fetal RPE cells (Fig. 2).[20]

AMD, a blinding disease that affects 30–50 million elderly people worldwide, is associated with RPE loss.[21,22] Approximately 30% of adults over the age of 75 show signs of AMD and around 10% of patients develop the late stage of disease.[23,24] AMD primarily exists in two forms: dry (atrophic or nonexudative) and wet (neovascular or exudative). Early in the course of disease, there is cellular dysfunction without cell death. In late-stage disease, AMD is characterized by extensive cell death. In the case of dry AMD, loss of vision develops due to loss of the RPE, photoreceptors, and/or choriocapillaris; this can lead to patches of geographic atrophy (GA) which result

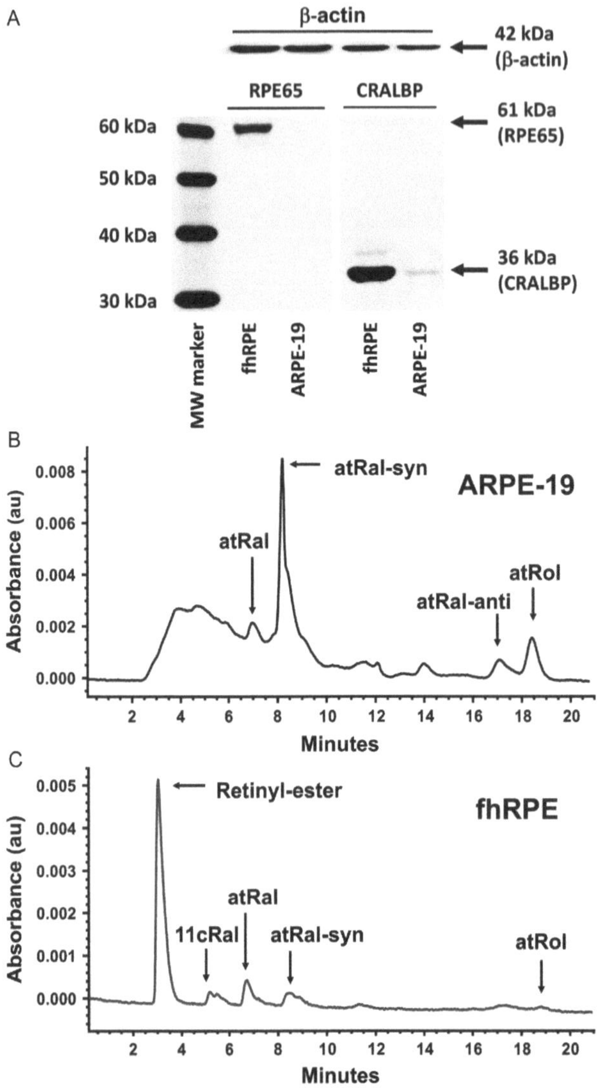

Figure 2 Retinoid metabolism in monolayer RPE cultures. Expression of RPE visual cycle proteins (A) showing immunoblots against RPE65 and CRALBP in both fhRPE and ARPE-19 cells. While fhRPE cells abundantly express both proteins, ARPE-19 cells express only low quantities of CRALBP. β-Actin indicated approximately equal loading of the lanes. Functional retinoid metabolism was missing from ARPE-19 cells (B); but was present in fhRPE cells (C) as shown by the corresponding retinoid profiles (aldehydes indicated both as unconverted and converted to oximes by hydroxylamine) taken by monitoring the HPLC at 360 nm 2 days postadministration of 5 μmol/L all-*trans* retinal (atRal) in phosphatidylcholine (PC) vesicles. The profile for ARPE-19 cells (B) was dominated by unconverted all-*trans* retinal, while the profile for fhRPE cells (C) and was dominated by retinyl esters and 11-*cis*-retinal (11cRal), indicating functional retinoid metabolism. The data are representative of three independent experiments. *With the full permission of all authors of the original publication, Fig. 3 of Ref. 20 has been included here.*

clinically in paracentral scotomas.[25] Replacement of dying RPE cells with new healthy RPE cells may be a useful treatment for AMD. Further, models of *in vitro* cell cultures of RPE from patients with AMD/GA will likely be useful for studying RPE-related AMD pathogenesis.

5. INDUCED PLURIPOTENT STEM CELL-DERIVED RETINAL PIGMENT EPITHELIUM

There have been many clinical challenges in the transplantation of intact sheets and suspensions of primary RPE cells in patients with retinal disorders.[26–31] Stem cell engineering allows us to overcome these obstacles inherent in this treatment through the use of iPSCs, which can then be differentiated into RPE using well-established differentiation protocols. Stem cells are an ideal replacement source for these lost or damaged cells, since stem cells have a significant ability to proliferate *in vitro*—prior to transplantation—and *in vivo*—after subretinal transplantation.

The use of iPSC-derived RPE as a research tool and potential therapeutic for cell transplantation therapy holds great potential for the treatment of retinal degeneration. For example, patients with advanced GA in AMD will require repopulation of the atrophic area with new cells, namely, new RPE.[32] In principle there are several potential sources for donor RPE, including adult RPE, fetal RPE, immortalized RPE (i.e., ARPE-19), and ESCs. These sources are limited and can often vary in terms of quality and expansion capacity.[26,33–35] As noted above, there are also ethical implications as well as the problem of a limited reservoir of donor cells.[36] There have been many important studies demonstrating the use of ESCs for the treatment of GA including a Phase 1 clinical trial in humans started in 2011 by Advanced Cell Technology, Inc., for the treatment of retinal degenerative disorders.[26,37] This is the first FDA-approved trial for the treatment of macular degeneration using RPE derived from human ESCs that have been injected into the subretinal space of these patients. One caveat is that these patients are on an immunosuppressive regimen for the duration of their treatment.

The use of iPSC technology offers a source of individualized or patient-specific cells that can then be differentiated into iPSC-derived RPE. Many research groups have efficiently differentiated iPS cells into RPE using established protocols.[38–41] These cells morphologically and functionally behave like native RPE. Figure 3 demonstrates the differentiation process of establishing iPSC-derived cell lines. Cultures are established within 45–60 days and tested for their ability to produce RPE-specific markers such

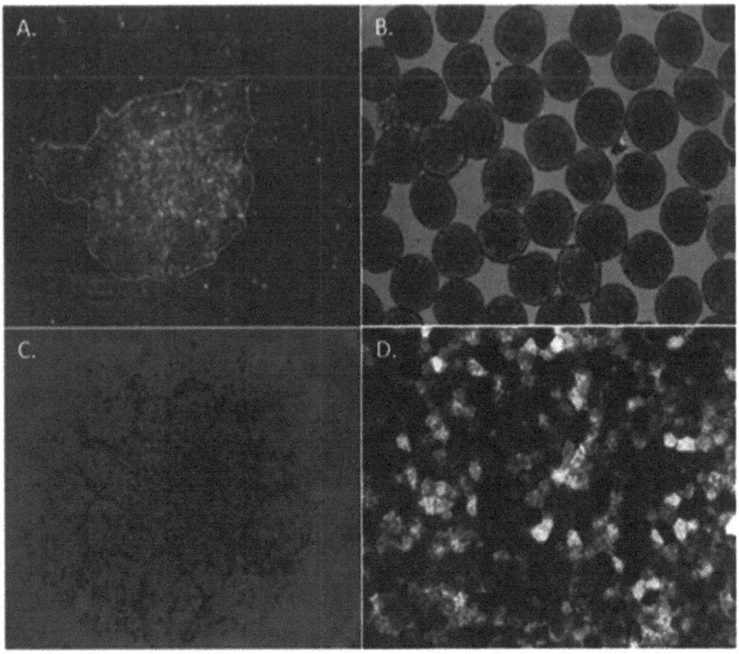

Figure 3 Differentiation of induced pluripotent stem cells toward an RPE fate. Undifferentiated iPSC colony at day 0 (A). Embryonic bodies form by day 7 (B), and eventual formation of neural aggregates (C) by day 14. Pigmented monolayer of iPSC-derived RPE form by day 45 postdifferentiation (D).

as tight junction protein zona occluden-1 (ZO-1) (Fig. 4). Functional testing, such as integrity of the monolayer is analyzed (Fig. 5). When treating a disorder such as GA, this approach offers the advantage of potentially circumventing the problem of immune rejection after transplantation in the area of disease as well as being a reservoir for large numbers of cells for replacement. The feasibility of this approach has been demonstrated in animal models of retinal degeneration.[39,42]

IPSC-derived RPE can also provide a valuable research tool and platform to understand disease. Through the generation of iPS cells from patients with specific diseases such as AMD and Best disease, genetic studies and models to express particular disease phenotypes can be developed which can then be used to understand the pathophysiology of disease and determine the efficacy of therapeutic interventions.[13,43,44] These models can help to understand human inherited diseases,[12] and can serve as a tool for drug discovery and pharmacological toxicity testing of therapeutic agents, what is often referred to as personalized medicine.[13,45,46]

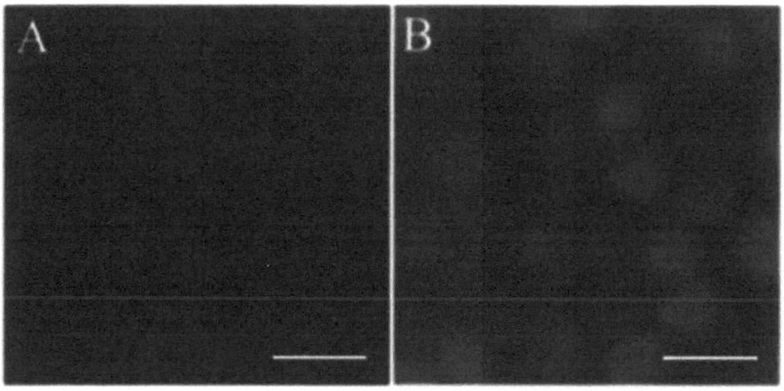

Figure 4 Epithelial expression of RPE cell markers. Immunofluorescent staining of RPE marker ZO-1 (A) in pigmented IPS-derived RPE cells. (B) DAPI image of same cells. Scale bars = 50 μm.

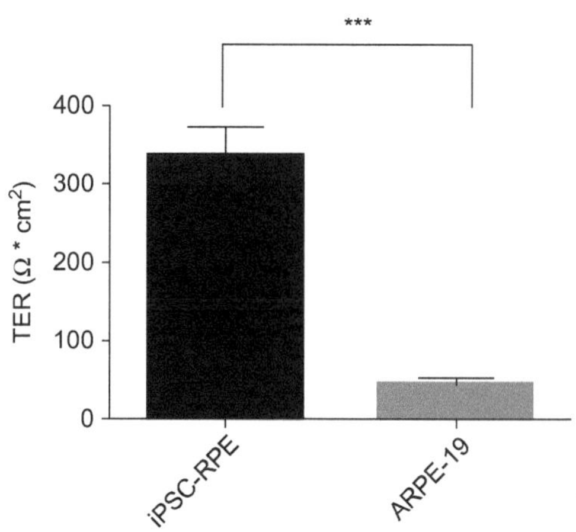

Figure 5 Transepithelial resistance (TER) of iPSC-derived RPE. Monolayer permeability was assessed by transepithelial resistance measurements in polarized iPSC-RPE cells. Transepithelial resistance was tested at an average of $335 \pm 36\ \Omega * cm^2$ for iPSC-RPE compared with ARPE-19 cells which tested at an average resistance of $46 \pm 73\ \Omega * cm^2$.

6. RETINOID PROCESSING IN INDUCED PLURIPOTENT STEM CELL-DERIVED RETINAL PIGMENT EPITHELIUM

When considering the use of iPSCs derived as a therapeutic option for individuals with retinal degenerations, it is vital that these cells are able to perform and restore visual cycle regeneration.[47,48] *In vitro* assays have been developed to verify visual cycle function in iPSC-derived RPE cultures. While many research groups have efficiently developed iPSC-RPE cells, only a limited number have demonstrated the production of retinoids in these cultures.

There are different systems that can induce the production and analysis of retinoids. Maeda *et al.* have demonstrated the ability of iPSC-RPE cells to express visual cycle enzymes both *in vivo* and *in vitro*.[48] These cells are able to produce visual cycle proteins such as RPE65, lecithin retinol acetyltransferase (LRAT), and cellular retinaldehyde-binding protein (CRALBP).[48] It was shown that the cellular environment and culture conditions may play a major role in the maintenance of visual cycle proteins in these cultures. Muniz *et al.* have also conducted studies on visual cycle production by iPSC-RPE cultures. In these studies, iPSC-RPE cultures were able to uptake all-*trans* retinol, synthesize retinyl esters, and release 11-*cis*-retinaldehyde into the culture media after analysis by high-performance liquid chromatography (HPLC).[47] Similar experiments have been conducted by Ablonczy *et al.* in human fetal cultures to assess retinoid metabolism.[20] Briefly, iPSC-RPE cells are cultured to confluence, forming cobblestone morphology and allowed to produce pigment. Cultures are treated with all-*trans* retinal in phosphatidylcholine (PC) vesicles. Media is then collected, cells lysed, and HPLC analysis follows. Separation of retinoids is performed and quantified by comparison of retention times and absorption properties with pure retinoid isomeric standards.[20,49] As shown in the HPLC profile in Fig. 6, iPSC-RPE monolayers treated with all-*trans* retinol form detectable amounts of 11-*cis*-retinal. These results suggest the ability of these cultures to process retinoids may be similar to that of native RPE. Given the ability of these cells, it is clear that iPSC-RPE could be a potential therapeutic tool for transplantation.

7. FUTURE DIRECTIONS

iPSCs are a promising technology as a research tool and potential reservoir of cells for the treatment of many diseases, but challenges do remain. For example, IPSCs derived from the affected patient contain the predisposing mutation that initially caused the disease. This can provide a challenge when assessing use of these particular cells given their dysfunction, but it also provides a unique opportunity for disease models. Use of technology such as clustered regularly interspaced short palindromic repeats (CRISPR) has also begun to change the way disease mutations are approached. This technology has the ability to correct mutations directly within the cell allowing for precise and targeted gene therapy.[50] However,

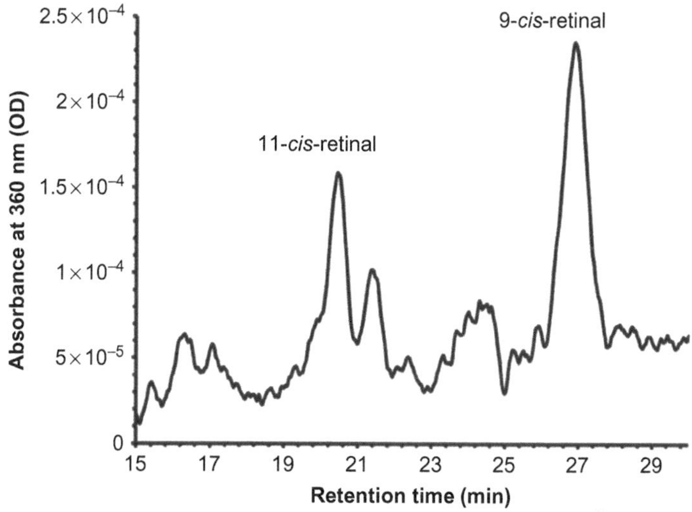

Figure 6 Retinoid metabolism in iPSC-RPE monolayers. iPSC-RPE were cultured in 6-well plates until a confluent and pigment monolayer was observed. Retinoid profiles were taken by monitoring the HPLC at 360 nm, 2 days postadministration of 5 μmol/L all-*trans* retinol (atRol) in 1% bovine serum albumin. iPSC-RPE cultures express 11-*cis*-retinal indicating functional retinoid metabolism. No hydroxylamine was used in these experiments, therefore oximes are not detected. Note that significant quantities of the more stable 9-*cis* isomer are also formed with the administration of all-*trans* retinol, and that the solvent system/column used in this experiment was different from that in Fig. 2 to better resolve the retinal isomers.

challenges remain in efficient RPE generation. Optimizing protocols to provide homogenous populations in sufficient numbers suitable for transplantation will have to be addressed. Better understanding of iPSC technology and refining the methodology of iPSC-derived RPE generation will create new inroads in regenerative medicine and in treatment of retinal degenerations.

ACKNOWLEDGMENTS

Supported in part by NIH grant EY19065 (Z. A.) and an unrestricted grant to MUSC-SEI from Research to Prevent Blindness, New York, NY; and Foundation Fighting Blindness, Columbia, MD. The authors would also like to thank Dr. Rosalie Crouch for critically reading this chapter.

REFERENCES

1. Thomson JA, Itskovitz-Eldor J, Shapiro SS, et al. Embryonic stem cell lines derived from human blastocysts. *Science*. 1998;282:1145–1147.

2. Cumano A, Paige CJ, Iscove NN, Brady G. Bipotential precursors of B cells and macrophages in murine fetal liver. *Nature*. 1992;356:612–615.
3. Lantz CS, Huff TF. Differential responsiveness of purified mouse c-kit+ mast cells and their progenitors to IL-3 and stem cell factor. *J Immunol*. 1995;155:4024–4029.
4. Takahashi K, Yamanaka S. Induction of pluripotent stem cells from mouse embryonic and adult fibroblast cultures by defined factors. *Cell*. 2006;126:663–676.
5. Liao J, Cui C, Chen S, et al. Generation of induced pluripotent stem cell lines from adult rat cells. *Cell Stem Cell*. 2009;4:11–15.
6. Shimada H, Nakada A, Hashimoto Y, Shigeno K, Shionoya Y, Nakamura T. Generation of canine induced pluripotent stem cells by retroviral transduction and chemical inhibitors. *Mol Reprod Dev*. 2010;77:2.
7. Liu H, Zhu F, Yong J, et al. Generation of induced pluripotent stem cells from adult rhesus monkey fibroblasts. *Cell Stem Cell*. 2008;3:587–590.
8. Takahashi K, Tanabe K, Ohnuki M, et al. Induction of pluripotent stem cells from adult human fibroblasts by defined factors. *Cell*. 2007;131:861–872.
9. Okita K, Nakagawa M, Hyenjong H, Ichisaka T, Yamanaka S. Generation of mouse induced pluripotent stem cells without viral vectors. *Science*. 2008;322:949–953.
10. Yamanaka S. A fresh look at iPS cells. *Cell*. 2009;137:13–17.
11. Robinton DA, Daley GQ. The promise of induced pluripotent stem cells in research and therapy. *Nature*. 2012;481:295–305.
12. Jin ZB, Okamoto S, Mandai M, Takahashi M. Induced pluripotent stem cells for retinal degenerative diseases: a new perspective on the challenges. *J Genet*. 2009;88:417–424.
13. Song M, Paul S, Lim H, Dayem AA, Cho SG. Induced pluripotent stem cell research: a revolutionary approach to face the challenges in drug screening. *Arch Pharm Res*. 2012;35:245–260.
14. Tang PH, Kono M, Koutalos Y, Ablonczy Z, Crouch RK. New insights into retinoid metabolism and cycling within the retina. *Prog Retin Eye Res*. 2013;32:48–63.
15. Kiser PD, Golczak M, Maeda A, Palczewski K. Key enzymes of the retinoid (visual) cycle in vertebrate retina. *Biochim Biophys Acta*. 1821;2012:137–151.
16. Baylor DA, Burns ME. Control of rhodopsin activity in vision. *Eye*. 1998;12(Pt. 3b): 521–525.
17. Okada T, Ernst OP, Palczewski K, Hofmann KP. Activation of rhodopsin: new insights from structural and biochemical studies. *Trends Biochem Sci*. 2001;26:318–324.
18. Lamb TD, Collin SP, Pugh Jr EN. Evolution of the vertebrate eye: opsins, photoreceptors, retina and eye cup. *Nat Rev Neurosci*. 2007;8:960–976.
19. Lamb TD, Pugh Jr EN. Dark adaptation and the retinoid cycle of vision. *Prog Retin Eye Res*. 2004;23:307–380.
20. Ablonczy Z, Dahrouj M, Tang PH, et al. Human retinal pigment epithelium cells as functional models for the RPE in vivo. *Invest Ophthalmol Vis Sci*. 2011;52:8614–8620.
21. Defoe DM, Ahmad A, Chen W, Hughes BA. Membrane polarity of the Na(+)-K+ pump in primary cultures of Xenopus retinal pigment epithelium. *Exp Eye Res*. 1994;59:587–596.
22. Tuo J, Grob S, Zhang K, Chan CC. Genetics of immunological and inflammatory components in age-related macular degeneration. *Ocul Immunol Inflamm*. 2012;20:27–36.
23. Prasad PS, Schwartz SD, Hubschman JP. Age-related macular degeneration: current and novel therapies. *Maturitas*. 2010;66:46–50.
24. Klein R, Chou CF, Klein BE, Zhang X, Meuer SM, Saaddine JB. Prevalence of age-related macular degeneration in the US population. *Arch Ophthalmol*. 2011;129:75–80.
25. Jager RD, Mieler WF, Miller JW. Age-related macular degeneration. *N Engl J Med*. 2008;358:2606–2617.
26. Schwartz SD, Hubschman JP, Heilwell G, et al. Embryonic stem cell trials for macular degeneration: a preliminary report. *Lancet*. 2012;379:713–720.

27. Binder S, Krebs I, Hilgers RD, et al. Outcome of transplantation of autologous retinal pigment epithelium in age-related macular degeneration: a prospective trial. *Invest Ophthalmol Vis Sci.* 2004;45:4151–4160.
28. Algvere PV, Berglin L, Gouras P, Sheng Y. Transplantation of fetal retinal pigment epithelium in age-related macular degeneration with subfoveal neovascularization. *Graefes Arch Clin Exp Ophthalmol.* 1994;232:707–716.
29. Kaplan HJ, Tezel TH, Berger AS, Del Priore LV. Retinal transplantation. *Chem Immunol.* 1999;73:207–219.
30. MacLaren RE, Bird AC, Sathia PJ, Aylward GW. Long-term results of submacular surgery combined with macular translocation of the retinal pigment epithelium in neovascular age-related macular degeneration. *Ophthalmology.* 2005;112:2081–2087.
31. Lappas A, Weinberger AW, Foerster AM, Kube T, Rezai KA, Kirchhof B. Iris pigment epithelial cell translocation in exudative age-related macular degeneration. A pilot study in patients. *Graefes Arch Clin Exp Ophthalmol.* 2000;238:631–641.
32. Huang Y, Enzmann V, Ildstad ST. Stem cell-based therapeutic applications in retinal degenerative diseases. *Stem Cell Rev.* 2011;7:434–445.
33. Song MK, Lui GM. Propagation of fetal human RPE cells: preservation of original culture morphology after serial passage. *J Cell Physiol.* 1990;143:196–203.
34. Gamm DM, Wright LS, Capowski EE, et al. Regulation of prenatal human retinal neurosphere growth and cell fate potential by retinal pigment epithelium and Mash1. *Stem Cells.* 2008;26:3182–3193.
35. Maminishkis A, Chen S, Jalickee S, et al. Confluent monolayers of cultured human fetal retinal pigment epithelium exhibit morphology and physiology of native tissue. *Invest Ophthalmol Vis Sci.* 2006;47:3612–3624.
36. West EL, Pearson RA, MacLaren RE, Sowden JC, Ali RR. Cell transplantation strategies for retinal repair. *Prog Brain Res.* 2009;175:3–21.
37. Pan CK, Heilweil G, Lanza R, Schwartz SD. Embryonic stem cells as a treatment for macular degeneration. *Expert Opin Biol Ther.* 2013;13:1125–1133.
38. Buchholz DE, Hikita ST, Rowland TJ, et al. Derivation of functional retinal pigmented epithelium from induced pluripotent stem cells. *Stem Cells.* 2009;27:2427–2434.
39. Carr AJ, Vugler AA, Hikita ST, et al. Protective effects of human iPS-derived retinal pigment epithelium cell transplantation in the retinal dystrophic rat. *PLoS One.* 2009;4:e8152.
40. Singh R, Phillips MJ, Kuai D, et al. Functional analysis of serially expanded human iPS cell-derived RPE cultures. *Invest Ophthalmol Vis Sci.* 2013;54:6767–6778.
41. Hirami Y, Osakada F, Takahashi K, et al. Generation of retinal cells from mouse and human induced pluripotent stem cells. *Neurosci Lett.* 2009;458:126–131.
42. Li Y, Tsai YT, Hsu CW, et al. Long-term safety and efficacy of human-induced pluripotent stem cell (iPS) grafts in a preclinical model of retinitis pigmentosa. *Mol Med.* 2012;18:1312–1319.
43. Singh R, Shen W, Kuai D, et al. iPS cell modeling of Best disease: insights into the pathophysiology of an inherited macular degeneration. *Hum Mol Genet.* 2013;22:593–607.
44. Yang J, Li Y, Chan L, et al. Validation of genome-wide association study (GWAS)-identified disease risk alleles with patient-specific stem cell lines. *Hum Mol Genet.* 2014;23:3445–3455.
45. Ferrer M, Corneo B, Davis J, et al. A multiplex high-throughput gene expression assay to simultaneously detect disease and functional markers in induced pluripotent stem cell-derived retinal pigment epithelium. *Stem Cells Transl Med.* 2014;3:911–922.
46. Inoue H, Yamanaka S. The use of induced pluripotent stem cells in drug development. *Clin Pharmacol Ther.* 2011;89:655–661.
47. Muniz A, Greene WA, Plamper ML, et al. Retinoid uptake, processing, and secretion in human iPS-RPE support the visual cycle. *Invest Ophthalmol Vis Sci.* 2014;55:198–209.

48. Maeda T, Lee MJ, Palczewska G, et al. Retinal pigmented epithelial cells obtained from human induced pluripotent stem cells possess functional visual cycle enzymes in vitro and in vivo. *J Biol Chem.* 2013;288:34484–34493.
49. Groenendijk GW, De Grip WJ, Daemen FJ. Quantitative determination of retinals with complete retention of their geometric configuration. *Biochim Biophys Acta.* 1980;617: 430–438.
50. Li HL, Nakano T, Hotta A. Genetic correction using engineered nucleases for gene therapy applications. *Dev Growth Differ.* 2014;56:63–77.

CHAPTER TWENTY-NINE

Potential Role of Exercise in Retinal Health

Machelle T. Pardue*, Micah A. Chrenek*, Robin H. Schmidt*, John M. Nickerson*, Jeffrey H. Boatright*,†,1

*Department of Ophthalmology, Emory University School of Medicine, Atlanta, Georgia, USA
†Center for Visual and Neurocognitive Rehabilitation, Atlanta VA Medical Center, Decatur, Georgia, USA
1Corresponding author: e-mail address: jboatri@emory.edu

Contents

Abstract

For many patients suffering vision loss due to retinal degeneration, the potential exists for therapeutic intervention to halt or delay disease progression. Proposed molecular, pharmacological, and surgical treatments are expensive and complicated. Finding low-cost interventions to sustain vision and thereby quality of life is vitally important. This chapter reviews findings from animal model and human subject studies indicating that physical exercise has direct, beneficial effects on regions of the central nervous system and is protective against neurodegenerative disease, including recent data from animal models showing similar effects for retina and vision. Potential local and systemic mechanistic pathways for exercise-induced retinal neuroprotection are discussed.

Globally over 8 million people are blind or significantly visually impaired due to retinal degenerative (RD) diseases such as age-related macular degeneration (AMD)[1] and inherited retinal degenerations such as retinitis pigmentosa.[2,3] Only a small percentage of these patients are predicted to develop complete blindness (no light perception).[1–3] Thus, in the majority of retinal degeneration patients, the potential exists for therapeutic

Progress in Molecular Biology and Translational Science, Volume 134
ISSN 1877-1173
http://dx.doi.org/10.1016/bs.pmbts.2015.06.011
© 2015 Elsevier Inc.
All rights reserved.

intervention to halt disease progression. Clinical trials suggest that these patients may respond to molecular, pharmacological, and surgical approaches. However, outcomes have been modest and variable, and these treatments remain expensive, experimental, or complicated.[4,5] Finding low-cost interventions to sustain vision and thereby quality of life is vitally important. Numerous animal models and human subject studies demonstrate that physical exercise can have direct, beneficial effects on regions of the central nervous system (CNS) and is protective against neurodegenerative disease (reviewed in Refs. 6–10). As the retina is an organ of the CNS, the question arises: *Might physical exercise, a low-cost, noninvasive intervention, benefit retina and vision?*

1. EXERCISE IS NEUROPROTECTIVE IN HUMANS AND ANIMALS

In humans, physical exercise is protective and rehabilitative for many neuronal functions and structures (reviewed in Ref. 6). Treadmill training and other exercise regimens improve or maintain motor and nonmotor functions in Alzheimer's disease (AD),[11] Parkinson's disease (PD),[12,13] amyotrophic lateral sclerosis (ALS; Lou Gehrig's disease),[7] and stroke.[14–20] Exercise also benefits cognitive function in healthy aged and young individuals, suggesting a role in maintenance of neuronal structures.[21,22] Thus, exercise is beneficial in disease and health, affecting several functions associated with disparate neuronal structures (e.g., substantia nigra in PD, cerebellum in stroke, cortex and spinal cord in ALS, cortex and other brain structures in AD) in both young and aged people. Often these beneficial outcomes are accompanied by increased levels of circulating brain-derived neurotrophic factor (BDNF; reviewed in Ref. 23). Treadmill running greatly enhances recovery from peripheral nerve injury in mice and rats, an effect mediated by BDNF.[24–29] Centrally, the rehabilitative effects of exercise are well documented in models that span stroke to PD to normal aging,[7–10] including in CNS structures that are not considered regenerative, such as substantia nigra and basal ganglia,[30–32] corticostriatal synapses,[33] and cerebellum.[34]

2. EXERCISE MAY BE BENEFICIAL TO RETINA AND VISION

Reports on the beneficial effects of exercise to the visual system in humans are limited to epidemiology studies. Lower incident AMD risk is associated with higher levels of vigorous exercise (running) for middle-aged adults[35] and moderate exercise (walking) in people over 75 (but not younger).[36] Increased activity and exercise are associated with better visual outcomes in patients who have ongoing AMD,[37] whereas decreased activity is associated with AMD precursors.[38] Thus, clinical data suggest that exercise may have preventive and rehabilitative effects on visual outcomes. How this may occur is unknown. As to clinical relevance, studies reveal that patients are highly complaint to tailored exercise regimens (mean age of 70[39–43]) and that RD patients are nearly 100% compliant in following tailored prescriptions from retina specialists.[41] This suggests that RD patients may benefit more from exercise than other patient groups as they may adhere to exercise "prescriptions."

3. EXERCISE PROTECTS RETINA AND VISION IN ANIMAL MODELS OF RETINAL DISEASE

Animal studies indicate that exercise has direct benefits to retina and vision. While enriched environment testing paradigms that include voluntary exercise (e.g., running wheels) inhibit vision function loss and retinal degeneration in the rd10 mouse,[44,45] we recently reported that modest treadmill exercise alone protects retinas of mice undergoing light-induced retinal degeneration (LIRD), preserving retinal morphology and function.[46] Similarly, swimming protects mouse retinal ganglion cells following elevated intraocular pressure, a model of glaucoma.[47] Treadmill exercise also protects cells of the inner nuclear layer in streptozotocin-injected rats, a model of diabetic retinopathy.[48] In both of these latter studies, exercise suppresses neuroinflammation and apoptotic pathway constituents in retina tissue, including markers for mitochondrial stress.[47,48]

4. MECHANISMS THAT MAY MEDIATE EFFECTS OF EXERCISE ON RETINA AND VISION

How does the retina respond to a whole-body activity that can be stressful and beneficial to systemic physiology? It seems reasonable to suppose that exercise may have protective effects on visual outcomes in disease and during healthy aging through biological and molecular mechanisms that are similar to those observed in the brain in response to exercise. Here, we discuss the potential role for exercise effects on mitochondria and trophic factors.

5. BDNF MEDIATES EFFECTS OF EXERCISE IN HUMAN AND ANIMAL MODELS

Often beneficial effects of exercise in CNS and PNS structures are accompanied by increased levels of circulating BDNF (reviewed in Ref. 23). Treadmill running greatly enhances recovery from peripheral nerve injury in mice and rats, an effect mediated by BDNF.[24–29] Numerous studies, both animal and human, indicate that the effects of aerobic exercise are mediated by BDNF acting via TrkB receptor activation, and that, indeed, it may be the major neurotrophic mediator of the effects of exercise in substantia nigra, cortex, hippocampus, neuromuscular junctions, and other neuronal structures (for reviews, see Refs. 49,50). The most recent systematic, critical review of the clinical research literature indicates that aerobic exercise, whether long-term endurance training or short-term acute exercise, increases circulating BDNF levels in healthy and chronically ill humans and that strength training does not have this effect (reviewed in Ref. 51). Assessment of arterial-to-internal jugular venous differences in BDNF levels demonstrates that the brain is the predominant source of this BDNF.[52]

Animal studies provide a much more detailed exploration of the role of BDNF in mediating the effects of exercise. In particular, animal studies show that aerobic exercise increases BDNF levels in many neuronal structures, not just in the hippocampus (where it is arguably most studied), and that in addition to localized elevation of BDNF, several of these studies demonstrate that suppression of BDNF expression or blockade of BDNF receptor activation and transduction precludes the beneficial effects of exercise. For example, aerobic exercise is protective in stroke models, and this is accompanied by elevated BDNF in cortex, hippocampus, and striatum.[53] Sciatic nerve regeneration is prevented if endogenous BDNF is bound by antibodies.[54] Suppression of BDNF levels in amygdala by siRNA or suppression of BDNF signal transduction by kinase inhibitors prevents the beneficial effects of treadmill and wheel running on learning and memory tasks in mice.[55] Aged rodents also respond to exercise. For instance, stress-induced BDNF induction in the hippocampus is impaired in aging, leading to decreased synaptic plasticity and memory and cognition deficits in rats.[56–59]

Voluntary or involuntary exercise in aged (24-month-old) rats completely reverses this, with restored or preserved BDNF induction and memory capabilities.[56,60]

6. THE POSSIBLE ROLE OF BDNF IN EXERCISE AND RETINAL NEUROPROTECTION

Endogenous BDNF is one of the several trophic factors released in retina upon injury[61] and is elevated in retinas from AMD patients,[62] suggesting that BDNF is an important player in the natural response to stress and disease in retina. The same enhanced environment paradigm described above that includes wheel running and that is so protective in the rd10 mouse model of retinal degeneration[44,45] results in elevated BDNF levels in retina.[63,64] Certainly providing BDNF to the retina is beneficial to retinal photoreceptors and neurons since treatment with locally delivered, exogenous BDNF[61,65–67] or transgenically overexpressed BDNF[68] is protective in rat and mouse models of chronic and acute vision loss. Systemic treatment with small molecule agonists of the BDNF TrkB receptor is protective in the LIRD model of retinal degeneration.[69,70] Similarly, intravitreal injection of agonistic TrkB antibodies protects retinal ganglion cells in rodent models of glaucoma.[71,72] Thus, it appears that BDNF or its agonists would be excellent candidates for retinal therapeutics. However, the major drawback to BDNF as a therapeutic compound is its short half-life of as little as 3 h in neuronal tissues.[68,73] Interventions that chronically elevate endogenous BDNF, such as exercise, should avoid this shortcoming and so provide benefit. It is of note in this regard that our work demonstrating that treadmill running protects against retinal degeneration and vision loss in mice exposed to toxic light indicates a role for BDNF.[46] Exercise increased BDNF levels in serum, hippocampus, and retina.[46] Critically, treatment of mice with ANA-12, a TrkB antagonist specifically designed to cross the blood–brain barrier,[74] prevented the protective effects of exercise. These data demonstrate that exercise protects retina and vision function via a mechanism that requires BDNF TrkB signal transduction.[46] How BDNF mediates exercise-induced photoreceptor cell protection remains to be elucidated. The mechanistic pathway may involve multiple retinal cell types as BDNF and TrkB localize to subpopulations of photoreceptor cells, Muller cells, ganglion cells, and amacrine cells.[66,75,76]

7. FROM MUSCLE TO RETINA: SYSTEMIC AND LOCAL PATHWAYS?

Aside from mechanistic pathways within the retina (or other brain regions), it remains unclear how whole-body exercise benefits specific CNS targets such as photoreceptor cells. Indeed, similar lack of detail exists in the more-studied effects of exercise on cognition, learning, and other CNS functions and neuroprotection associated with various brain regions. Though circulating levels of BDNF increase following exercise,[46,77] BDNF probably does not cross the blood–brain barrier,[78] and hence probably does not cross the blood–retina barrier. Thus, it seems likely that the protective effects of exercise are in some way mediated by systemic signaling that results in local upregulation of BDNF. Wrann and colleagues recently reported data that suggest such a mechanism. They found that mice allowed to run on wheels for 30 days had elevated levels of hippocampal BDNF and FNDC5, a protein that is also upregulated in muscle before being cleaved and secreted into the circulation as irisin following endurance exercise.[79,80] *In vivo*, peripheral delivery of FNDC5 increased blood irisin and increased hippocampal Bdnf expression.[79] In cultured neurons, overexpression of FNDC5 increased Bdnf gene expression and increased cell viability, whereas RNAi-mediated suppression of FNDC5 had opposite effects.[79] Neuronal culture experiments also revealed that overexpression of Pgc1alpha upregulated Fndc5, whereas shRNA-mediated suppression of ERRalpha (a Pgc1alpha cofactor) blocked this effect. RNAi-mediated suppression of Pgc1alpha resulted in decreased Fndc5 expression,[79] suggesting that Fndc5 gene expression is regulated by a Pgc1alpha/ERRalpha complex.[80]

These data suggest to the authors and others[80] local and systemic pathways (represented in Fig. 1) whereby exercise induces Pgc1alpha expression with subsequent upregulation of Fndc5 that is then cleaved and secreted as irisin into the extracellular space (autocrine model) and into the circulation (systemic model). Though the experimental data are not complete enough to definitely exclude additional extracellular and circulating exercise factors,[80] the findings that peripheral delivery of FNDC5 is sufficient to induce CNS expression of Bdnf suggest that a circulating form of FNDC5, possibly irisin, travels to CNS targets (including BDNF-producing retina cells) and in a Pgc1alpha-dependent manner, directly and in addition with local FNDC5 pathway, enhances BDNF production with subsequent TrkB activation. It may be that similar pathways (Fig. 1) mediate exercise-induced retinal neuroprotection that we[46] and others[47,48] observe.

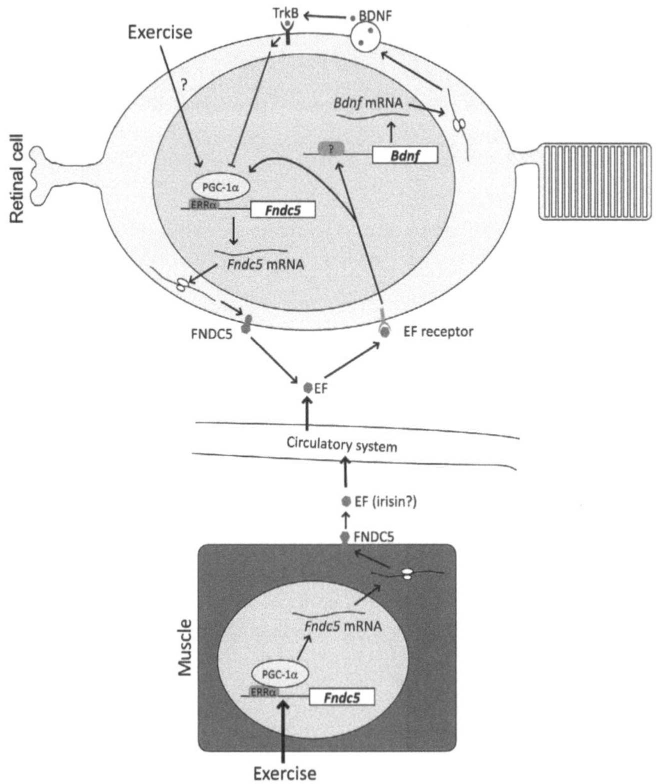

Figure 1 Pathways mediating exercise-induced retinal neuroprotection. Figure represents possible local and systemic pathways whereby exercise induces Pgc1alpha expression with subsequent upregulation of Fndc5 that is then cleaved and secreted as irisin into the extracellular space (autocrine, retinal cell model) and into the circulation (systemic model that includes muscle, circulation, and retinal cell). Though the experimental data are not complete enough to definitely exclude additional extracellular and circulating exercise factors (EFs),[80] the findings that peripheral delivery of FNDC5 is sufficient to induce CNS expression of Bdnf suggest that a circulating form of FNDC5, possibly irisin, might travel to retina targets (including BDNF-producing retina cells) and in a Pgc1alpha-dependent manner, directly and in addition with local FNDC5 pathway, enhance BDNF production. The elevated BDNF activates TrkB receptors on photoreceptor cells (shown), Muller cells, ganglion cells, or amacrine cells,[66,75,76] resulting in retinal neuroprotection. *Figure modified from Xu* et al.[80]

Note added in proof: The identification of irisin as a circulating form of FNDC5 that mediates the effects of exercise on CNS targets is controversial.[81]

ACKNOWLEDGMENTS

This work was supported by NIH R01EY016470, R01EY021592, R01EY014026, P30EY06360; The Abraham and Phyllis Katz Foundation, an unrestricted award to the

Department of Ophthalmology, Emory University from Research to Prevent Blindness, Inc.; Rehabilitation Research and Development Service Veterans Affairs Research Career Scientist Award; and Atlanta Veterans Administration Center of Excellence in Vision and Neurocognitive Rehabilitation. Parts of this work were presented at ARVO May 2013 and 2014.

REFERENCES

1. Jonas JB, Bourne RR, White RA, et al. Visual impairment and blindness due to macular diseases globally: a systematic review and meta-analysis. *Am J Ophthalmol.* 2014;158(4):808–815.
2. De l'Aune WR, Williams MD, Welsh RL. Outcome assessment of the rehabilitation of the visually impaired. *J Rehabil Res Dev.* 1999;36(4):273–293.
3. Grover S, Fishman GA, Anderson RJ, et al. Visual acuity impairment in patients with retinitis pigmentosa at age 45 years or older. *Ophthalmology.* 1999;106(9):1780–1785.
4. Fahed DC, Ghazi NG, Jabbour NM, Fahed CD, Fahed JC, Salti HI. Impact of AREDS in a developing country 5 years after publication of the study. *Eur J Ophthalmol.* 2011;21(1):67–72.
5. Zhang K, Hopkins JJ, Heier JS, et al. Ciliary neurotrophic factor delivered by encapsulated cell intraocular implants for treatment of geographic atrophy in age-related macular degeneration. *Proc Natl Acad Sci USA.* 2011;108(15):6241–6245.
6. Smith PJ, Blumenthal JA, Hoffman BM, et al. Aerobic exercise and neurocognitive performance: a meta-analytic review of randomized controlled trials. *Psychosom Med.* 2010;72(3):239–252.
7. McCrate ME, Kaspar BK. Physical activity and neuroprotection in amyotrophic lateral sclerosis. *Neuromolecular Med.* 2008;10(2):108–117.
8. Klaus F, Amrein I. Running in laboratory and wild rodents: differences in context sensitivity and plasticity of hippocampal neurogenesis. *Behav Brain Res.* 2012;227(2): 363–370.
9. Sahay A, Scobie KN, Hill AS, et al. Increasing adult hippocampal neurogenesis is sufficient to improve pattern separation. *Nature.* 2011;472(7344):466–470.
10. Lau YS, Patki G, Das-Panja K, Le WD, Ahmad SO. Neuroprotective effects and mechanisms of exercise in a chronic mouse model of Parkinson's disease with moderate neurodegeneration. *Eur J Neurosci.* 2011;33(7):1264–1274.
11. Vreugdenhil A, Cannell J, Davies A, Razay G. A community-based exercise programme to improve functional ability in people with Alzheimer's disease: a randomized controlled trial. *Scand J Caring Sci.* 2012;26(1):12–19.
12. Filippin NT, da Costa PH, Mattioli R. Effects of treadmill-walking training with additional body load on quality of life in subjects with Parkinson's disease. *Rev Bras Fisioter.* 2010;14(4):344–350.
13. Cruise KE, Bucks RS, Loftus AM, Newton RU, Pegoraro R, Thomas MG. Exercise and Parkinson's: benefits for cognition and quality of life. *Acta Neurol Scand.* 2011;123(1):13–19.
14. Sacco RL, Gan R, Boden-Albala B, et al. Leisure-time physical activity and ischemic stroke risk: the Northern Manhattan stroke study. *Stroke.* 1998;29(2):380–387.
15. Liang W, Lee AH, Binns CW, Zhou Q, Huang R, Hu D. Habitual physical activity reduces the risk of ischemic stroke: a case-control study in southern China. *Cerebrovasc Dis.* 2009;28(5):454–459.
16. Krarup LH, Truelsen T, Gluud C, et al. Prestroke physical activity is associated with severity and long-term outcome from first-ever stroke. *Neurology.* 2008;71(17): 1313–1318.
17. Stroud N, Mazwi TM, Case LD, et al. Prestroke physical activity and early functional status after stroke. *J Neurol Neurosurg Psychiatry.* 2009;80(9):1019–1022.

18. Quaney BM, Boyd LA, McDowd JM, et al. Aerobic exercise improves cognition and motor function poststroke. *Neurorehabil Neural Repair.* 2009;23(9):879–885.
19. Reisman DS, McLean H, Bastian AJ. Split-belt treadmill training poststroke: a case study. *J Neurol Phys Ther.* 2010;34(4):202–207.
20. Rose D, Paris T, Crews E, et al. Feasibility and effectiveness of circuit training in acute stroke rehabilitation. *Neurorehabil Neural Repair.* 2011;25(2):140–148.
21. Kramer AF, Hahn S, Cohen NJ, et al. Ageing, fitness and neurocognitive function. *Nature.* 1999;400(6743):418–419.
22. Tomporowski PD, Davis CL, Miller PH, Naglieri JA. Exercise and children's intelligence, cognition, and academic achievement. *Educ Psychol Rev.* 2008;20(2):111–131.
23. Voss MW, Vivar C, Kramer AF, van Praag H. Bridging animal and human models of exercise-induced brain plasticity. *Trends Cogn Sci.* 2013;17(10):525–544.
24. Thompson NJ, Sengelaub DR, English AW. Enhancement of peripheral nerve regeneration due to treadmill training and electrical stimulation is dependent on androgen receptor signaling. *Dev Neurobiol.* 2014;74(5):531–540.
25. Boeltz T, Ireland M, Mathis K, et al. Effects of treadmill training on functional recovery following peripheral nerve injury in rats. *J Neurophysiol.* 2013;109(11):2645–2657.
26. Wilhelm JC, Xu M, Cucoranu D, et al. Cooperative roles of BDNF expression in neurons and Schwann cells are modulated by exercise to facilitate nerve regeneration. *J Neurosci.* 2012;32(14):5002–5009.
27. English AW, Wilhelm JC, Sabatier MJ. Enhancing recovery from peripheral nerve injury using treadmill training. *Ann Anat.* 2011;193(4):354–361.
28. English AW, Cucoranu D, Mulligan A, Sabatier M. Treadmill training enhances axon regeneration in injured mouse peripheral nerves without increased loss of topographic specificity. *J Comp Neurol.* 2009;517(2):245–255.
29. Sabatier MJ, Redmon N, Schwartz G, English AW. Treadmill training promotes axon regeneration in injured peripheral nerves. *Exp Neurol.* 2008;211(2):489–493.
30. Smith BA, Goldberg NR, Meshul CK. Effects of treadmill exercise on behavioral recovery and neural changes in the substantia nigra and striatum of the 1-methyl-4-phenyl-1,2,3,6-tetrahydropyridine-lesioned mouse. *Brain Res.* 2011;1386: 70–80.
31. Wu SY, Wang TF, Yu L, et al. Running exercise protects the substantia nigra dopaminergic neurons against inflammation-induced degeneration via the activation of BDNF signaling pathway. *Brain Behav Immun.* 2011;25(1):135–146.
32. Vucckovic MG, Li Q, Fisher B, et al. Exercise elevates dopamine D2 receptor in a mouse model of Parkinson's disease: in vivo imaging with [(1)F]fallypride. *Mov Disord.* 2010;25(16):2777–2784.
33. Cepeda C, Cummings DM, Hickey MA, et al. Rescuing the corticostriatal synaptic disconnection in the R6/2 mouse model of Huntington's disease: exercise, adenosine receptors and ampakines. *PLoS Curr.* 2010;2. pii: RRN1182. http://dx.doi.org/10.1371/currents.RRN1182.
34. Ben-Ari S, Ofek K, Barbash S, et al. Similar cation channels mediate protection from cerebellar exitotoxicity by exercise and inheritance. *J Cell Mol Med.* 2012;16(3): 555–568.
35. Williams PT. Prospective study of incident age-related macular degeneration in relation to vigorous physical activity during a 7-year follow-up. *Invest Ophthalmol Vis Sci.* 2009;50(1):101–106.
36. Gopinath B, Liew G, Burlutsky G, Mitchell P. Physical activity and the 15-year incidence of age-related macular degeneration. *Invest Ophthalmol Vis Sci.* 2014;55(12):7799–7803.
37. Knudtson MD, Klein R, Klein BE. Physical activity and the 15-year cumulative incidence of age-related macular degeneration: the Beaver Dam Eye Study. *Br J Ophthalmol.* 2006;90(12):1461–1463.

38. Munch IC, Linneberg A, Larsen M. Precursors of age-related macular degeneration: associations with physical activity, obesity, and serum lipids in the inter99 eye study. *Invest Ophthalmol Vis Sci*. 2013;54(6):3932–3940.
39. Subramanian U, Hopp F, Mitchinson A, Lowery J. Impact of provider self-management education, patient self-efficacy, and health status on patient adherence in heart failure in a Veterans Administration population. *Congest Heart Fail*. 2008;14(1):6–11.
40. Olayiwola JN, Sobieraj DM, Kulowski K, St Hilaire D, Huang JJ. Improving diabetic retinopathy screening through a statewide telemedicine program at a large federally qualified health center. *J Health Care Poor Underserved*. 2011;22(3):804–816.
41. Hochstetler BS, Scott IU, Kunselman AR, Thompson K, Zerfoss E. Adherence to recommendations of the age-related eye disease study in patients with age-related macular degeneration. *Retina (Philadelphia, Pa)*. 2010;30(8):1166–1170.
42. Kulkarni SV, Damji KF, Buys YM. Medical management of primary open-angle glaucoma: best practices associated with enhanced patient compliance and persistency. *Patient Prefer Adherence*. 2008;2:303–314.
43. Tsai JC. A comprehensive perspective on patient adherence to topical glaucoma therapy. *Ophthalmology*. 2009;116(suppl 11):S30–S36.
44. Barone I, Novelli E, Piano I, Gargini C, Strettoi E. Environmental enrichment extends photoreceptor survival and visual function in a mouse model of retinitis pigmentosa. *PLoS One* 2012;7(11):e50726. http://dx.doi.org/10.1371/journal.pone.0050726, Epub 2012 Nov 28. PubMed PMID: 23209820; PubMed Central PMCID: PMC3508993.
45. Barone I, Novelli E, Strettoi E. Long-term preservation of cone photoreceptors and visual acuity in rd10 mutant mice exposed to continuous environmental enrichment. *Mol Vis*. 2014;20:1545–1556. eCollection 2014. PubMed PMID: 25489227; PubMed Central PMCID: PMC4225138.
46. Lawson EC, Han MK, Sellers JT, et al. Aerobic exercise protects retinal function and structure from light-induced retinal degeneration. *J Neurosci*. 2014;34(7):2406–2412.
47. Chrysostomou V, Kezic JM, Trounce IA, Crowston JG. Forced exercise protects the aged optic nerve against intraocular pressure injury. *Neurobiol Aging*. 2014;35(7): 1722–1725.
48. Kim DY, Jung SY, Kim CJ, Sung YH, Kim JD. Treadmill exercise ameliorates apoptotic cell death in the retinas of diabetic rats. *Mol Med Rep*. 2013;7(6):1745–1750.
49. Zoladz JA, Pilc A. The effect of physical activity on the brain derived neurotrophic factor: from animal to human studies. *J Physiol Pharmacol*. 2010;61(5):533–541.
50. Lista I, Sorrentino G. Biological mechanisms of physical activity in preventing cognitive decline. *Cell Mol Neurobiol*. 2010;30(4):493–503.
51. Knaepen K, Goekint M, Heyman EM, Meeusen R. Neuroplasticity—exercise-induced response of peripheral brain-derived neurotrophic factor: a systematic review of experimental studies in human subjects. *Sports Med*. 2010;40(9):765–801.
52. Seifert T, Brassard P, Wissenberg M, et al. Endurance training enhances BDNF release from the human brain. *Am J Physiol Regul Integr Comp Physiol*. 2010;298(2): R372–R377.
53. Ke Z, Yip SP, Li L, Zheng XX, Tong KY. The effects of voluntary, involuntary, and forced exercises on brain-derived neurotrophic factor and motor function recovery: a rat brain ischemia model. *PLoS One*. 2011;6(2):e16643.
54. Zhang JY, Luo XG, Xian CJ, Liu ZH, Zhou XF. Endogenous BDNF is required for myelination and regeneration of injured sciatic nerve in rodents. *Eur J Neurosci*. 2000;12(12):4171–4180.
55. Liu YF, Chen HI, Wu CL, et al. Differential effects of treadmill running and wheel running on spatial or aversive learning and memory: roles of amygdalar brain-derived neurotrophic factor and synaptotagmin I. *J Physiol*. 2009;587(pt 13):3221–3231.

56. Barrientos RM, Frank MG, Crysdale NY, et al. Little exercise, big effects: reversing aging and infection-induced memory deficits, and underlying processes. *J Neurosci.* 2011;31(32):11578–11586.
57. Barrientos RM, Frank MG, Hein AM, et al. Time course of hippocampal IL-1 beta and memory consolidation impairments in aging rats following peripheral infection. *Brain Behav Immun.* 2009;23(1):46–54.
58. Frank MG, Barrientos RM, Watkins LR, Maier SF. Aging sensitizes rapidly isolated hippocampal microglia to LPS ex vivo. *J Neuroimmunol.* 2010;226(1–2):181–184.
59. Godbout JP, Chen J, Abraham J, et al. Exaggerated neuroinflammation and sickness behavior in aged mice following activation of the peripheral innate immune system. *FASEB J.* 2005;19(10):1329–1331.
60. Aguiar Jr AS, Castro AA, Moreira EL, et al. Short bouts of mild-intensity physical exercise improve spatial learning and memory in aging rats: involvement of hippocampal plasticity via AKT, CREB and BDNF signaling. *Mech Ageing Dev.* 2011;132(11–12):560–567.
61. Wen R, Cheng T, Song Y, et al. Continuous exposure to bright light upregulates bFGF and CNTF expression in the rat retina. *Curr Eye Res.* 1998;17(5):494–500.
62. Kolomeyer A, Sugino I, Zarbin M. Characterization of conditioned media collected from aged vs. young human eye-cups. *Invest Ophthalmol Vis Sci.* 2011;52(8):5963–5972.
63. Landi S, Cenni MC, Maffei L, Berardi N. Environmental enrichment effects on development of retinal ganglion cell dendritic stratification require retinal BDNF. *PLoS One.* 2007;2(4):e346.
64. Landi S, Sale A, Berardi N, Viegi A, Maffei L, Cenni MC. Retinal functional development is sensitive to environmental enrichment: a role for BDNF. *FASEB J.* 2007;21(1):130–139.
65. Unoki K, LaVail MM. Protection of the rat retina from ischemic injury by brain-derived neurotrophic factor, ciliary neurotrophic factor, and basic fibroblast growth factor. *Invest Ophthalmol Vis Sci.* 1994;35(3):907–915.
66. Rohrer B, Korenbrot JI, LaVail MM, Reichardt LF, Xu B. Role of neurotrophin receptor TrkB in the maturation of rod photoreceptors and establishment of synaptic transmission to the inner retina. *J Neurosci.* 1999;19(20):8919–8930.
67. Duncan JL, Paskowitz DM, Nune GC, et al. Retinal damage caused by photodynamic therapy can be reduced using BDNF. *Adv Exp Med Biol.* 2006;572:297–302.
68. Okoye G, Zimmer J, Sung J, et al. Increased expression of brain-derived neurotrophic factor preserves retinal function and slows cell death from rhodopsin mutation or oxidative damage. *J Neurosci.* 2003;23(10):4164–4172.
69. Jang SW, Liu X, Pradoldej S, et al. N-acetylserotonin activates TrkB receptor in a circadian rhythm. *Proc Natl Acad Sci USA.* 2010;107(8):3876–3881.
70. Jang SW, Liu X, Yepes M, et al. A selective TrkB agonist with potent neurotrophic activities by 7,8-dihydroxyflavone. *Proc Natl Acad Sci USA.* 2010;107(6):2687–2692.
71. Bai Y, Xu J, Brahimi F, Zhuo Y, Sarunic MV, Saragovi HU. An agonistic TrkB mAb causes sustained TrkB activation, delays RGC death, and protects the retinal structure in optic nerve axotomy and in glaucoma. *Invest Ophthalmol Vis Sci.* 2010;51(9):4722–4731.
72. Hu Y, Cho S, Goldberg JL. Neurotrophic effect of a novel TrkB agonist on retinal ganglion cells. *Invest Ophthalmol Vis Sci.* 2010;51(3):1747–1754.
73. Fukumitsu H, Ohtsuka M, Murai R, Nakamura H, Itoh K, Furukawa S. Brain-derived neurotrophic factor participates in determination of neuronal laminar fate in the developing mouse cerebral cortex. *J Neurosci.* 2006;26(51):13218–13230.
74. Cazorla M, Premont J, Mann A, Girard N, Kellendonk C, Rognan D. Identification of a low-molecular weight TrkB antagonist with anxiolytic and antidepressant activity in mice. *J Clin Invest.* 2011;121(5):1846–1857.

75. Cellerino A, Kohler K. Brain-derived neurotrophic factor/neurotrophin-4 receptor TrkB is localized on ganglion cells and dopaminergic amacrine cells in the vertebrate retina. *J Comp Neurol.* 1997;386(1):149–160.
76. Di Polo A, Cheng L, Bray GM, Aguayo AJ. Colocalization of TrkB and brain-derived neurotrophic factor proteins in green-red-sensitive cone outer segments. *Invest Ophthalmol Vis Sci.* 2000;41(12):4014–4021.
77. Rasmussen P, Brassard P, Adser H, et al. Evidence for a release of brain-derived neurotrophic factor from the brain during exercise. *Exp Physiol.* 2009;94(10):1062–1069.
78. Zhang Y, Pardridge WM. Conjugation of brain-derived neurotrophic factor to a blood-brain barrier drug targeting system enables neuroprotection in regional brain ischemia following intravenous injection of the neurotrophin. *Brain Res.* 2001;889(1–2):49–56.
79. Wrann CD, White JP, Salogiannnis J, et al. Exercise induces hippocampal BDNF through a PGC-1alpha/FNDC5 pathway. *Cell Metab.* 2013;18(5):649–659.
80. Xu B. BDNF (I)rising from exercise. *Cell Metab.* 2013;18(5):612–614.
81. Albrecht E, Norheim F, Thiede B, Holen T, Ohashi T, Schering L, et al. Irisin – a myth rather than an exercise-inducible myokine. *Sci Rep* 2015;5:8889. http://dx.doi.org/10.1038/srep08889.

CHAPTER THIRTY

The Biology of Retinoblastoma

Pia R. Mendoza, Hans E. Grossniklaus[1]
Department of Ophthalmology, Emory University School of Medicine, Atlanta, Georgia, USA
[1]Corresponding author: e-mail address: ophtheg@emory.edu

Contents

Abstract

Retinoblastoma, the most common primary intraocular cancer of childhood, is a malignancy arising in the developing retina. Tumor formation usually begins with mutation in both alleles of the retinoblastoma tumor suppressor gene RB1, followed by a series of other genetic alterations that correlate with the clinical stage and pathologic findings of the tumor. Analysis of sporadic and heritable retinoblastoma led to the development of Knudson's Two-Hit Hypothesis. The tumor suppressor RB1 gene codes for the retinoblastoma protein which is a key regulator of cellular replication via its binding to the E2F family of transcription factors and chromatin remodeling proteins. Studies of preclinical models of retinoblastoma in the form of transgenic mice and xenograft animal models have significantly contributed to the development of effective therapies for this disease. Research on retinoblastoma has paved the way toward understanding many of the mechanisms in cancer genetics.

1. INTRODUCTION

Retinoblastoma is the most common primary cancer of the eye in children[1] but compared to other pediatric cancers is quite rare representing only 3–4% of all childhood malignancies.[2,3] Retinoblastoma has an

Progress in Molecular Biology and Translational Science, Volume 134
ISSN 1877-1173
http://dx.doi.org/10.1016/bs.pmbts.2015.06.012
© 2015 Elsevier Inc.
All rights reserved.

incidence rate of 1 in 14,000–20,000 livebirths and has no gender predilection. Two-thirds of cases are diagnosed before the age of 2 years, and 90% are diagnosed before the age of 5 years.[2] Retinoblastoma is a malignant tumor of the developing retina that arises from primitive retinal stem cells or cone precursor cells. Tumorigenesis in retinoblastoma is classically thought to be due to inactivation of both alleles of the retinoblastoma susceptibility gene RB1, the first tumor suppressor gene to be identified. Research on retinoblastoma has been and continues to be pivotal in many of the landmark discoveries in cancer biology. Innovations in molecular genetics and targeted therapy have led to remarkable improvements in survival rates in children with this disease.[4]

2. GENETICS AND MOLECULAR BIOLOGY OF RETINOBLASTOMA

2.1 Clinical Genetics of Retinoblastoma

Retinoblastoma presents in two distinct clinical forms: a heritable form (40% of cases) in which tumors are bilateral and multifocal, presenting at an earlier age; and a nonheritable form (60% of cases) in which tumors are unilateral and unifocal, presenting at a later age. In 1971, Knudson proposed the Two-Hit Hypothesis, explaining that two mutational events in a developing retinal cell lead to tumor formation in retinoblastoma.[5] In hereditary retinoblastoma, patients have an inherited germline mutation (first "hit") followed by an acquired second mutation (second "hit") as their retinas develop. In nonhereditary retinoblastoma, there are two distinct mutations acquired in both alleles of the retinoblastoma gene in somatic cells of the developing retina (see Fig. 1). Hereditary retinoblastoma is transmitted in an autosomal dominant fashion and only 6–10% of patients have a positive family history of retinoblastoma. Individuals with heritable retinoblastoma or heterozygous carriers of a cancer-predisposing retinoblastoma gene mutation are at increased risk of developing secondary malignancies such as osteosarcomas, soft tissue sarcomas, and melanomas.[6] Some patients exhibit reduced expressivity, such as in 15% of unilateral, nonfamilial cases that actually have germline mutations. There are also reports of incomplete penetrance as well as mutational mosaicism.[7]

2.2 The RB1 Gene

The retinoblastoma susceptibility or RB1 gene, localized and cloned in 1986, was the first human tumor suppressor gene identified.[8] It is located

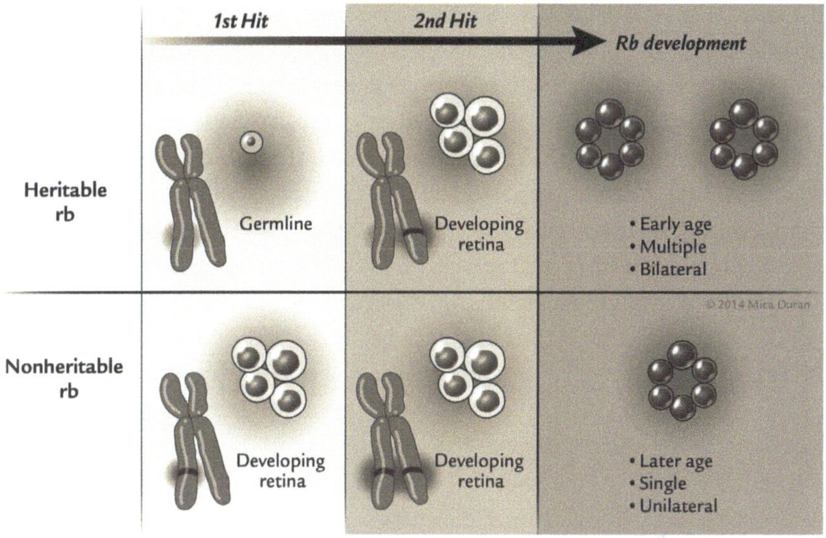

Figure 1 Inheritance in retinoblastoma. Two mutational events (two hits) in a developing retinal cell lead to tumor formation in retinoblastoma. In heritable retinoblastoma, there is an inherited germline mutation (first "hit") followed by an acquired second mutation (second "hit") in the developing retina. These tumors are usually bilateral, multiple, and occur at an early age. In nonheritable retinoblastoma, there are two distinct mutations acquired in both alleles of the retinoblastoma gene in somatic cells of the developing retina. These tumors are usually unilateral, single, and occur at a later age.

on chromosome 13q14, spans 180 kb, and has 27 exons. The 5′ region of the gene contains a CpG island, which is normally unmethylated. All 27 exons contribute to the 2.7-kb open reading frame coding for a 110-kDa nuclear phosphoprotein known as the retinoblastoma protein or Rb. The Rb protein, composed of 928 amino acids, functions as a regulator of cellular replication. The protein has an N-terminal region, a central pocket domain with A and B boxes, and a C terminal region. Exons 13–21 code for the binding site within the functional pocket domain of the protein formed by interactions of the A and B boxes. Most clinically significant mutations result in amino acid insertion, deletion, or substitution within this pocket domain.[9] Rb suppresses transcription by binding to chromatin remodeling proteins BRG and histone deacetylases (HDACs) via the LxCxE binding site.[10] It also restricts cell division by binding to the E2F family of transcription factors and preventing E2F from targeting gene products required for the transition of the cell through the G1 to the S phase of the cell cycle.[11] Rb can only bind to E2F when it is hypophosphorylated; cyclin–cyclin-dependent kinase (CDK) complexes that phosphorylate Rb cause it to dissociate from E2F which in turn facilitates transcription. In retinoblastoma, mutation of RB1 leads to dysfunction or absence of the Rb protein, resulting in constant activity of E2Fs and continuous instigation of the cell cycle, causing tumor formation. When Rb is dephosphorylated via p16, HDAC and BRG are recruited to the Rb-E2F complex and cause chromatin structure alteration, thus preventing access to transcription leading to exit from the cell cycle. When Rb is phosphorylated, HDAC and BRG are released, allowing progression through the cell cycle. Under stress and excess proliferation, hyperphosphorylation changes the tertiary structure of Rb, releasing E2Fs and leading to apoptosis. Thus, Rb plays an important role in deciding the fate of a cell through either exit of the cell cycle or apoptosis. Mutations resulting in abnormal Rb function or inhibition of Rb by oncoproteins produced by tumor-causing viruses (SV40, adenovirus, human papillomavirus) negate Rb's ability to mediate exit from the cell cycle or apoptosis and lead to uncontrolled cell division, a hallmark of cancer (see Fig. 2).

There is a wide array of genetic variations linked to RB1, including chromosomal rearrangements, large exonic deletions, hypermethylation of the gene promoter region, small length mutations, and single nucleotide substitutions. The majority of germline mutations that have been identified in familial RB are nonsense or frameshift mutations within exons 2–25, resulting in absence of Rb or loss of Rb function. Molecular genetic testing of RB1 in white blood cells and tumor cells can identify a germline or somatic mutation in individuals with a predisposition to retinoblastoma. If a cancer-predisposing RB1 mutation is identified in the proband's white blood cells, mutation analysis can determine the genetic status of at-risk siblings and offspring. Testing for loss of heterozygosity in tumors involves comparative genotyping of polymorphic loci within and flanking RB1 in DNA from peripheral blood and tumor that can reveal that loss of the normal allele (hemizygosity) with or without duplication (homozygosity) of the mutant allele constitutes the somatic mutation.[7]

2.3 Secondary Mutations in Retinoblastoma

Biallelic loss of RB1 function is required for retinoblastoma tumor development; however, additional mutational events (three or more "hits") are needed for tumor progression.[13] Loss of both copies of RB1 may lead to the benign variant of retinoblastoma, called retinoma or retinocytoma, and progressive genomic instability then leads to the frank malignancy retinoblastoma.[14] RB1 plays a central role in maintaining genomic stability and

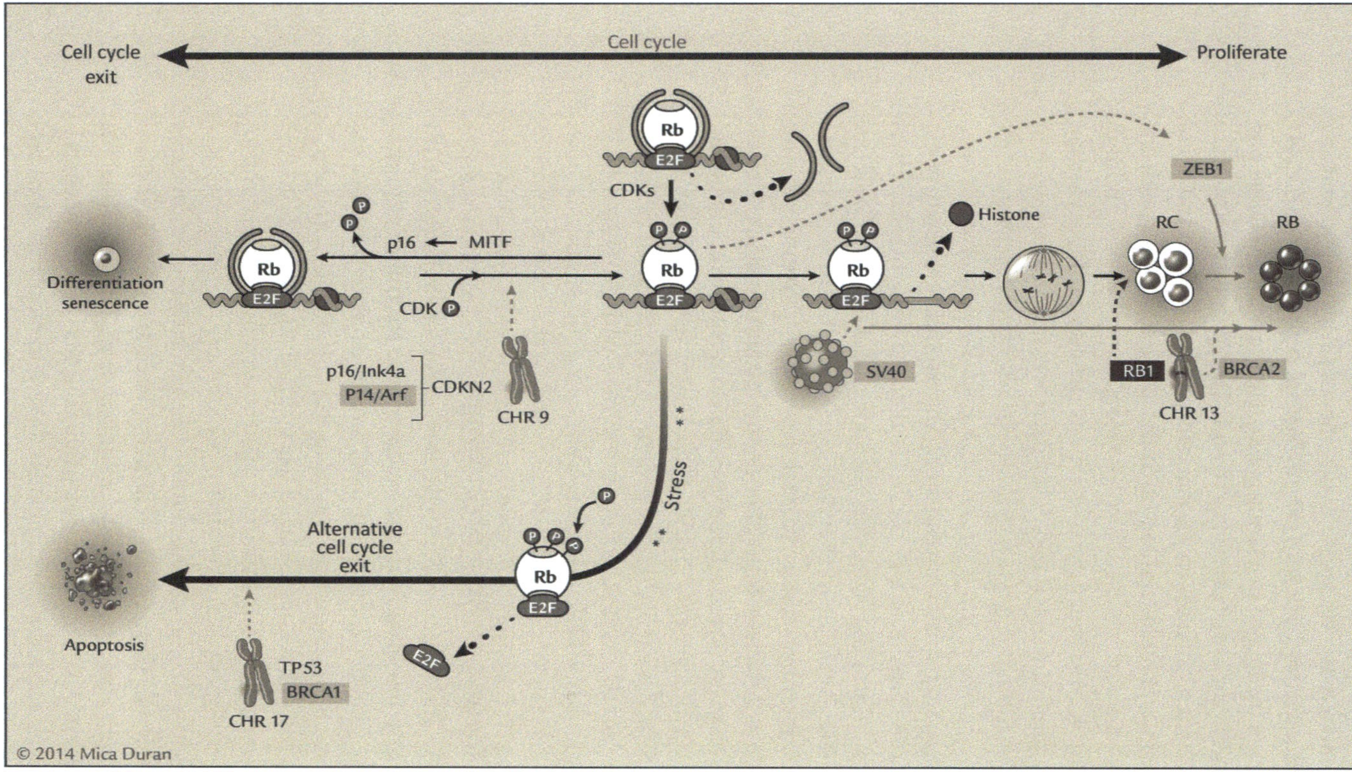

Figure 2 Role of the Rb protein in progression to retinoblastoma. Rb recruits proteins HDAC and BRG1 to E2F sites, resulting in chromatin structure alterations and prevention of access to transcriptional machinery. Phosphorylation by CDKs releases HDAC and BRG1, in turn releasing histone from DNA, thus allowing access to transcriptional machinery and resulting in cell division. Mutations in both alleles of Rb result in proliferation/retinocytoma formation. Further genetic events (such as p16/Ink4a, P14/Arf, or BRCA2 mutations; SV40 interactions with Rb; or ZEB1 activation by inactivated pRb) lead to progression to retinoblastoma. The cell cycle may be exited under normal conditions when Rb is hypophosphorylated via MITF induction of p16, allowing HDAC and BRG1 to rebind to Rb and suppress transcription. Alternative cell cycle exit occurs under stress or hyperproliferation. Rb is hyperphosphorylated, releases E2F, and undergoes tp53-mediated apoptosis which may be suppressed by BRCA1 mutations. Thus, RB1/Rb mutations initiate the process; however, subsequent mutational events involving chromosomes 9, 13, and 17 are needed for progression to retinoblastoma. *Modified from Harbour.*[12]

inactivation of the RB1 gene could lead to chromosome instability, allowing secondary and tertiary mutations in key cancer pathways to be rapidly acquired. Approximately two-thirds of retinoblastoma tumors have MDM4/MDM2 amplification leading to inactivation of the p53 pathway.[15] The p53 pathway is responsible for inducing cell cycle exit and apoptosis when a cell encounters deregulated proliferation or "oncogenic stress." It should be noted that the p53 pathway is directly connected to the RB pathway through the p14/Arf tumor suppressor gene, which is regulated by RB1 through E2F3.[16] In addition, a small proportion of tumors develop in the context of normal RB1 and amplification of MYCN has been identified in such cases.[17]

Subsequent alterations in genes such as BRCA2 on chromosome 13, CDKN2 on chromosome 9, and TP53/BRCA1 on chromosome 17 contribute to retinoblastoma tumor progression. Breast cancer-related genes BRCA1 and BRCA2 are involved in DNA damage response pathways and may directly affect the function of Rb. Loss of p16/INK or p16/INK hypermethylation may alter cell cycle exit. RB1 inactivation upregulates ZEB protein, a transcriptional repressor of E-cadherin (see Fig. 2). E-cadherin is a cell-to-cell adhesion molecule that when downregulated allows for tissue invasion and cell transmigration or metastasis, two other hallmarks of cancer.[18] Structurally related retinoblastoma-like genes RBL1 and RBL2 code for related proteins p107 and p130 that share the same function as Rb. The Rb protein together with p107 and p130 constitutes the retinoblastoma family of proteins.

2.4 Preclinical Models of Retinoblastoma

With the discovery that mutations in the RB1 gene are responsible for the development of retinoblastoma, researchers ventured to model retinoblastoma in mice.[19] One of the first preclinical models of retinoblastoma was a viral oncoprotein-induced murine model using the SV40 T antigen.[20] These transgenic mice have since been used in several studies to test alternative treatments for retinoblastoma, such as periocular injection of sustained-release nanoparticles containing carboplatin.[21] Heterozygote $RB1^{+/-}$ mice developed pituitary and thyroid tumors; however, retinoblastoma was not observed. Surprisingly, chimeric mice made with $RB^{-/-}$ embryonic stem cells did not develop retinoblastoma, whereas $RB^{-/-}$; $p107^{-/-}$ mice did develop retinoblastoma suggesting that p107 can suppress retinoblastoma in mice. Studies on the role of the RB family in mice and humans showed that when RB1 is inactivated in the developing retina, the retinoblastoma-related RL1/p107 gene is upregulated in a compensatory manner.[22] This is probably due to a direct role of Rb in suppressing p107 expression in the mouse through conserved E2F binding sites in the p107 promoter. Using Chx10-Cre;$Rb^{Lox/Lox}$;$p107^{-/-}$ mice, the first knockout mouse model of retinoblastoma was created by inactivating the RB1 gene in the retinal progenitor cells of $p107^{-/-}$ mice.[23] Several other models using other Cre transgenic lines have since been developed. The notable difference between humans and mice is the expression of p107, since there is no compensatory increase in p107 following RB1 gene inactivation in the human retina.[24]

One of the limitations of these genetic murine models of retinoblastoma is that they develop retinal tumors in a completely mutant retina, whereas in humans the tumor arises from a mutant cell in an otherwise normal retina. Moreover, it has not been demonstrated that the mouse tumors recapitulate all the subsequent genetic mutations found in retinoblastoma. In contrast, orthotopic xenograft models have several advantages that complement the deficiencies in the genetic models.[25] In these xenograft models, cells from cultured human retinoblastoma cell lines derived from enucleated eyes are injected into the vitreous or subretinal space of mice, rats, or rabbits. They have well-characterized genetic lesions that occur in human disease and they develop intraocular tumors that invade the retina and form vitreous seeds closely resembling human retinoblastoma. These models have been extensively used in preclinical studies of novel methods of delivering chemotherapy or other agents for the treatment of retinoblastoma. Intravitreal and suprachoroidal injection of topotecan using microneedles in a rabbit model of retinoblastoma with vitreous seeds are also currently being developed in our laboratory. The main advantage of using rabbits over mice is their larger eye size, making it more suitable for targeted drug delivery experiments.[26]

Preclinical models that faithfully recapitulate the human disease are required for rare childhood cancers such as retinoblastoma because there are not enough patients for large-scale multicenter clinical trials. Moreover, animal models of retinoblastoma were instrumental in identifying secondary mutations and developing targeted therapy for this disease.

3. CLINICAL FEATURES OF RETINOBLASTOMA

The most common presenting clinical signs of retinoblastoma are leukocoria (white pupil) (60%) followed by strabismus (misaligned eyes)

(20%). The tumor arises from the retina and forms an elevated, white mass that may or may not be associated with vitreous or subretinal tumor seeds. When the tumor outgrows its vascular supply, it undergoes ischemic necrosis and may contain areas of dystrophic calcification, resulting in a chalky white or cottage cheese appearance. Tumor growth patterns include endophytic, in which the tumor grows into the vitreous; exophytic, in which the tumor grows under the retina toward the choroid; mixed (both endophytic and exophytic); and diffuse infiltrating in which a plaque is formed on the retinal surface. The benign variant retinoma/retinocytoma may clinically appear as a gray-translucent lesion or as regressed retinoblastoma with calcification, and both retinoblastoma and retinoma lesions can be found in the same eye. The diagnosis of retinoblastoma is usually established by examination of the retina of the eye using indirect ophthalmoscopy. Fundus imaging, ultrasonography, and magnetic resonance imaging are used to support the diagnosis and stage the tumor. The clinical evaluation for retinoblastoma includes the International Classification of Retinoblastoma[27] and staging using the American Joint Committee on Cancer Tumor Node Metastasis system.[28] Retinoblastoma can spread to the anterior segment via the vitreous/subretinal space; to the orbit through the sclera; to the cranial cavity through the optic nerve; and to distant sites hematogenously through the choroid.

As mentioned earlier, retinoblastoma presents in two distinct clinical forms: (1) a hereditary form in which tumors are bilateral (both eyes affected) and multifocal, usually presenting at an earlier age; and (2) a nonhereditary/sporadic form in which tumors are unilateral (only one eye affected) and unifocal, usually presenting at a later age. Trilateral RB is an entity wherein patients have bilateral retinoblastoma and a third primary intracranial tumor[29] which may be a primitive neuroectodermal tumor, a pinealoblastoma, or ependymoblastoma. A novel subset of unilateral nonfamilial cases of retinoblastoma that lack mutations in the RB1 gene but are driven by MYCN amplification ($RB1^{+/+}MYCN^{A}$) have been described. Patients that were $RB1^{+/+}MYCN^{A}$ were diagnosed at a younger age (median 4–5 months) and clinically have large, invasive tumors.[17]

4. PATHOLOGY OF RETINOBLASTOMA

One of the most debated topics is the cell of origin in retinoblastoma. Retinoblastoma is derived from cells of neuroectodermal origin from the inner layer of the optic cup. The presumed cell of origin is the retinoblast, or primitive retinal progenitor or precursor cell (RPC), that has the capability of differentiating into multiple retinal cell types including horizontal, amacrine, Muller, and photoreceptors. Other studies show that human retinoblastoma exhibits properties of a cone precursor cell.[30] Recent evidence from gene expression profiling suggests that the cell of origin for poorly differentiated retinoblastoma is the primitive RPC and for well-differentiated tumors is the cone precursor cell[31] (see Fig. 3).

Retinoblastoma has been microscopically characterized as a proliferation of small round cells with darkly staining basophilic nuclei and scant cytoplasm, high nuclear to cytoplasmic ratios, tumor necrosis, apoptosis, and prominent mitotic activity. The most important histologic features predictive of metastasis and survival (so-called high-risk features) are invasion of

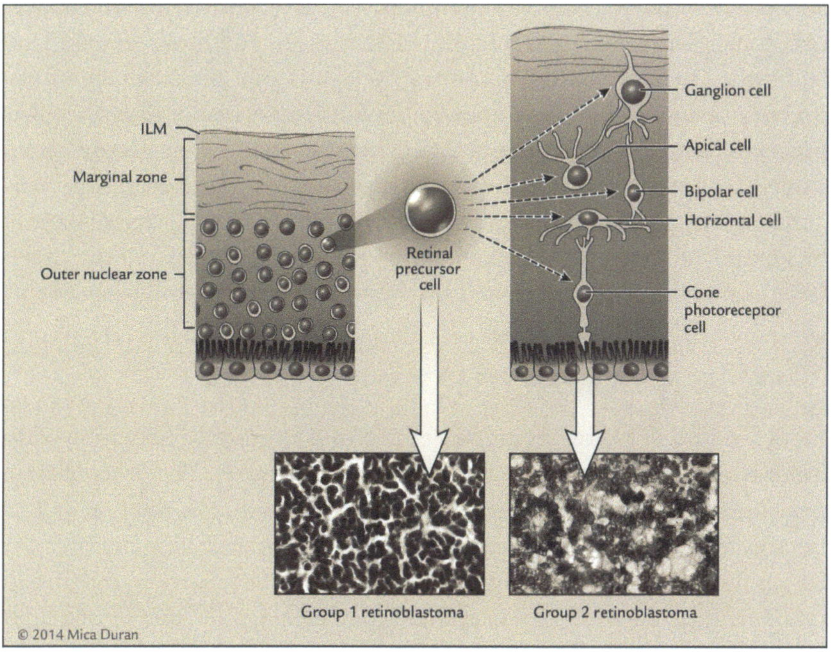

Figure 3 Cell of origin of retinoblastoma. Primitive retinal precursor cells (RPCs) are present in the outer nuclear zone of the developing retina and have the capability of differentiating into multiple retinal cell types. Gene expression profiling has shown two general phenotypic groups of retinoblastoma. If mutational events occur in an RPC before further differentiation, Group 1 retinoblastoma occurs. Group 1 tumors are poorly differentiated and express RPC genes. If mutational events occur in a cone photoreceptor cell, Group 2 retinoblastoma occurs. Groups 2 tumors may exhibit photoreceptor differentiation such as rosette or fleurette formation and express cone-associated genes.

the optic nerve beyond the lamina cribrosa, particularly if tumor is present at the surgical margin; and massive invasion of the choroid measuring 3 mm or greater in diameter especially if with extension into or beyond the sclera or orbit. The grading of cellular anaplasia and nuclear morphometry has recently been shown to have prognostic significance in the absence of high-risk histologic features.[32] The benign variant retinocytoma consists of well-differentiated cells often with associated calcifications. Retinocytoma cells have small, round, bland nuclei and abundant cytoplasm. Necrosis and mitotic figures are typically absent.[33]

The degree of differentiation in retinoblastoma is determined by the development of rosettes and fleurettes, which has been confirmed by electron microscopy and tissue culture. The least differentiated is the neuroblastic Homer Wright rosette, which can also be observed in other embryonal central nervous system tumors such as pineoblastoma and medulloblastoma. The Flexner–Wintersteiner rosette and the more differentiated fleurette are two formations mimicking the cytological and architectural features of photosensory rod and cone cells, which are evidence of further photoreceptor differentiation. The central portion of the Homer Wright rosette is filled with neuropil, in comparison to the clear center of the Flexner–Wintersteiner rosettes attributable to a recapitulation of the external limiting membrane of the retina. Fleurettes have well-developed photoreceptor outer segments in their lumen and are abundant in retinocytoma (see Fig. 4).

5. MANAGEMENT OF RETINOBLASTOMA

The goal of treatment in retinoblastoma is to ensure elimination of the tumor while minimizing collateral injury to other tissues. The order of treatment priority is (1) preservation of life, (2) preservation of the eye, and (3) preservation of vision. Treatment depends on tumor stage, number of tumor foci (unifocal, unilateral multifocal, or bilateral), localization and size of tumors within the eye, presence of vitreous seeding, and age of the child. Treatment options include enucleation, cryotherapy, laser photocoagulation, systemic or ocular chemotherapy, radiation therapy through episcleral plaques or rarely external beam, or a combination of these. Systemic chemotherapy for tumor chemoreduction is commonly given to patients with bilateral/heritable disease and a regimen of chemotherapy agents vincristine, carboplatin, and etoposide is used. Chemoreduction is usually followed by local therapies to consolidate the tumor such as thermotherapy/laser photocoagulation for small posterior tumors, cryotherapy for anterior tumors, and

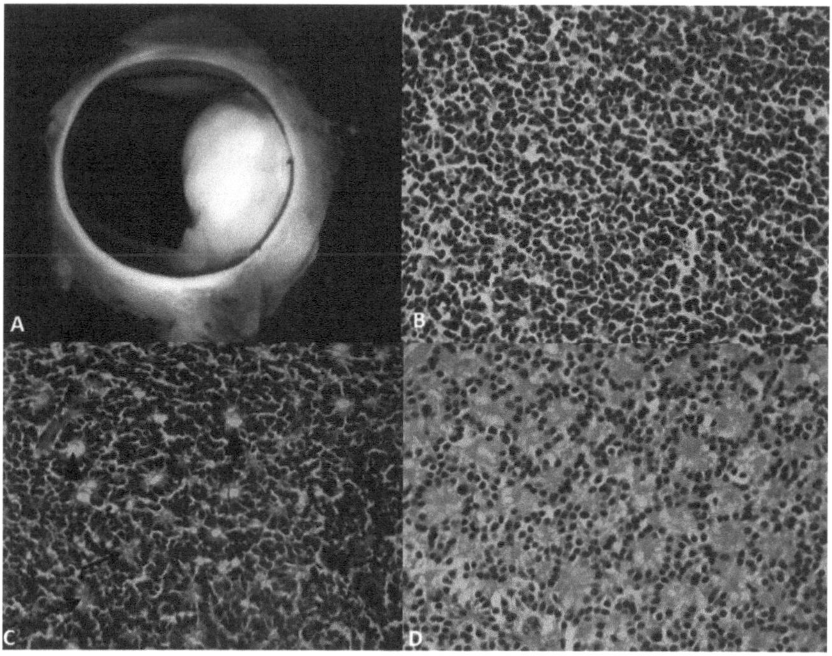

Figure 4 Pathology of retinoblastoma. (A) Retinoblastoma tumors grossly appear white, fleshy, or cottage cheese-like. (B) Retinoblastoma is a small, round, blue cell tumor forming sheets of cells. (C) Retinoblastoma may exhibit rosette formation, including Homer Wright rosettes with neuropil in the lumen (arrows) or Flexner–Wintersteiner rosettes with a clear lumen (arrowheads). (D) Retinocytoma is composed of cells with small nuclei, abundant cytoplasm, and fleurettes with photoreceptor outer segments. (B–D: Hematoxylin and eosin, 25 ×.)

brachytherapy for larger tumors. Eye-salvaging approaches using laser or cryotherapy sometimes combined with local ocular chemotherapy are recommended for unilateral/nonheritable, early-stage disease. Enucleation of eyes with unilateral disease without extraocular extension is an effective treatment with a cure rate of 85–90%. Adjuvant systemic chemotherapy is utilized for eyes with optic nerve invasion beyond the lamina cribrosa or massive choroidal invasion. Treatment of extraocular or orbital disease includes radiation and chemotherapy. If possible, any radiation (including X-ray, CT scan, and external beam) should be avoided in individuals with heritable retinoblastoma to minimize the lifetime risk of developing second cancers.

Treatment failures are attributable to persistence of vitreous seeds, subretinal seeds, intraretinal tumor, or tumor that is unresponsive to chemotherapy.[34] Some have presumed that well-differentiated tumors are less likely to respond to treatment, as cells are not cycling.[35] Periocular injections of carboplatin have been used for the treatment of vitreous seeds but have largely been abandoned owing to lack of success and toxicity. Intra-arterial chemotherapy as an eye-preserving modality has been effective in many cases with large tumors associated with vitreous seeds. It involves infusion of a chemotherapeutic agent (melphalan, topotecan, or carboplatin) into the ophthalmic artery. Although drug injection through the sclera into the vitreous was initially discouraged due to fear of tumor spread along needle tracks, techniques for safe and effective intravitreal injections have been developed to treat vitreous seeds. Melphalan or topotecan is injected using a 32-gauge needle and the needle is frozen with a cryoprobe as it is withdrawn from the eye.[36]

Emerging treatment modalities include gene therapy and sustained drug release platforms. Intravitreal injections of an adenovirus vector containing herpes simplex thymidine kinase gene have been used in patients with bilateral retinoblastoma with vitreous seeds.[37] A sustained-release episcleral exoplant has also been developed for targeted intraocular drug delivery.[38] Major advances such as the understanding of the molecular biology of the tumor and the development of targeted therapy have greatly improved survival to 97% in developed countries.[39]

REFERENCES

1. Broaddus E, Topham A, Singh AD. Incidence of retinoblastoma in the USA: 1975–2004. *Br J Ophthalmol*. 2009;93:21–23.
2. Young JL, Smith MA, Roffers SD, et al. Retinoblastoma. In: Ries LAG, Smith MA, Gurney JG, et al., eds. *Cancer Incidence and Survival Among Children and Adolescents: United States SEER Program 1975–1995*. Bethesda: National Cancer Institute; 1999:73–78.
3. Parkin DM, Stiller CA, Draper GJ, Bieber CA. The international incidence of childhood cancer. *Int J Cancer*. 1988;42:511–520.
4. Grossniklaus HE. Retinoblastoma: fifty years of progress. The LXXI Edward Jackson memorial lecture. *Am J Ophthalmol*. 2014;158:875–891.
5. Knudson Jr AG. Mutation and cancer: statistical study of retinoblastoma. *Proc Natl Acad Sci USA*. 1971;68:820–823.
6. Kleinerman RA, Yu CL, Little MP, et al. Variation of second cancer risk by family history of retinoblastoma among long-term survivors. *J Clin Oncol*. 2012;30:950–957.
7. Lohmann DR, Gallie BL. Retinoblastoma. In: Pagon RA, Adam MP, Ardinger HH, et al. eds. *GeneReviews*. Seattle: University of Washington; 1993–2015:1–22.
8. Friend SH, Bernards R, Rogelj S, et al. A human DNA segment with properties of the gene that predisposes to retinoblastoma and osteosarcoma. *Nature*. 1986;323:643–646.
9. Harbour JW. Overview of RB gene mutations in patients with retinoblastoma: implications for clinical genetic screening. *Ophthalmology*. 1998;105:1442–1447.
10. Lee JO, Russo AA, Pavletich NP. Structure of the retinoblastoma tumour-suppressor pocket domain bound to a peptide from HPV E7. *Nature*. 1998;391:859–865.

11. Lee C, Chang JH, Lee HS, Cho Y. Structural basis for the recognition of the E2F transactivation domain by the retinoblastoma tumor suppressor. *Genes Dev.* 2002;16:3199–3212.
12. Harbour JW. Eye cancer: unique insights into oncogenesis. The Cogan lecture. *Invest Ophthalmol Vis Sci.* 2006;47:1736–1745.
13. Corson TW, Gallie BL. One hit, two hits, three hits, more? Genomic changes in the development of retinoblastoma. *Genes Chromosomes Cancer.* 2007;46:617–634.
14. Dimaras H, Khetan V, Halliday W, et al. Loss of RB1 induces non-proliferative retinoma: increasing genomic instability correlates with progression to retinoblastoma. *Hum Mol Genet.* 2008;17:1363–1372.
15. Laurie NA, Donovan SL, Shih CS, et al. Inactivation of the p53 pathway in retinoblastoma. *Nature.* 2006;444:61–66.
16. Aslanian A, Iaquinta PJ, Verona R, Lees JA. Repression of the Arf tumor suppressor by E2F3 is required for normal cell cycle kinetics. *Genes Dev.* 2004;18:1413–1422.
17. Rushlow DE, Mol BM, Kennett JY, et al. Characterisation of retinoblastomas without RB1 mutations: genomic, gene expression, and clinical studies. *Lancet Oncol.* 2013;14:327–334.
18. Di Fiore R, D'Anneo A, Tesoriere G, Vento R. RB1 in cancer: different mechanisms of RB1 inactivation and alterations of pRb pathway in tumorigenesis. *J Cell Physiol.* 2013;228:1676–1687.
19. Jacks T, Fazeli A, Schmitt EM, et al. Effects of an Rb mutation in the mouse. *Nature.* 1992;359:295–300.
20. Windle JJ, Albert DM, O'Brien JM, et al. Retinoblastoma in transgenic mice. *Nature.* 1990;343:665–669.
21. Kang SJ, Durairaj C, Kompella UB, et al. Subconjunctival nanoparticle carboplatin in the treatment of murine retinoblastoma. *Arch Ophthalmol.* 2009;127:1043–1047.
22. Robanus-Maandag E, Dekker M, van der Valk M, et al. p107 is a suppressor of retinoblastoma development in pRb-deficient mice. *Genes Dev.* 1998;12:1599–1609.
23. Zhang J, Schweers B, Dyer MA. The first knockout mouse model of retinoblastoma. *Cell Cycle.* 2004;3:952–959.
24. Donovan SL, Schweers B, Martins R, et al. Compensation by tumor suppressor genes during retinal development in mice and humans. *BMC Biol.* 2006;4:14.
25. Dyer MA. Biology of retinoblastoma. In: Rodriguez-Galindo C, Wilson MW, eds. *Retinoblastoma.* New York: Springer; 2010:1–9.
26. Kang SJ, Grossniklaus HE. Rabbit model of retinoblastoma. *J Biomed Biotechnol.* 2011;2011:394730.
27. Murphree AL. Intraocular retinoblastoma: the case for a new group classification. *Ophthalmol Clin North Am.* 2005;18:41–53.
28. Edge SB, Byrd DR, Compton CC, et al., eds. Retinoblastoma. In: *AJCC Cancer Staging Manual.* 7th ed. New York: Springer; 2010:560–568.
29. Bader JL, Meadows AT, Zimmerman LE, et al. Bilateral retinoblastoma with ectopic intracranial retinoblastoma: trilateral retinoblastoma. *Cancer Genet Cytogenet.* 1982;5:203–213.
30. Xu XL, Fang Y, Lee TC, et al. Retinoblastoma has properties of a cone precursor tumor and depends upon cone-specific MDM2 signaling. *Cell.* 2009;137:1018–1031.
31. Kapatai G, Brundler MA, Jenkinson H, et al. Gene expression profiling identifies different sub-types of retinoblastoma. *Br J Cancer.* 2013;109:512–525.
32. Mendoza PR, Specht CS, Hubbard GB, et al. Histopathologic grading of anaplasia in retinoblastoma. *Am J Ophthalmol.* 2015;159:764–776.
33. Margo C, Hidayat A, Kopelman J, Zimmerman LE. Retinocytoma. A benign variant of retinoblastoma. *Arch Ophthalmol.* 1983;101:1519–1531.

34. Hwang CK, Aaberg Jr TM, Chevez-Barrios P, et al. Residual intraretinal retinoblastoma after chemoreduction failure. *Arch Ophthalmol.* 2012;130:246–248.
35. Demirci H, Eagle Jr RC, Shields CL, Shields JA. Histopathologic findings in eyes with retinoblastoma treated only with chemoreduction. *Arch Ophthalmol.* 2003;121:1125–1131.
36. Munier FL, Soliman S, Moulin AP, et al. Profiling safety of intravitreal injections for retinoblastoma using an anti-reflux procedure and sterilisation of the needle track. *Br J Ophthalmol.* 2012;96:1084–1087.
37. Subramanian N, Navaneethakrishnan S, Biswas J, et al. RNAi mediated Tiam1 gene knockdown inhibits invasion of retinoblastoma. *PLoS One.* 2013;8:e70422.
38. Edelhauser HF, Rowe-Rendleman CL, Robinson MR, et al. Ophthalmic drug delivery systems for the treatment of retinal diseases: basic research to clinical applications. *Invest Ophthalmol Vis Sci.* 2010;51:5403–5420.
39. Abramson DH. Retinoblastoma in the 20th century: past success and future challenges the Weisenfeld lecture. *Invest Ophthalmol Vis Sci.* 2005;46:2683–2691.

CHAPTER THIRTY-ONE

The 11-*cis* Retinal Origins of Lipofuscin in the Retina

Leopold Adler IV, Nicholas P. Boyer, Chunhe Chen, Zsolt Ablonczy, Rosalie K. Crouch, Yiannis Koutalos
Department of Ophthalmology, Albert Florens Storm Eye Institute, Medical University of South Carolina, Charleston, South Carolina, USA

Note: This chapter is available in its entirety online at http://dx.doi.org/10.1016/bs.pmbts.2015.07.022

Progress in Molecular Biology and Translational Science, Volume 134
ISSN 1877-1173
http://dx.doi.org/10.1016/bs.pmbts.2015.07.022
© 2015 Elsevier Inc.
All rights reserved.

INDEX

Note: Page numbers followed by "*f*" indicate figures and "*t*" indicate tables.

D

E

L

M

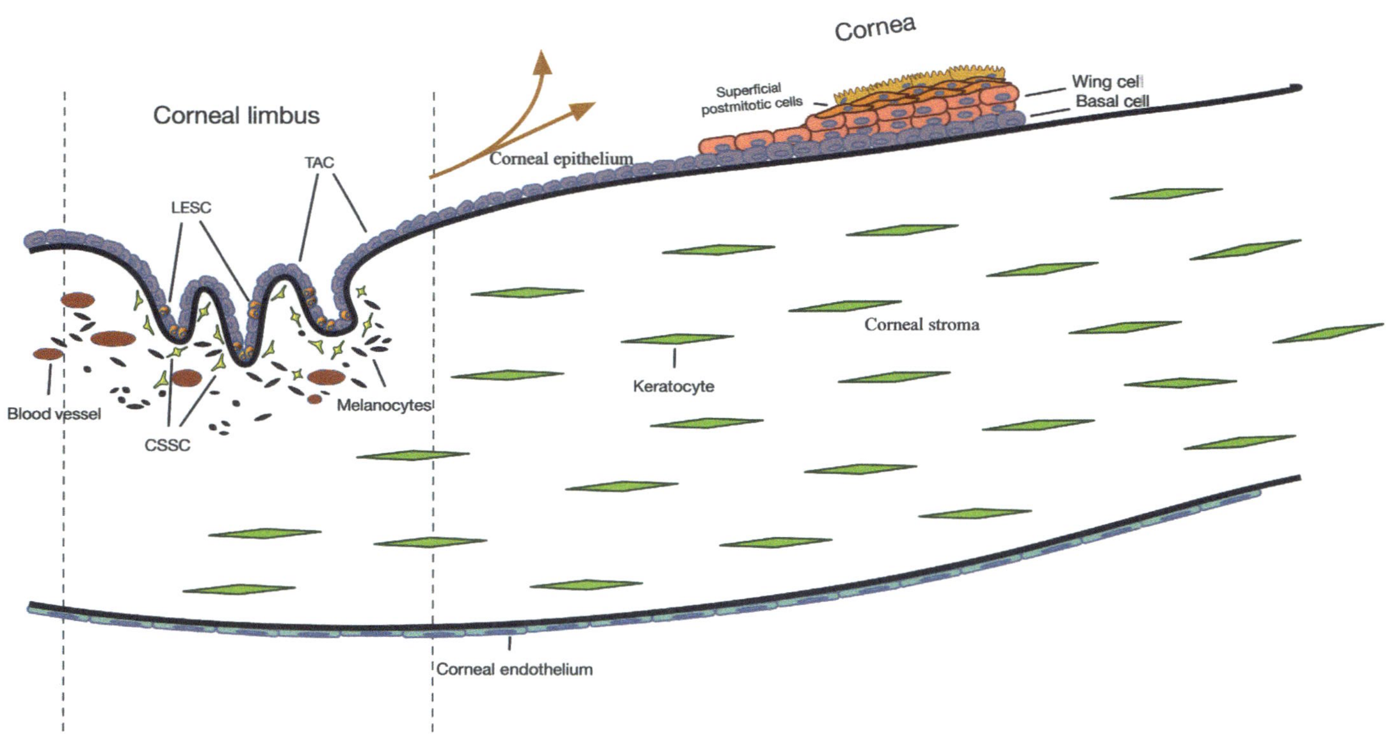

Andrew J. Hertsenberg and James L. Funderburgh, Figure 1 The cornea is composed of three cellular layers: the epithelium, stroma, and endothelium. The vascular limbal region is located at the peripheral cornea and is bordered by the conjunctiva—this region is the proposed niche for stem cell populations in each layer. LESC, limbal epithelial stem cell; TAC, transit-amplifying cell; CSSC, corneal stromal stem cell.

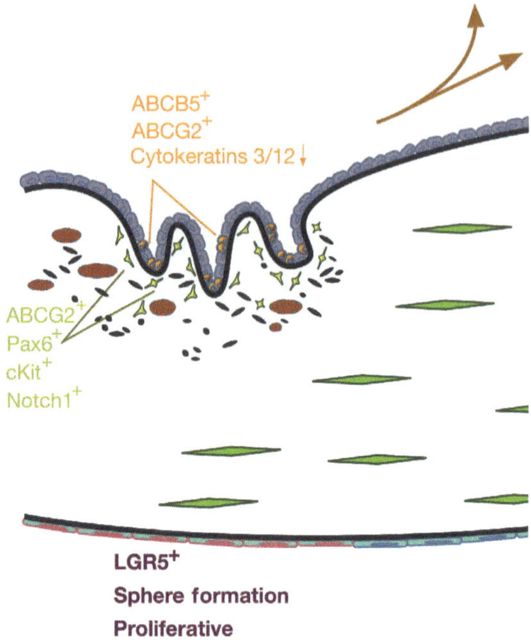

Andrew J. Hertsenberg and James L. Funderburgh, Figure 2 Stem cell markers/characteristics. Orange, limbal epithelial stem cell (LESC) markers, expression generally compared to central epithelial cells; green, human corneal stromal stem cell markers; and purple, proposed markers and characteristics corneal endothelial progenitor cells.

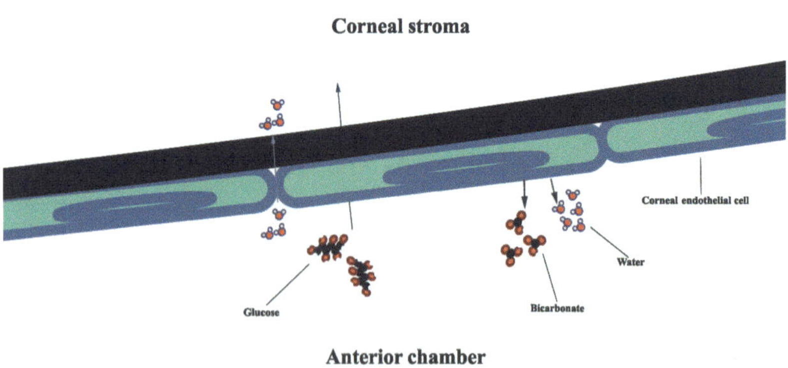

Andrew J. Hertsenberg and James L. Funderburgh, Figure 3 The corneal endothelium serves as a leaky barrier between the corneal stroma (anterior) and anterior chamber (posterior). Water passively moves from the ac into the stroma, while protein and other nutrients (such as glucose) are actively transported by corneal endothelial cells. To pump water from the stroma into the anterior chamber, bicarbonate is actively pumped out of the endothelial cells (along with other ions) and water follows.

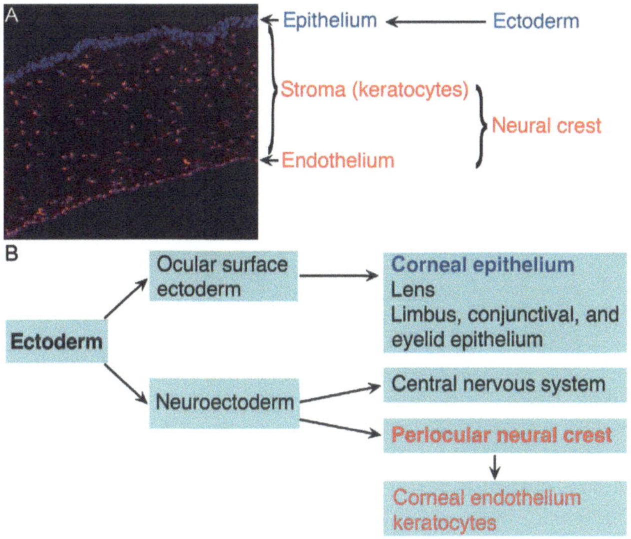

Peter Y. Lwigale, Figure 1 Ectodermal origin of the three cellular layers of the cornea. (A) Cross section through E8 chick cornea showing the ectoderm-derived corneal epithelium and neural crest cell-derived stromal keratocytes and corneal endothelium. (B) Schematic diagram showing the derivatives of the cranial ectoderm that contribute to the formation of the corneal epithelium and the neural crest cell progenitors of the corneal endothelium and keratocytes.

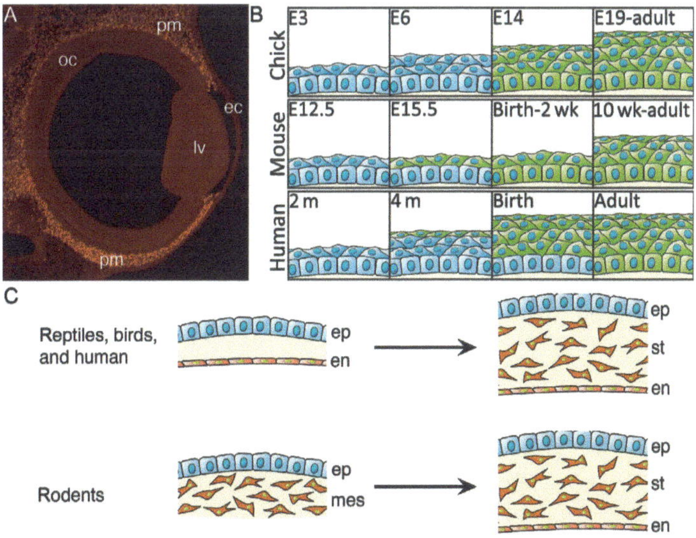

Peter Y. Lwigale, Figure 2 Cornea development. (A) Cross section through an E3 chick eye at the rudimentary stage showing the optic cup (oc) and lens vesicle (lv) being surrounded by the neural crest-derived periocular mesenchyme (pm), and the presumptive corneal epithelium (ec). (B) Schematic showing stages of stratification of the corneal epithelium and keratin expression in chick, mouse, and human. (C) Schematics showing the different modes of neural crest cell migration between the corneal epithelium (ep) and lens during the formation of the corneal endothelium (en) and stroma (st). mes, neural crest mesenchyme.

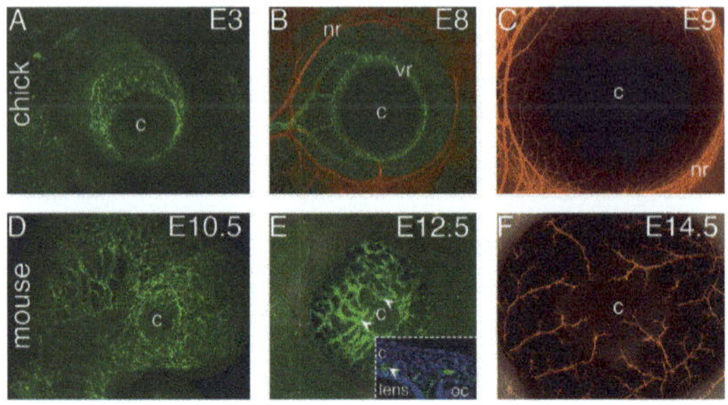

Peter Y. Lwigale, Figure 3 Neurovascular development in the anterior eye. (A and B) Whole-mount Tg(tie1:H2B:eYFP) transgenic quail eyes at E3 and E7 showing that angioblasts and forming vasculature avoid the developing cornea. (B and C) Formation of the pericorneal nerve ring (nr) and innervation of the corneal periphery at E9. (D and E) Endomucin staining of E10.5 and E12.5 mouse embryos showing the vascularization of the anterior eye. Inset in (E) shows that the mouse cornea is not vascularized although it forms adjacent to the pupillary membrane (arrowheads). (F) Projection of sensory nerves into the corneal periphery of an E14.5 mouse embryo. c, cornea; nr, pericorneal nerve ring; vr, vascular ring; oc, optic cup.

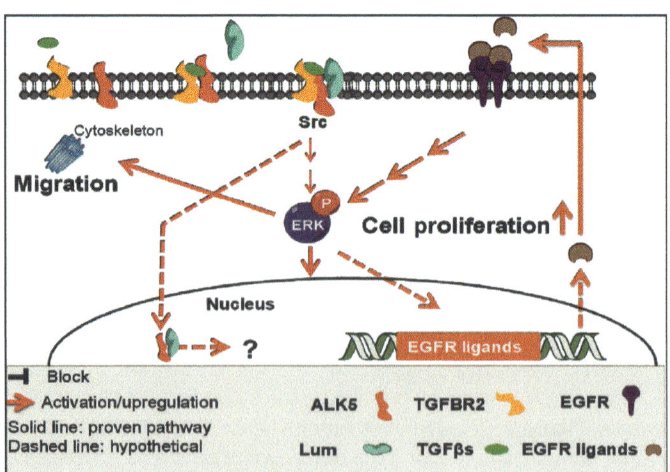

Chia-Yang Liu and Winston Whey-Yang Kao, Figure 2 Lumican binds to TβRI (ALK5) to activate EGF ligands in corneal epithelial cells.

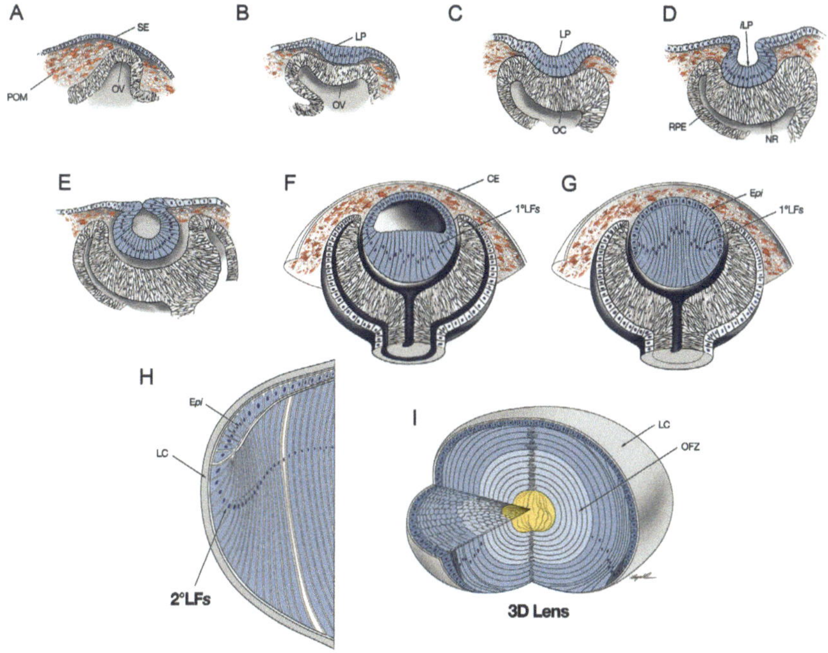

Ales Cvekl *et al.*, Figure 1 Embryonic lens development (mouse). (A) E9.0 (prospective lens ectoderm–surface ectoderm). (B) E9.5 (lens placode). (C) E10 (lens placode at the beginning of invagination). (D) E10.5 (transition from invaginating lens placode to lens pit). (E) E11 (open lens vesicle). (F) E12.5 (onset of primary lens fiber cell differentiation). (G) E13.5–E14.5 (completion of primary lens fiber cell elongation–secondary lens fiber cell formation). (H) Secondary lens fiber cell differentiation as a driving force of continuous lens growth. Apical–basal polarity of lens epithelial and lens fiber cells is indicated. Anterior lens epithelium, ALE; corneal epithelium, CE; invaginating lens placode, iLP; invaginating lens placode/pit, iLP/p; lens capsule, LC; lens epithelium, Epi; lens placode, LP; neuroretina, NR; optic vesicle, OV; periocular mesenchyme, POM; primary lens fibers, 1° LFs; prospective lens ectoderm, PLE; retinal pigmented epithelium, RPE; secondary lens fibers, 2° LFs; surface ectoderm, SE. *From Cvekl and Ashery-Padan.*[1]

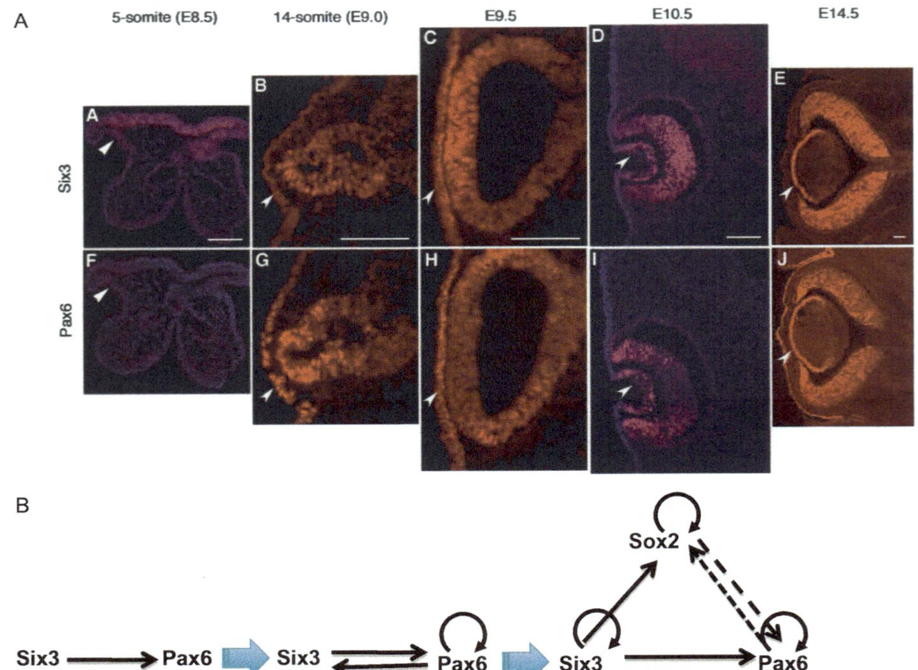

Ales Cvekl *et al.*, Figure 2 Expression of Pax6 and Six3 during early stages of lens formation and regulatory mechanisms involving Pax6, Six3, and Sox2. (A) Six3 is expressed first at the surface ectoderm (E8.5), followed by Pax6 (~E8.75). (B) Onset of Pax6 expression in the presumptive lens ectoderm requires Six3. At E9.0, Pax6 and Six3 cross-regulate each other and autoregulation of Pax6 is being established. At E9.5, expression of Sox2 in the invaginating lens placode is dependent on Six3 and transitionally also on Pax6. *Images from Liu* et al.[88]

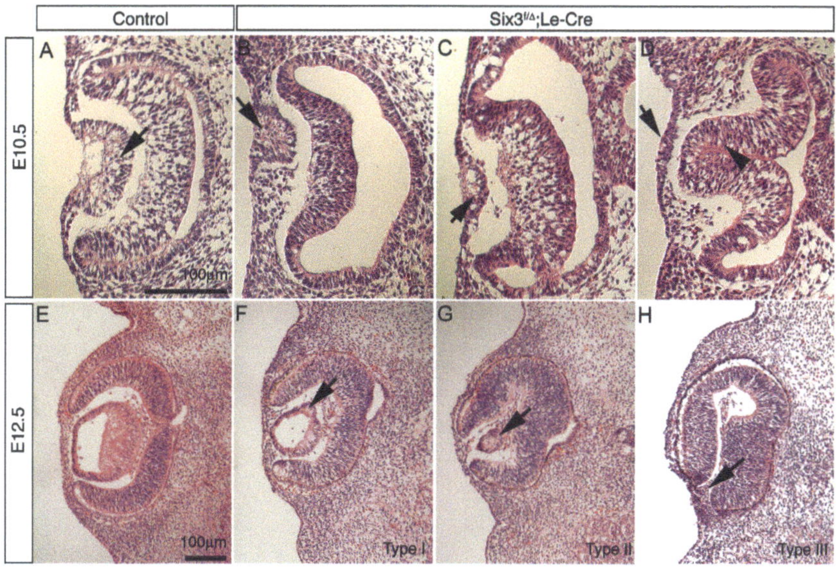

Ales Cvekl *et al.*, Figure 3 Depletion of Six3 in the presumptive lens ectoderm disrupts lens morphogenesis. Conditional inactivation of Six3 with Le-Cre causes a range of lens defects (A–H); in the most severe case the lens does not form (D, H).[88]

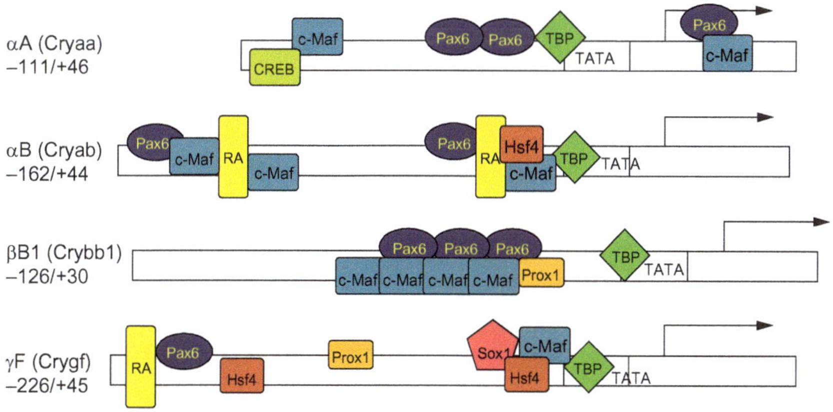

Ales Cvekl *et al.*, Figure 4 Interaction of DNA-binding transcription factors with a representative panel of vertebrate crystallin promoters and enhancers. Four crystallin promoters are shown: mouse αA-, mouse αB-, chicken βB1-, and mouse γF-crystallins. Transcription factors c-Maf, CREB, Hsf4, Pax6, Prox1, RA-responsive nuclear factors (e.g., RARβ/RXRβ), Sox1, and TBP are indicated by different shapes and colors.

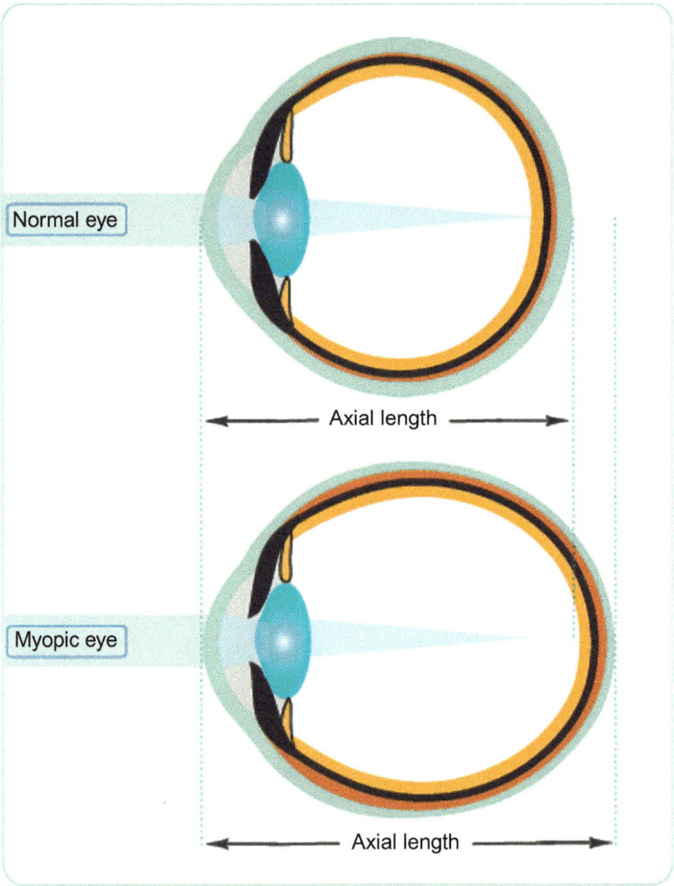

Yan Zhang and Christine F. Wildsoet, Figure 1 Schematic diagram illustrating the principal gross anatomical differences between human emmetropic and myopic eyes, the latter typically being longer and more prolate, with a longer vitreous chamber.

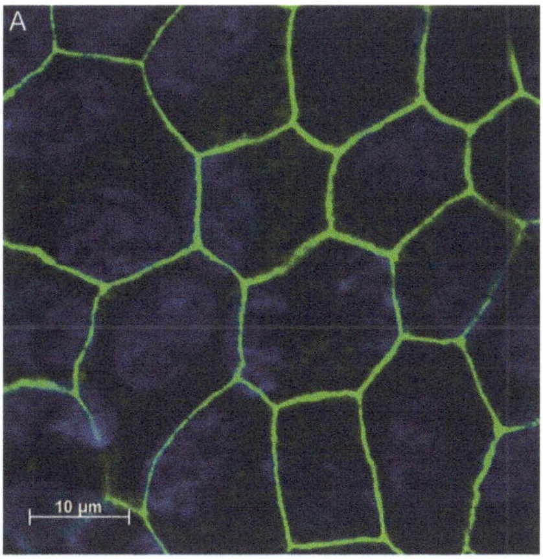

Yan Zhang and Christine F. Wildsoet, Figure 4 (A) Cultured human fetal RPE cells. Tight junctions, shown stained for ZO-1 (zonula occludens-1, in green, ensure that the RPE functions as an effective barrier between the retina and the choroid, with exchange tightly regulated through ion channels and transporters.

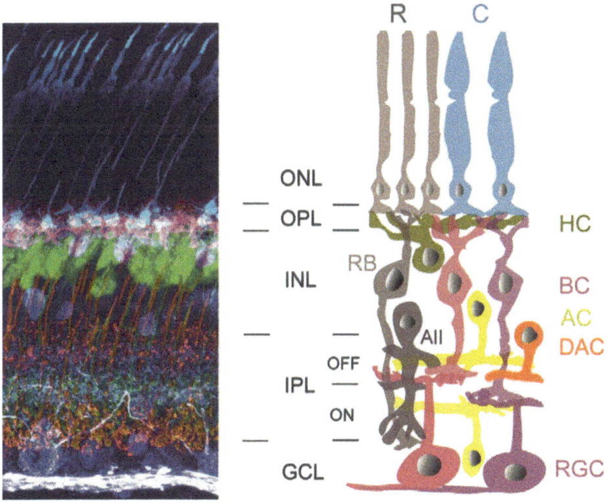

Ranjay Chakraborty and Machelle T. Pardue, Figure 1 Retinal anatomy and circuitry. During visual processing, the output from the rod and cone photoreceptors in the outer nuclear layer (ONL) is decomposed into a number of different parallel information channels by synapsing in the outer plexiform layer (OPL) to different cells in the inner nuclear layer (INL; bipolar, amacrine, and horizontal cells). The output from these inner retinal cells is sampled by different retinal ganglion cells in the inner plexiform layer (IPL). Finally, depending on the type of information, specific ganglion cells in the ganglion cell layer (GCL) transmit the signal to higher visual structures in the brain. R, rod; C, cone; RB, rod bipolar cell; BC, cone bipolar cell; HC, horizontal cell; AC, amacrine cell; AII, AII amacrine cell; DAC, dopaminergic amacrine cell. *Image modified from Ref. 4.*

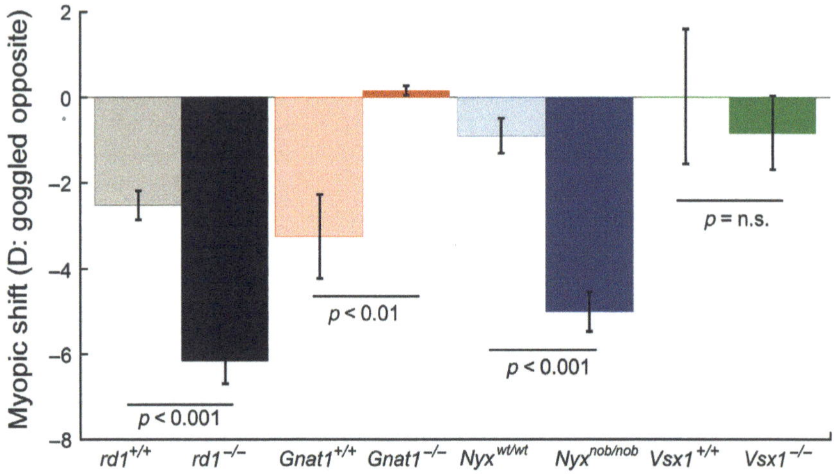

Ranjay Chakraborty and Machelle T. Pardue, Figure 2 Myopic shift (goggled opposite eye) across several mutant mouse strains with photoreceptor or ON/OFF retinal pathway defects after 2 weeks of form deprivation. Note that the mutant mice are on different background strains and thus, the WT controls are shown for each strain. *Rd1* $^{-/-}$ mice and *Nyx* $^{nob/nob}$ mice show significantly greater myopic shift than their WT controls. In contrast, *Gnat1* $^{-/-}$ mice with nonfunctional rods did not show a shift with form-deprivation myopia. *Vsx1* $^{+/+}$ mice are on a 129SV background and do not respond to form deprivation. The loss of *Vsx1* did not change the susceptibility to form-deprivation myopia.

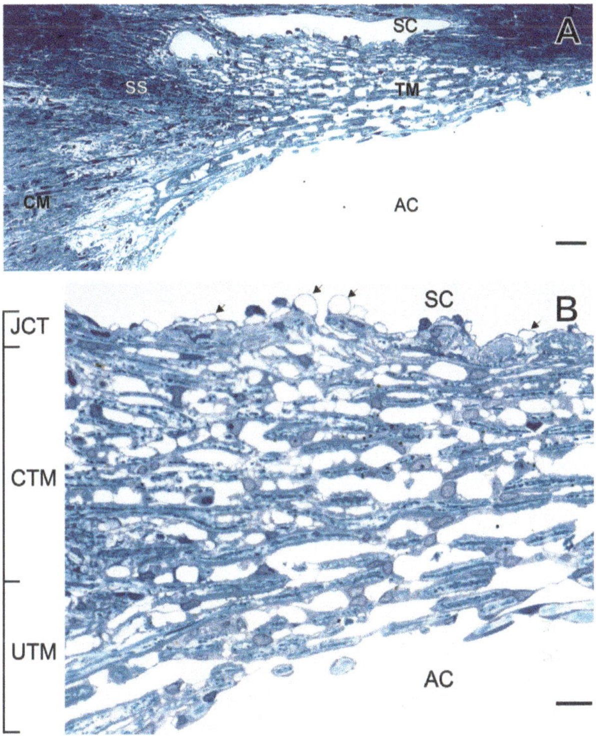

Ernst R. Tamm *et al.*, Figure 1 (A) Light micrograph of a meridional section through the trabecular meshwork. (B) is a magnification of (A). In the plane of this section, Schlemm's canal (SC) has two lumens that are separated by a septum. TM, trabecular meshwork; SS, scleral spur; CM, ciliary muscle; AC, anterior chamber; JCT, juxtacanalicular tissue; CTM, corneoscleral trabecular meshwork; UTM, uveal trabecular meshwork. Arrows in (B) point to giant vacuoles in the inner wall endothelium of SC. Magnification bars: 20 μm (A), 5 μm (B). *From Ref. 11.*

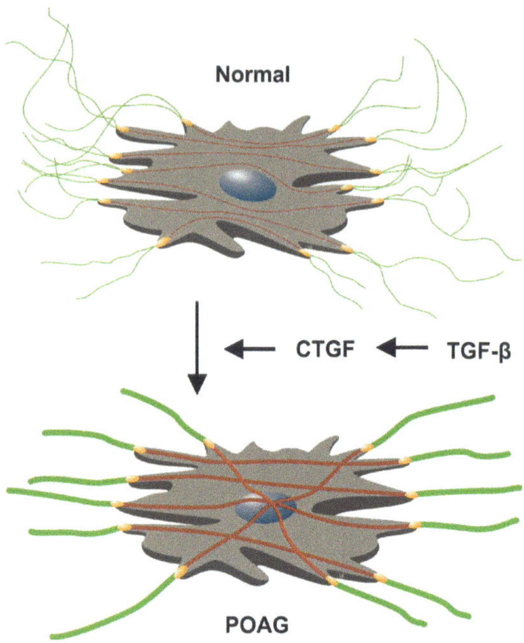

Ernst R. Tamm *et al.*, Figure 4 Schematic drawing of the actin cytoskeleton and the fibrillar extracellular matrix of cells in the juxtacanalicular tissue. The force generated by the actin cytoskeleton (red) is transmitted to the fibrillar extracellular matrix (green) by integrin-based cell-matrix contacts (yellow). The high activity of TGF-β signaling in POAG causes, via the downstream mediator CTGF, a switch to a myofibroblast-like phenotype including an augmentation of the actin cytoskeleton and its directly associated extracellular fibrillar matrix. Overall, the changes cause an increase in TM rigidity and aqueous humor outflow resistance. *From Ref. 38.*

FSC
www.fsc.org
MIX
Papier | Fördert
gute Waldnutzung
FSC® C083411

Kolibri 360 GmbH
Ferdinand-Jühlke-Straße 7
99095 Erfurt, Deutschland
produktsicherheit@kolibri360.de